INTERMEDIATE BOTANY

Other Works by L. J. F. Brimble

All published by Macmillan

TREES IN BRITAIN
SCHOOL COURSE OF BIOLOGY
PHYSIOLOGY, ANATOMY AND HEALTH

★

SCIENCE IN A CHANGING WORLD

(*University of Cairo Press*)

A DISPLAY OF DAFFODILS IN DOROTHY'S FIELD, RYDAL.
Rich with the memories of Wordsworth, the field is named after the poet's sister.

INTERMEDIATE BOTANY

the late L. J. F. BRIMBLE

B.Sc. (LONDON AND READING), F.R.S.E., F.R.S.A., F.L.S.

FORMERLY LECTURER IN THE UNIVERSITY OF MANCHESTER
AND IN THE UNIVERSITY OF GLASGOW
FORMER EDITOR OF *NATURE*

FOURTH EDITION

Entirely Revised and Rewritten in collaboration with

S. WILLIAMS

PH.D. (GLASGOW), D.Sc. (MANCHESTER), F.R.S.E.

SOMETIME SENIOR LECTURER IN THE UNIVERSITY OF GLASGOW

and with the assistance of

G. BOND

PH.D., D.Sc. (GLASGOW), F.R.S.

PROFESSOR OF BOTANY IN THE UNIVERSITY OF GLASGOW

First Edition 1936
Second Edition 1939
Reprinted 1942
Third Edition 1946
Reprinted 1949
Fourth Edition (with S. Williams and G. Bond) 1953
Reprinted 1955, 1957, 1960, 1962, 1969, 1971, 1973, 1977, 1978, 1980

Published by
THE MACMILLAN PRESS LTD
*Associated companies in Delhi Dublin
Hong Kong Johannesburg Lagos Melbourne
New York Singapore and Tokyo*

ISBN 0 333 08403 9 (hard cover)
0 333 23335 2 (paper cover)
Printed in Hong Kong

PREFACE TO FOURTH EDITION

This book was first published in 1936 and has passed through three editions (and two reprints), during which certain modifications and changes were made. The time has now come for a completely new edition in the light of botanical advances during the past fifteen years and the needs of present-day school and university students. This has now been done, the original author having been fortunate in obtaining the collaboration of Dr. S. Williams and Dr. G. Bond.

The text has been rewritten and much new matter added. In the preparation of the Second Edition help was afforded by Dr. J. Ramsbottom, O.B.E. (formerly Keeper of Botany in the British Museum (Natural History)), Prof. R. Ruggles Gates, F.R.S. (formerly Research Fellow in Biology, Harvard University), Dr. M. A. H. Tincker (formerly of the Royal Horticultural Society's Gardens), and Prof. F. W. Sansome (University College, Ibadan, Nigeria) ; their suggestions are still incorporated in this new edition.

Many illustrations have been replaced by better ones, and additional illustrations included. Most of the line figures were drawn by L. J. F. Brimble; some new figures were drawn by Dr. S. Williams and Mrs. F. W. Williams; and Dr. G. Bond has supplied some useful photographs.

The new edition of *Intermediate Botany* is now brought up to date, and is suitable for candidates for the General Certificate of Education (Advanced Level) and for candidates for first-year University examinations such as those of Intermediate Science, Arts, Pharmacy, Agriculture, Horticulture and Medicine.

London 1952   L. J. F. BRIMBLE

CONTENTS

CHAPTER I

LIVING THINGS

Botany is a branch of biology—the scientific investigation of **organisms** or living things. To-day, a knowledge of botany is essential in very many industries, occupations and vocations. A knowledge of the structure (**morphology**) including internal structure (**anatomy**) of plants is necessary to all who are seriously interested in their cultivation—for example, farmers, gardeners, planters (tea, rubber, sugar, etc.); the function of plants and their life processes (**physiology**) to the same type of person and to the physician; the structure of fossil plants (**palaeobotany**) to the geologist and others interested in the past and present structure of the earth; the diseases of plants (**plant pathology**) to all cultivators ; and the breeding of plants (**genetics**) to those who are concerned with obtaining new or better types of fruits, flowers or vegetables. To the botanist, of course, all branches of the subject are important. After having acquired a general knowledge of botany, the romance of plant life will reveal itself, and it will then be realised what an important science botany is, and has been, since man learned to make use of plants and their products for his pleasure, sustenance, health and general well-being.

In the study of plants, whether they are in a condition of health or disease, it is necessary to remember that they are living things. It may seem easy to define what are living things, because the average living thing is so *obviously* alive; but actually to do so with scientific accuracy is really very difficult. All that can be said is that living things possess certain characteristics in common, and these provide the power which results in the phenomenon of what is called life.

On the other hand, certain familiar objects, such as a piece of glass or a bar of iron, have no life, and are said to be non-living. A dead plant or animal could conceivably be placed in the latter category. Actually, however, dead things are not classified with non-living things; for the latter has no life, but more important still, it never did possess life. On the other hand, a dead thing, though it no longer possesses life, did so at one time. There are, therefore, three groups of matter from the biologist's point of view : **living** (or **animate**), **dead,** and **non-living** (or **inanimate**).

ANIMATE AND INANIMATE

Men of science and philosophers, from very early times, have attempted to define life, and even to prove that *all* living things possess souls. **Aristotle,** the great philosopher, who lived in the fourth century, B.C., and may be looked upon as the ' father of biological research ', spent a considerable time trying to establish the presence of souls in plants. But for many centuries it was not realised that to define life is practically impossible. In fact, even today, it cannot definitely be said what life is. As science students, all we can do is accept it as such, to take life for granted, and examine its features so far as is scientifically possible. The main differences between animate and inanimate, to be considered now, were emphasised chiefly by **Claude Bernard**, the nineteenth-century French physiologist, known chiefly for his work on the function of the liver and the pancreas, and on other features of human physiology.

One difference between the living and the non-living is the power of growth. Nearly all living things grow during some part of their life, whereas non-living things do not, at any rate in the biological sense. Non-living things such as crystals grow by the deposition of new material on their already-existing surfaces (**accretion**). Growth in living things takes place not only by this method, but also by the deposition of new molecules *interspersed* between those already present (**intussusception**). But many living things cease growth long before they die. Nevertheless, growth, through internal action, may be looked upon as being characteristic of living things.

The next difference is more diagnostic. It is well known that living things possess the power of producing new living things, that is, young ones like themselves. There are a number of different ways in which this is done, and the process is called **reproduction.** Not a single non-living thing can do this. So the power which living things possess of reproducing themselves seems to be an excellent distinction; but, it is not infallible, for there are exceptions. Many individual living things do not possess the power of reproduction, for example, most mules, and are then said to be **sterile.**

There is a further difference between living and non-living things. If an animal burns itself, it quickly moves from the source of heat. On the other hand, a non-living thing like a bar of iron does not. Also, most plants, especially green plants, need light; in fact, it is necessary to them. If such a plant were put in a darkened place, with just a little light coming from one point, its shoot would slowly bend towards the source of light. In these two cases, it is said that the animal **responds** to the heat and the plant responds to the light (see Chap. XX). The light and the heat are referred to as **stimuli.** Thus living things respond to stimuli. This phenomenon is known as **irritability.**

All life is in constant chemical communication with its environment. For example, all living things—both plant and animal—respire, usually by absorbing oxygen and, after certain chemical reactions have taken place, giving off carbon dioxide. But this process is so complicated, and varies so much in different organisms, that it is impossible to accept its occurrence as a criterion of distinction from the non-living in much of which analogous processes occur, such as, combustion (see Chap. XVII).

Protoplasm, the fundamental basis of life (see Chap. IX) is constantly undergoing physical and chemical changes. Life, therefore, is the resultant of these constantly occurring changes.

The apparent gap between the animate and the inanimate may possibly be occupied by **viruses** (p. 320), substances or organisms which cause plant and animal diseases, and by **bacteriophages** which attack and destroy bacteria. Both are excessively small, smaller than some organic molecules, but both reproduce themselves in the cells of the body attacked. But it is difficult, to be sure whether they are animate or inanimate.

PLANTS AND ANIMALS

There are two great groups into which living things may be classed: **plants** and **animals**. Both the plant and animal kingdoms are very extensive. It is customary, therefore, to regard the science of life under two comprehensive heads, namely, botany, the study of plants, and zoology, the study of animals.

In spite of this subdivision into botany and zoology, however, it is by no means easy to define the terms *plant* and *animal*. Just as a clear line of demarcation cannot be drawn between the living and the non-living, so also it may be said that *Nature knows no hard and fast lines of distinction between plants and animals*.

There are about 335,000 different kinds of plants, ranging from the microscopically small to the huge forest trees; and the animal kingdom is much larger even than this.

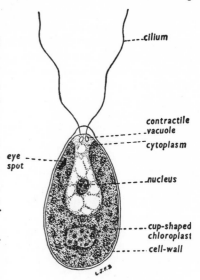

FIG. 1. *Chlamydomonas*, A UNI-CELLULAR PLANT (× 1500).

Many plants and animals are so small that they can only be seen with the aid of a microscope. Fig. 1, for example, is a diagram of a complete plant, highly magnified. This plant thrives in fresh water, and is called *Chlamydomonas*. It can be seen only with the aid of a microscope, and is clearly different in size and form from, for example, an elm tree.

Consider certain differences between plants and animals, realising that not one is absolutely diagnostic in that not one is true of *every* living plant and, conversely, of *every* living animal. It might be suggested that an animal such as a dog has a head, legs and tail, whereas a plant has not. The latter is true of all plants; but the former certainly is not true of all animals, as in the case of a jelly-fish or an oyster. Again, it might be said that plants have roots, stems, leaves and flowers, but that animals have not ; but there are thousands of plants which do not possess such organs, for example, *Chlamydomonas* (Fig. 1). So one could go on in this fruitless effort to find an *absolute* distinction between plants and animals, that is, one which involves all known cases. Nevertheless, there are certain differences which involve the majority of cases.

First, most animals are free to move about from place to place (**locomotion**), whereas the majority of plants are stationary. The result is that most plants must obtain their food without moving; but most animals are able to go in search of it. Yet there are many exceptions. Sponges and sea anemones, for example which are animals, are normally stationary; on the other hand, *Chlamydomonas*, a plant, can swim freely in the water.

Plants, in contradistinction to animals, are usually branched. This is clearly seen in the case of trees, shrubs and herbs; but there are many exceptions in other plants, such as *Chlamydomonas* and bacteria. Few animals are branched; in fact, most of them are of a definite shape. This is another good distinction between the two groups.

In most plants, each cell of which they are composed (see Chaps. IX and X) is surrounded by a firm wall, whereas in animals cell-walls seldom exist.

Also, the regions of growth in most plants are usually at, or near, the extremities of their organs, and growth at such extremities goes on, either continuously or periodically, through the life of the plant, giving a *continued embryology*. On the other hand, in animals, growth in length is seldom restricted to the ends of organs, and ceases in nearly all cases, as in man, long before death occurs.

NUTRITION

The main difference between plants and animals lies in the way in which they obtain their food.

All living things must have food in order to carry out the manifold processes necessary for their existence. Plants and animals absorb food-stuffs into their bodies. The entire process of taking in food, whether it be in plants or in animals and absorbing it into the system of the organism, is called **nutrition**.

All food-stuffs are chemical compounds, some simple and others very complex. The chief elements which unite to give the hundreds of different compounds in foods are: carbon, oxygen, hydrogen and nitrogen. Other elements are found in some foods, such as sulphur and phosphorus. Moreover, still further elements must be available, often in very minute quantities ; but these are frequently confined to certain kinds of plants and animals only. Just a trace is necessary, but this trace must be there if the organism is to remain healthy. Such elements are called **trace elements** or **micro-elements**.

Most elements are comparatively easy to obtain, since they are present in the soil and the air. Once extracted from these sources, they are built up by various complex chemical processes into the different forms of food. The manner in which the elements unite gives the clue to the difference between the majority of plants and animals. The compounds present in food-stuffs are not formed by the spontaneous union of the elements. The elements will not unite except in the presence of the green colouring matter which is so characteristic of green leaves, young stems, etc. This colouring matter is a mixture of several chemical compounds (see Chap. XV) and is called **chlorophyll**. *In very few cases are foods manufactured in Nature from the raw elements without the help of chlorophyll.* Therefore *the essential organs of food manufacture are foliage leaves* and, in some cases, other parts of plants which are green.

Thus, by the virtue of possesssing green leaves containing chlorophyll, plants can manufacture food-stuffs from the raw elements. On the other hand, animals cannot, for they do not possess chlorophyll. Yet animal food is essentially similar to plant food. Since, therefore, animals cannot manufacture food for themselves, they must perforce depend upon plants for it. Some animals depend directly upon plants for their food; for example, the rabbit, which in the wild state lives entirely on the vegetation around it. On the other hand, some animals eat very little plant food; for example, the dog, lion and many snakes. They consume other animals and animal products, such as bone, meat and milk. But such animals still ultimately depend upon plants for this food. For example, a cat can live almost entirely on milk, but the milk comes from the cow, and the food present in that milk has been built up from the grass and other plants which the cow had previously eaten. Thus we have two types of animals from the point of view of nutrition : those which depend upon plants directly, like the rabbit ; and those which depend upon plants indirectly, like the dog. There is a third type such as man, which depends upon plants directly by consuming them, and also indirectly by consuming other animals.

Here, therefore, lies the chief point of difference between the two groups. Plants manufacture their own foods ; animals take theirs already manufactured. The former is called **holophytic** nutrition, and the latter **holozoic**.

This distinction in mode of nutrition is not a fundamental one, for there are numerous exceptions. Many plants, such as the mushroom, are not green and contain no chlorophyll. They therefore are not holophytic (see Chap. XXIV). Also a few animals are holophytic. Such animals are all microscopic ; but they contain chlorophyll.

The following points of difference may therefore be said to apply to most plants and animals :

PLANTS	ANIMALS
1. The majority of plants are stationary. Only a few are capable of moving about from place to place (locomotion), for example *Chlamydomonas*.	**1.** Most animals are capable of locomotion. Some, however, live a sedentary life, for example, sponges and sea anemones.
2. The cells of most plants are surrounded by a cell-wall (see Chaps. IX and X).	**2.** The cells of most animals are not enclosed in a cell-wall.
3. Most plants are branched.	**3.** Most animals are of a definite shape, and are not branched.
4. The regions of growth in the majority of plants are at, or near, the extremities of organs, and growth occurs throughout life.	**4.** The regions of growth in animals are seldom localised at the extremities of organs, and further growth usually becomes impossible long before death.
5. Most plants manufacture their foodstuffs from the raw elements (holophytic nutrition).	**5.** Most animals must absorb their foodstuffs already manufactured (holozoic nutrition). They are incapable of manufacturing food from its elements.

CHAPTER II

THE PLANT KINGDOM

IT is clear that plants vary, to a considerable degree, in size and form. But though it is obvious that many plants differ from each other, it is just as clear that others resemble each other very closely. They are therefore classified into groups and sub-groups according to their structural affinities or differences.

SEED PLANTS AND NON-SEED PLANTS

Many plants bear seeds ; on the other hand, many other plants do not. The total number of *different* kinds of seed-bearing plants on the earth is approximately 120,000.

Plants, therefore, may be classified first into two great groups ; those which bear seeds (**Phanerogams** or **Spermophyta**) and those which do not (**Cryptogams**).· This classification, however, like all classifications, cannot be absolutely rigid ; for intermediate types, especially of past ages but now extinct, are known.

The plant kingdom can be subclassified into five divisions, three of which contain non-seed plants, and two, seed plants. The five divisions can be arranged roughly in order of complexity.

(1) **Thallophyta.** The simplest kinds of plants, which, however, are not necessarily the smallest, are found growing chiefly in the sea, lakes, rivers, ponds, and also on land usually in damp situations. They constitute the group Thallophyta. They are usually simple in structure because they can obtain their raw food materials from the water in which they are usually submerged. Since the water completely surrounds them, high morphological specialisation is scarcely necessary. An example of the Thallophyta is *Chlamydomonas* (Fig. 1). Another is the green, powdery growth seen on damp wood and on the bark of trees. This green powder is actually composed of millions of the small, microscopic plant called *Pleurococcus*. This plant is roughly spherical in shape as seen in Fig. 2. *Pleurococcus* will not thrive unless plenty of water is available.

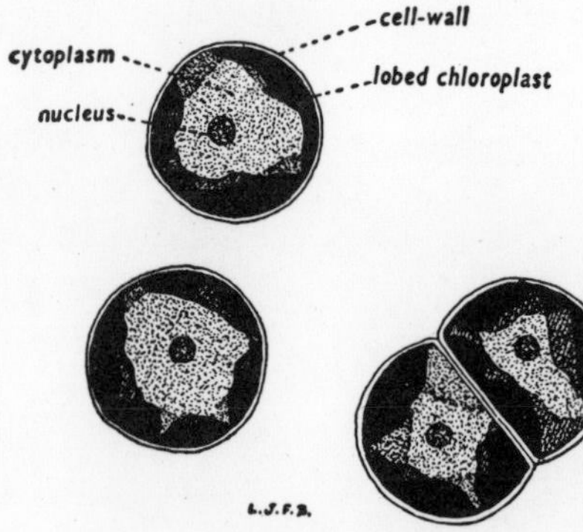

FIG. 2. *Pleurococcus*, A THALLOPHYTE.
(× *c*.840)

Another plant, simple in structure, and belonging to the same group, is the one responsible, together with other plants, for the green scum seen floating on the surface of stagnant ponds and pools. If a little of this scum be taken and placed in a glass of water, it will be seen to be composed of threads or filaments. Many of these filaments are plants called *Spirogyra* (Fig. 3).

Next we come to those plants which, though comparatively simple in structure, are much more complicated than those so far considered in this group. They are

6

the seaweeds. Some seaweeds are very familiar, for example, the sea wrack (*Fucus serratus*) (Fig. 6). It appears to be formed of stems and leaves ; but actually it is not, for although those structures are different in shape, their internal structure is very similar, whereas the internal structures of true stems and leaves differ considerably. There are three groups of seaweeds, the **green,** the **brown,** and the **red.** The green seaweeds are found in very shallow water, and are very often exposed to the atmosphere. The brown seaweeds either grow right under the water, or between tide marks, so that they become exposed at low tide.

Some of the brown seaweeds are very large, as for example, some of the giant kelps (for example *Nereocystis*) which, growing in some fifty feet of water, attain a length of up to 50-60 feet. A few types are not attached, such as the Gulf weed, *Sargassum*, which floats in the sea by means of gas bladders and which multiplies by simple fragmentation of the plant body (thallus, p. 307). Certain forms of this plant are to be found in the Sargasso Sea, an area 250,000 square miles in extent, east of the West Indies. The red seaweeds are the least common of the three groups and the majority of them grow in deeper water than the other two, under conditions of reduced light intensity. Some, however, grow between the tide marks and particularly in rock pools.

One would think that the brown and the red seaweeds, since they are not green, could not be holophytic ; but actually they are, for they do contain chlorophyll. Brown seaweeds contain chlorophyll together with a brown pigment **fucoxanthin,** which masks the green colour. Red seaweeds contain chlorophyll together with red pigments, **phycoerythrin** and **phycocyanin.**

All the plants so far considered are placed in the sub-group of the Thallophyta called **Algæ.** The other sub-group of the Thallophyta is the **Fungi.** This group contains a great variety of plants. Many of them are of microscopic dimensions, whereas others are comparatively large and conspicuous (Fig. 4). Some fungi are useful to man ; for example, the common edible mushroom (Fig. 4) and the truffle, a fungus which is an article of diet chiefly on the Continent. A curious feature of this plant is that the edible portion, like the rest of the plant, grows below the surface of the soil (Fig. 4). Many fungi are very poisonous, such as certain toadstools or ' seats of death '.

Other familiar fungi are the moulds which grow on decaying plant and animal material such as damp bread, jam, old leather and farmyard manure (Fig. 4). Certain fungi are useful in flavouring various forms of cheese ; others are of inestimable value medicinally (p. 339). Other fungi, closely related to the moulds, grow on living plants and animals, thus causing disease (Fig. 4).

Closely related to the fungi is a great group of plants called **Bacteria.** All the bacteria are of microscopic size. They may be spherical, 10d-shaped or spiral (Fig. 5). They are of outstanding importance. Some are beneficial to man and

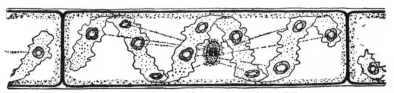

FIG. 3.   *Spirogyra*, A THALLOPHYTE (× *c.* 300).

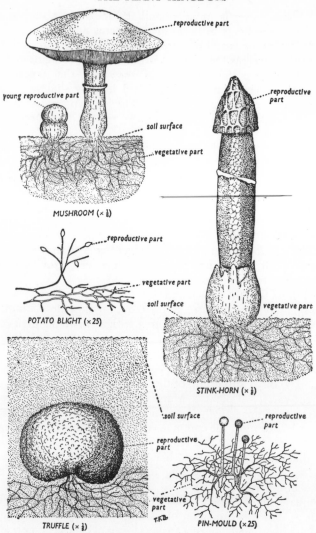

FIG. 4. TYPES OF FUNGI (THALLOPHYTA).

the lower animals, others are useless but harmless, whereas others are very harmful. To get some idea of the size of bacteria, the number of times that the diagram (Fig. 5) is enlarged should be carefully noted.

The most important feature of fungi and bacteria is that they do not contain chlorophyll although many fungi are brilliantly coloured. Therefore the nutrition of these plants cannot be holophytic (see Chap. XXIV).

(2) **Bryophyta**. The Bryophyta form the second group of the non-seed-bearing plants. This group contains the mosses, **Musci,** and the liverworts, **Hepaticae**, which are much more complicated in structure than the Thallophyta (Fig. 6). The majority of liverworts grow in damp situations or actually in fresh water. The plant body is called a thallus, and is similar in appearance

to a green seaweed. Mosses, on the other hand, have a simple stem and leaf structure.

(3) **Pteridophyta.** Still more complicated in structure than the Bryophyta are the Pteridophyta, which constitute the third group of Cryptogams. This group contains plants with a definite stem and root structure and a variety of leaf forms. Such plants are like flowering plants, but differ from them in that they do not in any circumstances bear flowers. All the ferns belong to this group. Well-known examples are the common bracken (*Pteridium aquilinum*), which grows on heaths and in open woods, the shield or male fern (*Dryopteris filix-mas*) (Fig. 6), which grows in damp, shaded places such as woods and the banks of secluded streams, and the hart's tongue fern (*Phyllitis scolopendrium*), which grows on moist banks and on walls. In tropical countries,

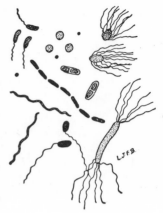

FIG. 5. TYPES OF BACTERIA (× 1500).

especially in the forests, ferns attain a great size, even to the extent of being tree-like (Fig. 7). Altogether, throughout the world, from equator to Arctic regions, there are about 7,000 different ferns which belong to the class **Filicales.**

Other members of the group Pteridophyta are the horsetails (*Equisetum* species) (Fig. 6). These form a very ancient family which is fast dying out. In early geological times, especially during the Coal Age of the Carboniferous Period, the horsetails dominated the flora, some of them attaining the size of trees. The living species, of which several are common in Great Britain, are nearly all small plants, although one growing in the American tropics may reach a height of thirty feet. These form the class **Equisetales.**

Clubmosses (Fig. 6), which must not be confused with the true mosses, are also Pteridophytes. Examples of these are found on heaths and moors in Great Britain. They form the class **Lycopodiales.**

(4) **Gymnosperms.** The Gymnosperms are the first group of the Phanerogams. Nearly all of them are trees. They bear flowers and seeds; but, unlike the next group of flowering plants such as the rose, dandelion, and pea, *the seeds are borne naked* and exposed instead of being enclosed in a case. For this and other reasons they are placed in a separate division. By far the most important examples of gymnosperms are the conifers, such as pines (*Pinus* species), spruces (*Picea* species), and larches (*Larix* species)). They are so called, because their seeds are borne collectively in **cones** (Fig. 8). They are nearly all large plants and, the group contains the largest and the longest-lived examples of plants that exist to-day.

For example, the redwood (*Sequoia*) or giant tree of California is so big, that in one case a carriage drive has been cut through the trunk, without killing the tree (Fig. 9). One of the largest of these redwood trees is still growing on the south side of San Francisco Bay. It must be of very great age, although the exact date of its birth is not known. Its history was traced in 1931, and it was found that its earliest records go back to the first Spanish explorers of that region. In 1769 Gaspar da Pabola camped beneath it, and in 1777 it was already a very large tree with a height of 137.5 feet and a trunk with a diameter of 15 feet. Now the diameter is 23 feet.

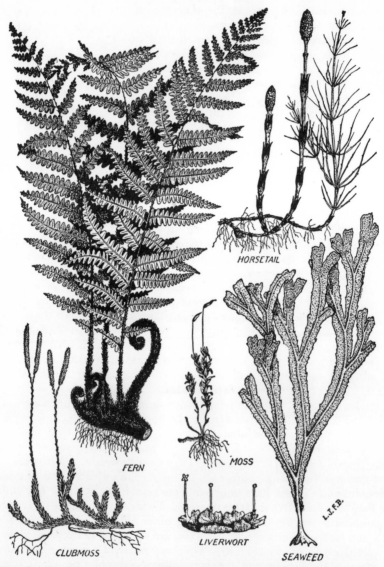

FIG. 6. TYPES OF CRYPTOGAMS: THALLOPHYTA (SEAWEED), BRYOPHYTA (MOSS, LIVERWORT), AND PTERIDOPHYTA (CLUBMOSS, HORSETAIL AND FERN).

(5) **Angiosperms.** The fifth and final division contains the most popular and well-known of all the plants. They are the flowering plants which bear *seeds enclosed in some kind of case* (Chap. XXI).

Thus the plant kingdom is grouped in five divisions, namely : Thallophyta, Bryophyta, Pteridophyta, Gymnosperms, and Angiosperms.

An interesting thing to note in connexion with this classification is that apart from the fungi, which, as will be seen later, form a very exceptional group, all the Thallophyta live either entirely submerged in fresh or salt water, or in very damp situations. The Bryophyta and the Pteridophyta live chiefly in very damp situations or entirely submerged. On the other hand, nearly all the Gymnosperms and Angiosperms live on dry land.

The classification of plants is represented in the following table :

Division		*Class*
THALLOPHYTA.	1.	**Algae**—Simple, green plants. Many microscopic, also green scum, seaweeds, etc.
	2.	**Fungi**—Simple, non-green plants. Yeasts, moulds, mildews, blights, smuts, mushrooms, bacteria, etc.
	3.	**Lichens**—Algae and fungi growing together for mutual benefit.
BRYOPHYTA.	1.	**Musci**—Mosses. Green plants with stems and leaves, but no true roots.
	2.	**Hepaticae**—Liverworts. Usually thalloid organisms.
PTERIDOPHYTA.	1.	**Filicales**—Ferns. Green plants with stems, leaves and roots, and a vascular system.
	2.	**Equisetales**—Horsetails. Green plants with stems, leaves and roots, and a vascular system.
	3.	**Lycopodiales**—Clubmosses, etc Green plants with stems, leaves and roots, and a vascular system.
GYMNOSPERMS.	1.	**Coniferales**—Mostly cone-bearing plants with naked seeds.
	2.	**Gnetales**—Tropical plants.
	3.	**Cycadales**—Tropical plants.
	4.	**Ginkgoales**—Tropical plants.
ANGIOSPERMS.	1.	**Monocotyledons**—Flowering plants with seeds borne enclosed in a case. One cotyledon.
	2.	**Dicotyledons**—Flowering plants with seeds borne enclosed in a case. Two cotyledons.

Dr. K. Biswas

FIG. 7.   TREE FERNS.

Harold Bastin

FIG. 8. OLD CONES OF LARCH, A GYMNOSPERM.

There seems thus to be a gradual transition from simple to complex and from aquatic to terrestrial plants. In this respect, plants are similar to animals. There is reason for believing that all types of life in the first place began in the sea ; and throughout the ages which stretch over millions of years, as the various organisms became more complex, they invaded the land, and over millions of years became modified according to their change in environment. To-day there exist types representing all stages of this process in both the plant and the animal kingdoms.

Plants and animals living in the sea (**aquatic**) and those living on the land (**terrestrial**) are very familiar; but there are examples of plants and animals which have 'got more or less half-way' between sea and land. Animals of this type, such as the frog, which live part of their time in water and part of their time on dry land, are said to be **amphibious**. Most Bryophyta and Pteridophyta, at one time in their life-history must have liquid water available, and at another time dry conditions, especially during the process of reproduction. They may therefore be said to be amphibious (see Chaps. XXV and XXVI).

Moreover, as more is learned by the student about plant (and animal) form, structure and habit, many observations will come to light which support the view that throughout the evolution of life there has been a slow invasion of land by organisms from the sea.

TYPES OF ANGIOSPERMS

The typical angiosperm is composed of roots, stems, leaves and flowers ; but it is very clear that these structures vary considerably in dimensions, shape, colour

E.N.A.

FIG. 9. A ROADWAY CUT THROUGH THE TRUNK OF A LIVING BIG TREE OR GIANT
REDWOOD (*Sequoia*).

This is at Wawona, California. The tunnel is 10 ft. high and 9½ ft. across at the base.

Canon Lonsdale Rugg

FIG. 10. ELM TREE IN VICTORIA PARK, BATH.

and so forth. These different shapes are often so obvious and familiar that a cursory glance at them is often sufficient to identify the plant. Nevertheless, in spite of these differences, many plants possess certain features in common, and for that reason they can be divided into groups, according to their vegetative structure.

Some plants grow to a great size. The stem becomes thick and woody, and is called a **trunk** or **bole**. These plants are called **trees** (Fig. 10).

A much greater number of plants vary considerably in height from a few inches to several feet, for example, buttercups, primroses, grasses, foxgloves and sunflowers. Although they all contain a certain amount of wood, this varies, though it never amounts to as much as is present in a tree. For example, the dandelion stalk contains very little wood, and is therefore much softer and juicier than the sunflower stem which contains a much greater amount of wood. All these plants are called **herbs**. They are probably the most successful plants, considered from the point of evolution (Chap XXVIII).

Intermediate between the trees and the herbs are the **shrubs.** They are bushy plants with many branches and are all very woody. Unlike the trees, they do not possess a trunk, that is, the main stem is not so pronounced. Examples of shrubs are the privet, bramble, box and gorse.

LENGTH OF LIFE

The length of life of organisms varies within extremely wide limits. The shortest-lived creatures, are, of course, the microscopic ones. For example, some bacteria can reproduce themselves within half an hour of their origin. In the animal kingdom, there is great diversity. Many house-flies complete their life-history within ten days, whereas elephants have been known to live for more than a hundred years. Among the oldest living animals are the Galapagos turtles, some of which have been known to live to be two hundred years of age.

But the oldest things which are living to-day are certain plants. Yet a very large number of herbs carry out the whole of their life-history in one season. That is the seed germinates, the plant develops and finally produces its own seeds all within the space of a year. Such plants are therefore called **annuals.** Wheat (*Triticum vulgare*), barley (*Hordeum vulgare*), and the common poppy (*Papaver rhoeas*) are examples of annuals.

Some plants can complete their life-history so quickly, that the offspring which they produce can also complete their life-history in the same season. Thus two generations are produced in one season. Sometimes even three, or four, generations are produced within a season. Such plants are called **ephemerals**, because their life is so ephemeral or short-lived. The groundsel (*Senecio vulgaris*) is an ephemeral, and its quick growth makes it an objectionable weed in the garden.

Some plants take two years to complete their life-history, for example, the foxglove (*Digitalis purpurea*) and the beet (*Beta vulgaris*). They develop vegetatively in the first year and produce seeds in the second, and are therefore called **biennials.** Both annuals and biennials are characterised by the fact that they die after producing one lot of fruit ; they are therefore said to be **monocarpic**. Some monocarpic types, however, have a much longer life-history. For example, the century plant (*Agave*) lives purely vegetatively for 20-100 years ; it then flowers, and after setting seed, dies.

Many plants go on living year after year, sometimes producing seeds every season and sometimes only once in several seasons. These are **perennials** and are of two distinct types. In some, such as the michaelmas daisy (*Aster* species) and solomon's seal (*Polygonatum multiflorum*), the aerial shoot dies down each year after flowering ; but the underground parts remain alive and send up new flowering shoots in the following season. Such plants are known as **herbaceous perennials.** Contrasted with these are the shrubs and trees in which the aerial parts of the plant grow on year after year, producing an annual crop of fruits ; these are called **woody perennials**. Since both these types of perennial plant bear flowers and fruit many times they are said to be **polycarpic.**

Prof. C. J. Chamberlain

FIG. 11.   THE BIG TREE OF TULE.

The woody perennials, particularly the trees, often live for many years.   The biggest, and possibly the oldest tree in the world, is the cypress called the Big Tree of Tule, still growing in Mexico (Fig. 11).   Its trunk is 50 feet in diameter where it begins branching.   It was a big tree when the ancient pyramids of Egypt were being built.   The General Sherman Tree in the Sequoia National Park, California, is another well-known veteran—probably 3,500 years old.   The Fortingall Yew in Perthshire is considered to be one of the veterans of European vegetation.

EVERGREEN AND DECIDUOUS PLANTS

Perennials in temperate regions seldom maintain the same appearance through-out the year.   The majority of British trees, for example, shed all their leaves in autumn.   For this reason they are said to be **deciduous**.   Those trees and shrubs which do not shed their leaves for the winter months are **evergreen**.   Among the best-known evergreens of temperate regions are : most of the conifers except larch (*Larix decidua*) and deciduous cypress (*Taxodium distichum*) ; spurge laurel (*Daphne laureola*), ling or heather (*Calluna vulgaris*), cross-leaved heath (*Erica tetralix*), ivy (*Hedera helix*) mistletoe (*VViscum album*), the periwinkles (*Vinca minor* and *V. major*), many grasses, holly (*Ilex aquifolium*), sweet bay (*Laurus nobilis*), Japanese laurel (*Aucuba japonica*), rhododendrons (*Rhododendron* species), and a host of exotic shrubs. But it must be realised that although an evergreen bears leaves all the year round it *does* shed its leaves. It does so continuously, but never all at one time, and each leaf remains from one to several seasons.

Leaf-fall (see Chap. XIV) is clearly related to seasonal climate. The leaves of deciduous trees are shed just before winter sets in, and that is why plants growing in the low-lying, humid, tropics, where there is little seasonal variation in climate, are nearly all evergreen. Some plants may be said to be partially evergreen, that is, although they are never leafless like true deciduous plants, they bear more leaves in the warm season than in the cold. Privet (*Ligustrum vulgare*) is an example of this.

PRACTICAL WORK

Draw and describe typical examples of the plant kingdom. In the case of large plants, such as shrubs and trees, it is useful to take photographs, provided a written description is given at the same time.

There are many common types of each plant group ; suggestions with regard to types are given below :

1. Collect some of the green encrustation from the bark of a tree and mount in water and examine under the microscope (for directions, see p. 119). The spherical green structures are plants of the Alga, *Pleurococcus*.

2. Examine a plant of *Fucus* as an example of a brown seaweed. Note the disk-shaped or branched holdfast ; the stalk or stipe ; and the expanded lamina showing a thickened mid-rib and wings. Note also the indentations at the tips of the thallus where the growing points are situated.

3. Mount a small fragment of mould from old bread or decaying fruit and examine under the microscope. Note the filaments which constitute the vegetative body of this fungus.

4. Examine the thallus of *Pellia* or *Marchantia* as an example of a liverwort. Note that the general structure is very like the lamina of *Fucus* but that it is attached to the soil by means of hair-like rhizoids.

5. Examine any woodland moss, such as *Mnium* or *Catharinea*. Note the stem, small leaves, and the rhizoids by which the plant is attached to the soil.

6. Examine, in the field if possible, plants of bracken and the male fern. Examine also plants of the horsetail (*Equisetum arvense*). Note the presence of true roots, stem and leaves in these examples of Pteridophytes.

7. Observations should be made in the general structure of a pine tree (*Pinus sylvestris*) as an example of a Gymnosperm. Examine a cone and note the seeds lying naked on the cone scales.

8. Examine the general structure of any available flowering plants. Note that the seeds are always enclosed in a fruit, the characteristic feature of Angiosperms.

CHAPTER III

HISTORICAL SURVEY

Man's earliest interest in plants was purely utilitarian. Primitive man was naturally interested in the plants he had found good to eat, and in those he had found poisonous. At first, when man was a nomadic hunter, these plants would be wild ones, chosen presumably by a process of trial and error ; but quite early some of the plants were deliberately grown and agricultural practices started. In Europe, the age of agriculture probably began ten to twelve thousand years ago, and there is good evidence that Neolithic man grew wheat, barley and millet for food, and flax to provide the fibres from which a crude cloth was woven. It has been suggested that accidental or sacrificial scattering of wild food fruits over ground turned up during the burial of the dead might have led to the observation that an exceptionally good crop grew on the exposed soil and hence to the deliberate cultivation of the ground and the sowing of seeds.

Apart from this interest in plants as a source of food, early man seems to have discovered that decoctions of certain plants would alleviate pain and heal wounds or have a narcotic effect. Such knowledge tended to become the monopoly of the ' medicine man ', or ' witch doctor ', and it may be noted that such is still the case in many primitive tribes even today.

This interest in medicinal plants was, in fact, the line along which botany progressed for a considerable period. In early Greek and Roman civilisations the medicine man was replaced by collectors of plants which had real or supposed medicinal virtues and by pharmacists, men who compounded the medicines from the collected drugs. The necessary careful examination and identification of the medicinal plants naturally led to accurate observation of their external form and it was from such observation that the science of botany was born (Fig. 12). The first exponent was **Theophrastus** (born c. 370 B.C.), a pupil of Plato and Aristotle. In *The History of Plants* (nine books) and *The Causes of Plants* (six books) he dealt mainly with the economic uses of plants but also recognised a division of plants into annuals, biennials and perennials and a distinction between woody and herbaceous types. The early philosophers were, however, much concerned with abstract ideas, as, for example, those of Aristotle already mentioned in Chapter I. This speculative botany prevented the development of botany as a science, since ideas and wide generalisations were promulgated without any sound basis of observed facts. In any science, the collection and grouping of facts must come first ; only then can sound conclusions be drawn.

There followed a period of many centuries during which little progress was made, although the *Materia Medica* of **Dioscorides** (A.D. 64) was important in so far as it recognised the chief groups of flowering plants. Printed herbals, books describing medicinal plants of real or imaginary virtue, began to appear about 1470 ; but the descriptions of plants were poor and the illustrations were often more curious than accurate.

Botany as a science was reborn in the early part of the sixteenth century,

From Petrus Crescentius's ' Opus Ruralium Commodarum '
FIG. 12.   A MEDIEVAL HERB GARDEN.

largely through the work of a group of herbalists who have been called ' The German Fathers of Botany '.   These were **Brunfels** (1464-1534), **Fuchs** (1501-1566), **Bock** (1498-1554) and **Cordus** (1515-1544).   Their works are noteworthy for the excellence of their illustrations which are for the most part accurate and naturalistic representations of the plants described.   Apart from the listing and illustrating of medicinal plants, various interesting observations on the structure and distribution of plants were made.   For example, Bock carefully noted the localities in which the plants occurred and Cordus observed the root nodules on the roots of the Leguminosae (p. 455).

Many herbals in addition to those mentioned were published between 1470 and 1670 ; but towards the end of the sixteenth century and in the early years of the seventeenth century, botany became established as a pure science developing along various lines and divorced from the purely medical aspect of the herbals. The history of these developments will be considered separately.

PLANT CLASSIFICATION

Although many of the herbals were mere catalogues and descriptions of medicinal plants, some of the writers of these works attempted to arrange the plants in groups, that is to classify them.   Early groupings were very broad ones, for example, into trees, shrubs and herbs ; but some of the best herbalists saw the desirability of associating together " such plants as Nature seems to have linked together by similarity of form "

More scientific attempts to classify plants were made in the later years of the sixteenth century. For example, **Caesalpino,** an Italian professor, classified plants on the basis of the characters of the fruits and seeds. Other attempts were made using leaf characters as the criterion. **John Ray** (1628-1715), of Cambridge, published an important work, the *Historia Plantarum Generalis,* in which the number of seed leaves or cotyledons in the embryo was used as a guiding character, thus founding the two great classes of flowering plants known as the Dicotyledons and Monocotyledons (p. 12). The first Flora of the British Isles was also published by Ray in 1690.

Many other botanists were concerned in efforts to classify the rapidly increasing number of known plants. It was finally **Linnaeus** (1708-78), an outstandingly acute observer, who brought order into the system of classification. He devised what he himself recognised to be a very artificial scheme of groupings based largely on the number of stamens and styles. He did indeed begin a more natural classification, but he did not live to complete this. The main contribution of Linnaeus to botanical science was the establishment of the **Binomial System of Nomenclature. Bauhin** had earlier already recognised the importance of naming plants in a uniform manner and instituted the practice of giving each plant a double name consisting of two Latin words, for example *Solanum tuberosum* (potato), the first being the name of the genus and the second the name of the species. This practice was established by Linnaeus, and it may be noted that a large proportion of our native plants were first described and named by him, as is indicated by the fact that their names in any Flora are followed by an abbreviated form of his name, for example *Ranunculus bulbosus* Linn.

There followed a period of rapid development with the work of de **Jussieu** (1748-1836), **Pyrame de Candolle** (1778-1841), whose system of classification, based largely on the nature of the perianth, was almost universally accepted until 1860, and **Robert Brown** (1773-1858). Then came the association of two famous British botanists, **George Bentham** (1800-84) and **Sir J. D. Hooker** (1817-1911). The *Handbook of the British Flora* was first published by Bentham in 1858. It was revised by Sir Joseph Hooker in 1886 and again by **Dr. A. B. Rendle** in 1937, and it is still one of the most widely used British Floras.

Most of the early systematists believed in the doctrine known as the **Constancy of Species,** which held that each type of plant was unalterable and that it had remained the same in form and structure since the Creation. Doubts about the truth of this doctrine were, however, entertained, notably by **C. Darwin** and **Wallace** (p. 398). Both indeed came to the conclusion that species are variable, that there is a constant struggle for existence and that favourable variations will tend to survive, thus giving rise to new species. This is the process of **Natural Selection** (p. 398). In 1859, Darwin's *Origin of Species by Natural Selection* was published, and this book had a profound effect on the views of biologists everywhere. It came to be generally accepted that species are not constant and that the plants and animals of today are in fact the product of long-continued evolutionary processes. Efforts to classify the plant kingdom took on a different aspect now that it was realised that the classification should aim at constructing a genealogical table and not merely a directory of plants. It has, however, proved to be very difficult, and indeed it may well remain impossible, to draw up a scheme which will truly reflect the blood relationships of the various groups of plants. It is relatively easy in many cases to group together related species. For example, *Ranunculus acris, R. bulbosus,* and

R. repens are, even to a non-botanist, obviously buttercups closely related to one another. This similarity is due to the fact that they arose from a common ancestor. A more careful examination is necessary to determine that species of the genus *Ranunculus* are related to other genera such as *Trollius* and *Anemone*, but even here the resemblance is marked and it seems quite reasonable to group them in a family, the RANUNCULACEAE, the assumption being that all the genera in this family had a common ancestor at some time in the distant past. Families which resemble one another are grouped into orders ; orders can be grouped into classes and classes into phyla but it becomes more and more difficult to trace the evolutionary connexions between these larger groups.

Modifications of the scheme of classification for the flowering plants are still being suggested and new lines of evidence sought, such as for example the cytology and genetics of the various species.

This brief survey has been confined to the classification of the flowering plants. Indeed, it was not until well on into the nineteenth century that much attention was paid to the lower plants.

MORPHOLOGY AND ANATOMY

The work of the early herbalists necessarily involved a meticulous examination of the external form of the plants with which they were concerned. The mass of observations thus accumulated laid a sound foundation for the understanding of the various plant organs and the knowledge concerning these expanded with the progress of classification. The general acceptance of the idea of evolution after 1859 brought a new point of view to bear on the numerous modifications of leaf, stem and root and such modifications were interpreted, often without experimental evidence, as adaptations to special needs, evolved in the effort to succeed in the struggle for existence. The origin and significance of such adaptations, as, for example, those characterising aquatic plants or climbing plants, is still a debatable question.

The investigation of the internal structure or anatomy of plants commenced much later than that of the external morphology. Early ideas scarcely advanced beyond the view of Theophrastus, who regarded the plant body as being composed of sap, veins and flesh. It was not until the invention of the compound microscope at the end of the sixteenth century that further progress could be made.

Robert Hooke (1635-1703) was one of the first people to look at a variety of objects under the microscope. Among the objects investigated were charcoal and cork which he described as being " all perforated and porous, much like a honey-comb ". He referred to the cavities as " cells " and this term is still used, though now with a wider significance. Hooke's results were recorded in his famous book, *Micrographia*.

Contemporary with Hooke were two observers of greater scientific merit : **Malpighi** (1628-94), an Italian professor of medicine, and **Nehemiah Grew** (1641-1712), a British physician. Both carried out important pioneer work on plant anatomy and published their work almost simultaneously. In his classical work, *The Anatomy of Plants*, Grew dealt with the structure and distribution of tissues and speculated on their functions ; he observed the important breathing pores of the plant, the stomata, and demonstrated the chloroplasts.

These early investigators tended to regard the cell as being essentially a cavity surrounded by a wall.   The contents of the cell, which we now know to be the essential living part, were not noted until 1772.   In 1831, **Robert Brown**, already mentioned as an outstanding systematist, observed the rotation of the cell contents in the hairs on the stamens of *Tradescantia* and also observed the nucleus which was to prove to be such an important structure.   Robert Brown is also to to be remembered as the first observer of the oscillation of minute granules in the cell contents, a movement now known as ' Brownian movement '.   The cell contents were also observed by **von Mohl** (1805-72).   He insisted that these were the same as the substance of animal cells, and, in 1844, he applied the term **protoplasm**, already coined by **Purkinje** in 1839, to the plant cell contents.

The origin and development of the plant cells and tissues was not accurately known for a long time but the work of **von Mohl, Schleiden** (1804-81) and **Nageli** (1817-91) established the **Cell Theory**, an important generalisation to the effect that all cells, whatever their ultimate structure and function, arise from the division of pre-existing cells.   By the middle of the nineteenth century the cell theory was well established and protoplasm was recognised as the physical basis of life.

Then followed a period of investigations into the nature of protoplasm and, in particular, attention was directed to the structure and behaviour of the nucleus. The recognition of chromosomes and their behaviour as the vegetative nucleus divides, the process of mitosis (p. 107) and their different behaviour in the divisions giving rise to spores in meiosis (p. 111) was based on the work of many investigators.

Then, when Mendel's laws of inheritance were rediscovered about 1900 it became clear that the chromosomes represented the physical basis of these laws. In this way the science of cell structure and particularly nuclear structure (**cytology**) was linked with the science of **genetics**.

More recent work has added much to our detailed knowledge of these matters, and at the present time efforts are being made to elucidate the biochemistry and biophysics of the cell contents and particularly of the nucleus.   Although the goal is as yet infinitely distant, the aim is to elucidate the mechanism of heredity, and, perhaps, eventually the physical nature of life itself, in terms of chemistry and physics.

Since the establishment of the cell theory one hundred years ago, our knowledge of tissue systems in plants has been vastly extended and there is now available a fairly complete picture of these.   Nevertheless, many points still await investigation, particularly those relating to the differentiation of tissues from the growing points and to minute structures difficult to observe under the ordinary microscope. The electron microscope is proving a useful tool in the investigation of minute structural detail.

The mass of anatomical detail accumulated during the last century has many applications of economic importance.   For example, a detailed knowledge of wood anatomy plays a vital part in dealing with timber and other forest products ; and, as another example, a meticulously accurate and wide knowledge of anatomy is needed in the science of pharmacognosy for the identification of impurities in samples of plant organs used in the preparation of drugs.

Finally, it may be mentioned that at the present time there is a revival of interest in what is termed **experimental morphology,**—an attempt to investigate by means of experiments the causes underlying the form, structure and position of plant organs, for example, the factors underlying the origin of leaves and the definite arrangement of these in various plants.

PHYSIOLOGY

The development of classification, morphology and anatomy as outlined above depended upon increasingly accurate observation and methods of observation. Knowledge concerning the functioning of plant organs and tissue, that is **plant physiology,** must however, be obtained by carefully devised and accurately observed experiments.

It was not until the beginning of the eighteenth century that an experimental approach to plant physiology was initiated. Theophrastus had recognised the absorbent functions of the root system and the circulatory system of the stem ; but he was completely ignorant concerning the nutrition of plants. For many centuries little progress beyond these crude ideas was made and even Grew and Malpighi, who contributed so much to the advance of anatomy, were content with mere speculations as to the functions of the various parts they described.

It was left to **Stephen Hales** (1677-1761) to replace speculation with purely experimental methods and his contribution to physiology will be mentioned again later. The development of our knowledge concerning some of the main aspects of this subject will now be considered.

Photosynthesis

Knowledge concerning photosynthesis (p. 175) could not be gained until after Black had discovered carbonic acid in 1754 and Priestley had confirmed in 1774 the discovery of oxygen. In fact the unravelling of the nutrition of plants has depended throughout its history on progress in chemistry and physics.

The work of **Priestley** (1733-1804), **Ingen-Housz** (1730-99), **de Saussure** (1767-1845) and **Boussingault** (1802-87) led to the establishment of the fact that plants take in carbon dioxide and give out oxygen during the process. It was of course natural that the gaseous interchange involved should be the first part of the process to be investigated. It was not until some time later that the products of photosynthesis were investigated. It was about 1860 that **Sachs** demonstrated that starch is, as he said, the first visible product, though it may be noted that von Mohl had much earlier suspected this to be the case. The general reaction of the photosynthetic process was thus made clear. Many investigations since the time of Sachs have, however, shown that a hexose sugar is the first carbohydrate formed and that a conversion to starch follows.

The mechanism and course of photosynthesis were now investigated by a number of workers ; but the final solution of these problems has not even now been reached. A few of the major results may be noted. The composition of chlorophyll was determined by **Tsvett**, who used chromatographic methods, and by **Willstätter** and **Stoll** using the principle of differential solubility. (p. 175)

The intermediate stages in the elaboration of the hexose sugar produced have been the subject of much discussion. In 1870 **Baeyer** suggested that formaldehyde might be the intermediate substance—a view that has been widely accepted. In recent years the use of ' marked atoms ' or tracers, such as the radioactive isotope of carbon, ^{14}C (p. 181), has opened up promising methods of tracing the intermediate substances which have for so long escaped detection.

Water Relations

Knowledge concerning the mode of entrance of water into the plant and its subsequent passage through the plant hardly advanced from the time of Theophrastus until the beginning of the eighteenth century. Grew was aware that the roots absorb water and the leaves give off water vapour ; but the first well-established facts came from the experiments of Hales (p. 185). These demonstrated the passage of water through the wood of the stem into the leaves. There remained, however, the problem as to the nature of the forces causing the ascent of the water even to the top of a tall tree. Very diverse views have been put forward on this question. Some of them regard the activity of the living cells of the wood as the important factor, but these seem to have been disproved by experiments showing that the movement of water still takes place even if the living cells are killed. Contrasted with the ' vital theories ' of the ascent of sap are the ' physical theories ' which dispense with the activity of the living cells as a factor in the upward movement of water. The most generally accepted view is that put forward by **Dixon** and **Joly** in 1895 and known as the ' cohesion theory ' (p. 188).

The solution of the problem as to how water enters the root hair and passes into the wood of the root followed from the work of **Graham** on colloids and crystalloids (p. 98) and that of **Pfeffer** and **de Vries** on osmotic phenomena (p. 123). Recent work, however, suggests that factors other than osmosis may be concerned.

Mineral Requirements of the Plant

The earlier physiologists had no knowledge of the importance of inorganic mineral substances to the life of the plant. It was **de Saussure** (1767-1845) who first proved by quantitative methods that the mineral substances are essential for the growth of plants. It was, however, the work of **Sachs** and **Knop** about 1860 which demonstrated the importance of nitrogen, sulphur, phosphorus, potassium calcium, magnesium and iron, the method used being that of ' water cultures ' (p. 59). Subsequent work has shown this list to be incomplete. In 1905 **Bertrand** showed that small quantities of manganese had a favourable effect on growth and later experiments have shown this element to be essential. More recently, zinc, boron, copper and molybdenum have been added to the list of essential elements. Since these substances need only be present in dilutions of the order of one in five millions they are described as trace elements or micro-elements (p. 58). Much research is being carried out on these nowadays.

The inability of green plants to absorb nitrogen from the atmosphere was demonstrated by Boussingault more than a century ago ; at the same time, by means of sand cultures he demonstrated that plants obtain their nitrogen from the soil in the form of nitrates. The process of nitrification in the soil (p. 140) was the subject of much research towards the end of the nineteenth century and **Winogradsky** succeeded in isolating the bacteria concerned. The peculiar case of the Leguminosae (p. 138) also attracted much attention, and the solution of the problem was largely due to **Hellriegel** and **Willfarth**, who in 1866 published the results of their experiments on peas and other leguminous plants. The organism concerned with the fixation of nitrogen in the root nodules was first isolated by **Beijerinck** in 1888.

As indicated above, the basic facts about the nutrition of plants are now known. Space will not permit a consideration of the history of our knowledge concerning the synthesis of proteins and fats (pp. 71 and 74) ; but the ways in which the plant uses its foods may, however, be briefly considered.

Respiration

A considerable part of the food elaborated by the plant is utilised in the formation of new cells at the growing points ; the remainder is largely used in giving the energy necessary for the growth of the plant. This energy is provided by the process of respiration (p. 197).

As far back as the time of Ingen-Housz and de Saussure the gaseous interchange, that is the intake of oxygen and the giving out of carbon-dioxide, was understood. **Dutrochet** in 1837 established the fact that respiration in plants is fundamentally the same as that in animals although it is masked during the day by the gaseous interchange associated with photosynthesis. It was not, however, until about 1868 that **Sachs** established the vital fact that the respiratory process is the source of energy for the plant's growth and movement. The type of respiration known as anaerobic respiration (p. 206) was first dealt with by **Pasteur** in 1861. Although the general nature of this process was understood thus early, many later investigators have been concerned with the nature of the substances broken down in the process, the effect of external factors on the process and with the chemical changes involved.

Enzymes

The history of enzyme action began with the work of **Dubrunfaut**, who in 1830 prepared an extract from germinating barley which had the power of converting starch to sugar. A little later, in 1833, **Payen** and **Persoz** prepared a substance from a similar extract which also brought about this reaction, and to this substance the term diastase (p. 93) was applied. Other enzymes were rapidly identified and it became clear that these substances play a vital part in all plant metabolism. Much, however, remains to be learnt about them, and at present their chemical constitution is not known with any exactitude.

Movement in Plants

Plants show many curvatures and movements, some of them due to internal causes, others induced by external stimuli. The ones most fully investigated are the tropisms (p. 230)—the curvatures of plant organs due to the action of unilateral stimuli such as gravity (geotropism) and light (phototropism).

Many speculations were made by early writers such as **Cordus, Magnus,** and **Caesalpino** ; but it was not until 1806 that **Knight** showed by experimental methods that the vertical growth of primary stems and roots is due to gravity (p. 231). It is impossible to trace here the full story of further developments, but some of the more important advances may be mentioned.

With regard to geotropism, **Ciesielski** showed that it is the tip of the root which acts as the perceptor of the stimulus of gravity since decapitated roots did not show geotropic curvature. Later it was shown that the tips of stems act in the same way. The mechanism of this perception is still not fully established ; but the work of **Haberlandt** and **Němec**, and, more recently, that of **Hawker**, has led to the widely held ' Statolith Theory ', statoliths being starch grains or other solid cell inclusions which fall under the influence of gravity to the lowermost side of the cells containing them. The excitation caused by the geotropic stimulus then passes to the growing region further back where the actual curvature takes place. The work of **Boysen Jensen** and of **Went** within the last thirty years has clearly established the fact that the stimulus is conducted by the movement of a hormone or growth substance, the differential distribution of which causes the differential growth in the growing region which is the immediate cause of the geotropic curvature.

Research on phototropism has followed a parallel course. The experiments of **Charles** and **Francis Darwin** established the fact that the tips of coleoptiles and stems constitute the seat of perception of the stimulus of light. As early as 1910 Boysen Jensen was able to show that phototropic curvature is brought about by the movement of a definite chemical substance and later Went was able to establish the hormone theory in relation to phototropism.

Other tropisms such as chemotropism (p. 234) and hydrotropism (p. 234) have, so far, received no adequate explanation. But some nyctinastic movements, such as the sleep movements of certain leaves (p. 234), have been shown to be due to changes in growth substance content leading to differential growth-rates as between the upper and lower surfaces of petioles or other parts showing the characteristic curvatures. In the case of the sensitive plant (*Mimosa pudica*) (p. 235), which has been investigated by numerous workers, the shock stimulus is also transmitted by means of a hormone.

THE CRYPTOGAMS

Little attention was paid by the herbalists to the lower plants, and indeed it was not until the beginning of the nineteenth century that the main subdivisions of these were recognised.

Knowledge concerning the structure and life-histories of the mosses and ferns was the first to accumulate. Various workers demonstrated the sexual organs of these plants and these observations were vastly extended and correlated by **Hofmeister** (1824-77) in his famous *Vergleichende Untersuchungen* (1849). This work demonstrated for the first time the important generalisation that there exists in Bryophytes, Pteridophytes, Gymnosperms and Angiosperms a fundamentally similar rhythmic life-cycle. In the hundred years since the publication of Hofmeister's results many workers have filled in and extended the picture of the mosses and ferns. There may be mentioned only a few ; **Bower, Campbell** and **Goebel** have all contributed greatly to our knowledge and understanding of these groups. More recently, cytological studies and the application of experimental methods have begun to throw light on the classification and on the causes underlying the organisation of these plants.

Even as late as 1850, knowledge concerning the Algae and Fungi was in a chaotic state. Many thousands of species had been named and figured but little was known about their structure and life-history ; some workers indeed still thought that simple Algae like *Chlamydomonas* (Fig. 1) arose by spontaneous generation. Hofmeister's classical elucidation of the structure and life-history of the higher Cryptogams, however, stimulated similar investigations on the lower groups. The accurate observations of **Thuret, Bornet, de Bary, Pringsheim, Oltmanns** and many others have made clear the structure and complete life-histories of the main algal groups.

It was not until the early decades of the nineteenth century that Fungi were scientifically investigated. From very early times the Fungi had been merely objects of wonder and superstition, ' the superfluous moisture of earth and trees and of rotten wood ', as one herbalist put it.

Useful observations on the development of the larger fungi, such as the mushroom (Fig. 276) were made as far back as 1729; but it was not until a hundred years later that accurate knowledge about the structure and reproductive methods of the Fungi began to accumulate. It was, however, **de Bary**, who securely laid the foundations of **mycology**, the science relating to Fungi. In addition to observing the Fungi in their natural habitat, he grew them in culture and so was able to follow their complete development and reproductive methods. He also gave the first adequate description of the way in which Fungi are able to penetrate the interior of plants thus causing disease and sometimes the death of the plant. These observations paved the way for the many subsequent investigations into the plant diseases which are of such great economic importance in agriculture, forestry and horticulture, and which form the subject matter of plant pathology.

Mention must also be made of the many recent investigations of the antibiotic substances produced by Fungi, such as penicillin and streptomycin, which are playing such an important part in modern medical practice.

PALAEOBOTANY

Classification rests mainly on a careful comparison of all the characters of living plants, and, as has been pointed out, it is difficult to trace the evolutionary connexions between the larger groups. Direct evidence of such connexions can only be obtained from observations on the plants of former geological ages, the remains of some of which are present as fossils in the various rock strata. **Palaeobotany** is the branch of botany dealing with such fossils. Botanists of the sixteenth and seventeenth centuries regarded fossils either as *lusus naturae* or as the relics of organisms which had existed before the Flood. It was not until the end of the eighteenth century that the significance of the fossils was fully realised. **Brongniart** published a *History of Fossil Plants* in 1828 and **Lindley and Hutton** began to publish their classical *The Fossil Flora of Great Britain* in 1831. Since then, many investigators have described fossil plants from the most ancient to the most recent rocks. The fossil record is very incomplete; but, nevertheless, many types of plant, some of them completely different from living types, have been described, and these have helped to fill in the picture of evolutionary progress in the plant kingdom.

GENETICS

Until **Camerarius** (1665-1721) discovered the true nature of the stamens and carpels of a flower no scientific investigation of the transmission of parental characters to the offspring was possible. A little later **Kölreuter** was able to show that hybrids resulted from the pollination of the stigma of one flower by the pollen from another flower. From then until the latter half of the nineteenth century plant breeding was mainly an empirical commercial practice. In 1865, however, **Mendel**, an Augustinian monk of Brunn, Austria, recorded the results of his classical experiments on garden peas. These, together with more recent developments in the science of plant breeding are dealt with in Chap. XXVIII.

ECOLOGY

The term plant ecology is usually applied to a study of plants as they exist in their natural habitats. This aspect of botany has only been studied in relatively recent times, the stimulus having come from the publication of **Professor Warming's** *Ecological Plant Geography* in 1896. Since then, great progress has been made not only in the analysis of the communities in which plants live but also in an understanding of the so-called habitat factors, such as the nature of the soil and climate.

CHAPTER IV

THE FLOWERING PLANT

A typical, simple flowering plant, such as a common buttercup (*Ranunculus acris*), wallflower (*Cheiranthus cheiri*) or rosebay willowherb (*Chamaenerion angustifolium*) is clearly composed of various parts or organs. An organ is that part of a plant or animal which is distinguished by its shape and has certain particular functions to perform.

The organs of a typical flowering plant may be divided into two systems, namely that which normally grows below the soil called the **root** system, and that which grows above the soil called the **shoot** system (Fig. 14).

THE ROOT

The root performs several functions, the chief two of which are to anchor the plant firmly in the soil, and to absorb water and substances dissolved in it from the soil, for the use of the plant. Many plants store food and the stores are kept in various parts of the plant; the root is the store-house in many cases.

If a wallflower plant be dug up from the soil and thoroughly washed, it will be seen that the root system is composed of a main root or **tap root**, which grows vertically downwards. This gives off branch or lateral roots, which grow out obliquely from the tap root and these branch roots are further branched. Near the tips of many of the roots may be seen a tuft of very fine, white hairs called **root hairs**, to which many soil particles will be adhering. The main or tap root is the first which grows out from the seed. The branches appear later (Fig. 13).

Differing from the tap root system is that system of roots, equally as common, where the first root, instead of persisting as the main root, withers away, and a series of roots more or less equal in size takes its place. This can be seen in the case of grasses, and is called a **fibrous root** system (Fig. 15).

Such fibrous roots arise from the bases of the stems and are therefore not produced by the first root. A root which grows from any part of the plant other than the original root system is said to be **adventitious** and most fibrous root systems are of this type. Although adventitious roots usually arise from stems, they may arise from leaves (Fig. 35). Cut ends of stems of the geranium, carnation, fuchsia and willow,

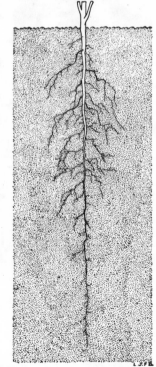

FIG. 13. A TYPICAL YOUNG TAP ROOT SYSTEM.

29

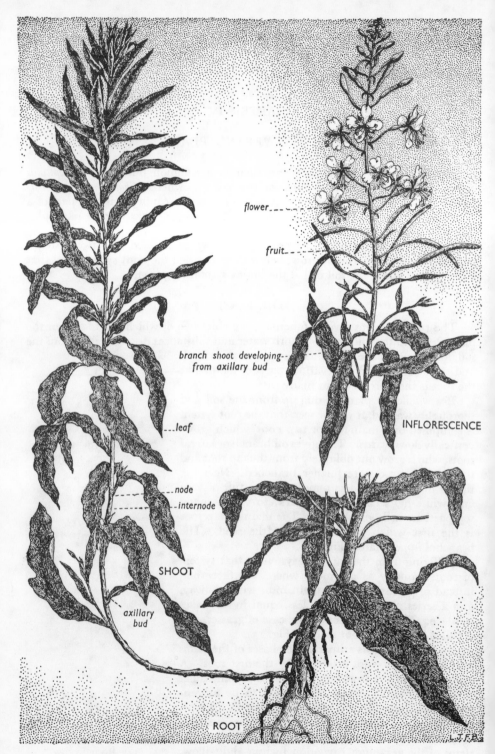

flower

fruit

branch shoot developing--
from axillary bud

INFLORESCENCE

...leaf

...node
...internode

SHOOT

axillary
bud

ROOT

L.J.F.B.

FIG. 14. ROSEBAY WILLOWHERB.

when placed in water or the soil, often give off such
adventitious roots.

THE SHOOT

The shoot is composed of stems which bear leaves
and flowers. In the case of the buttercup, willowherb or wallflower, the shoot is composed of a central
axis or main **stem** which terminates in a bud, which
is really a telescoped, embryonic shoot. Wherever
this **terminal** bud is a leaf bud only, it is capable
of·opening out and continuing the growth in length
of the main stem. If, however, the terminal bud
contains a flower, once it has opened and flowered,
growth in length *in that direction* ceases.

BUDS

On the side of the stem the leaves are borne (Fig.
14). That part of the stem to which a leaf is
attached is called a **node**. The leaf is usually given
off from the node at an oblique angle ; seldom at
right angles. The angle which the leaf makes with
the stem is called the **axil**, and borne in the axil
frequently is a bud which is called the **axillary**
bud, as distinct from the terminal bud. These
buds often grow out and produce branch stems
(Fig. 14).

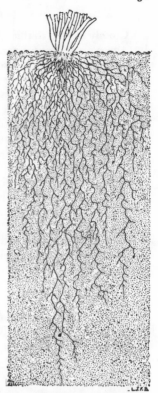

FIG. 15. A TYPICAL FIBROUS
ROOT SYSTEM.
Composed of adventitious roots
of a grass.

That part of the stem between one node and
the adjacent node is called the **internode**. Usually
the internodes of the stem are cylindrical and circular in cross section, though
some are angled, or flanged or even square in cross section. In some plants,
such as the poppy and buttercup, the internodes are often covered with fine
silky **hairs**.

Buds are young, undeveloped shoots. If they are leaf buds only, they finally
grow out to produce branch stems bearing leaves. If they are flower buds, on the
other hand, they finally produce the flower or flowers, and then their growth ceases.
The production of a flower always stops vegetative growth in that direction.
Some of the axillary buds produce axillary branches, but others do not develop
and are said to be **dormant**. In exceptional circumstances, however, dormant
buds will awaken and actively develop branches. This is so in the case of the
severing of a terminal bud. If it is severed or injured so that it cannot itself
develop, then some of the dormant axillary buds develop. If the main trunk of a
tree, such as the common oak (*Quercus robur* and *Q. petraea*) or elm (*Ulmus procera*)
be cut down or broken, a dense outgrowth of branches develops from the base of
the shoot. These branches are produced from dormant and adventitious buds.
This is called **tillering** In some gardening processes, such as nipping off the
tips of runner beans, etc., the axillary buds are thus stimulated to develop more
quickly.

POLLARDING AND COPPICING

Closely related to this effect are processes known as **pollarding** and **coppicing**; but these cases are not concerned with dormant buds, but with the development of adventitious buds. These latter are buds which arise in unusual positions on the stem, (that is they are not axillary), or they may sometimes develop on roots or on leaves. The process of pollarding is fairly common in Britain. It consists in lopping off the top of a tree and leaving just the trunk or bole. This is some-times done with willow trees (*Salix* species) growing along the banks of rivers (Fig. 16). When the top is lopped off, adventitious buds arise usually near the edge of the cut end of the trunk in relation to the wound-healing tissue called **callus**. They then develop into branch shoots.

Coppicing is very similar to pollarding except that, instead of lopping the tree at the top of the trunk, it is lopped off at the base, near the ground. This is done to a considerable extent in Great Britain with hazel (*Corylus avellana*), alder (*Alnus glutinosa*) and osier willow (*Salix viminalis*). It produces very bushy trees, and when growing closely together they form a very thick copse. The practice is common in forestry and on large estates. By its means, rods for making hurdles, crates, hop-poles and baskets are obtained. The resulting copses sometimes act as coverts for game.

GROWTH IN THICKNESS

Many plants, especially woody perennials, cannot continue growth in length unless this is accompanied by growth in thickness, else the plant becomes so long

Harold Bastin

FIG. 16. POLLARDAL CRACK WILLOWS IN WINTER.

and slender as to be unable to remain upright. Such plants, as they grow in length, also grow in thickness.

Nearly all stems and roots contain a certain amount of wood, and when these begin to thicken they do so chiefly by the formation of more wood in those organs. This growth in thickness is called **secondary thickening** (Chap. XIII). Trunks of trees, which are nothing but thickened stems, are thus composed chiefly of wood produced by the process of secondary thickening. The characteristic unbranched appearance of the trunk is due to the loss of numerous branches which were present in the younger sapling stages. As the tree increases in size, the lower branches, which are shaded by the upper ones, grow only feebly and soon die. The process of secondary thickening leads to their being broken off and their bases are soon buried in the mass of secondary wood.

THE TRUNK

If the trunk or a thick branch of a tree be cut across, the cut surface will reveal to what extent the deposition of wood has taken place in the stem. The wood forms the chief part of the thickened stem, and it may be seen to be composed of two sections—that in the centre, which is often of a dark colour, called the **heart** wood, and a lighter portion of wood surrounding it, called the **sap** wood. It is only through the latter that the sap, composed of water and dissolved food materials, passes up to the rest of the shoot. The heart wood, whose dark coloration is due to the presence of resins, tannins and wood dyes, gives mechanical rigidity to the trunk. This dead section of the trunk is frequently protected from rotting by the formation of antiseptic substances which prevent the decaying activities of attacking fungi. This is well exemplified in pollarded willows ; these trees do not produce toxins in their heart wood which consequently easily tends to rot on exposure.

Every year during the growing season, which is from spring to autumn in temperate countries, the tree develops a thin layer of wood, which is added to the outside of the already existing woody cylinder. No new wood is formed during the winter. As will be seen in Chapter X, the wood is composed of a series of tubes, and when the stem is cut across these tubes present circular ends. If the cut end of the trunk be examined with a pocket lens, the circular ends of the tubes will clearly be seen. The tubes which are produced at the beginning of the growing season, that is, during the spring, are much larger than those produced towards the end of the season, that is, the autumn. Then growth ceases during the winter. The following spring sees more large tubes produced. Therefore each season's growth is distinct, giving the complete cut section of the trunk an appearance of concentric rings (Fig. 17)—evidence of the seasonal rhythmic growth which that woody trunk or branch has undergone in the past.

Each ring represents one year's formation of wood, and the rings are therefore called **annual rings**. Furthermore, some annual rings are wider than others, thus showing that during that year the tree experienced very good growing conditions. By counting the number of annual rings, which is a comparatively easy matter, it is possible to tell the age of the tree. That for example, in Fig. 17 is about forty years old. One cross section of the trunk of a Californian redwood tree, which can be seen in the British Museum (Natural History), South Kensington, shows 1335 annual rings. A section of a redwood tree in the United States displays rings going back more than two thousand years.

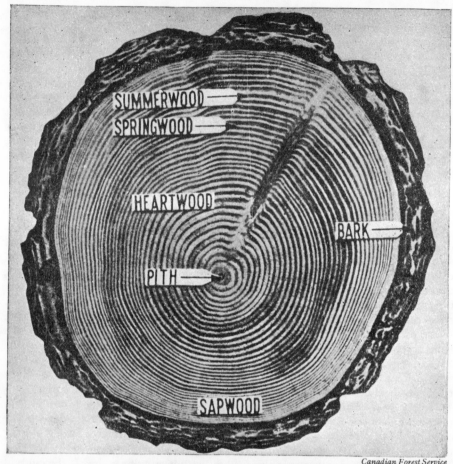

Canadian Forest Service

FIG. 17.   TRANSVERSE SECTION OF A CONIFEROUS LOG.

By examining comparatively the growth of the annual rings of a single trunk, it is possible to deduce, at any rate roughly, what kind of weather that tree has experienced during the years of its existence. This idea has been carried still further by an American, **Dr. A. E. Douglas**, who examined the trunks of some fossil trees which must have been alive several million years ago. From his results he was able to deduce certain facts of meteorological and geological interest about those times when the trees were growing. Dr. Douglas has also made some interesting discoveries concerning climates of more recent times by adopting similar methods. For example, he has been able to show that in the United States there was a great drought which commenced in 1276 and lasted for twenty-three years.

CORK AND BARK

Surrounding the woody cylinder in the thickened plant stem are two other important layers, both of which, though not so extensive as the wood, are visible to the naked eye. Immediately surrounding the wood is a very thin layer called

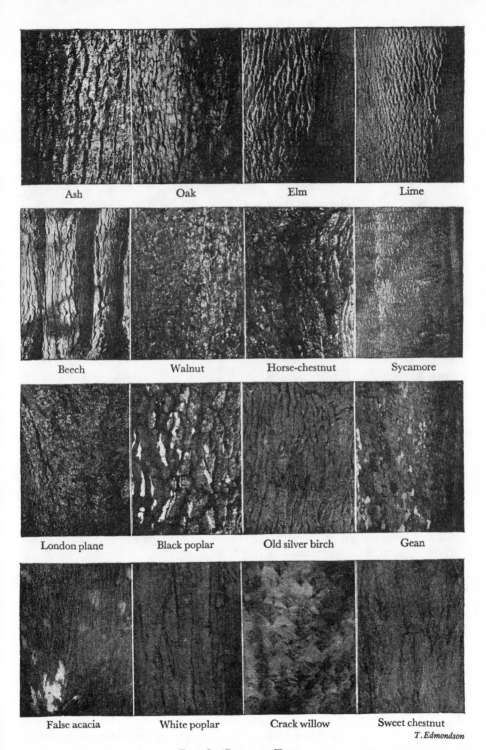

Ash	Oak	Elm	Lime
Beech	Walnut	Horse-chestnut	Sycamore
London plane	Black poplar	Old silver birch	Gean
False acacia	White poplar	Crack willow	Sweet chestnut

T. Edmondson

FIG. 18. BARKS OF TREES.

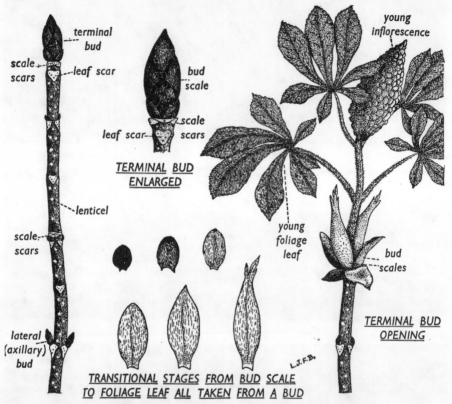

TRANSITIONAL STAGES FROM BUD SCALE
TO FOLIAGE LEAF ALL TAKEN FROM A BUD

FIG. 19. HORSE-CHESTNUT TWIG IN THE RESTING STAGE AND
JUST BEGINNING DEVELOPMENT.

the **phloem** down which the food, manufactured by the green leaves, passes to the
rest of the plant (see Chap. XIII). Outside this is the **bark**. This is a composite
tissue but consists largely, and sometimes only, of **cork** cells, the walls of which
consist of **suberin** (p. 70). This layer varies considerably in thickness and
texture. Many trees therefore can be identified by their bark (Fig. 18). The
bark of the elm, for example, is thick and very rough, whereas that of the beech
is thinner and smoother (See Chap. XIII). The bark of some trees, such as birch
(*Betula* species), and plane (*Platanus* species) is so thin that periodically it flakes off
in irregular sheets.

THE TWIG IN WINTER

The examination of a twig during the winter reveals other interesting characteristics of the plant. For example, the scars of the leaves which have been shed
may clearly be seen. They are either crescent- or horseshoe-shaped, and on
them even the small scars of the veins which passed up into the leaf may be seen
(Fig. 19).

A twig in winter affords a splendid opportunity of examining buds, as, for
example, those of the horse-chestnut (*Aesculus hippocastanum*). The bud is often

covered with a glutinous material as a
protection against excessive moisture
which would get into the heart of the bud
and make it rot. The outermost **bud
scales** are boat-shaped, and are used
for protecting the tender foliage leaves
inside. If all the scales and leaves are
dissected from such a bud, it will be seen
that there is no line of real distinction
between the bud scales and the young
foliage leaves, but that there is a gradual
transition from one to the other (Fig. 19).

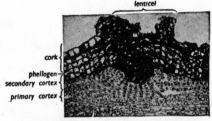

FIG. 20. PART OF SECTION THROUGH A
SECONDARILY THICKENED STEM OF ELDER.
Passing through a lenticel (× 20).

Bud scales, like foliage leaves, leave their scars after they have fallen. These
small **scale scars**, left after the terminal bud has opened in the spring, collectively
form a ring round the stem (Fig. 19). A similar ring of scale scars will be seen
further down the twig ; this indicates the position occupied by the terminal bud
a year ago. Other such rings occur at intervals down the length of the twig and
it is clear therefore that the distance between one ring of scale scars and the next
represents the amount of growth in length of that twig, for óne year.

Other marks on the young twig are small dots all over the surface, scarcely
bigger than a pin's point. They are usually lighter in colour and are composed
of microscopic particles of cork, very loosely packed so that the whole area is
porous. As will be seen later, nearly all plants take in (**inhale**) oxygen and give
off (**exhale**) carbon dioxide, and so do nearly all animals. In the case of land-
living plants and animals, this interchange is a gaseous one.

Now such gases cannot pass through a continuous layer of cork cells quickly
enough, and such a layer has already been seen to surround a thickened stem ;
hence the spongy areas in the form of dots on the twig's surface. They are called
lenticels, and form a passage for gaseous interchange (Figs. 19 and 20). As will
be seen, however, in Chapter XIV, lenticels are not the most important organs
present in the plant for gaseous interchange.

THE LEAF

Leaves vary considerably in size and shape in different plants. There is the
comparatively small leaf of the privet, and in contradistinction to this there is the
banana leaf which is roughly the same shape, but hundreds of times larger, being
anything from one to three yards long.

The different shapes of leaves are best learned by collecting and drawing them.
One should be able to recognise a **simple** privet leaf or tulip leaf with its smooth
margin, the elm leaf with its serrated margin, the oak leaf with its deeply indented
margin, and the **compound** wild rose (*Rosa canina*) and horse-chestnut leaves, etc.,
(Fig. 21). Compound leaves have their blades divided into **leaflets**. If these
are arranged along the length of a central axis or rachis, as in the wild rose, the
leaf is described as being a **compound pinnate** one. If, however, the leaflets
radiate from the top of the petiole as in the horse-chestnut, the leaf is said to be
a **compound palmate** one. Although the shape of the leaf is usually characteristic of the plant, it frequently varies, sometimes even in one and the same
plant. For example, it can be seen that the lowest leaves of a groundsel plant

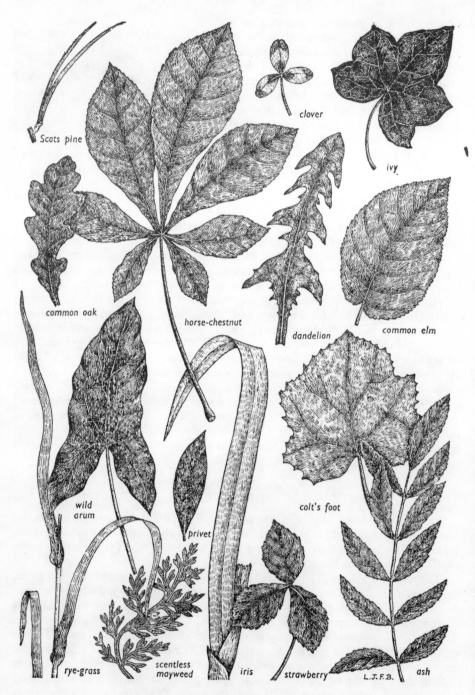

Scots pine

clover

ivy

common oak

horse-chestnut

dandelion

common elm

wild
arum

privet

colt's foot

rye-grass

scentless
mayweed

iris

strawberry

L.J.F.B.

ash

FIG. 21. TYPES OF FOLIAGE LEAVES.

(*Senecio vulgaris*) have only slightly notched leaves, but higher up on the plant the leaves become more deeply notched. Ivy (*Hedera helix*) also shows a considerably heterogeneity of leaf-form, though usually from one plant to another. Usually where the ivy plant is closely adpressed to a wall or tree-trunk or well exposed to the light, its leaves are deeply lobed in a palmate fashion : on the other hand, where an ivy branch grows upright, its leaves are usually heart- or lance-shaped.

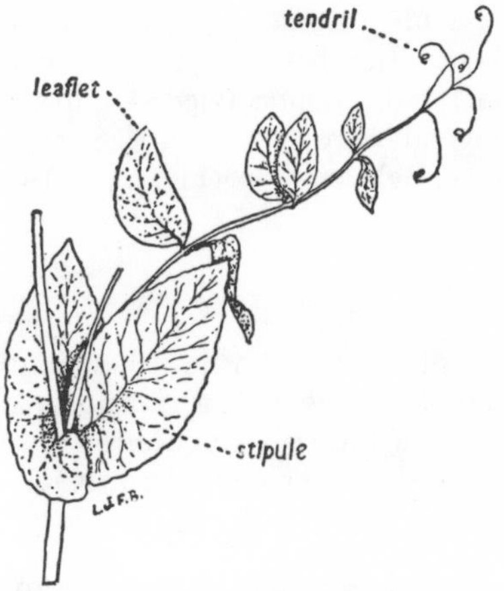

The typical leaf is composed of a **leaf-stalk** or **petiole** which widens at its base, where it joins the node, into what is called the **leaf-base**. Sometimes the leaf-base is merely a thicker structure, as in the case of the horse-chestnut, or it may become elongated into a sheath as in the buttercup. In grasses, this sheath folds over the stem for a considerable distance.

FIG. 22. COMPOUND PINNATE LEAF OF GARDEN PEA, SHOWING LARGE STIPULES.

The main flattened portion of the leaf is called the **leaf-blade**, or **lamina**. Sometimes the leaf-blade is joined on to the leaf-base, without the intermediate petiole, in which case the leaf is said to be **sessile**.

Very commonly, wing-like outgrowths arise at the base of the petiole one on each side, and have the appearance of leaflets. These may be seen in the rose (*Rosa* species) and still better in the garden pea (*Pisum sativum*) (Fig. 22). These outgrowths vary considerably in shape, size and function and are called **stipules**. In many plants stipules are absent, when the leaf is said to be **exstipulate**, for example, in the privet. In the beech, the mature leaves show no stipules but the young leaves in the bud do possess them. Furthermore, pairs of stipules, devoid of any leaf blade, act as the protective scales on the outside of the buds.

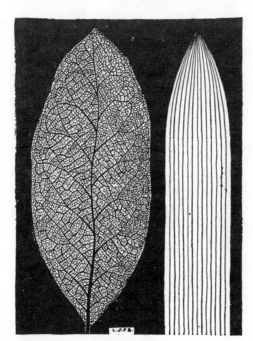

FIG. 23. VENATION OF LEAVES.
Left, *Rhododendron* (net) ; right, top of *Iris* (parallel).

It is in the lamina that leaves usually show their great diversity of form. On the lamina, thicker lines called **veins** are visible. These are channels for conducting water to, and food from, the leaf (see Chap. XIV). Often there is one main vein. Then this vein gives off finer branch veins, then these branch veins give off still finer ones, and so forth. This venation is said to be **net** or **reticulate**. In other cases, especially the blade-like leaves of grasses, lilies, irises, etc. (in fact, most monocotyledons), instead of having a main vein with branches, there are several veins, equal in size, running parallel to each other. This is known as **parallel** venation (Fig. 23). In some cases of net venation there are several main veins.

THE FLOWER

Flowers are borne either singly or in groups called **inflorescences** (Fig. 13). The flower is such a complicated structure, and so important from the point of view of the reproduction of the plant, that it will be deferred for detailed consideration later (see Chap. XXI).

PRACTICAL WORK

GENERAL DIRECTIONS

In the study of plant life, practical work is even more important than theoretical study from a book ; but it is essential that the practical work should be carried out thoroughly. One should try to find out things for oneself during such work, and then use the text of the book as a means of understanding what has already been observed. Practical work should never be treated merely as a supplement to theoretical study.

No practical work should be done without keeping a permanent record of it, either by diagrams, written description, or both. Such records should be made at the time, either in the laboratory or in the field. Diagrams, etc., can be ' touched up ' afterwards, in the class-room or study ; but things should never be put aside.

Written description should always be clear, full, and to the point. Diagrams, too, should be very clear. Although it is desirable to bring in artistic effect in order to make the diagrams neat and attractive, such effect must invariably be subsidiary to clarity, and the portrayal of the true facts. For this reason, it is usually desirable to use line diagrams. Avoid shading and, so far as is possible, use pencil and not ink. Colours, either by means of crayons or wash, should be used very sparingly. Great care should be taken over the diagrams. They should always be very large—much larger than those used to illustrate the text of this book.

A diagram is useless unless it is fully labelled. One should be able to revise one's work from the diagrams, and this would be impossible if they were not labelled. The best method of labelling is the one chiefly used in this book. Labelling by means of letters, with an explanatory legend at the bottom of the diagram, is not desirable, since constant reference to the legend is tiring and unnecessary.

No diagrams should be copied from the book. Those in the book are given as a help in following the text, and also a *guide* during practical work. But they must be used *only* as a guide. Draw exactly what is seen, and then, with the the help of the book, label carefully.

A written description is also a splendid means of driving home important points. Where the study of plant structures is concerned, the written description is not essential, provided good drawings are made ; yet it is well worth while. In experimental work the written description of the experiment, method, results and conclusions, is essential.

---

1. Make a through examination of the external features of a herbaceous flowering plant. Good specimens for this purpose are, the common buttercup (*Ranunculus acris*), wallflower (*Cheiranthus cheiri*), and the groundsel (*Senecio vulgaris*). Note its division into root and shoot, and carefully observe the structure of the stem, leaf, axillary and terminal buds, and flower. Note also the nodes and internodes. Make a fully labelled drawing of the plant and, if time will permit, write a description of the plant.

2. Make enlarged drawings showing the details of the most important parts of the plant, such as the root system, node, leaf, etc.

3. Examine and draw a seedling, noting especially the root hairs. A convenient seedling to use is that of the mustard (*Brassica nigra*), for this can easily be cultivated by sowing some seeds on damp blotting paper or cotton wool, and the seedlings, which develop in the course of a few days, can be examined without handling them.

4. Examine, draw, and describe various root systems ; tap root (wallflower, groundsel) fibrous (a grass) and adventitious (water cress).

It is also interesting to watch the development of adventitious roots on the twig of the willow (*Salix* species). The twigs should be gathered late in the winter, and placed in a vessel water. In a few weeks, the buds will open, when their progress can be watched with profit, and adventitious roots will appear on the stem. Keep a dated record of the progress of the shoot, and draw examples of the adventitious roots.

5. Examine and draw a section across a tree trunk. It is usually possible to do this in the field, where felling has recently taken place ; otherwise, the sawn end of a fairly thick branch may be used. The end could then be smoothed and polished, making the structures more easily seen. Note especially, sap wood, heart wood, annual rings, phloem and bark.

6. Make a careful study of the twigs of various trees in winter. The following are suggested as suitable types :

. (*a*) Horse–chestnut (*Aesculus hippocastanum*). Note : the terminal bud ; the leaf scars with associated axillary buds ; the ring of bud scale scars ; the lenticels. Dissect a terminal bud. Arrange the scales and young foliage leaves in series, noting any transitional types between them. The scales are leaf bases.

Some examples will show two apparently terminal buds with a large scar between them. The scar indicates the position occupied by an inflorescence in the previous spring. The formation of an inflorescence by the terminal bud leads to the growth of the branch being carried on by the two axillary buds immediately below.

Examine and draw stages in the opening of the buds in spring.

The sycamore (*Acer pseudoplatanus*) twig shows similar features.

(*b*) Lime (*Tilia* species). Note the same general features as in the horse–chestnut. Carefully examine the apparently terminal bud and note that in addition to a leaf scar there is another small scar. This latter was formed by the withering and falling off of the original terminal portion of the shoot. The apparently terminal bud is therefore axillary.

Dissect a bud and arrange the constituent parts in a series. Note the pair of outer scales followed by pairs of inner scales which have a small foliage leaf between them. The bud scales are in fact stipules.

Note the opening of the buds in the spring and the fact that the stipules soon fall from the foliage leaves.

Beech (*Fagus sylvatica*) and oak (*Quercus* species) also possess stipular bud scales.

7. Examine, draw and describe different types of foliage leaves. Some examples may be taken from those illustrated in Fig. 21, but there are many other types. Note also the stipules of certain leaves such as the rose, pea, etc. Parallel and net venation can easily be examined in many common types of leaf.

CHAPTER V

SPECIAL FORMS AND FUNCTIONS OF PLANT ORGANS

All plant organs have a general function to perform ; for example, the general function of the green leaf in food manufacture. But many have *special* functions and are modified accordingly.

MODIFICATIONS OF THE ROOT

The plant very often manufactures more food than it can utilise immediately. The excess food is stored for future use.

In many plants, food is stored in the roots. In order to form such a store-cupboard, in certain cases adventitious roots become very thick and fleshy. Such fleshy roots are called **root tubers.** They can be diagnosed as roots in that, like normal roots, they do not bear leaves, they have the internal structure of the root, and at their tips they have a protective cap called the **root cap**, which all roots, but no other organs, possess (see Chap. XII).

Root tubers are seen in the *Dahlia* and lesser celandine (*Ranunculus ficaria*) (Fig. 24) (inulin being stored in the former and starch in the latter), and in many orchids. In all these cases, the root tubers are adventitious roots, which have been developed in relation to axillary buds at the base of the shoot. The aerial

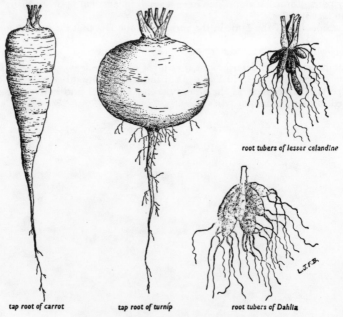

root tubers of lesser celandine

tap root of carrot   tap root of turnip   root tubers of Dahlia

FIG. 24. MODIFICATIONS OF THE ROOT.

42

shoots of the plant die down at the end
of the growing season but the axillary
buds and the tubers remain alive through
the winter. At the beginning of the next
growing season (spring), the buds develop
rapidly into new aerial shoots at the expense of the food stores in the tubers and
new fibrous absorptive roots arise from
the base of the shoots. The process is now
repeated so that late in the growing season
one can see the depleted old tubers and
new ones developing. Thus new plants
are formed from purely vegetative organs
without the intervention of sex. This form
of reproduction is said to be **vegetative**.

J. Schmidt

Swollen tap roots (which are not root
tubers but swollen true roots) are seen in
the carrot (*Daucus carota*) and cultivated
parsnip (*Peucedanum sativa*) the former of
which stores food in the form of sugar, and

FIG. 25. RESPIRATORY ROOTS OF *Sonneratia
alba* IN A MANGROVE SWAMP.

the latter starch. By some botanists, the swollen part of the turnip (*Brassica rapa*)
is considered to be the hypocotyl (p. 355), and the main root continues below.

In all these cases, the swollen part is produced during the first year of growth,
and food is stored in it. With the approach of winter, the foliage withers, but no
flowers have yet been formed. The following year, new leaves and also flowers are
eventually produced (followed by seeds, at the expense of the food stored in the
storage organ which gradually shrivels). So these plants are biennials.

Special modifications are shown by the roots of many tropical orchids which
grow, as **epiphytes**, on the trunks or branches of forest trees. These plants possess
aerial roots, which hang down in the moist atmosphere but do not reach the
soil. These roots are able to absorb rain water, and, perhaps, even condense
water vapour, since they are surrounded by a sheath of dead, empty cells with
pores in their walls. This sheath, or **velamen**, acts like a sponge.

All roots, like other living structures, must have air for gaseous interchange.
This is usually possible, for there is plenty of air in the small spaces between soil
particles. However, in some cases, for example, mangrove swamps, the soil
becomes so water-logged, or perhaps even submerged in water, that the aeration
of the soil is impossible. In such cases, many plants have exceptional root modifications. Some of these roots, instead of growing downwards, as roots usually do,
grow upwards into the air. That part of the root exposed to the air becomes covered
with lenticels, to facilitate gaseous interchange. The roots are therefore **breathing** or **respiratory roots** (Fig. 25).

It is well-known that to tear ivy away from a wall or tree-trunk to which it is
clinging often demands considerable force. In this case, the stem gives off hundreds
of adventitious roots which cling tenaciously to the support. Such roots are
called **climbing roots** (Fig. 26). These are, like some roots, sensitive to light
(p. 233) and grow away from it. They also secrete a glutinous substance by which
they adhere to their support. The plant, however, obtains its water and dissolved
mineral substances in a normal manner by means of a tap root in the soil.

Another relatively common modification is the development of **prop roots** which serve to keep the shoot erect. In the maize plant (*Zea mays*) (Fig. 27), adventitious roots arise from successively higher levels on the tall but relatively slender stem and grow obliquely downwards into the soil, thus acting as guy-ropes to keep the stem upright. Some of the tropical mangroves have a similar system of much stouter prop-roots, sometimes called **stilt roots**, which serve to keep the plant erect in the unstable mud of the swamp. A remarkable development of prop-roots is seen in the banyan tree (Fig. 28). The widely spreading branches are supported only by the numerous stout adventitious roots.

Buttress roots, which serve the same function as prop-roots, are found at the base of many tropical forest trees and, less well developed, at the base of some of our native trees. These roots are flattened in the vertical plane and radiate outwards from the base of the trunk.

MODIFICATIONS OF THE STEM

Most stems store a certain amount of food for the plant ; but in some instances they become swollen and succulent in relation to this function. Many cacti are of this character. The stems are sometimes cylindrical and sometimes slightly flattened, though always thick and juicy. These stems store chiefly water, as might be expected in such desert plants. Here they are also green, and carry out the functions of the green leaf. The leaves of such plants during the course of evolution have become modified into sharp spines, which protect the plants against browsing animals (Fig. 29).

The normal stem of a plant grows vertically into the air, giving off its branches obliquely upwards. There are, however, many exceptions. Some are recumbent on the surface of the ground and others grow actually under it.

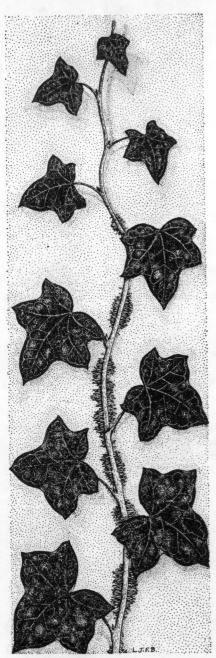

FIG. 26. CLIMBING SHOOT OF IVY.
Note adventitious roots.

One such type, in the potato (*Solanum tuberosum*), is very familiar. The edible portion of the potato is not a root, since it bears shoots, as may be seen when it is stored for several months in a warm, dark place, for it then ' shoots out '. These shoots develop from the ' eyes ' of the potato, which are really young buds in the axils of insignificant scale leaves. The edible portion of the potato is therefore an **underground stem** and, since it is very swollen, is called a **stem tuber**. Branch stems which develop from the axillary buds of the basal leaves of the plant grow along underground and towards their tips they swell to form the tubers (Fig. 30). This explains why one can ensure a maximum supply of tubers by earthing up the growing potato plant, thus covering those axillary buds which will in due course grow out as underground stems. The food stored in this swollen organ is chiefly starch, with a certain amount of protein. The potato tuber differs from that of the *Dahlia* in that the former is a modified stem, whereas the latter is a modified root (p. 43). The stem structure of the potato tuber can be easily deduced from certain diagnostic features. Actually it is the swollen end of a branch stem. In its early stages it bears leaves in the form of small scales; as it gets older and larger these scales wear off, leaving only their scars (' eyes '). The ' eyes ' are arranged in a spiral fashion, being fairly close together at the ' rose ' end of the potato, and becoming placed further apart in the region of the tuber's greatest growth. In the axils of the scale leaves are axillary buds which are able to develop and produce shoots. Roots, on the other hand, *never* bear leaves. Also the internal structure of a stem tuber is similar to that of a normal stem (Chap. XIII).

The potato tuber is not only used for food storage. As has already been seen, seeds are a means of the reproduction of the plant. But they are not the only method available to the plant. All methods of reproduction in flowering plants other than that by seeds are called **vegetative reproduction**.

The potato tuber is a means of vegetative reproduction. The gardener never uses the real potato seeds for the production of new crops ; he uses the tubers,

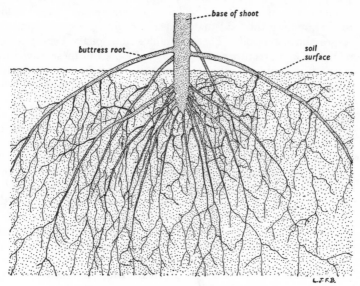

FIG. 27.　PART OF THE ROOT SYSTEM OF MAIZE.
Note the prop roots.

FIG. 28. A BANYAN TREE IN CAIRO.   *E.N.A*

~aved from the plants of the previous season. The true seed of the potato is used only by plant breeders who are after new varieties.

Whereas the seeds usually begin growth by the production of roots, the potato tuber, once it has been placed in the soil, begins its growth by the production of new shoots. These grow out from the buds present in the ' eyes '. The shoot goes on growing, getting the food that is necessary for its growth from the tuber, until it emerges above the soil. Then it unfolds new foliage leaves, which, since they are green, can manufacture a new supply of food. But these leaves must have a supply of water and mineral salts from the soil. This becomes possible in that those parts of the stems of the shoots which are below the soil give off adventitious roots. Thus the new plant becomes established in the soil.

The cultivated crocus (*Crocus purpureus*, etc.) and the autumn crocus (*Colchicum autumnale*) present another example of a stem modified for food storage ; but here it is the main stem of the plant. The swollen food-storing structure in this case is called a **corm** (Fig. 30). If a crocus corm is examined in the autumn, it will be seen to consist of a short, much-thickened stem with one to several large buds near the top and a ring of adventitious roots round the base. It is covered by thin membranous scales, each of which completely surrounds the stem ; if these are removed, circular scars are left and small axillary buds are exposed. The growing point of the stem was used up during the last season in the production of one or more flowers, whose withered stalks still remain at the top of the corm. An examination of one of the large axillary buds from the upper end of the corm will show that it consists of a short stem bearing protective scale leaves on the outside, then several yellowish foliage leaves, and, in the centre, one or more young flowers (Fig. 33).

Such a corm examined in the spring will show that each of these large buds has given rise to green foliage leaves and flowers, the whole shoot being surrounded at its base by the now enlarged scale leaves. Food manufactured by the leaves is passed into the short stem bearing them. The stem swells and thus forms a new corm on top of the old one, which by this time is withering away. The bases of the scale leaves shrivel to form fibrous sheaths round the base of the new corm. When the foliage leaves die, their bases shrivel and form the fibrous sheaths round the top of the corm. Axillary buds develop in relation both to the scale leaves and the foliage leaves. The buds in the axils of the scale leaves remain small; but one or several buds in the axils of the foliage leaves, that is, those near the top

E.N.A.

FIG. 29. GIANT CACTI GROWING IN THE DESERT NEAR TUCSON, ARIZONA.

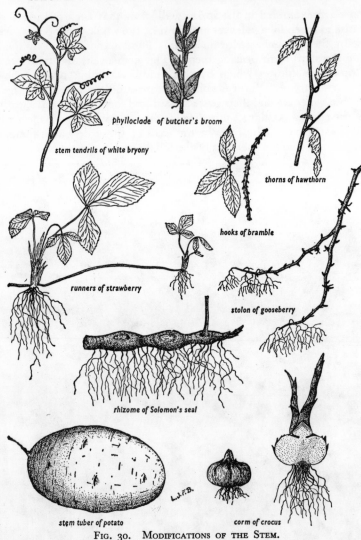

phylloclade of butcher's broom

stem tendrils of white bryony

thorns of hawthorn

hooks of bramble

runners of strawberry

stolon of gooseberry

rhizome of Solomon's seal

stem tuber of potato

corm of crocus

FIG. 30.   MODIFICATIONS OF THE STEM.

of the new corm, enlarge and these contain the rudiments of the shoots of the following year.

Several new corms will thus have been formed on top of the old one.   Each will be similar to the parent and capable of repeating the same cycle of development.   The new corms are naturally placed at a higher level in the soil than the old one.   They send out stout roots which later contract in length and thus pull the new corms to a lower level.   Such roots are said to be **contractile.**

After the plant's activity has ceased in late spring, the new corms remain in the soil, and thus the plant tides itself over the winter—as a corm, resting in the soil.   This habit of resting during inclement weather is called **perennation.** The food stored is ready as a supply for the next season's outburst of growth.   It is stored in the form of starch.

There are examples of stems which grow more or less horizontally beneath the soil, and may or may not be swollen. They have the same structure as the normal stem, but of course cannot contain chlorophyll like many of the latter. They are cylindrical, and are composed of nodes and internodes. The leaves borne at the nodes are usually reduced to insignificant, colourless, tissue-like scale leaves.

These underground stems are called **rhizomes**. At the nodes, adventitious roots are given off, so that from the point of view of getting supplies of water and other substances from the soil, the stem could go on growing indefinitely. The buds in the axils of some of the scale leaves develop, and send up branch shoots which finally open out as normal shoots in the air and bear foliage leaves and flowers.

These foliage leaves thus ensure a supply of manufactured food, and therefore the rhizome can grow to considerable lengths in the soil, being independent of the original roots and shoots for supplies. Sedges possess such rhizomes, so also does the couch-grass (*Agropyron repens*). With these rhizomes, the plants spread rapidly. The iris (*Iris pseudacorus*) and Solomon's seal (*Polygonatum multiflorum*) also possess rhizomes which are, however, swollen throughout their length (not like the potato, only in part), and store food (Fig. 30).

In other cases, instead of the stem growing along beneath the soil, it grows along, in a recumbent position, on the surface of the soil. In some cases, such a trailing, **creeping stem** is a main stem. This is seen in the moneywort (*Lysimachia nummularia*). The stem creeps along the soil, giving off

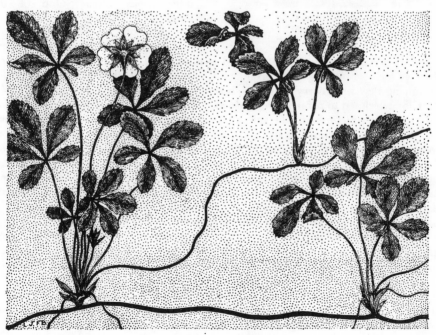

Fig. 31. Creeping Cinquefoil.
Note the runners producing new plants.

FIG. 32.   HOUSELEEK.
Note the offsets.

adventitious roots at its nodes. Therefore, if the stem be severed, both parts can live.

Often the creeping stem is a branch, formed from an axillary, not the terminal, bud. Such stems are known as **runners**, and examples are those of the wild strawberry (*Fragaria vesca*), creeping cinquefoil (*Potentilla reptans*) (Fig. 31) and the creeping buttercup (*Ranunculus repens*). Such runners bear a series of young plants, each supplied with adventitious roots, along their length. In the creeping buttercup each young plant arises from an axillary bud on the runner, the main growing point continually adding to the length of the runner, thus making this plant an objectionable weed. The runner of the strawberry (Fig. 30) develops in a rather different way. It gives rise at its tip to a young plant with a rosette of leaves and supplied with adventitious roots. One of the axillary buds of this new plant grows out as a continuation of the runner and this terminates in a second young plant. This process may be repeated a number of times. This is, therefore, another means of vegetative reproduction, and is used artificially by gardeners for producing new strawberry plants in the cultivated varieties.

Closely related to runners are **stolons**. These are normally growing branches which are not recumbent like those of the strawberry, but owing to their great length, bend over and touch the soil. Where a node touches the soil, adventitious roots are given off and the axillary bud at that node grows out to produce a shoot. Thus is a new plant established by another method of vegetative reproduction. Stolons are present in the blackberry (*Rubus fruticosus*), currants (*Ribes* species) and the gooseberry (*Ribes uva-crispa*) (Fig. 30) and the gardener helps this vegetative method of producing new plants by fixing the stolon to the soil by means of a small staple. The process is called **layering**.

The name **offset** is given to very short axillary runners which turn up at their ends and produce new plants from their apices. The houseleek (*Sempervivum tectorum*) (Fig. 32), an inhabitant of the roofs of old houses, old walls and on cultivated rockeries, is a case in point.

Sometimes branch stems are arrested in their growth in length. In place of the usual terminal bud at the end of the branch, the stem forms a sharp point called a **thorn**. This is very evident in the case of the hawthorn or may (*Crataegus monogyna* and *C. oxyacanthoides*), and it is easy to prove that the thorn is really a branch since it arises from a bud in the axil of a leaf, and sometimes the thorn itself is divided into one or two nodes and internodes towards its base, bearing foliage leaves at the nodes (Fig. 30).

Hooks differ from thorns in that, instead of being modified *complete* stems like the latter, they are formed by the modification of a *part* of the stem only, that is, the outer tissues. These hooks are used for climbing as in the case of the rose, especially the rambler rose and wild rose, blackberry (Fig. 30) and raspberry. Hooks may be found on any part of the stem, on the node, internodes and even on the leaf petioles.

Stems may become flattened. Sometimes they become so flattened and green that they resemble leaves and actually carry on the function of leaves, which is chiefly food manufacture. The butcher's broom (*Ruscus aculeatus*) is a case in point. There are several means of proving that these leaf-like structures are really stems. For example, rarely do true leaves produce buds and flowers on themselves. Yet on this structure, small buds will be seen half-way up the mid-rib, in the axil of a small scale leaf. Later these buds open out to form flowers. Also these flattened stems arise, as they should do, since they are branch stems, in the axils of the leaves, which here are reduced to small scales. Therefore these structures, in spite of their leaf-like appearance and function, are stems. They are called **phylloclades** (Fig. 30).

In favourable climates where luxuriant vegetation is found, such as those parts of the tropics where rain is plentiful, there is clearly a great struggle among the crowded plants to get plenty of light and air. Trees are very favoured in this war of Nature, and millions of smaller plants are choked out during the course of a year. But some plants succeed, although they are not trees. They develop very long, thin stems by means of which they scramble or climb up the larger plants, especially the trees. For example, some twine around the stems of the support like a runner bean does (see Chap. XIII). Others climb by adventitious roots like the ivy, and others by hooks like the rose.

Some plants climb by means of **tendrils**. There are several kinds of tendrils, but some of these are modified stems, and are called **stem tendrils**. In plants like the white bryony (*Bryonia dioica*), the twining tendrils are modified axillary branch stems (Fig. 30).

MODIFICATIONS OF THE LEAF

Just like stems and roots, leaves frequently become modified for special purposes.

One familiar example is found in the **bulbs** of certain lilies, the bluebell (*Endymion nonscriptus*), onion (*Allium cepa*), and tulip (*Tulipa* species) (Figs. 33 and 34). The function of the bulb is similar to that of the tuber and corm. In the case of the bulb, however, the bulk of the fleshy part is not swollen stem, but swollen, fleshy, colourless leaves. In some examples, these swollen food-storing structures are modified complete leaves (for example, tulip), while in others (for example, onion, Fig. 34) they are the swollen bases of the foliage leaves of the preceding season ; both types occur in the bulb of the daffodil. The best way to examine a bulb is to cut it vertically in half. It is then seen to be a shoot, which is composed of stem, leaves and terminal bud. The stem has become modified into a flat, bun-shaped structure which gives off adventitious roots around its edges. From the centre of the upper surface the terminal bud is given off and finally grows out to produce the foliage leaves and the flowers.   The lateral leaves are very thick and colourless and store the food, except the outermost ones, which are thin, brownish in colour and membranous for protection.   In the axils of some of the thick, modified leaves, axillary buds may be found.

After the flowers have died down, the foliage leaves often become larger.   Then as they manufacture food, some of this is passed down to the axillary bud or buds in the axils of the scales of the old withering bulb.   These buds then swell to store the food and thus form new bulbs.   In Nature these perennate beneath the soil until the following year.   This explains why after the flowers of tulips, daffodils, etc. (and, in the case of corms, the crocus) have died off, the leaves should not be cut but should remain until they too wither, if new strong bulbs are to result.

Bulbs are another means of vegetative reproduction and perennation.   The food reserve in the tulip is starch, and in the onion it is sugar.

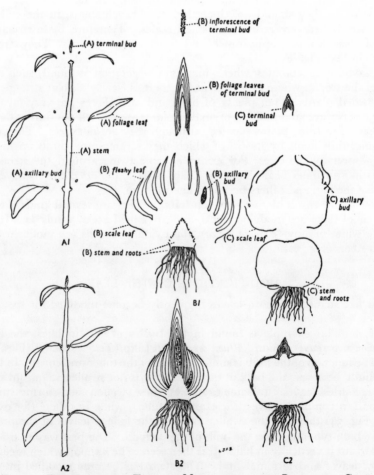

FIG. 33.   DIAGRAM OF THE TYPES OF MODIFICATIONS IN A BULB AND A CORM.
A1, normal foliage shoot dissected ; B1, bulb dissected ;
C1, corm dissected; A2, B2, C2, the same organs assembled.

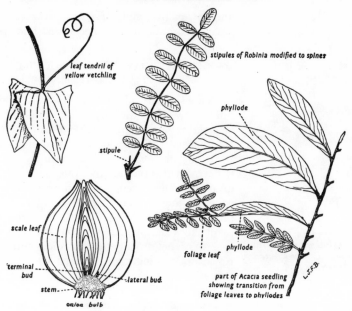

stipules of Robinia modified to spines

leaf tendril of yellow vetchling

phyllode

stipule

scale leaf

phyllode

foliage leaf

terminal bud

lateral bud

stem

part of Acacia seedling showing transition from foliage leaves to phyllodes

L.J.F.B.

onion bulb

FIG. 34. MODIFICATIONS OF THE LEAF.

Leaves may also become modified to produce **leaf tendrils**. In the garden pea, only the terminal leaflets become so modified (Fig. 22). The normal functions of the leaf are performed by the unmodified lower leaflets and by the enlarged stipules. In the yellow vetchling (*Lathyrus aphaca*), the whole leaf is modified into a tendril, except the stipules which are enlarged and carry on food manufacture (Fig. 34).

Leaves are sometimes reduced to sharp **spines** in order to form a means of protection against animals. Only a part of the leaf may be so modified, as seen on the edges of the holly (*Ilex aquifolium*). In some plants the whole leaf becomes modified into a spine, as in the case of the gorse (*Ulex europæus*) and cacti. In the gorse other spines are produced also by branch stems, so that arguing logically we should find long spines arising from the axils of other spines ; and this is exactly what is found. In the false acacia or locust (*Robinia pseudoacacia*) the spines are modified stipules so that there are two spines given off at the base of the petiole (Fig. 34).

The Australian acacias (*Acacia* species) show an extraordinary kind of leaf modification. The first leaves of a seedling acacia have a compound pinnate lamina but later leaves have their petiole flattened in the vertical plane and the leaf blade is reduced. Finally, the plant produces leaves which have a relatively large and flattened petiole, called a **phyllode**, but no lamina (Fig. 34). The absence of lamina and the vertical position of the phyllodes, placed edge on to the sun, are modifications which reduce the risk of over-exposure to light and heat in tropical and sub-tropical habitats. Mature plants grown in moist conditions sometimes produce leaves of the seedling type with the lamina developed to a varying degree.

Some leaves become modified to protect the young undeveloped foliage leaves

and flowers in a bud. They are the scale leaves and have already been examined in the bud (p. 37).

VEGETATIVE REPRODUCTION IN THE GARDEN

Some of the plant modifications already considered are often used by the horticulturist, for the artificial vegetative production of new plants. There are, however, others which are used in horticulture.

If a short length of leafy stem is severed at a node from the parent and the lower end inserted in moist sand, then adventitious roots sometimes arise in the proximity of the cut surface. This process is described as 'taking a cutting'. Many plants, for example, pansy, *Pelargonium*, privet, *Buddleja*, may be propagated by such cuttings. The process of rooting may be aided by the use of hormones (p. 227, Fig. 168).

FIG. 35. PART OF A *Begonia* LEAF BEARING ADVENTITIOUS ROOTS AFTER CULTIVATION ON A WARM, DAMP, SOIL.

Begonia illustrates a curious method adopted by horticulturists for the production of new plants. It is somewhat similar to the method of cuttings, but in this case the cut surface is that of the leaf and not the stem. If a *Begonia* leaf be cut either in the petiole or the lamina, the cut surfaces, if kept damp by, say, placing on a damp soil, will give rise to adventitious buds and roots (Fig. 35). The young plants which develop from such buds soon become independent and the old leaf dies.

Still more artificial methods of vegetative reproduction used by the gardener are those of **budding** and **grafting** (see Chap. XIII).

GARDEN VARIETIES OF VEGETABLES AND FRUITS

The new plants produced by methods of vegetative propagation possess exactly the same hereditary characters as the parent from which they were derived. It is for this reason that many garden varieties of plants are grown only from such vegetative organs and not from seeds, which on account of the mixed origins of the parent, frequently give rise to a very variable progeny. The strawberry may be considered as an example. The variety Royal Sovereign is propagated vegetatively by means of its runners and all the Royal Sovereign plants in existence to-day are, in a sense, parts of one original seedling which possessed the characteristics of this variety.

The term **clone** is applied to the vegetatively produced progeny of a single plant. Each of the numerous varieties of the strawberry is therefore a clone. The varieties of the potato (for example, Arran Pilot) are clones propagated by tubers. Varieties of apple (for example, Cox's Orange Pippin) are likewise clones propagated by grafting.

PRACTICAL WORK

1. Examine and make a drawing of the root system of the lesser celandine. Note the fibrous, normal roots, and the club-shaped root tubers. Examine plants at various seasons of the year and trace the origin and mode of development of the tubers. In a similar way, study the much larger specimens of root tubers in the *Dahlia*.

2. Tear a spray of ivy from an old wall, or the trunk of a tree. Note how difficult it is to do this without breaking the stems of the ivy, thus showing how tenaciously the adventitious roots cling to their support. Examine and draw these adventitious, climbing roots, and state why they must be of adventitious origin.

3. Examine the subterranean portion of a potato plant, looking especially for stem tubers. Note their position. Examine a tuber in detail, looking for the features already described, and state the reasons for distinguishing between this type of tuber and that of the *Dahlia* or lesser celandine. Make a drawing of a tuber enlarged to show details.

4. Examine a crocus corm in the autumn. Note the flattened swollen stem ; the adventitious roots ; the membranous scales encircling the stem ; the axillary buds, one or more of which near the top of the corm are strongly developed. Cut longitudinally through the middle of a corm and passing through one of the large buds. Draw the cut surface and note particularly the structure of the bud with its axis, scale leaves, foliage leaves and central flowers. Test the cut surface of the stem with iodine.

Grow a number of corms and trace the development of the flowering shoot and of the new corms.

Examine also corms of *Montbretia* and note that the old corms are more persistent than in the crocus. Note also that some of the axillary buds produce widely spreading underground stems which terminate in new corms.

5. Examine and draw the external appearance of a tulip bulb. Cut it longitudinally through the middle and note the stem ; the outer membranous and inner fleshy scale leaves; the large central bud containing the rudiments of foliage leaves and flower. Carefully dissect a bulb and note the presence of buds in the axils of the scale leaves. Cut a bulb transversely and note the arrangement of the scale leaves. Investigate the nature of the food reserves. Compare the tulip bulb with those of *Narcissus* and onion. Grow a number of bulbs and investigate the origin of new bulbs.

6. Examine part of the rhizome system of couch-grass (*Agropyron repens*) and the aerial shoots which it bears. Note that at each node there is a scale leaf with an axillary bud or branch. Adventitious roots also arise at the nodes. Note carefully the ways in which the rhizome system becomes progressively more extensive.

Examine the stouter rhizomes of Solomon's seal and iris. Note the presence of scale leaves, axillary buds, and adventitious roots. Note carefully the position of the aerial shoots and the way in which further growth of the rhizomes is carried on. The rhizome of Solomon's seal should be examined at intervals throughout the year.

7. Examine some specimens of creeping stems. There are several good examples which are common, such as moneywort (creeping Jenny), *Abronia*, ground ivy, etc. Note the long, recumbent habit of the stem, the absence of scale leaves (compare this with the rhizome), the position of the adventitious roots, etc. Make comparative drawings of the types collected.

8. Examine the structure of the strawberry plant. Either cultivated or wild strawberry will serve the purpose. Look for and draw examples of runners, and the formation of new plants.

9. Make a study of examples of stolons. Note the curved nature of the stems and how the adventitious roots are given off from that part of the stem which touches the ground. Note exactly where the new adventitious shoots are formed. The currant and gooseberry are good material for this purpose.

10. Note the general character of the woody stem and the leaves in the hawthorn. Look for the thorns, and note especially their position on the stem by means of diagrams. Look also for larger examples, which, themselves, will be seen to be bearing foliage leaves. Compare the structure of a twig of gorse with that of the hawthorn.

11. Study the hooks on the stem and petioles of the bramble or rose. Compare them with thorns, and explain why the hooks cannot be modified branch shoots. Study carefully their shape, as compared with thorns.

12. Examine the phylloclades of butcher's broom, especially with relation to the very reduced leaves. Fully explain why we know that these also are modified branches.

13. Examine the phyllodes of *Acacia*. Note the buds or branches in the axils of the phyllodes indicating that these are modified leaf structures.

14. Study, draw and describe examples of stem tendrils. Good examples are : the twining tendrils of white bryony, the adhesive tendrils of the Virginian creeper, etc.

15. Examine specimens of leaf tendrils. Note by their position and relation to other leaf structures, why they are leaf modifications, and from this point of view, make comparisons with stem tendrils. Note the difference between the leaf tendril of the garden pea, or sweet pea and that of the yellow vetchling.

CHAPTER VI

FOOD OF LIVING THINGS

All living things must have an ample supply of food if they are to remain healthy. If their food supply is inadequate they soon show signs of malnutrition. They also become more susceptible to disease. In many plants and animals surplus food is stored within the tissues and is available for use should adverse conditions develop (Chap. V). In general, the food of plants and animals is fundamentally similar.

The most important food-stuffs may be classified into five great groups: **water, mineral salts, carbohydrates, proteins** and **fats**. Some others will be mentioned later.

WATER

All plants and animals require water. It is an important constituent of protoplasm, the fundamental basis of life (see Chap. IX). Apart from this connexion, however, water is very essential for various other purposes. The need for water is as great as the need for any other food-stuff in plants and animals. It is of interest to note that the other foods can be withheld from man for more than thirty days and not cause permanent injury, yet starvation due to the absence of water can take place within three or four days.

Many plants have to transport their manufactured food-stuffs from one place in their bodies to another, and they are always carried in solution in water. The transport of food throughout the plant is called **translocation** (Chap. XVI).

Also, all the chemical changes which occur in the living cell (and there are many) do so in solution in water, and many chemical reactions in the cell actually involve molecules of water. Food manufacture, digestion, germination and growth are important processes which demand the presence of water. The amount of water required varies with the organism. In man, sixty to seventy per cent of his body-weight is made up of water. In dry seeds, the water content is very low ; that of some seaweeds is as much as ninety-five per cent ; cabbage ninety-two per cent and lettuce ninety-four per cent.

Some plants or plant organs such as lichens and seeds can withstand desiccation for a long time.

Submerged plants, such as *Chlamydomonas* and seaweeds, absorb water all over their surface. Terrestrial plants absorb their water from the soil through their roots ; and in the case of some epiphytes, the water is condensed on the aerial roots, which hang suspended in the air (see p. 43), and absorbed in that way.

MINERAL SALTS

Until comparatively recently it was not realised how important inorganic mineral substances are to life. Man for example must have certain inorganic substances, such as sodium, iron, calcium and phosphorus salts. Common salt and iron are in the blood, phosphorus in the brain and nerves, and calcium in the bone.

Plants require many elements for their foods, etc. Hydrogen and oxygen are obtained from water in the soil. Carbon is absorbed from the air (see Chap. XV). All the rest are obtained from solution in the soil. Nitrogen is required for the formation of proteins. Phosphorus is required for certain proteins ; magnesium is one of the elements present in chlorophyll. Iron, calcium and potassium are necessary to a plant for various reasons.

These ten elements so far mentioned are known to be necessary to all living green plants and are therefore called **essential** elements. But certain other elements are now being found to be necessary for the healthy well-being of some, if not all, plants. They are, however, required only in very minute amounts. For example, boron, zinc, copper, aluminium, molybdenum, sodium, chlorine, silicon and manganese appear to be required. It is possible that manganese, copper and zinc may enter into certain types of enzyme activity (Chap. VIII). If boron is entirely absent, the growing points of shoot and root tend to die and there is often disintetegration of tissues in other parts of the plant. All these elements, traces of which are required (though we have yet to learn much about them) are called **micro-elements** or **trace elements**.

It must be borne in mind that the essential feature is that these micro-elements must be present in traces only. For example, in 1927 it was shown that broad beans growing in a nutrient solution entirely free of boron failed to complete their development ; concentrations of 0.1 to 40 parts per million restored development to normality ; whereas concentrations of 200 parts or more per million had an adverse effect.

All the elements, with the exception of hydrogen, oxygen, and carbon, are normally absorbed from the soil in the form of salts which pass into the root dissolved in the soil water. In the case of submerged plants, there is normally sufficient quantity of the mineral salts in the sea- or fresh-water (whatever the case may be) for the plant.

The relative amounts of these chemical substances that are absorbed by a plant are not necessarily proportional to their concentrations in the external solution. For example, in sea-water there is a high percentage of common salt (about 2.7 per cent) but an infinitesimal amount of iodine. Yet brown seaweeds absorb a much greater ratio of iodine, and have, indeed, provided a commercial source of that element. Thus plants have an ability to accumulate certain salts or ions.

Information on the nature of the minerals absorbed by a plant is obtained by chemical analysis of the tissues. One way to analyse plants for mineral matter is to obtain the plant ash. The plant material, after drying, is heated in a crucible until all the volatile material is driven off. In this way all organic compounds containing the carbon, hydrogen, oxygen and much of the nitrogen are driven off, and the inorganic compounds remain. These are qualitatively and quantitatively analysed by the usual chemical methods.

The percentages of various elements (in this case in the dry matter) found present in certain maize plants after they had reached maturity and had ceased to absorb solutes from the soil are given in the following table (data from Prof. E. C. Miller) :

Oxygen	-	-	-	44·431	Sulphur	-	-	-	0·167
Carbon	-	-	-	43·569	Iron	-	-	-	0·083
Hydrogen		-	-	6·244	Silicon	-	-	-	1·172
Nitrogen		-	-	1·459	Aluminium		-	-	0·107
Phosphorus		-	-	0·203	Chlorine	-	-	-	0·143
Potassium		-	-	0·921	Manganese		-	-	0·035
Calcium		-	-	0·227	Undetermined elements		-	0·933
Magnesium		-	-	0·179					

The organic materials are clearly the products of the living protoplasm, and until the beginning of the nineteenth century it was concluded that such compounds were the products of a ' vital force '. It was considered impossible to prepare organic substances such as sugars, amino-acids, malic acid, etc., by the usual chemical means. A force peculiar to life called a vital force was believed necessary for such synthesis. But this **vitalistic theory** was disproved in 1828 by **Wöhler** who for the first time prepared the *organic* substance urea, $CO(NH_2)_2$, from the *inorganic* compound ammonium cyanate, NH_4CNO. Since that date, many organic compounds have been prepared from inorganic ones.

EXPERIMENTAL METHODS

Further information on the mineral nutrition of plants is obtained by means of experiments of various types.

Field experiments, like those carried out, for example, at the Rothamsted Experimental Station, have been giving important information for well over a century. That Station was founded in 1843 by **Sir J. B. Lawes** for carrying out field experiments, following the example of **J. B. Boussingault**, the French agricultural chemist. These took the form of attempts in the field to discover the effects of various substances or **fertilisers** on plant growth, by adding them to the soil. But Lawes's most important work was on what is now called **superphosphate**, an artificial manure obtained by treating calcium phosphate with sulphuric acid. He patented this process in 1842, and thus initiated the artificial manure industry. The phosphate used to-day, for treatment with sulphuric acid in order to produce superphosphate, is ground rock phosphate. Some results of testing the effects of artificial manures, in field experiments, are shown in Fig. 36.

Greater control of mineral uptake is possible if the plants are grown in pots filled with a rooting medium which is, to begin with, free of available mineral matter. Often a purified sand is used and to this is added a **culture solution** containing selected mineral salts. Alternatively a solid rooting medium can be dispensed with and the plants grown with their roots immersed in the culture solution alone. This is the **water culture** method. Most plants will make good growth under these conditions, provided that suitable mineral salts are supplied.

There are many prescriptions for a culture solution. One of the first proposed was that of **W. Knop** in 1865. This consists of 1000 c.c. distilled water, 0·8 gm. calcium nitrate, 0·2 gm. magnesium sulphate, 0·2 gm. acid potassium phosphate, 0·2 gm. potassium nitrate, and a few drops of ferric phosphate or ferric chloride solution. Detailed instructions for performing water culture experiments will be found in the practical work.

Knop's solution contains the elements hydrogen, oxygen, calcium, nitrogen, magnesium, sulphur, potassium, phosphorus, iron and chlorine. Most plants will

Rothamstead Experimental Station

Fig. 36. Some Results of Field Culture Experiments.
Produce from the Broadbalk Wheat Field, Rothamsted, on which wheat has been
grown every year since 1843, without exception. Plot 3, no manure since 1839 ; Plots 6, 7
and 8, artificial fertilisers since 1843, including superphosphate, sulphate of potash, and
sulphate of ammonia (the last-named applied at the rate of 200 lb., 400 lb. and 600 lb. per
acre, respectively.

grow and remain healthy in that solution. By modifying the composition of the
solution the effect of the absence of any of the elements can be studied. Thus by
substituting salts of sodium for those of potassium, the latter element can be
omitted from the culture solution. This results in markedly reduced growth and
in the development of brown unhealthy areas on the leaves. The omission of
calcium leads to interference in root development and causes distorted and yellowish leaves. Yellowness in the leaves (**chlorosis**), due to interference with chlorophyll formation, is particularly noticeable when iron is absent from the solution.
This might be thought to indicate that iron is a constituent of chlorophyll, but
actually this is not the case (p. 175).

Certain results of water culture experiments are illustrated in Fig. 37.

Many experiments on the application of water culture methods to the commercial production of crop plants, such as the tomato, and of flowers, as an alternative
to conventional cultivation in soil, have been carried on, often with successful results.
The application of this cultural method to crop production is known as **hydro-
ponics.**

FERTILISERS AND MANURES

Mainly through the absorptive action of plants and as a result of leaching
(loss of salts in drainage water), the amount of mineral nutrients present in a soil

tends to become reduced. In most natural plant communities there are, however, compensating processes at work. Fallen leaves and other plant and animal remains give rise to a **humus** and from this, mainly by the action of bacteria and fungi, the mineral matter is eventually released again into the soil and is again available for plant uptake.

In agricultural and horticultural cultivation this is not the case, since the plants or parts of them are removed from the soil for use. In this case, the nutrients are replaced by manuring. Natural manure, such as leaf mould or animal dung, is added to the soil. Artificial manures, which contain nitrogen, phosphorus, potassium, etc., are also used to a great extent in modern farming and gardening.

Dr. G. Bond

FIG. 37. WATER CULTURES OF BARLEY.
Left to right : complete Knop's solution ; Knop's solution minus potassium ; Knop's solution minus nitrogen ; Knop's solution minus phosphorus.

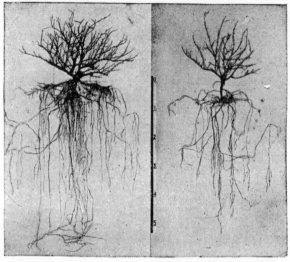

East Malling Research Station

FIG. 38.   EIGHT-YEAR-OLD WHITESMITH GOOSEBERRY PLANTS.
On the right, not manured in year of planting ; on the left,
manured with farmyard manure in year of planting.

The best-known nitrogen-ous fertilisers are Chile salt-petre and ammonium sul-phate, the latter being obtained from gasworks or manufactured by a process in which nitrogen from the air is heated under pressure with hydrogen, so that the two elements combine. For general plant cultivation, mixtures of artificials are often used, some of them being obtained commerci-ally already mixed.   For example, 'Growmore' is a general-utility fertiliser con-taining ammonium sul-phate (one part), potassium sulphate (two parts) and superphosphate of lime three parts).   It is recom-mended that this fertiliser should be applied at the rate of 42 lb. for every 300 square yards.

In spite of the great advantages of artificial manure in crop production, natural manure is frequently more successful with many plants.   The effect of ordinary farmyard manure on a gooseberry plant is shown in Fig. 38.   The superiority of farmyard manure for wheat is illustrated graphically in Fig. 39.

ROTATION OF CROPS

Since some plants require more of certain elements than others it would in general seem advisable not to grow the same kind of annual plant on the same soil in two successive seasons.   In the second season, another type of crop should be planted, and in the third another, and so on.   Such a changing-round of crops is known as the **rotation of crops**, and has been practised in agriculture since before Roman times.   By this rotation of crops, one crop is alternated with a ' recuper-ative ' crop.   In one season a grain crop is grown, and in the next a root crop.

What is called the ' four-course system ' is used extensively in agriculture. In this system, the rotation is wheat one year, then roots (such as swedes) in the next, then barley and then clover.

CARBOHYDRATES

Carbohydrates are common constituents of both plants and animals.   The name *carbo-hydrate* was given in the first place because it was discovered that these substances contain the elements carbon, hydrogen and oxygen, the last two being present in the same proportion as they are in water, so that they could be repre-sented by the general formula $Cn(H_2O)m$.   In a few carbohydrates, however, the elements do not conform to this proportion.

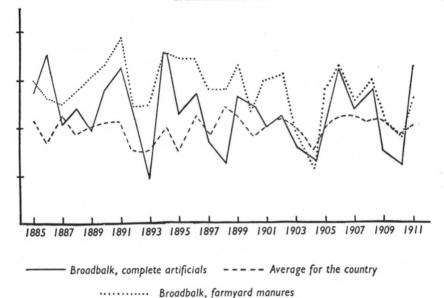

——— Broadbalk, complete artificials   - - - - - Average for the country

············ Broadbalk, farmyard manures

FIG. 39. YIELDS OF WHEAT, IN BUSHELS PER ACRE, FROM THE BROADBALK FIELD, ROTHAMSTED.

Manured with complete artificial manures and with farmyard manures respectively. These are compared with the average yield for the whole country between the dates indicated.

Carbohydrates are of the utmost importance to plants and man. They are the primary products in photosynthesis (Chap. XV). They are the main basis of respiration. Besides being a normal constituent of protoplasm, many exist as food reserves such as **sugars**, **starches**, and occasionally in the plant cell-wall as **cellulose**. In animals, they seldom exist other than in the cell protoplasm.

Carbohydrates may be roughly classified into two groups, namely **sugars** and **non-sugars**. All of them have high molecular weights, but those of non-sugars are much higher than those of the sugars. Sugars are usually sweet to the taste and soluble in water, whereas the non-sugars are not.

An important reaction in plant and animal physiology is that in which the elements of water are added to a molecule, with the result that such a molecule becomes split up into two or more simpler molecules. This process, which is also common among inorganic chemical reactions, is called **hydrolysis**. For example, some inorganic salts will decompose when dissolved in water, and produce a free acid and a base, with the fixation of the water. Thus, a solution of potassium cyanide will hydrolyse and produce hydrocyanic acid and potassium hydroxide :

$$KCN + H_2O \leftrightarrows HCN + KOH$$

The reverse reaction is known as **condensation** ; and this also is common in physiological processes.

The classification of carbohydrates can be best understood from a consideration of their products of hydrolysis.

The simplest carbohydrates are certain sugars, which may therefore be looked upon as the unit for classification, since they are so simple that they cannot be hydrolysed. For this reason, they are referred to as **monosaccharides**, and most of them may be represented by the formula, $C_6H_{12}O_6$. Those sugars which will split up into two monosaccharide molecules on hydrolysis are called **disaccharides**, and can be represented by the formula $C_{12}H_{22}O_{11}$. Hydrolysis of a disaccharide is demonstrated in the following equation :

$$C_{12}H_{22}O_{11} + H_2O \rightarrow 2C_6H_{12}O_6$$

Certain **trisaccharides** and **tetrasaccharides** also occur in plants; but they are by no means so important as mono-and di-saccharides.

All the foregoing saccharides are sugars. Still more complex saccharides, however, occur. They are not sugars and are either incapable of dissolving in water, or form only colloidal, but not true, solutions in water (see Chap. IX). Starch is an example. Since, on hydrolysis, such saccharides give a large number of monosaccharide molecules, they are referred to as **polysaccharides**. The molecule of a polysaccharide is usually very large (see p. 68), but their hydrolysis can be represented in general by the following equation :

$$(C_6H_{10}O_5)n + nH_2O \rightarrow nC_6H_{12}O_6$$

Another type of polysaccharide present in plants is the **pentosan**. There are several types of pentosan, some of which occur in wood and straw. This polysaccharide is called a pentosan because it gives a monosaccharide, which, instead of containing six carbon atoms, contains only five. Therefore, there are two types of monosaccharides, namely those which contain six carbon atoms, **hexoses**, and those which contain only five carbon atoms, **pentoses**.

The classification of the carbohydrates according to the products of their hydrolysis may be represented thus :

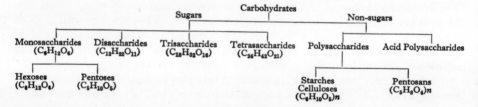

SUGARS

Sugars are very common in plant and animal tissues. They are all soluble in water giving crystalloidal solutions sweet to the taste. For our purpose, the consideration of one or two will be sufficient ; but it must be realised that there is quite a large number of them.

One of the most important sugars is a monosaccharide named **glucose**. It is a white crystalline substance with the formula $C_6H_{12}O_6$. It is present as a food reserve in the fruit of the grape, the onion bulb and onion seed, and various leaves and roots of many plants. It is usually present in these cases in solution in the cell-sap of the vacuole (see p. 131). But it also occurs in smaller quantities in nearly all plants.

Glucose is also present in the human blood. When a person suffers from the illness called sugar diabetes, there is excess of this sugar which is given off in the urine. The malady can be diagnosed by testing the urine. By medical men, this sugar is therefore called ' diabetic sugar '.

The monosaccharides contain hydroxyl groups (–OH). Besides these groups, some contain an aldehyde group (–CHO), whereas others include a ketone group (>C:O). Glucose contains an aldehyde group.

The various types of monosaccharides, like many other organic compounds, exhibit the phenomenon of **isomerism.** Where such organic chemical compounds are concerned, it is possible for several *different* compounds to possess the *same* chemical elements in the *same* percentage composition. Therefore, the same *molecular* formula will represent several different compounds ; but the organic groups of which such compounds are built up, are different ; hence the *structural* formulae are different. For example, the molecular formula $C_3H_6O_2$ represents three substances, ethyl formate, methyl acetate and propionic acid. On the other hand, when the structural formulae of these three compounds are examined, it will be seen that the grouping of the atoms in each molecule is different. They are : ethyl formate, $H.CO.O.C_2H_5$; methyl acetate $CH_3.CO.O.CH_3$; propionic acid, $C_2H_5.CO.OH$. This structural difference between the molecules of compounds which have the same percentage composition is called **structural isomerism.**

Another hexose is important, and is interesting in this connexion in that it is a structural isomer of glucose. It is called **fructose**, and often occurs with glucose in plant organs, especially in green leaves, sweet·fruits and in certain storage organs such as bulbs. Both glucose and fructose are hexoses, and may therefore be represented by the molecular formula $C_6H_{12}O_6$.

Their structural isomerism, however, is due to the fact that, whereas they both contain five hydroxyl groups, glucose contains an aldehyde group (**aldohexose**), and fructose a ketone group (**ketohexose**). The structural formula for each may be represented as follows, though it should be noted that under many conditions the molecules of the sugars assume a ring form as a result of a modification of the straight-chain formulae shown here :

CHO	CH$_2$OH
CHOH	CO
CHOH	CHOH
CHOH	CHOH
CHOH	CHOH
CH$_2$OH	CH$_2$OH
Glucose	*Fructose*

Another important form of isomerism is that known as **stereoisomerism** in which the constituent groups of the molecule are arranged *differently in space*. For example, **galactose**, another hexose which occurs in the combined state in plant organs, but rarely free, and is an important constituent of brain tissue in animals, is a stereoisomer of glucose. If the structural formula for the latter be written in further detail, in which nearly all the valence bonds (four for carbon, two for oxygen and one for hydrogen) are indicated, then this formula compared with the structural formula for galactose, the stereoisomerism or different arrangement in space of the groups of atoms, will at once become apparent :

```
        CHO                    CHO
         |                      |
    H—C—OH                 H—C—OH
         |                      |
   HO—C—H                  HO—C—H
         |                      |
    H—C—OH                 HO—C—H
         |                      |
    H—C—OH                  H—C—OH
         |                      |
      CH₂OH                  CH₂OH
```
$$\begin{array}{cc} CHO & CHO \\ H-C-OH & H-C-OH \\ HO-C-H & HO-C-H \\ H-C-OH & HO-C-H \\ H-C-OH & H-C-OH \\ CH_2OH & CH_2OH \end{array}$$

Glucose   *Galactose*

Another stereoisomer of glucose which is known to exist is **mannose**, though it is not a common sugar.

A very familiar form of stereoisomerism in organic chemistry is that known as **optical isomerism**. This form is very common among the natural products of plants and animals. As in the former case of stereoisomerism, optical isomerism is due to the asymmetric nature of the carbon atoms in the molecule.

The carbon atom is tetravalent, therefore, in order to saturate it, four other atoms or groups of atoms must be joined to it. In space, the carbon atom can be pictured as being in the centre of a tetrahedron with the four other atomic groups situated at the angles. If the carbon atom is asymmetric, that is, if it has four *different* atomic groups satisfying its valencies, then it is possible to conceive of two tetrahedra, the one being the mirror image of the other. This is represented in Fig. 40. The difference between two such mirror images is indicated *physically* in the different ways in which they turn the plane of *polarised light*. Polarised light is that which, by means of a Nicól prism, is made to vibrate in one plane only. In the case of optical isomers, one mirror image is able to turn the plane of polarised light through a definite measurable angle, to the right, and is therefore said to be **dextro-rotatory**. The other mirror image turns the polarised light to the left, and is therefore said to be **laevo-rotatory**.

The sugars exhibit optical isomerison. Naturally occurring glucose is dextro-rotatory, and an alternative name for it is, therefore, **dextrose**. (This same sugar is sometimes also called **grape sugar**, since it is the constituent sugar in the fruit of the grape.) The Fructose is laevo-rotatory. It is, therefore, sometimes called **laevulose**.

Hexoses are easily oxidised to form organic acids, usually in stages. For example, glucose can be oxidised to gluconic acid $(CH_2OH)\,(CHOH)_4.COOH)$,

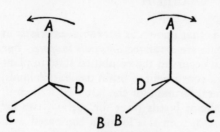

FIG. 40. TETRAHEDRAL ARRANGEMENT OF OPTICAL ISOMERS.

and this acid further oxidised to glucuronic acid $(COOH)\,(CHOH)_4(CHO)$ and this further to saccharic acid $(COOH.(CHOH)_4.COOH)$. Carried still further the final oxidation ends in carbon dioxide and water.

In the presence of weak alkalies, glucose and fructose are interconvertible.

A much more familiar sugar is **cane-sugar** or **sucrose**. This sugar is very common in nearly all the organs of plants, especially the leaves. It is present in varying amounts, but in some plants the percentage is high enough to make it worthy of extraction. In the sugar-cane, the sucrose content of the stem is 11 to 18 per cent. In the root of the sugar-beet it is 10 to 25 per cent, and it is also high in the birch and maple. Sucrose, glucose and fructose are often present together, as indicated in the following table, which gives the percentage carbohydrate content of certain fruits, etc. (data from Widdowson and McCance, 1934).

Carbohydrate / Fruit	Glucose	Fructose	Sucrose	Starch
Apple	1·7–2·2	5·0–7·2	2·4–3·6	0·2–0·35
Banana	5·8	3·8	6·6	3·0
Cherry	4·7–5·5	6·1–7·2	0	0
Date	32·0	23·7	8·2	0
Grape	8·2	7·2–8·0	0	0
Lemon	1·4	1·4	0·4	0
Orange	2·5	1·8	4·2	0
Pear	2·2–3·5	6·0–7·0	1·0	0
Strawberry	2·6	2·3	1·3	0
Tomato	1·6	1·2	0	0·02

Sucrose is a disaccharide with the molecular formula $C_{12}H_{22}O_{11}$. It can be hydrolysed to form two monosaccharides according to the equation :

$$C_{12}H_{22}O_{11} + H_2O \rightarrow C_6H_{12}O_6 + C_6H_{12}O_6$$

Sucrose   Water   Glucose   Fructose

In the laboratory, such hydrolysis can be produced by warming a solution of sucrose with some dilute hydrochloric acid. All three of these sugars are optically active ; but a remarkable phenomenon accompanying the hydrolysis of this disaccharide is that, whereas sucrose is dextro-rotatory, the solution of glucose and fructose produced is laevo-rotatory. This is easily explained, however. The specific rotation of dextro-rotatory sucrose is $+66·5°$, that of the dextro-rotatory glucose, $+52·3°$, and that of the laevo-rotatory fructose $-93°$. On the hydrolysis of sucrose, the glucose and fructose are produced in equimolecular quantities, as indicated in the foregoing equation. Therefore, the specific rotation of the mixture of sugars produced is

$$\left(\frac{-93 + 52·3}{2}\right)° = circa - 20°,$$

that is, laevo-rotatory. Because of this inversion of specific rotation during the hydrolysis of sucrose, the resulting mixture is often referred to as **invert sugar.**

Many considerations suggest that the opposite reaction, namely, the condensation of glucose and fructose to give sucrose frequently occurs in the plant ; but there is doubt about the mechanism of the reaction.

NON–SUGARS

Non-sugars or polysaccharides are characterised by their high molecular weight and insolubility in water. Starches and celluloses are two examples. The former are used for food-storage in the plant, whereas the latter are used chiefly in the formation of the plant's skeleton

Starches are insoluble in water and therefore cannot exist in solution in the cell-sap as do the sugars. The starch molecule is built up from glucose molecules and may be represented as $(C_6H_{10}O_5)n$. By various methods, it has been shown that n has a value of some hundreds, so that the molecular weight of starch is very high.

Starch is present in leaves, stems, roots and storage organs, such as the potato tuber and tulip bulb. Some plants, such as the onion, however, do not contain starch, except in certain cells such as the guard cells ; but contain a great deal of sugar. The main sources of commercial starch are cereal grains, for example, maize (containing 70 per cent starch), and potato tubers (containing 20 per cent starch). Rice grains (after polishing) contain a high percentage of starch, namely, 85 per cent.

In the plant, starch is manufactured by the process of photosynthesis, and therefore appears periodically in the **chloroplasts** present in green organs (see Chap. XIV). Chloroplasts contain the green colouring matter chlorophyll, and occur only in those organs exposed to sunlight. Starch which is produced in organs not thus exposed, such as a potato tuber, is produced by granules similar to chloroplasts, with the exception that they do not possess chlorophyll (Fig. 41). These granules are called **leucoplasts.** If leucoplasts are exposed to the light, they develop chlorophyll, and thus become converted into chloroplasts. That is why potato tubers turn green on exposure to the light.

Starch takes the form of microscopic grains in the plant, embedded in the cell cytoplasm. These grains vary considerably in shape and structure, so much so, that from a microscopic examination of them it is often possible to determine from which plant they have been extracted. They vary in size too, from about 0·002 mm. to 0·17 mm. in diameter. Those produced by leucoplasts are usually much larger than those produced by chloroplasts.

Starch grains are not merely homogeneous masses. They are formed of several layers or **stratifications ;** thick, dense layers alternating with thin, less dense ones. The centre around which these layers are deposited is called the **hilum.** Sometimes two starch grains begin development near each other, then as the

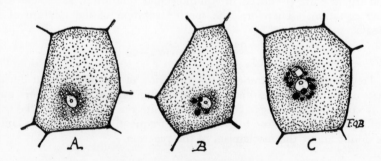

FIG. 41. CELLS OF A YOUNG POTATO

A, with minute leucoplasts surrounding the nucleus ; B, some leucoplasts already forming starch, stained darkly with iodine ; C, further starch formation and one cubic protein crystal (x220).

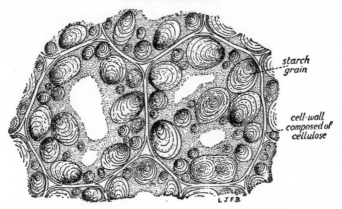

FIG. 42. STARCH GRAINS IN CELLS OF A POTATO TUBER.

successive layers are deposited, the growing grains touch each other and the succeeding layers are deposited around both grains, thus forming a **compound** starch grain, with two hila. Sometimes there are numerous hila and the components of such a compound grain are frequently angular as a result of mutual pressure. The normal grain with only one hilum is called a **simple** starch grain (Figs. 42 and 43).

The most important economic use of starch to man is the same as in the plant, that is, as a food. It is the fundamental food present in many articles of diet such as the potato tuber, flour (with proteins), haricot beans (with proteins), peas (with proteins), arrowroot, sago and tapioca (Fig. 43).

There is evidence that the starch grains of many plants contain two different forms of starch. One is named **amylopectin**, and this may form a kind of skeletal network which encloses the other called **amylose**.

The starches are polysaccharides and can either be partially or completely hydrolysed. Unlike sucrose, starch, on *complete* hydrolysis, produces only one monosaccharide, namely, glucose. The reaction may be illustrated by the following equation :

$$(C_6H_{10}O_5)n + nH_2O \rightarrow nC_6H_{12}O_6$$
Starch   Water   Glucose

In view of the large size of the starch molecule, it is natural that hydrolysis should take place in stages, and that certain definite intermediate compounds should be produced.

The first stage in the starch hydrolysis brought about by boiling with dilute acid is probably the production of another polysaccharide, **soluble starch**, which also is a condensation product of glucose, though not containing so many molecules of glucose as starch does. Further hydrolysis gives **dextrin**, which can be represented by the formula $(C_6H_{10}O_5)n$, where n represents a smaller number than that in the formula for soluble starch. Further hydrolysis produces the disaccharide, maltose, $C_{12}H_{22}O_{11}$, and this sugar, on still further hydrolysis, yields glucose.

Cellulose and its related substances are of great importance to the plant, and also economically to man. It is the chief constituent of the wall of plant cells, although in special cases, such as the vessels of the xylem, other substances are added to the original cellulose (see below). Cellulose is chemically a carbohydrate built up, like starch, from glucose, so that the empirical formula $(C_6H_{10}O_5)n$ can be assigned to it. One estimate is that about 200 glucose molecules are required to make one cellulose molecule, though much higher figures have also been obtained. On complete hydrolysis glucose is yielded.

The cotton plant (*Gossypium*), a member of the mallow family, Malvaceae, is the chief source of cotton in industry. Cotton is composed chiefly of cellulose, which is present in the cell-walls of certain very long cells which grow on the seeds of the plant, forming a very hairy covering (Figs. 44 and 45). The cotton fibre contains 91 per cent cellulose, 8 per cent water and 1 per cent other substances.

Cellulose is not present in man, and it is extremely rare in other animals. A substance closely related to it, however, known as **tunicin**, is present in certain tunicates and insects.

Hemi-cellulose is another type of cellulose, showing certain chemical differences from the normal form. It is deposited as a food reserve in the seeds of various plants (for example, lupin, date-palm) and this leads to the cell walls being greatly thickened. This reserve food is utilised during germination, the hemi-cellulose being converted to sugars under the influence of the enzyme cytase (Chap. VIII).

Wood is formed of cellulose united with a substance called **lignin** and also other substances. This forms the basis of the structure of vessels, tracheids, fibres and stone cells (Chap. X). The chemical nature of lignin is still not perfectly understood. **Suberin** and **cutin** are substances of fatty or waxy nature which

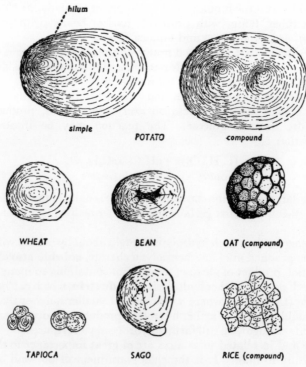

FIG. 43. VARIOUS FORMS OF STARCH GRAINS.

appear in some cell-walls, and render them more or less impermeable to water. Suberin is the chemical basis of cork, and cutin that of cuticle, such as is found on the outer epidermal walls of leaves (see p. 166).

Gums are chemically very complicated carbohydrates. Some gums will dissolve in water to form adhesive substances. Others will not dissolve, but will absorb water and form a jelly. All are laevo-rotatory. Hydrolysis is difficult, but once it occurs, mixtures of sugars (chiefly hexoses and pentoses) and uronic acid are obtained. **Gum arabic** is obtained from a plant (*Acacia senegal*) which grows in the Sudan, and is used in finishing silks and other textiles, and a host of other processes.

Inulin is another polysaccharide of complex composition, condensed from fructose. It occurs as a food reserve in some plant organs, such as the *Dahlia* root tuber (p. 24).

FIG. 44. FLOWERS AND FRUITS OF THE COTTON PLANT.

In the bottom right-hand corner is an opened fruit with its seeds covered with long hairs.

PROTEINS

Proteins are the most important constituents of protoplasm ; they therefore occur in all living parts. As protoplasmic constituents, proteins occur either in colloidal solution or in a very viscid colloidal state (see Chap. IX).

The term ' protein ' embraces a large variety of substances occurring in the plant and animal kingdoms. There is quite a high percentage of proteins in milk. The ' white ' of an egg is composed mainly of protein and water, the chief protein being called **albumin**. The lean part of meat is composed chiefly of different proteins, so also is the flesh of fish.

In plants, proteins occur either as solids or in the colloidal state as reserves in almost any organ, but especially in seeds and vegetative organs of reproduction such as the potato tuber. In the latter the reserve proteins are present chiefly in those parenchymatous cells immediately beneath the epidermis or skin (Fig. 41).

After Figuier

FIG. 45. SINGLE SEED OF COTTON WITH SUPERFICIAL HAIRS.

Solid proteins are sometimes amorphous and sometimes crystalline. A special kind of protein reserve is present in the seed of the castor-oil plant (*Ricinus communis*), Brazil ' nut ' (*Bertholletia excelsa*) and in other seeds. In this case, several proteins combine to form an ovoid structure called an **aleurone grain** (Fig. 46). Under the microscope, this structure appears as an egg-shaped matrix in which is embedded two smaller structures. One of these forms a crystal shape. This is composed of protein, and is called the **crystalloid**. (This term must not be confused with the term which is used in contradistinction to the term colloid in Chap. IX). The other structure within the matrix of the aleurone grain is spherical in shape, and is therefore called the **globoid**. This is not a protein, but is composed of the double phosphate of calcium and magnesium. The matrix in which the crystalloid and the globoid are embedded is composed of protein which is different from the protein of the crystalloid in that it is chemically much simpler. Surrounding the whole aleurone grain is a thin·skin which is formed of another different protein.

In the potato, the proteins are chiefly solid, forming cubical crystals (Fig. 41). In pea and bean seeds, which are particularly rich in protein food reserve, the aleurone grains contain neither globoids nor crystalloids. The same is true of the protein reserves of cereal grains, such as wheat and maize.

Proteins are all of a very high molecular weight, values as high as 400,000 having been obtained in some cases. They are all laevo-rotatory and contain the elements carbon, hydrogen, oxygen, nitrogen and sulphur. Some contain phosphorus. The percentage composition of the majority of proteins is: carbon, 50–55 ; oxygen 20–25 ; nitrogen, 15–20 ; hydrogen, 6·5–7·5 ; sulphur, 0·3–5.

Proteins yield amino-acids (R.CH (NH$_2$).COOH) on hydrolysis. There is a large number of different amino-acids present in plant and animal proteins. Examples are : glycine (a–amino acetic acid), CH$_2$NH$_2$.COOH ; aspartic acid (a–amino succinic acid), COOH.CH$_2$.CH(NH$_2$).COOH ; glutamic acid (a–amino glutaric acid), COOH.CH$_2$.CH$_2$.CH(NH$_2$).COOH ; cystine (Di–β–thio a– amino propionic acid)

$$\text{S.CH}_2.\text{CH(NH}_2).\text{COOH}$$
$$\text{S.CH}_2.\text{CH(NH}_2).\text{COOH}$$

and tryptophane (β–indole a–amino propionic acid)

$$\text{C.CH}_2\text{CH(NH}_2).\text{COOH.}$$

The proteins contain –COOH and –NH$_2$ groups, and since the former is an acid group and latter an alkaline, they can act either as acids or as bases. They are therefore amphoteric electrolytes (see Chap. IX).

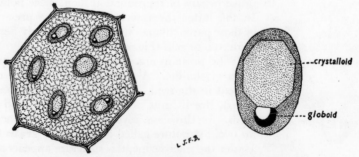

FIG. 46.   LEFT, ALEURONE GRAINS IN A CELL OF THE SEED OF THE CASTOR-OIL PLANT
(× 550)   RIGHT, A SINGLE ALEURONE GRAIN (× 1800).

There are many different proteins ; but a great deal more has yet to be learned about them. Until recently, they were roughly classified according to their physical properties, but since the products of the hydrolysis of many of them have been elucidated, a rough chemical classification is now possible, as in the following table :

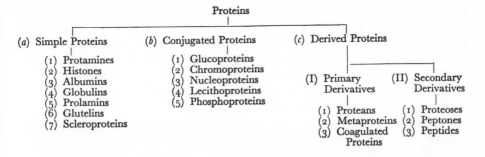

Proteins

(a) Simple Proteins	(b) Conjugated Proteins	(c) Derived Proteins
(1) Protamines	(1) Glucoproteins	
(2) Histones	(2) Chromoproteins	
(3) Albumins	(3) Nucleoproteins	(I) Primary Derivatives (II) Secondary Derivatives
(4) Globulins	(4) Lecithoproteins	
(5) Prolamins	(5) Phosphoproteins	(1) Proteans (1) Proteoses
(6) Glutelins		(2) Metaproteins (2) Peptones
(7) Scleroproteins		(3) Coagulated (3) Peptides
		Proteins

Protamines are the simplest proteins known. They usually exist as salts in combination with *nucleic acid*. They have been isolated from the sperm of fish, but have not been definitely extracted from plant organs. **Histones** are more complex. These are unknown in plants, but are present in blood. **Albumins** are present in plants and in animals. The most familiar example is egg albumen. They are also present in pea seeds and the grains of wheat and maize. **Globulins** are insoluble in water, but soluble in salt solutions. Examples are present in blood serum, and in many seeds of plants including pea, bean, lupin, etc. In fact, globulins are the most common form of protein reserves in plants. Albumins and globulins are the only forms of proteins which are coagulated by heat. **Prolamins** and **glutelins** are exclusively plant proteins. They are common protein reserves in cereals. Together they form the substance known as *gluten*, which, together with starch, forms the chemical basis of flour. **Scleroproteins** are unknown in the plant kingdom, but are well known in animals, being represented by *keratin* in hair, horn, etc., and by *gelatin*.

Conjugated proteins, on hydrolysis, yield a simple protein together with a substance of a different nature, sometimes known as the *prosthetic* group. **Nucleoproteins** are present in the nuclei of all cells (see Chap. X). Here the prosthetic group is nucleic acid. **Lecithoproteins** are formed by the condensation of a protein with *lecithin*, a substance related to fats. They are found in plants and animals, especially in egg yolk. **Chromoproteins** are exemplified by *haemoglobin*, the colouring matter of the blood. This same pigment has been shown to be present in the root nodules of leguminous plants (p. 139) ; but in general the chromoproteins are of uncertain occurrence in plants. **Glucoproteins** are present in some plant mucilages.

Derived proteins are best understood once the actual structure of proteins has been grasped.

The units from which proteins are built are the **amino-acids**. These, as has already been seen, contain acid and basic groups. They can therefore unite and condense, thus forming what are called **peptides**, which contain the –CONH– group. By further condensation, polypeptides and eventually proteins are formed.

Hydrolysis of a simple protein gives a mixture of different amino-acids together with ammonia.   That of conjugated proteins gives a mixture of different amino-acids, ammonia and, of course, the prosthetic group.

The protein molecule can be modified by the action of acids, alkalis, heat, partial hydrolysis, etc.   In this way, the derived proteins are obtained.   Primary protein derivatives retain most of their protein characteristics.   Most seeds contain **proteans**, which are produced by the action of an acid upon a protein.   **Meta-proteins** are obtained from the first stage of hydrolysis of proteins.   **Coagulated** proteins are the result of the action of heat, heavy metals or alcohol.   **Secondary proteins** do not retain so many of their protein characteristics, since they are all further stages in the breaking down of the original proteins by hydrolysis.   Their formation by this process is best understood from the following sequence :

Proteins→Metaproteins→Primary proteoses→Secondary
Proteoses→Peptones→Polypeptides→Amino-acids.

Proteoses are found in many plant juices.   So also are **peptones**, though these are confined to plant tissues already rich in simple proteins, such as in certain seeds.

FATS AND OILS

There is no fundamental difference between a fat and an oil.   Both names are only general ones, a fat denoting a solid and an oil denoting a liquid, under normal conditions of temperature, etc.

Each name denotes a group of many substances which, apart from being important constituents of living protoplasm, form food reserves in plants and animals.

All fats contain the elements carbon, hydrogen and oxygen, but not in the same proportion as they are present in the carbohydrates.

If an acid reacts with a base, a salt is produced.   For example, hydrochloric acid reacts with caustic soda (a base) to produce sodium chloride, a salt :

$$HCl + NaOH \rightarrow NaCl + H_2O.$$

In organic chemistry, a similar reaction takes place between acids and alcohols, the latter containing one or more –OH groups, like a base.   The resulting compound, corresponding to the salt in the inorganic reaction quoted, is called an **ester**.   There is quite a large number of alcohols, but each contains the –OH group, for example, ethyl alcohol, C_2H_5OH, and methyl alcohol, CH_3OH.

Acetic acid will react with ethyl alcohol to produce an ester known as ethyl acetate :

$$CH_3COOH + C_2H_5OH \rightarrow CH_3COOC_2H_5 + H_2O$$
Acetic   *Ethyl*   *Ethyl*   *Water*
acid   *alcohol*   *acetate*

Fats are all esters.   The alcohol concerned is glycerin, or what is more correctly termed **glycerol**, which is a trihydric alcohol, since it contains three hydroxyl (–OH) groups.   Its formula is $CH_2OH.CHOH.CH_2OH$.   Fats are not produced by the esterification of *any* acid by glycerol.   There are definite acids which produce fats, and they are therefore referred to as the **fatty acids**.   Some of the important fatty acids are **formic acid**, $HCOOH$; **acetic acid**, CH_3COOH ; **butyric acid**, C_3H_7COOH ; **oleic acid**, $C_{17}H_{33}COOH$; **stearic acid**, $C_{17}H_{35}COOH$; **palmitic acid**, $C_{15}H_{31}COOH$.

Fats which naturally occur in plants and animals are mixtures of esters produced by the reaction of glycerol with different fatty acids. One example of a fat
is **tristearin**,

$$C_{17}H_{35}COOCH_2.C_{17}H_{35}COOCH.C_{17}H_{35}COOCH_2.$$

This is produced by the esterification of stearic acid, a fatty acid, by glycerol,
according to the following equation :

$$
\begin{array}{cccc}
& CH_2OH & C_{17}H_{35}COOCH_2 & \\
& | & | & \\
3C_{17}H_{35}COOH + & CHOH \rightarrow & C_{17}H_{35}COOCH & + 3H_2O \\
& | & | & \\
& CH_2OH & C_{17}H_{35}COOCH_2 & \\
\textit{Stearic acid} & \textit{Glycerol} & \textit{Tristearin} & \textit{Water}
\end{array}
$$

By a reversal of this type of reaction, fats can by hydrolysed into their constituent fatty acids and glycerol.

The specific gravity of all oils and fats is less than that of water, and they are
all insoluble in it ; but they are soluble in chloroform, ether, carbon disulphide and
carbon tetrachloride. The majority also make a translucent mark on paper.

Fats (or oils as they usually are referred to in plants, since they are often in the
liquid state) are often found as food reserves, usually in parenchymatous cells as
small droplets suspended either in the cytoplasm or the cell-sap. They occur in
all types of plants, from the seaweeds upwards. In angiosperms, fats and oils are
found chiefly in seeds ; for example, the Brazil ' nut ' (which is really a seed and not
a nut) contains nearly 70 per cent, and the almond contains about 55 per cent.

Olive oil is extracted from the fruit of the olive (*Olea europœa*) and is used as a
food, such as a medium in which to preserve other foods, and also in salad oils
and dressings. Inferior olive oil is extracted by dissolving it out in a fat solvent
such as carbon disulphide, and this oil is used in making certain kinds of soap.
Cotton-seed oil is extracted, as its name implies, from the seed of the cotton
plant (*Gossypium* species). After purification it is used in the manufacture of soap
and rubber substitutes. **Coco-nut oil** is obtained from the seed of the coco-nut
(*Cocos nucifera*). **Palm oil** is extracted from the fruits of the oil palm (*Elæis
guineensis*), and has the consistency of lard. It is used in the manufacture of
margarine. **Rape oil** is obtained from the seeds of *Brassica napus*, and is sometimes
used as an illuminant. **Linseed oil** is obtained by pressure from the seeds of the
flax plant (*Linum usitatissimum*). The residue left after pressure forms the well-
known oil-cake which is used as cattle food. Linseed oil, and also **walnut oil** and
poppy-seed oil, are used in oil paints. **Castor oil** is obtained by compressing
the seeds of the castor-oil plant (*Ricinus communis*). It is used as a medicine, and
is also a very valuable chemical in the dyeing industry.

The foregoing survey of the various types of substances which help to make up
the plant body gives an idea of the complicated chemical structure of plants.
Animals, including ourselves, are built up of similar substances (in many cases,
the same), therefore they too are extremely complicated organisms.

By consuming plants, animals obtain food-stuffs. The food, or **dietetic**, value
of plants depends on (*a*) their own chemical composition, and (*b*) what the animal
consuming them actually digests.

COMPOSITION OF FOOD PLANTS

Parts of many plants are fit for human consumption. A very large number of wild plants are nutritious ; in fact we still eat some of them though not so much to-day as in earlier times. Certain plants are cultivated on farm and garden as a food supply. Some are more valuable than others, but civilized man depends so much on the plants that he cultivates, that agriculture is a major world industry being continuously improved upon by scientific research. One branch of this research involves the analysis of those plant organs which are eaten either in the raw state or cooked in some form or other. The accompanying tables (reconstructed from data given in Prof. F. O. Bower's *Botany of the Living Plant*) give a very general idea of the percentage composition of well-known food plants :

VEGETATIVE PARTS (ROOTS, STEMS AND LEAVES)

Common Name	Botanical Name	Water	Proteins and other Nitrogenous Substance	Fats	Carbohydrates and Derivatives		Ash
					Sugars, Starches, etc.	Cellulose, Lignin	
Potato	*Solanum tuberosum*	74·98	2·08	0·15	21·01	0·69	1·09
Beetroot	*Beta vulgaris*	82·25	1·27	0·12	14·40	1·14	0·82
Parsnip	*Peucedanum sativa*	79·31	1·32	—	16·36	1·73	1·28
Onion	*Allium cepa*	85·99	1·68	0·10	10·82	0·71	0·70
Carrot	*Daucus carota*	86·79	1·23	0·30	9·17	1·49	1·02
Turnip	*Brassica rapa*	87·80	1·54	0·21	8·22	1·32	0·91
Cauliflower	*Brassica oleracea* variety	90·89	2·48	0·34	4·55	0·91	0·83
Celery	*Apium graveolens*	84·09	1·48	0·39	11·80	1·40	0·84
Spinach	*Spinacia oleracea*	88·47	3·49	0·58	4·44	0·93	2·09
Lettuce	*Lactuca sativa*	94·33	1·41	0·31	2·19	0·73	1·03

The origin of the potato (*Solanum tuberosum*) has been the subject of much controversy. Wild potato plants have been found in both Chile and Peru : it grows very commonly on the seashore of the archipelago of Southern Chile. It was supposed to have been brought to Spain by a monk during the sixteenth century, and thence passed east to Italy and north to France and Belgium. It is also supposed that the navigators Drake and Hawkins introduced the potato into England from South America during 1563 ; but the botanist **Sir Joseph Banks** (1763–1820) claimed that what they brought was the sweet potato—a totally different plant. Recent research, however, suggests that Banks was wrong.

It is certain that Sir Walter Raleigh brought the potato from North Carolina to Ireland and planted it on his estate near Cork in 1586, though by that time the plant had already arrived in Continental Europe. More recent research into its early history shows that it was actually under cultivation in South America so early as A.D. 200 : it was often used as the motif for the Indian pottery of those regions at that time. It now seems clear that there were two original areas of cultivation, namely, Chile and Peru together with Bolivia.

About 80 per cent of the organic food stored in the potato tuber is starch (p. 68) ; just under the skin is a layer containing a store of proteins.

The beetroot is a sub-species of the wild beet (*Beta vulgaris*), an inhabitant of the sandy shores of Britain and spread over most areas to the Mediterranean

Sea. The swollen root stores cane sugar (about 14 per cent). The sugar beet (*B. vulgaris* var. *ropa*) contains an even higher percentage.

The parsnip (*Peucedanum sativa*) is the cultivated variety of another native plant, the wild parsnip, an inhabitant of waste places. It has been cultivated in Britain since the time of the Romans. It stores food in the form of starch in its swollen tap root (p. 43).

The onion (*Allium cepa*), though closely related to the wild garlics of Britain, certainly originated in the wild state in western Asia. It has been cultivated for centuries ; and was used as a condiment by the Greeks, Romans and Egyptians. It is even depicted on some Egyptian monuments, and one species in that ancient country was awarded divine honours. The fleshy leaves of the bulb contain a high percentage of sugar. A close relative is the shallot (*A. ascalonium*), a vegetable known from the beginning of the Christian era.

The carrot (*Daucus carota*) has been developed under cultivation from its wild progenitor which is native to Britain, growing in pastures. The swollen tap root of this plant is of two-fold value, for it contains not only a high percentage of sugar but also a colouring substance known as **carotene** which is the precursor of vitamin A (p 89).

The turnip (*Brassica rapa*) stores a certain amount of carbohydrates and proteins in its swollen hypocotyl. This plant, together with the cauliflower (*B. oleracea* variety), is a member of the very large cabbage family (Cruciferae) all members of which are edible and a comparatively high number of which are cultivated and supply food for man.

The celery (*Apium graveolens*) is derived from a wild form which grows in the salt marshes of Britain and is native to much of Europe and western Asia. Its food value is not high.

Spinach (*Spinacia oleracea*) belongs to the same family as the beetroot. It is not known in the wild state, having been introduced into Europe from western Asia during the sixteenth century. Its leaves have a comparatively high protein and vitamin content.

Lettuce (*Lactuca sativa*) was originally introduced probably from Asia. As a salad, lettuce was known to the ancient Greeks and Romans, and it has a high vitamin content.

LEGUMES

Common Name	Botanical Name	Water	Proteins and other Nitrogenous Substances	Fats	Sugars, Starch, etc.	Cellulose, Lignin	Ash
					Carbohydrates and Derivatives		
Broad bean (seed only, dry)	*Vicia faba*	13.49	25·31	1·68	48·33	8·06	3·13
French bean (whole fruit, green)	*Phaseolus vulgaris*	88·75	2·72	0·14	6·60	1·18	0·61
Green pea (seed only)	*Pisum sativum*	78·44	6·35	0·53	12·00	1·87	0·81
Groundnut (seed only, dry)	*Arachis hypogaea*	7·71	31·12	46·56	9·39	2·16	3·06

The family Leguminosae is represented in most parts of the world. The fruit and above all the seeds, have a very high protein content.

The broad bean (*Vicia faba*) has been cultivated from times immemorial, having originally reached Europe from regions south of the Caspian Sea.

The French bean (*Phaseolus vulgaris*) is a comparatively late arrival in Europe from Mexico.

The garden pea (*Pisum sativum*) was introduced into Europe from western Asia during prehistoric times.

The groundnut, pea-nut or monkey-nut (*Arachis hypogaea*) probably originated in Brazil ;   but to-day it is cultivated in many other tropical areas.

Among other leguminous plants cultivated for human food or animal fodder are : various species of clover (*Trifolium*), black medick (*Medicago lupulina*), lucerne or alfalfa (*Medicago sativa*), sainfoin (*Onobrychis viciifolia*), soy bean (*Glycine soja* and *G, hispida*), pulses or grams (*Phaseolus mungo*) and lentil (*Ervum lens*).

CEREALS (GRAINS)

Common Name	Botanical name	Water	Proteins and other Nitro-genous Substances	Fats	Carbohydrates and Derivatives		Ash
					Sugars, Starch, etc.	Cellulose, Lignin	
Wheat	*Triticum* species	13·37	12·04	1·91	69·07	1·90	1·71
Rye	*Secale cereale*	13·37	10·81	1·77	70·21	1.78	2·06
Barley	*Hordeum* species	14·05	9·66	1·93	66·99	4·95	2·42
Oats	*Avena sativa*	12·11	10·66	4·99	58·37	10·58	3·29
Maize	*Zea mays*	13·35	10·17	4·78	68·63	1·67	1·40
Rice(unpolished)	*Oryza sativa*	11·99	6·48	1·65	70·07	6·48	3·33

Wheat (*Triticum* species) is one of the oldest of cultivated cereals. We are still not sure of the identity of the parent species of wheat, though it is known that Chinese records of it date back to 2700 B.C. The two main cultivated species are *T. aestivum* and *T. turgidum* ; among other species are *T. monococcum*, the small spelt, *T. dicoccum*, the emmer (a very old species) and *T. durum*, the hard wheat from which macaroni is made.

Rye (*Secale cereale*) is a cereal of more recent origin. It is a special feature of north European agriculture having probably originated there.

Barley (*Hordeum vulgare* and *H. distichum*) is a hardy cereal and can be success-fully cultivated in areas much further north than the wheat zones. It grows wild around the Caspian Sea.

Oats (*Avena sativa*) probably originated in eastern Europe, though it is not now found growing wild. It dates back to prehistoric times.

Maize or Indian corn (*Zea mays*) is cultivated throughout tropical and sub-tropical regions and in certain temperate areas. Though not now found in the wild state, it originated in Mexico and Peru.

Rice (*Oryza sativa*) is one of the most important food crops in the world, forming the staple diet of millions of people, especially in the East. It has derived from the wild form of India and Malaya.

PRACTICAL WORK

1. Determine the moisture content of various plant organs (for example, cabbage or other leaves, storage organs such as carrot or potato, or seeds) by heating carefully weighed samples (of about 10 gm.) in dishes in a hot-air oven at 95°–100°C. Material such as cabbage or carrot is shredded before the samples are taken, while seeds are ground in a mill. Leave the samples in the oven overnight, then cool in a desiccator and re-weigh. Replace the material in the oven for a further few hours and then weigh again, to make sure that all moisture has been driven off. Calculate the percentage moisture content of the original material.

The ash content of the dry matter so obtained may then be determined by weighing out a powdered sample of 1–2 gm. into a crucible which is then heated over a bunsen in a fume chamber, gently at first, and then strongly. Alternatively the crucible may be heated in a muffle furnace. Continue heating until only a whitish residue (the ash) remains. After cooling re-weigh the crucible and calculate the percentage ash content in the dry matter.

2. Prepare a culture solution from Knop's prescription. Also prepare the following solutions :

(*a*) Knop's solution, but using calcium sulphate and potassium sulphate instead of the corresponding nitrate ; thus omitting nitrogen from the solution.

(*b*) Knop's solution, but omitting potassium phosphate ; thus omitting phosphorus.

(*c*) Knop's solution, but using the corresponding sodium salts instead of the potassium salts mentioned ; thus omitting potassium.

(*d*) Knop's solution, but substituting sodium nitrate for calcium nitrate ; thus omitting calcium.

(*e*) Knop's solution, but substituting sodium sulphate for magnesium sulphate ; thus omitting magnesium.

(*f*) Knop's solution, omitting the ferric salt ; thus omitting iron.

(*g*) Knop's solution, but substituting magnesium nitrate for magnesium sulphate ; thus omitting sulphur.

A series of glass or earthenware jars, of capacity 1–2 litres, is required, the jars being fitted with waxed corks bored with holes to take the plants. One jar is filled with the complete Knop's solution, others with solutions (*a*) to (*g*). By means of cotton-wool, seedlings of barley, wheat or broad bean are fixed into the corks. Glass jars should be wrapped with black or brown paper to exclude the light. Label the jars and place them in a green-house or well-lit window and keep periodic records of the growth of the plants and any other observed effects, over several weeks.

3. Make a solution of glucose. This substance is obtainable at the chemist's and just a little is required, dissolved in water in a test-tube. Then test for glucose by means of Fehling's reagent.

This solution is made up in separate parts, Fehling's solution A and Fehling's solution B. A is prepared by dissolving 34·6 gm. of copper sulphate in 500 c.c. of distilled water. B is prepared by dissolving 175 gm. of Rochelle salt and 50 gm. of sodium hydroxide in 500 c.c. of distilled water. The complete reagent is prepared when required for use by mixing equal quantities of the A and B solutions.

To about half a test-tube full of glucose solution add a few drops of Fehling's reagent and warm the solution gently over a Bunsen burner, with constant shaking. Note that the blue colour gradually disappears and a bright red precipitate of copper oxide is formed. This indicates the presence of glucose. Fructose gives the same reaction.

4. Test various plant organs for glucose and fructose. This is done by extracting some of the juice from the plant organ to be tested. Cut the tissues into fine pieces, and then crush the material with a pestle and mortar. If little juice is expressed, add a few c.c. of water and continue to crush. Then perform Fehling's test on the plant extracts obtained.

The following are suggested plant materials for this test : various types of leaves ; carrot, turnip, dandelion, beet, etc. roots ; various stems, potato tubers ; various types of fruit such as apple, pear, plum, tomato, prunes, etc.; seeds ; and so forth.

5. Perform the Fehling's test on sucrose (cane–sugar). Note that there is no reaction to Fehling's reagent in this case.

Then hydrolyse a solution of sucrose in a test–tube by adding a few drops of dilute hydrochloric acid and boiling for a few minutes. Cool, and neutralise by means of caustic soda. Perform the Fehling's test on the resulting solution.

6. To a piece of household starch, add a little iodine solution. Iodine is scarcely soluble in water, so the solution is usually composed of iodine in potassium iodide. Note that the iodine on the starch gives a deep blue coloration. This is a well-known reaction of starch.

7. Test for starch in plant organs, by means of the iodine test. Test certain roots ; stems ; storage organs such as potato tubers, carrot and parsnip roots, onion and tulip bulbs ; apple, orange, banana fruit, etc. Test as many parts of plants as possible. In all cases, it will be necessary to add the iodine solution to a *cut* surface.

In the case of green leaves, to note the colour reaction, it is best to remove the chlorophyll first ; also, for reasons to be discussed later on, leaves should be gathered immediately after they have been exposed to several hours of daylight. To remove the green colour from a leaf, place it for a few minutes in boiling water. This kills the cells. Then immerse the leaf in alcohol in a test-tube. Place the test-tube in hot water, *but remove the flame.* The boiling point of alcohol is lower than that of water, so, provided the water is hot enough, the alcohol will boil. Note that the boiling alcohol removes the chlorophyll from the leaf, and becomes green in consequence.

When the leaf is white through the loss of its chlorophyll, immerse it in an iodine solution. If starch is present in it, it will assume a deep blue colour. Perform this test on several types of leaves, such as those of *Pelargonium, Hydrangea,* elm, hyacinth, etc. Always use thin leaves.

8. Examine some starch grains under the microscope. Suitable material is provided by the potato tuber. Cut a tuber and take a small scraping by means of a knife or scalpel. Enough to cover a pin's head will do. Mount this in water on a slide and examine under the high power of the microscope. By very careful focusing up and down, it will be possible to see the hilum and layers or stratifications.

Make drawings of some of the grains. Then draw some iodine solution beneath the cover-slip, and note the deep blue colour reaction.

Examine and draw other examples of starch grains such as those of the bean seed, rice grain, some of these being of compound type. (*See directions for microscope work in the practical work following Chap. X*).

9. Perform the Fehling's test on a colloidal solution of starch. Note the reaction is negative. Then hydrolyse a portion of the starch solution by boiling with equal volume of dilute sulphuric acid for about 10 minutes, stirring all the time. Test the solution periodically by applying the iodine test on a drop on a tile. Note the stages of colour changes. Finally neutralise, apply the Fehling's test and account for the result.

10. Place some cotton wool in a saucer or dish. Pour some iodine solution over it and note that it turns yellow. Then drain off the extra iodine solution and add some concentrated sulphuric acid. The cotton wool now turns a deep blue colour.

Cotton wool is almost pure cellulose, and this colour reaction is a good test for cellulose. Make a record of these results and perform a similar test on a slice of onion bulb. Explain why a similar reaction, though not so clear, takes place in this case.

11. To the cut surface of a piece of wood, apply a little colourless solution of aniline sulphate or chloride. Note the assumption of a bright yellow colour. This is well–known colour reaction for wood.

Now cut across the stems of certain herbaceous plants, and apply the aniline chloride to the cut surfaces. Look for any evidence of wood and record its position in the stem.

12. Prepare a solution of protein by shaking up the white of an egg in its own volume of water, or by shaking a small quantity of ground-up pea seeds with water for several minutes, after which the mixture is filtered and the filtrate retained. Then perform the following colour tests for proteins, keeping a careful record of the results :

(*a*) *Biuret reaction.* Add a little caustic soda to some protein solution in a test–tube. Then add a little copper sulphate solution. A violet colour results.

(*b*) *Millon's reaction.* To some of the protein solution add some Millon's reagent. The protein is precipitated and turns pink when heated.

Millon's reagent is prepared by dissolving some mercury in twice its weight of concentrated nitric acid. Heat should be applied to accelerate solution, the operation being performed in a fume cupboard. When the reaction has ceased, dilute the solution with twice its volume of distilled water.

(c) *Xanthoproteic reaction.* Add some concentrated nitric acid to the protein solution and warm cautiously. A yellow colour results. After thoroughly cooling the mixture under the tap add a little strong ammonia, when the colour is intensified to orange.

13. The protein tests can be carried out in a test tube on other plant material by using expressed juice, aqueous extracts, pieces of tissue, or on a microscopic slide on which thin slices or sections of the tissue are placed.

14. Using a piece of animal fat, perform the following tests :

(a) The fat makes a translucent mark on paper.

(b) It is not soluble in water, but is soluble in ether, chloroform and acetone.

(c) If a little solution of osmic acid be applied to it, a black coloration results.

15. Apply the osmic acid test to the cut surface of the castor-oil or the sunflower seed. Note the presence of the oil.

CHAPTER VII

SOME OTHER PLANT PRODUCTS

Apart from essential food materials, constituents of protoplasm, chlorophyll pigments, and other substances, some of which will be considered in due course, there are many other chemical substances manufactured by plants. Some of these substances are of economic importance, though their significance to the plants producing them is not always clear. None of them is common to all plants, but is peculiar either to one special plant or to a group of plants.

GLYCOSIDES

Very widely distributed among plants is a group of chemical substances called **glycosides** (sometimes known as **glucosides**). Typically they are formed by the chemical combination of glucose or another sugar with a non-sugar. When the glycoside is hydrolysed by enzymic or other means, the sugar is released, together with one or more other substances.

In the so-called **cyanogenetic glycosides** one of the products of hydrolysis is prussic acid. A few of these glycosides are given in the following table :

Glycoside	Source	Products of Hydrolysis
Amygdalin	Seeds of bitter almond, plum, peach and apricot	2 mols. glucose + benzaldehyde + HCN
Dhurrin	Millet seedlings	Glucose + p-hydroxybenzaldehyde + HCN
Prunasin	Bird cherry twigs	Glucose + benzaldehyde + HCN
Lotusin	*Lotus arabicus*	Glucose + lotoflavin (a yellow pigment) + HCN

These cyanogenetic glycosides have poisonous properties, and if present in sufficient amount they make the plant tissues poisonous to animals, an example being the leaves of the yew which contain a glycoside of this type.

Among other well-known glycosides are **sinigrin** and **sinalbin**. The former occurs in the seeds of the black mustard (*Brassica nigra*) and yields glucose, $KHSO_4$ and a pungent so-called mustard oil (allylisothiocyanate) on hydrolysis ; the latter occurs in the seeds of the white mustard (*Sinapis alba*) and yields glucose, acid sinapin sulphate and a mustard oil (p-hydroxy benzene isothiocyanate) on hydrolysis. The glycoside **salicin** occurs in the bark of willows and yields glucose and an alcohol, saligenin. **Coniferin** occurs in coniferous trees and yields glucose and coniferyl alcohol.

One glycoside is of historical interest. This is the glycoside called **indican**, which is present in several plants, including the **woad plant**. From this glycoside, the textile dye, **indigo**, used to be manufactured. At that time, of course, the woad plant formed a very important crop. Nowadays its cultivation for the production of the glycoside for indigo manufacture is almost a thing of the past, for the indigo dye is manufactured almost exclusively by chemical means. Actually woad (*Isatis tinctoria*) is a wild plant, growing on banks and especially around chalk pits. It is interesting to note that this was one of the plants used by the Ancient Britons for staining their bodies. On hydrolysis, indican yields glucose and indoxyl. This latter compound on oxidation yields **indigotin**, the indigo dye.

Saponins form another group of glycosides. These are present in quite a large number of plants of different families. They are noted for their highly poisonous properties in that they have the effect of dissolving the blood corpuscles. Several also affect the heart. One part of saponin in 100,000 parts of water forms a solution sufficiently strong to kill fish. In the Far East this property is made use of for catching fish. The saponin solution is placed in lakes and ponds, and the fish when killed float on the surface. There is no danger in this practice, since such a low concentration of the glycoside has no harmful effects on the person who eats the fish.

Saponins, when mixed with other materials, are used for medicinal purposes, especially in cases of chronic bronchitis.

On hydrolysis, the saponins yield such sugars as glucose, galactose, arabinose, pentoses, together with the so-called sapogenin group, which possesses the specific physiological properties.

The leaves of the foxglove (*Digitalis purpurea*) contain several active glycosides collectively known as **digitalin.** They form the basis of certain preparations used in medicine as diuretics and cardiac stimulants.

The significance to the plants producing them of the glycosides reviewed above is obscure, but many of them are of a bitter taste and may give protection against animals. It is also possible that they are of value to the plant as food reserves. Enzymes having the ability to hydrolyse, under suitable conditions, the glycosides often accompany the latter in the plant tissues (Chap. VIII).

SOME PLANT PIGMENTS

Few people can fail to be deeply impressed by the remarkable display of colour that plants often show. In the petals of flowers almost every conceivable shade of colour is represented.

Before proceeding to consider any of these, it will be interesting to examine those flowers, such as the snowdrop, many lilies, narcissi, etc., which are not coloured. The whiteness of some flowers is so pure as to be almost dazzling. Yet this is not due to any white chemical pigment present in the cells of the petals.

All the cells of the petals of white flowers are colourless and transparent. therefore, on casual consideration, petals instead of being white should be transparent and colourless like a sheet of glass. But the petal is not one continuous sheet of material. It is composed of hundreds of small, colourless cells (see Chap. X). If a sheet of glass be laid flat it naturally appears transparent. If, however, the sheet be crushed into a fine powder, the resulting layer of powdered glass would no longer be transparent, neither would it be colourless. It would be white. This is due to the irregular reflection or scattering of light ; and it is this light-scattering by the hundreds of cells of the colourless petal that makes the petal actually appear white.

The colours of petals, on the other hand, are caused by different coloured chemical pigments. Sometimes only one pigment is present. More often, however, several differently coloured pigments are present, and then the shade of colour produced in the flower is the result of the mixing of the colours.

The chemical nature of many flower pigments is fairly well understood to-day, though there is still a great deal to find out. Most of the pigments can be classified into the groups **flavones, xanthones** and **anthocyanins**.

FLAVONES AND XANTHONES

Both flavones and xanthones contain the heterocyclic nucleus γ-pyrone:

$$
\begin{array}{c}
O \\
HC \diagdown \diagup CH \\
HC \diagup \diagdown CH \\
CO
\end{array}
$$

In each, the pyrone nucleus is attached to a benzene ring. Xanthone has an extra benzene ring; whereas to the benzo-γ-pyrone complex in flavone a phenyl group is attached.

Some xanthones and flavones are present in parts of plants, especially petals, where they impart a yellow colour; but in many petals the concentration is so low that the petals remain white though they will turn yellow or greenish on the application of ammonia vapour.

Pure yellow flowers, such as daffodils and buttercups, owe their colour to the presence of yellow flavones and xanthones and also carotenoids, the latter being yellow pigments which also occur in association with chlorophyll in the green parts of plants. The colouring matter can be extracted by placing the dried petals in boiling alcohol, when the anthoxanthins and the carotene will dissolve in the alcohol.

ANTHOCYANINS

Of much more common occurrence are the anthocyanin pigments. These occur in many chemical forms and colours. In their pure state or as glycosides and in various combinations with each other, they are responsible for many of the beautiful shades of colour present in flowers, and other plant organs. By far the majority of natural anthocyanins are glycosides of flavonol derivatives.

Anthocyanins are responsible chiefly for the blue, red, purple and brown shades so familiar in Nature. There are many of them; therefore it is impossible to consider all separately. A few are, however, worthy of mention. For example, **pelargonin** is the anthocyanin responsible for the bright red colour of the petals of *Pelargonium* and *Geranium*, **delphinin** for the beautiful blue of delphiniums (*Delphinium* species) and monk's hood (*Aconitum anglicum*), **cyanin** for the striking cornflower blue (*Centaurea cyanus*), **œnin** for the blue-black of grape skins, etc. Other colours may be produced by the combined effect of anthocyanin with other pigments. Thus the colouring matter in brown wallflowers is a mixture of anthocyanin, carotenoids and xanthones.

The colour of many anthocyanins is affected by the reaction of the solution in which they are present. They are red in acid solution and change colour through purple to blue if the acidity is decreased, as the indicator litmus does in the laboratory. It was suggested that the changes that sometimes occur in the colour of a flower after it has been open for a few days were due to alterations in the reaction of the cell sap. Thus originally blue flowers of forget-me-not (*Myosotis palustris*) or delphinium sometimes turn pink. A similar explanation was suggested for the contrasts in flower colour often shown by different plants of the same variety or species of plant (for example, sweet pea). Investigation has indicated, however, that other factors besides cell sap reaction are involved.

While the significance of the anthocyanins in organs such as the beetroot or the leaf of the copper beech is obscure, there is no doubt that the bright colours of flowers serve in attracting the insects such as bees, wasps and butterflies, on which, as will be seen later, many flowers depend for the process of fruit and seed production.

In the autumn coloration of leaves, the yellows are due to carotene and xanthophyll and related pigments and the reds and purples to anthocyanins. The chlorophyll disappears as the season progresses, but the carotene and xanthophyll remain and anthocyanins are manufactured, chiefly from any foods left in in the leaf (Chap. XIV).

TANNINS

Tannins are another group of chemical substances which are widely distributed throughout the plant kingdom. They are all of an acid nature, being condensation products of certain phenolic acids and occurring in colloidal solution. The most important tannin is called simply **tannic acid**. Although tannins are distributed in many types of plants and in all plant organs, such as stem, roots, leaves and unripe fruits to which they give the familiar astringent properties, they usually occur in small quantities only. Tannins are, however, found in large quantities in two instances. One is the bark of certain trees, especially the oak ; the other is what is called a plant **gall**.

A gall is an abnormal development of plant tissues or organs and is usually caused by a parasitic insect, worm or fungus. The parasite penetrates the tissue of the plant, and then the cells of the plant around the parasite begin to divide very actively, thus causing a large swelling of tissue forming the gall. One type of gall often develops on the leaves of the willow, and assumes the form of large red swellings. The well-known ' *oak-apples* ' are also plant galls caused by an insect (Fig. 47). The so-called ' *witches' brooms* ' which take the form of closely-packed branches on birch, elm and fir trees, among others, are also a form of gall, in some cases caused by an insect parasite, and in others by a fungal parasite.

Many galls contain tannin and the oak-apple galls are particularly rich in this substance. The tannin contained in the ' nut-galls ' formed on the twigs of a species of oak is used in the manufacture of ink. It is extracted and when mixed with ferrous sulphate gives a blue-black liquid, which is the basis of ink. Tannins, especially those present in the bark of certain trees, are used extensively in the tanning of hides for leather-making.

Many plant tannins are used for medicinal purposes ; in fact, tannic acid is officinal in both the British and United States pharmacopœias. Certain ointments are prepared from them. For example, one is used for stopping intestinal bleeding, owing to its astringent properties. It is also used in the treatment of severe scalds and burns.

For various reasons, tannins are considered to be waste products of the plant.

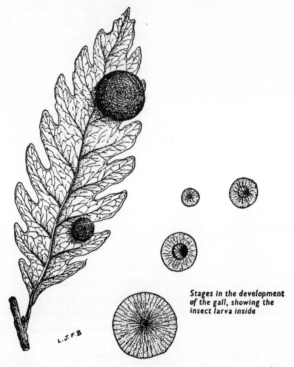

Stages in the development
of the gall, showing the
insect larva inside

FIG. 47. ' OAK-APPLE ' GALLS CAUSED BY AN INSECT PARASITIC ON THE OAK.

BEVERAGES

The three most important hot beverages to Great Britain are plant products, namely, tea, coffee and cocoa.

Tea is the prepared, dried leaves of the tea plant (*Camellia thea*), a shrub which grows to a height of 3–5 feet. The bronze colour of the leaves on drying is due to oxidation of the tannin present in the tea.

Coffee is the product of the coffee plant (*Coffea arabica*), a shrub which attains a height of about 20 feet. The so-called coffee ' beans ' are the seeds of the coffee fruit, which is actually a berry. The stimulating action of coffee is due to an alka-loid (see p. 87) called **caffeine**. This is also present in tea, though not to such a great extent. The aroma of coffee is due to an oil called **caffeone**.

Cocoa is produced from the seeds of the cacao tree (*Theobroma cacao*). It has a high nutritive value, and is mildly stimulating, since it contains a certain amount of the alkaloid caffeine, though not so much as coffee does.

DRUGS

A drug is any chemical substance or inorganic composition which is used either by itself or in combination with other substances as a medicine. In a particular way, the term is often applied to those substances which act as narcotics or poisons. Many drugs are of vegetable origin, and a few of them will be considered here.

ALKALOIDS

Some of the most important drugs belong to the group of substances known as the **alkaloids**, which are organic nitrogenous basic compounds. They may occur naturally in solution or in the solid state. They are not very widely distributed throughout the plant kingdom.

Most alkaloids contain the elements carbon, oxygen, hydrogen and nitrogen, though a few, for example nicotine, contain no oxygen. Most are colourless, crystalline solids, insoluble in water. All have a bitter taste and toxic or other physiological properties. They usually occur in the plant combined with an organic acid, such as tannic, malic, oxalic, etc., in the form of salts.

A very important alkaloid is **quinine**. It is used for preventing colds, for killing the malarial parasite in the blood of a patient suffering with that disease (in this respect, quinine proved a great boon during the Crimean War, and in all wars since), and for killing many of the bacteria which are responsible for diseases in man. Our knowledge of the medicinal uses of quinine is due chiefly to the work of **Prof. A. Binz** of Bonn, Germany. To-day, however, quinine has yielded place to **mepacrine** and **paludrine** as prophylactics for treating malaria.

Quinine is an alkaloid present in the bark of species of the *Cinchona* tree (Peruvian bark, etc.). The earliest record of the medical use of the cinchona bark dates back to 1638, when it was used effectively for curing the Countess of Chinchon, the wife of the Governor of Peru, of a fever. From this, the plant received its familiar name. At that time, cinchona trees grew only in South America ; but gradually the fame of its bark was spread throughout Europe, chiefly by the monks of the time, who were also the doctors. Then, in 1854, the Dutch Government obtained some trees from South America and transplanted them to Java. Similarly in 1859, the British Government transplanted some to India and Ceylon. To-day, cinchona plantations exist in all those countries, solely for the production of quinine. The world's chief supply, however, comes from Java.

Another alkaloid, very valuable to medical men as a drug, is **strychnine**, which is extremely poisonous. This was first discovered in 1818 in St. Ignatius's beans and other related plants, belonging to the genus *Strychnos*. Strychnine is obtained from the wood and the bark of the various species of *Strychnos*, especially *Strychnos nux-vomica*.

Cocaine is another alkaloid of great use to the doctor and the dentist. This alkaloid exists in the leaves of the **coca** plant (*Erythroxylum coca*). The plant is native to Bolivia and Peru ; but to-day it is cultivated, for the sake of the drug, in Java. Cocaine is a potent drug. It is used in surgical and dental practice because, when injected into the tissues of the body, it acts as a local anaesthetic, thus deadening the tissues to pain. Therefore, it may be injected for slight operations, such as some skin operations, or in the gums for extracting teeth. If taken internally, cocaine deadens any sensation of hunger in the stomach, with the result that the person can go a long time without feeling the need of food. Moderate doses of the drug produce a sensation of calmness and happiness. Heavy doses affect the nervous system. This leads to mental depression, physical laxity and often insanity.

The latex of the opium poppy, *Papaver somniferum*, contains a drug called **opium**. This contains several alkaloids. The drug is usually extracted from the fruit of the poppy. The only wild opium poppy grows along the northern shores of the Mediterranean Sea ; but it is cultivated in many parts of the world, such as France,

Germany, Turkey, Persia, China and United States. From the medicinal point of view, opium is very valuable owing to some of the alkaloids it contains. The chief alkaloid of medicinal value present in opium is **morphine**. This drug has a great pain-relieving power, and also induces deep sleep. No other plant drug known is so effective in relieving pain as morphine. For this reason it is the most common drug used for this purpose in medical practice.

The leaves of *Nicotiana tabacum* (the tobacco plant) contain an alkaloid called **nicotine**. Like the other alkaloids considered, in large doses it is a poison. Nicotine is still used in medicine, though not to any great extent. Its greatest value lies in its use to horticulturists. Certain mixtures, called **insecticides**, are used for treating crops attacked by insect parasites. The chief constituent of many insecticides is nicotine.

VITAMINS

Right up to the eighteenth century, one of the most dreaded diseases among sailors was a disease called **scurvy** ; for it almost always ended in death. It was so common that no mercantile or naval ship could go to sea for weeks at a time without, on its return, reporting the death through scurvy of a number of its crew. The same disease was prevalent in prisons and workhouses. The conditions under which sailors, prisoners and paupers lived were similar in at least one respect. That was the absence of fresh food. Naturally, in those days, since they had not the advantages of modern science, it was impossible for ships to take fresh fruit and vegetables on their long journeys.

So long ago as 1593, **Sir Richard Hawkins**, the famous English seaman, recognised that scurvy could be curtailed in its ravages if fresh fruit and vegetables were available ; but, realising how impossible this was on long journeys, he noted that to take an essence of fresh fruit in the form of orange or lemon juice (*Citrus* species) had a wonderfully beneficial effect. Nevertheless, it was not until two hundred years afterwards that the idea was followed up. The annual deaths in the Navy due to scurvy were appalling. In fact, the disease was almost as common among such people as the common cold is to-day. However, owing to the great physician, **Dr. James Lind**, who did so much brilliant work in connexion with hygiene in the British Navy, an Admiralty order was given, in 1795, that all ships should be supplied with lemon juice. From that day, scurvy disappeared from the Navy.

In India, Malaya, China and Japan, a disease which ravages the native population is one called **beriberi**. Now the natives of those countries live almost completely on rice (*Oryza sativa*). After much research work, it was discovered that the disease was more common among those people who ate polished rice, that is, rice from which the outer coating of the grain had been removed. It was then discovered that just as in the case of oranges and lemons, which must contain something which prevents scurvy, so do the fruit coats (pericarp) of the rice grain contain a substance which prevents beriberi.

Both these diseases, in other words, are due to a certain deficiency in diet. They are therefore called **deficiency diseases**.

It is clear from these and other observations that in addition to their ordinary dietetic value, certain foods, especially those of vegetable origin, contain accessory substances which help animals to develop correctly and to maintain good health.

These substances are called **vitamins**. The investigation of the chemical nature of vitamins and of their precise role in the development and metabolism of the animal body has occupied the attention of many biochemists and physiologists, and new information is constantly brought to light. In Great Britain, much of the pioneer work on vitamins was done by **Sir F. Gowland Hopkins** at the University of Cambridge, from 1906 onwards ; in Germany by **Stepp** from 1909 onwards ; and in the United States by **McCollum** and **Davis** from 1913 onwards. Up to the year 1922 only two vitamins were definitely known and were called **vitamin A** and **vitamin B**. Many more are now known, as will be seen below.

Vitamin A was not isolated in the pure state until 1937. Its absence from the diet leads to diseased conditions in the skin, the mucous membranes, the eye, and the nervous system. The so-called night blindness is an early symptom of deficiency. This particular vitamin is not present to any great extent in normal dietary articles, but is very abundant in fish-liver oils and occurs in smaller quantities in butter and eggs. The animal body is, however, able to synthesise the vitamin from the chemically very closely related substance carotene, which is a yellow pigment present in the green parts of plants, and also in the root of the carrot (p. 176). For this reason carotene is sometimes termed **provitamin A**.

What was originally called **Vitamin B** is now known to include several different vitamins. **Vitamin B_1** (chemically **aneurin** or **thiamin**) prevents beriberi, and is relatively abundant in yeast, whole-meal flour and lean meat. **Vitamin B_2** is chemically **riboflavin**. A deficiency of this leads to diseased conditions in the skin and eye. The vitamin is present in many vegetable foods, yeast being again a particularly rich source. Other vitamin substances classed as members of the vitamin B group include **nicotinic acid** (the anti-pellagra vitamin), **pyridoxine**, **pantothenic acid, biotin, folic acid** and **vitamin B_{12}** (occurring in liver), the last two being anti-anaemia factors.

Vitamin C (ascorbic acid) is the one which prevents scurvy. Its absence from the diet also leads to swelling of the gums and degeneration of the teeth, and inadequate healing of wounds ar ¹ ˹ actures. The vitamin occurs in nearly all fruits and vegetables, especially ri·.ı sources being rose-hips, black-currants, cabbage, and citrus fruits. Smalle amoun ₃s are present in the potato tuber and other vegetables.

Vitamin D includes several closely related ompounds which promote the proper development of the skeleton of animals and prevent or cure rickets. These compounds are derivatives of **sterols**, such as ergosterol, and the vitamins arise from such provitamins under the influence of irradiation with ultra-violet light. The naturally occurring form of the vitamin is distinguished as vitamin D_3. Animals, including man, can obtain the vitamin by consuming products derived from other animals, fish-liver oils, fish, eggs, and butter being relatively rich sources. Alternatively the vitamin can be actually formed in the animal body under the influence of the ultra-violet light from sunlight or from artificial sources, the radiation bringing about the conversion into the vitamin of sterols present in the skin. In this way the curative effect of sunlight on rickets is explained.

Vitamin E (concerned with the maintenance of fertility in animals), and **vitamin K** (promoting proper blood-clotting) have also been the subject of considerable investigation. They are present in foods of plant origin.

The diseased conditions to which reference has been made above result from long-continued deficiency of particular vitamins in the diet. A general re-

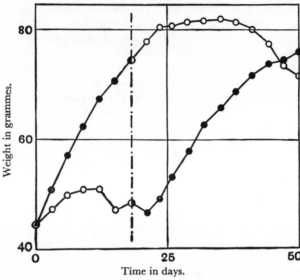

Time in days.

Fig. 48.   (For description see text.)

tardation of growth and development is frequently an earlier symptom.   This is
illustrated by an experiment carried out by Hopkins on rats, the results of
which are indicated in the accompanying graph (Fig. 48).   The lower curve
shows the weight of rats fed for the first eighteen days (the eighteenth is marked
with a dotted line) on a diet of purified protein, fats, carbohydrates, mineral salts
and water.   The upper curve shows the weight of another group of rats fed on the
the same diet except that a small amount of cow's milk was added.   After the
eighteenth day, the diets were reversed.   Note that the rats not supplied with milk
during the first eighteen days gained little weight and actually began to lose weight
by about the fifteenth day.   Those supplied with milk during the same period
gained weight quickly.   The reversal of diet resulted in the first group of rats now
starting to gain weight, whereas the second group eventually began to lose it.   The
beneficial effects of the addition of a small quantity of milk to the rats' diet are
attributed chiefly to the presence of vitamins A and D, together with smaller
amounts of the B and C vitamins, in milk.

    For some of the vitamins the reasons for their importance in the development of
the animal body and its maintenance in a state of good health are partly understood.   Several of the B vitamins or their immediate derivatives have been shown
to be the prosthetic groups of enzymes concerned in respiration (Chap. XVII).
The significance of the presence of these vitamin substances in plants may be the
same.   Vitamin C also has been suspected of having an importance in the respiratory processes of animals and plants.   The precise parts played by vitamins A and
D in the animal body remain rather obscure.

    From this brief survey it is clear that the synthetic action of plants in forming, as
constituents of their own metabolic machinery, these various vitamin or provitamin substances is of the greatest significance to the animal kingdom, and
illustrates further the dependence of the latter on the plant kingdom for nutrition.

CHAPTER VIII

ENZYME ACTION AND DIGESTION

The main purpose of digestion in man is to render the foods consumed soluble, so that later they may be absorbed by the tissues of the gut. The absorption involves the passage of substances through living cells, and, as will be seen later (Chap. XI), this is impossible unless such compounds are in true solution.

Some substances consumed by man, such as glucose, are already soluble, so that their absorption is a comparatively easy matter. In other cases, however, such as starch, the proteins in lean meat, fats, etc., the constituents are insoluble. They must, therefore, be broken down by chemical action into simpler substances before they can be absorbed. This breaking down process is called **digestion**. Most insoluble food-stuffs, such as polysaccharides, proteins and fats, will break down *spontaneously* into simpler substances when in solution in water, chiefly by a process of hydrolysis. But the process is far too slow for ordinary life processes. In digestion the break-down processes are accelerated by the organic substances produced by the protoplasm called **ferments** or **enzymes**.

The hydrolysis of sucrose into invert sugar (see Chap. VI) according to the following equation :

$$C_{12}H_{22}O_{11} + H_2O \rightarrow C_6H_{12}O_6 + C_6H_{12}O_6$$

can take place spontaneously, but very slowly, in aqueous solution. The addition of a little mineral acid and application of heat, however, greatly accelerate the reaction. In this reaction, the acid can be recovered unchanged after the complete hydrolysis of the sucrose to glucose and fructose has taken place. The acid is, therefore, a **catalyst**, and the acceleration is referred to as **catalysis**. In the plant, sucrose is hydrolysed to invert sugar (glucose and fructose) and the hydrolysis is again accelerated, but not by a mineral acid. An enzyme, **invertase**, is commonly the accelerating agent. An enzyme has been defined as ' a catalyst produced by living organisms ' (Bayliss).

To follow the passage of foods in man will give an idea of the importance and variety of enzymes.

Digestion in man begins in the mouth. Here the food is masticated and mixed with **saliva**, a secretion from the salivary glands. Saliva contains mucus, which acts as a lubricant to ease further passage along the alimentary canal ; and an enzyme, **ptyalin** (or **salivary amylase**), which accelerates the hydrolysis of starch into maltose which is soluble and can be absorbed.

The bolus of food then passes down the œsophagus to the stomach. There it is reduced to a semi-fluid, and is partially digested by the action of the enzymes in the *gastric juice*, which is secreted by glands in the wall of the stomach. The gastric juice contains hydrochloric acid and enzymes, the principal one of which is **pepsin**. The hydrochloric acid has no digestive action, but it facilitates the action of pepsin, which can act better in an acid medium (see p. 95). Pepsin accelerates the conversion of complex proteins into relatively simple peptides. Carbohydrates and fats present in the absorbed food are not digested in the stomach.

The food usually remains in the stomach for about four hours, after which it slowly passes into the duodenum, the beginning of the small intestine. Into the duodenum, a fluid is secreted by the pancreas. This fluid, known as the *pancreatic juice*, contains several enzymes notably **trypsin**, which breaks down proteins into peptides, **pancreatic amylase**, similar in activity to salivary amylase, and **lipase**, which hydrolyses fats into fatty acids and glycerol.

Into the duodenum, *bile* is also secreted by the liver. This has no digestive action, but renders the activity of the pancreatic juice easier, as will be seen later (p. 95). As the food passes through the small intestine, it is acted upon by the *intestinal juice*. This contains several enzymes, some of which are the same as those present in the pancreatic juice. Another enzyme present in this juice is **erepsin**, which converts peptides into amino-acids.

The food remains in the small intestine about four hours, after which it passes into the large intestine. While in the small intestine, it is digested, and the suitable substances, now made soluble, are absorbed into the blood stream to be circulated to various parts of the body. In the large intestine, the rest of the food remains 24–36 hours. Here, no digestion, and little absorption, take place ; but the food loses much of its water and undergoes partial decomposition by bacterial action. Some of the products of decomposition are absorbed here ; and the rest, which are waste products, are finally excreted through the rectum.

OCCURRENCE OF ENZYMES

Enzymes are present in all living parts of plants and animals. They exist in living cells in the form of colloids in the protoplasm itself (Chap. IX) and therefore cannot normally pass through the cell membranes (see Chap. XI). They are manufactured by the protoplasm.

Sometimes the enzyme is not present in the same part of the plant or animal as its **substrate**, that is, the substance the reactions of which it accelerates. Then the reaction can take place only after the enzyme and the substrate have been brought into contact with each other. This may happen naturally at certain stages in the development of the plant, or it may be brought about by injury. For example, when tissues (such as those of the bitter almond) containing cyanogenetic glucosides are crushed, an odour of prussic acid is produced since the enzyme and its glucoside substrate are now brought into contact.

There is some uncertainty regarding the chemical nature of enzymes. According to one view, many of them consist of a protein to which an active prosthetic group is attached. The chemical nature of the prosthetic groups of certain enzymes is known, and the presence of heavy metals such as iron and copper established. The term **co-enzyme** is often applied to a prosthetic group that is readily separable from the proteins.

CLASSIFICATION OF ENZYMES

Enzymes are classified according to the physiological reactions which they accelerate. Some of the more important enzymes occurring in plants are given in the tables below on pp. 93 and 94.

It should be noted that in some cases a process originally thought to be catalysed by a single enzyme has afterwards been shown to be the result of the action of two or more enzymes, each responsible for one stage in the process. Thus the final product of the hydrolysis of starch in the plant is glucose, and the name **diastase** was given to the single enzyme thought to be responsible. Further study has shown that the starch is first broken down to maltose, and the term diastase (or **amylase**) is now usually restricted to the enzyme concerned with this stage. The maltose is simplified to glucose by a second enzyme, **maltase**.

FIG. 49. STAGES (1-4) IN THE DIGESTION OF A STARCH GRAIN IN GERMINATING BARLEY.

Fig. 49 illustrates the dissolution of starch grains that occurs as a result of diastatic action during the germination of cereal grains. The alcoholic fermentation of sugars is another example where a transformation is now known to be achieved as a result of the action of a series of enzymes (see also p. 206). In this case the name **zymase** has been retained for the group of associated enzymes.

CLASSIFIED EXAMPLES OF ENZYMES

ENZYMES WHICH ACCELERATE HYDROLYSIS

	Enzyme	Plant Source	Substrate	Products
(1) *Fat-splitting*	Lipase	Oil-containing seeds	Fats	Fatty acids + glycerol
(2) *Carbohydrate-splitting*	Invertase	Green leaves, yeast	Sucrose	Glucose + fructose
	Maltase	Green leaves, cereal grains	Maltose	Glucose
	Diastase (Amylase)	Many plants	Starch	Maltose
	Cytase	Many plants	Reserve Cellulose	Sugars
(3) *Glycoside-splitting*	Emulsin	Bitter Almonds	Amygdalin	Glucose + benzaldehyde + prussic acid
	Myrosin	Mustard Seeds	Sinigrin	Glucose + mustard oil, etc.
(4) *Protein-splitting*	Protease Peptidase	Many plants Many plants	Proteins Peptones and Polypeptides	Amino-acids Amino-acids
(5) *Amide-splitting*	Urease	Soya bean and a few other plants	Urea	Ammonia + carbon dioxide
	Asparaginase	Barley	Asparagine	Aspartic acid + Ammonia

ENZYMES WHICH ACCELERATE FERMENTATION

Type of Fermentation	Enzyme	Plant Source	Substrate	Products
(1) *Alcoholic fermentation*	Zymase	Yeast and plants generally	Hexoses	Ethyl alcohol + carbon dioxide
(2) *Lactic acid fermentation*	In lactic acid bacteria	—	Hexoses	Lactic acid
(3) *Butyric acid fermentation*	In butyric acid bacteria	—	Hexoses	Butyric acid

ENZYMES WHICH ACCELERATE COAGULATION

Enzyme	Plant Source	Substrate
Rennin	–	Milk
Thrombin	–	Blood
Pectase	Fruits	Pectic substances

ENZYMES WHICH ACCELERATE OXIDATION AND REDUCTION

Enzyme	Plant Source	Substrate	Products
Oxidase	Many plants	Phenolic substances (+ oxygen)	Oxidised substrate
Peroxidase	Many plants	Phenolic substances (+ H_2O_2)	Oxidised substrate
Catalase	Many plants	H_2O_2	water + molecular oxygen
Dehydrogenase	Many plants	Various organic substances	Oxidised substrate + hydrogen (attached to H acceptor)
Reductase	Some plants	Nitrates	Nitrites

In addition to the enzymes listed in the tables, a great many more are now known to be present in plants. Research on them is very active, and enzymology is a rapidly developing field of scientific enquiry.

ACTION OF ENZYMES

Until the chemical and physical structures of enzymes are more clearly understood, the theory of their mode of action must perforce remain rather speculative. There are two suggestions, however, which may explain their catalytic activities. One is that the reacting compounds are adsorbed on to the colloidal particles of the enzyme, and thus brought into closer contact with each other. This makes reaction easier. The other suggestion is that the enzyme unites chemically with its substrate, forming a labile compound which can react more quickly than the substrate itself.

The action of enzymes is specific in that a given enzyme can act only on one substance or one type of substance. This specificity is probably connected with the molecular structure of the enzyme and its substrate, especially the space arrangement of the molecules of the substrate. Only a molecular structure of a certain form in the enzyme will fit that of the substrate. This relationship is like that existing between a lock and key, and gives strong support to that theory of enzyme action which postulates the formation of a loose chemical compound between enzyme and substrate.

FACTORS INFLUENCING ENZYMIC ACTION

Enzymic activity varies considerably with **temperature**. Within certain limits, the velocity of enzymic action is doubled with an increase of temperature by 10° C. All enzymes are destroyed at 100° C., and in most cases they cannot withstand a temperature above 60° C. At low temperatures, enzymes become inactive. There is a temperature at which the enzyme acts with the greatest velocity. This is known as the **optimum** temperature, and varies with each enzyme ; for example, for maltase it is 40° C.

Hydrogen ion concentration (that is, degree of acidity or alkalinity) also affects enzymic action. The *optimum* hydrogen ion concentration varies with the enzyme. For example, the hydrochloric acid secreted in the stomach of man merely serves to supply an acid medium for the chief enzyme of the gastric juice (pepsin) for the enzyme works better in acid medium. As soon as the food leaves the stomach, it meets the pancreatic fluid and the bile, both of which are alkaline. This is necessary since the pancreatic enzymes (trypsin, amylase and lipase) react only in neutral or slightly alkaline medium.

The rate of enzymic action also depends upon the **concentration of the enzyme and the substrate** ; but this is a complicated relationship. At the beginning of the process, the rate of the action varies directly as the concentration of the enzyme ; later, the substrate is naturally beginning to disappear, and, therefore, unless an unlimited supply of substrate is available, the process gradually slows down.

The **end products of the reaction** affect the rate of enzymic action. If such products are not removed as they are formed, they tend to retard the process. For example, the alcohol produced by the alcoholic fermentation of glucose by zymase eventually stops the process, unless the alcohol is removed.

Some substances, if present with enzymes, tend to prevent their activity. They are therefore called **inhibitors**. For example, salts of heavy metals, cyanides and alcohols, inhibit the action of most enzymes.

SYNTHETIC ACTION OF ENZYMES

The examples of enzyme action so far considered have been mostly of breakdown type, in which complex substances are resolved into simpler ones. It is highly probable that enzymes are also concerned in building-up, synthetic, transformations in the plant, though less is known about these enzyme activities. In some instances (for example, lipase) there is evidence that one enzyme is able to promote either the splitting of fats or their synthesis, depending on the concentrations of the substances concerned. In other cases, different enzymes may be concerned with up-grade and down-grade reactions. There is evidence that this is so with starch and sucrose (p. 181).

PRACTICAL WORK

1. Demonstrate the action of lipase in castor-oil seeds.

Shell about 10 gm. of the seeds and divide these into two portions. Pound one portion with 4 gm. of castor oil and 5 c.c. of water. Treat the second portion in the same way, but using 5 c.c. of $N/10$ sulphuric acid instead of the 5 c.c. of water. Allow both to stand for about one hour, then to each add about 25 c.c. of alcohol. Then titrate the free acid in each with a normal solution of caustic soda, using phenolphthalein as an indicator. Note that the second portion, which was pounded with acid, contains a much greater amount of free acid, showing that the activity of lipase present in the seeds has been accelerated in an acid medium.

2. Demonstrate the activity of diastase.

An active preparation of diastase can be obtained (a) by procuring taka-diastase from the chemist and preparing a 0·1 per cent solution of this in water, or (b) by germinating barley grains on damp filter paper until the shoots begin to emerge, and then pounding a handful of the seedlings with 50 c.c. water and filtering, when the filtrate will contain active diastase.

Prepare a 1 per cent solution of starch (p. 103) and place 5 c.c. in each of two test tubes. Add an equal quantity of the diastase extract, this, in the case of one of the tubes, being boiled before adding it to the starch. Periodically take a drop of the mixture from each tube and test its reaction with dilute iodine on a white tile. At first the blue–black starch colour will be obtained, but with the tube containing unboiled enzyme, this will soon be replaced by reddish colours, and finally no colour will be produced, indicating that the starch has now been converted to simpler substances, mostly to sugar. The temporary production of red colours is due to the formation of intermediate substances, termed dextrins. The mixture containing boiled enzyme will continue to give the starch reaction. The Fehling's test may be finally applied to each tube.

The above experiment will proceed faster if the tubes are placed in a water-bath at a temperature of 30°–40°C. A comparison can be made of the time required for the reaction to reach completion at room temperature and at the higher temperature mentioned.

3. Test for oxidase and peroxidase in plant tissues.

These enzymes may be detected by means of a 1 per cent solution of gum guaiacum in 60 per cent alcohol. Portions of various plant tissues are pounded with water in a mortar, and the extract decanted into a white dish and tested with the gum guaiacum. If oxidase is present a blue oxidation product of a constituent of the gum will be formed. If no colour is given a little hydrogen peroxide (5 vols.) should be added to the same mixture. The appearance of a blue colour now indicates that peroxidase is present.

Potato and carrot give the oxidase reaction; cabbage and turnip the peroxidase. Tissues containing oxidase often turn brown when cut surfaces are exposed to the air.

4. Test for zymase and catalase in yeast.

A mixture of fresh yeast, glucose and water in the proportion of 1 : 1 : 10 should be shaken up and placed in a saccharimeter or double test tube which is set at a temperature of 30°–40°C. An active formation of gas will soon commence, and by means of lime water this can be shown to be carbon dioxide. Ethyl alcohol will be present in the fermenting mixture and may be detected by suitable chemical test. The fermentation of the sugar is proceeding under the influence of the zymase complex of enzymes. The effect of using other sugars (sucrose, lactose) instead of glucose can be tested.

Yeast cells contain a very active catalase enzyme. Set up a double test tube with the inner tube filled with hydrogen peroxide (2 vols.) to which 1 gm. yeast is added. There is rapid evolution of a gas which, by testing with a glowing splint, can be shown to be oxygen.

CHAPTER IX

PROTOPLASM AND THE COLLOIDAL STATE

The establishment of the fact that the physical basis of life is a colourless, somewhat viscid fluid material was one of the outstanding scientific achievements of all time (Chap. III). This substance, **protoplasm**, always exists as small units known as **cells** and, in the case of plants, these cells are surrounded by a non-living cell-wall. Cells also contain non-living inclusions in the form of food materials and, sometimes, secretory or excretory products ; these are produced by the activity of the protoplasm. *Thus all living organisms may be said to be composed of protoplasm and its products.*

The physico-chemical nature of protoplasm is still far from being fully understood, for there are immense difficulties in the way of any investigations into this subject. It is obvious, for example that any chemical analysis will involve the death and consequent alteration of the protoplasm. Moreover, protoplasm is characterised by constant physical and chemical change and it is always associated with some of its non-living products. Nevertheless, it is possible to get a crude picture of protoplasm as an extraordinarily complex and highly organised physico-chemical system. One important feature of protoplasm is that it always exists in a *colloidal condition* so that it will be necessary to explain what is meant by this term before proceeding to an account of protoplasm itself.

THE COLLOIDAL STATE

Many substances, for example, common salt, dissolve freely in water, the substance dissolved being the **solute** and the water the **solvent.** In such true solutions the molecules of the solute are separated from one another and move freely in the solvent. Many of the molecules become dissociated, that is, they break up into two or more groups of atoms known as ions. If the solvent is evaporated, the solute crystallises out again. These substances which form true solutions are known as **crystalloids.** The individual molecules in a solution cannot be seen by any known device, for their dimensions are extremely small. For example, the smallest molecule (hydrogen) measures $0.1 \, \mu\mu$ and a medium-sized molecule (sugar) measures $0.7\mu\mu$ (1μ or micron is one thousandth of a millimeter ; $1 \, \mu\mu$ or millimicron is one millionth of a millimeter).

Suspensions, for example soot in water, and **emulsions**, for example, oil shaken up with water, are, like true solutions, two-phase systems. Each consists of a **continuous** phase, in this case water, and a **disperse** phase, the molecules of the solute, the particles of soot or the drops of oil. In suspensions and emulsions the dispersed particles are large molecular aggregates, those of a coarse clay suspension being larger than 3μ in diameter. Such aggregates can be seen under the high power of an ordinary microscope. The two phases of such suspensions and emulsions eventually separate under the influence of gravity.

There are, however, two-phase systems where the disperse phase consists of much smaller molecular aggregates (varying from $1\mu\mu$ to $100\mu\mu$ in diameter) which do not separate from the continuous phase under the influence of gravity. Such systems, which may at first sight appear to be true solutions, are called **colloidal solutions**, a name proposed by **Thomas Graham** in 1861 to distinguish them from crystalloidal solutions. The molecular aggregates in a colloidal solution naturally pass much more slowly (if at all) through a parchment-paper membrane than substances in true solution. By this means Graham was able to separate a mixture of two such solutions. The process is called **dialysis** (see practical work). This is an important property, since it follows that colloidal substances in the plant cannot pass through the cell membranes (see Chap. XI).

Certain inorganic substances, such as ferric hydroxide and gold, which are normally insoluble, can, by special methods of preparation, be obtained in colloidal solution in water. Many organic substances such as starch, most proteins, gums, etc., are present in the plant in a colloidal condition. The colloidal systems which concern biologists most are those in which the continuous phase is water ; but there is a variety of colloidal systems in which either or both phases may be formed from a liquid, solid or gas (with the exception of both phases being gaseous).

The particles of the disperse phase in a colloidal solution are too small to be seen under the highest power of an ordinary microscope. The smallest particle which can be made visible is not less than $200\mu\mu$ while even a very large colloidal particle measures no more than $100\mu\mu$. It is, however, possible to detect the presence of particles as small as $5\mu\mu$ by using an **ultra-microscope**, an instrument devised by **Professor R. Zsigmondy** towards the end of the nineteenth century. The ultra-microscope utilises the Tyndall effect. This effect is seen, for example, when a strong beam of light illuminates tobacco smoke in a partially darkened room and the particles of smoke scatter the light, thus giving the appearance of myriads of bright specks. In the same way, if a pencil of bright light is shone through a colloidal solution it appears milky if viewed at right angles to the light path. This appearance is due to the scattering of the light by the disperse particles. Under similar conditions a true solution appears uniformly clear. Under the ultra-microscope the colloidal solution is illuminated laterally against a dark background. Although the particles themselves are not seen, their presence can be detected as bright specks showing Brownian movement.

The two phases of a colloidal solution cannot be separated by the usual methods of filtration ; but it is possible to do so by **ultra-filtration** and, at the same time, to gain some idea of the actual size of the particles. In this method the colloidal solution is passed through a series of filters consisting of collodion membranes which can be prepared with pores of known size, varying from 1μ to $20\mu\mu$. Pores larger than the disperse particles will allow these to pass through, while smaller pores will prevent their passage. It is thus possible to estimate the size of the actual particles.

There are two main types of colloidal solutions or **sols** ; but the boundary between them is not very sharp. If the disperse phase consists of solid particles the sol is spoken of as a **suspensoid**, for example, the metallic sols mentioned earlier. This type may also be said to be **lyophobic** since neither phase dissolves in the other or **hydrophobic** since the disperse particles have little or no affinity for water. If, on the other hand, the disperse phase consists of liquid droplets, then the system is spoken of as an **emulsoid**, for example, starch solution, proteins, gums. These are said to be **lyophilic** since the two phases are more or less mutually soluble in each other or **hydrophilic** since the disperse particles absorb water.

Both suspensoids and emulsoids may, under certain conditions, set to a more rigid, jelly-like condition, to which the term **gel** is applied. An example of an emulsoid gel is gelatine. If a piece of gelatine is put into water it absorbs some of the water by imbibition and forms a gel which on heating forms a sol. On cooling this sol will set to a more rigid gel. The process is illustrated in Fig. 50. *A* depicts the heated colloidal sol ; *B*, the more rigid gel. If more water is added and heat again applied, the gel will again become a sol. Agar-agar, a mucilage derived from red seaweeds, behaves in much the same way. A 1 per cent solution in water, if heated and then allowed to cool sets into a firm gel which provides a suitable medium for the culture of fungi. In these gels the solid particles appear to form a sponge, the water being held as isolated droplets. Some gels show the phenomenon of **syneresis**, which is an extrusion of drops of the continuous phase. For example, a clot of blood gradually contracts and some of the serum is exuded. It has been suggested that this process underlies the secretion of chemical substances from living cells.

Some inorganic colloidal systems such as silica gels are of particular interest for comparison with protoplasm. The $Si(OH)_4$ molecules join end to end, losing water at each junction, and thus form long chains. When these chains lie parallel to one another the system is in a ' liquid crystal ' state, but when they are joined together in a random manner they form bundles or networks and the system becomes an elastic gel. Such gels liquefy when stirred—a phenomenon known as **thixotropy**.

Colloidal sols may be precipitated, as for example, the precipitation of certain proteins by the addition of ammonium sulphate (see practical work). Certain colloidal substances are precipitated by salts of heavy metals. That is why it is recommended to administer raw egg ' white ' to a person who has taken poison in the form of a salt of a heavy metal such as a mercury salt. The colloidal albumin and the mercury salt together form an insoluble precipitate, thus inactivating the poison. Colloids, on the other hand, can be protected from precipitation by the addition of certain salts, non-electrolytes, or another colloid. Like the sol-gel change, precipitation may be reversible or irreversible.

Coagulation of proteins, for example, white of egg, by heat or by reagents like alcohol is an irreversible change to a gel condition, but this is probably accompanied by an actual change in the proteins.

The stability of colloidal systems depends largely on the fact that each aggregate of the disperse phase carries an electric charge, thus acting in this respect

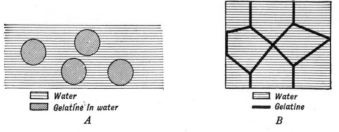

FIG. 50. DIAGRAMMATIC REPRESENTATION OF THE SETTING OF A HEATED COLLOIDAL SOL (A) TO A COOL, MORE RIGID GEL (B).

like a single true ion. In some examples, for example, ferric hydroxide and albumin, the charge is positive ; in others, for example, starch solution and oil emulsion, it is negative. Therefore if a potential difference be set up in such colloidal solutions, positively charged aggregates will move to the negative pole and negatively charged ones to the positive pole. This is known as **cataphoresis**. Owing to certain peculiarities of ionisation some colloidal substances, for example, proteins, act both as acids and bases and are therefore referred to as **amphoteric electrolytes**. The disperse aggregates are positively charged if the continuous phase is acid ; negatively charged if it is alkaline.

Some other characteristics of colloidal systems are of great importance in biology. It has already been mentioned that colloidal substances do not themselves diffuse, but it is necessary to realise also that they readily permit diffusion through them. Gels offer very little resistance to the diffusion of small molecules so that inorganic substances in solution pass very readily through colloidal gel membranes (see practical work).

An outstanding feature of colloidal systems is the very large surface of contact between the disperse phase particles and the continuous phase. The immense extent of this surface is suggested by the fact that if a cube with sides 1 cm. long and consequently a surface area of 6 sq. cm. is subdivided into particles with sides $10\mu\mu$ long (the size of an average colloidal particle) the surface area is increased a million-fold. It is clear therefore that when a substance goes into a colloidal condition the disperse particles will have an enormous surface in contact with the continuous phase. Their surfaces are the seat of surface energy and consequently chemical reactions will readily take place there. Moreover, other substances present in the continuous phase can concentrate on the surfaces of the aggregates of the disperse phase. This concentration on the surface is known as **adsorption** (see practical work). Thus many colloidal solutions have the property of adsorbing some substances to a marked degree, others to a less marked degree, and others not at all. This characteristic is referred to as **selective adsorption** and is extremely important in the physiology of plants and animals.

CHEMISTRY OF PROTOPLASM

Chemical analyses of protoplasm can only be carried out on dead protoplasm. A further difficulty is that the protoplasm always contains products of its own activity which are not essential and permanent constituents of the living substance. It is therefore impossible at present to elucidate the chemical constitution of living protoplasm ; but it is, nevertheless, useful to have some idea of the nature of the substances which are involved.

Lepeschkin analysed a slime fungus which is largely protoplasmic. His analysis showed that it consisted of 82·6 per cent water and the following substances (given in percentage of dry weight) :

Organic water-soluble substances (chiefly in the vacuoles)

Monosaccharides	14·2
Proteins	2·2
Amino-acids, purines, asparagine, etc.	24·3

Organic water-insoluble substances (the ground-mass of the protoplasm)

Nucleoproteins	32·3
Nucleic acid	2·5
Globulin	0·5
Lipoprotein	4·8
Fats	6·8
Phytosterol	3·2
Phosphatides	1·3
Polysaccharides, pigments.	3·4
Inorganic mineral substances	4·4

Apart from such quantitative analyses, much useful information has been obtained by cytochemical methods, that is by the use of stains which give strongly coloured compounds with the cell constituents. Suitable reagents are available for the identification of many of these. For example, the Feulgen test is specific for thymonucleic acid in the nucleus.

Water is obviously an extremely important constituent. Active protoplasm always consists of more than 75 per cent water. On the other hand, the protoplasm of resting spores and seeds may contain as little as 10 per cent, but the activity of these is not resumed until more water has been taken in.

Proteins, which may constitute 40–60 per cent of the dry weight, are essential constituents. Each organism appears to have its own characteristic proteins and the differences between the various species probably depend upon this fact. Simple proteins like albumins and globulins are constantly present in cytoplasm but many others are also present. Conjugate proteins appear to play an even more important part in the life of the protoplasm. Lecithoproteins and nucleoproteins are present in all cytoplasm and the nucleoproteins are particularly important constituents in nuclei. The enzymes which have such a vital role in all metabolism are also probably mainly protein in nature.

The true fats which occur in cells are probably mainly reserve food substances but more complex fatty compounds are permanent constituents of all protoplasm. The most important of these are the phosphorus-containing compounds known as phosphatides, phospholipins, lecithides and phospholipoids. Lecithins belonging to this group appear to play a major part in the formation and activity of protoplasmic membranes.

The inorganic mineral substances, though present in relatively small quantities as compared with the proteins and fatty substances, probably play an equally important part. Many of the essential elements (see p. 58) have been detected in protoplasm and each probably has a vital role.

PHYSICAL CHARACTERS OF PROTOPLASM

Living protoplasm appears under the microscope as a clear substance with non-living granules and globules distributed through it. Early theories as to its physical nature were based largely on dead and stained cells, but in the making of such preparations the proteins are coagulated and other changes take place. The appearance of such cells is therefore probably deceptive, since some at least of the features observed are **artefacts.**

More recent researches have been based on observations made on living cells by means of special techniques. The cells used are often grown in **tissue cultures**, that is cultures of plant and animal tissues grown aseptically in special fluids. Some of the more important new methods of observation may be mentioned. Fine structural details have been demonstrated by photomicrography using ultra-violet light which gives greater resolving power. Ultra-microscopic structures have been investigated by the use of polarised light. The use of a ' phase contrast microscope ' makes it possible to differentiate between cell structures which appear more or less homogeneous and transparent under ordinary illumination. Apart from these new methods of illumination, the use of vital stains has proved valuable. Such reagents colour various protoplasmic structures without killing them, so that their structure and behaviour can be more easily observed. Much important information has also been obtained by the use of the **micromanipulator**, an instrument by means of which it is possible to dissect, or to inject substances into, living cells under the high power of the microscope. Minute needles, scalpels and pipettes made of glass are used, and the precise and delicate manipulation of these is mechanically controlled.

All investigators agree that protoplasm is an extraordinarily complex and variable colloidal system with a number of different disperse phases. These latter consist largely of proteins and fatty materials and, since such substances are capable of forming emulsoid sols, it has been suggested that protoplasm is in this condition, with a continuous phase consisting of water with proteins and lipoids in solution. On the other hand, protoplasm often has a viscosity only two or three times that of water, and this suggests that the system is suspensoid rather than emulsoid.

There are, however, features which suggest that the protoplasmic structure is more complex than either the emulsoid or the suspensoid. Neither of these systems shows elasticity but protoplasm does. This feature has been demonstrated by inserting particles of nickel into protoplasm and then displacing them by an electro-magnet. When the current is cut off the particles tend to return to their original position. Similar properties are shown by systems like the silica gels, and it is suggested that in the case of protoplasm, the long chains of protein molecules form a sub-microscopic network or scaffolding. It may be noted that proto-plasmic gels often liquify when stirred with a micromanipulator needle just as do the silica gels.

Living protoplasm shows streaming movements round the cell. These are particularly obvious in the leaves of the Canadian pondweed (*Elodea canadensis*) (see practical work), but they probably occur in all living cells, the energy being provided by respiration. No complete explanation of this phenomenon has yet been given. It is suggested, however, that a folding and unfolding of the long protein molecules results in a rhythmic contraction and relaxation of the protoplasm leading to its movement. The rate of movement depends on the temperature. In *Elodea* it commences at $1\,^{\circ}$C., increases in speed as the temperature rises to $36\,^{\circ}$C. and ceases at $40\,^{\circ}$C.

A further complication of the protoplasmic system is indicated by the fact that many different chemical reactions proceed simultaneously in the same cell. This suggests that the protoplasm must be divided, sub-microscopically, into many compartments, possibly by films of gel consistency.

Whatever the precise nature of the protoplasmic system may prove to be, it is clearly colloidal in nature and shows many of the characteristics of this state. Thus :

(1) Protoplasm shows the phenomenon of selective adsorption, for example,

of the mineral substances absorbed by the roots (p. 131). It is also demonstrated by the adsorption of dyes (see practical work).

(2) Protoplasm is able to undergo a reversible change from the sol to the gel condition. The fluctuation between the two conditions seems to play an important part in the life of the cell, where it is probably brought about by the actions of enzymes and variations in acidity.

(3) Protoplasmic gels such as occur in spores and seeds take up water by imbibition (see practical work).

(4) Protoplasm will not itself diffuse through membranes but it allows many substances to pass through it very readily. This is characteristic of other colloidal gels such as gelatine (see practical work).

The above outline gives some idea of the complexity of the living substance. The most outstanding feature about it is, however, the fact that it is a delicately *organised* system showing a great variety of changes going on side by side and in orderly sequence. Very little is known about the nature of this organisation, and an enormous field of enquiry remains to be explored.

PRACTICAL WORK

1. Mount a complete leaf of Canadian pondweed (*Elodea canadensis*) in water on a slide and examine under the high power of the microscope. Note the small, green chlorophyll-containing granules (chloroplasts). Careful observation of some cells will reveal these moving, and since they are embedded in the protoplasm (cytoplasm), this indicates the circulation of the protoplasm within the cell-wall.

2. Make a colloidal solution of starch.
Shake 2 gm. of starch with a little cold water in a test-tube until it forms a paste. Boil 100 c.c. of water in a beaker, and while the water is boiling, slowly pour the paste from the test-tube, drop by drop, into it. After cooling, note that this colloidal solution is slightly opaque. Test the solution for starch by adding a few drops of iodine solution.

3. Separate a colloid from a crystalloid by the process of dialysis.
Glass cups for making dialysers with parchment paper membranes are easily obtainable (Fig. 51). On the other hand, to make a dialyser, soak a Soxhlet thimble in a 5 per cent solution of collodion in glacial acetic acid. Drain off the excess collodion and then thoroughly wash in water.

Then make a 2 per cent colloidal solution of gelatine, and to this add some common salt. Place this mixture of colloidal and true solutions in a dialyser, and immerse it in a beaker of water. After about a day, test the liquids in the dialyser and the beaker for gelatine and for sodium chloride. The former can be tested for with tannic acid, which precipitates gelatine out of solution. The latter can be tested for by adding silver nitrate, which reacts with any sodium chloride present to produce a precipitate of insoluble silver chloride. (Instead of using a colloidal solution of gelatine, one of starch will do, in which case the iodine test should be applied).
Note that whereas the crystalloid passes through the membrane, the colloid does not.

4. Examine the irreversible coagulation of egg 'white' by warming some in a test-tube placed in boiling water.
Then examine the reversible precipitation of a colloid, again using egg 'white'. The colloidal nature of egg 'white' is due chiefly to the presence of the protein albumin. Mix the 'white' of an egg with 100 c.c. of water. Filter by means of a filter pump. To about half of the clear solution thus obtained, gradually add some powdered ammonium sulphate, shaking all the time. As the sulphate approaches saturation, the albumin will be precipitated

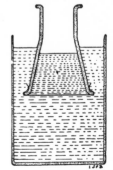

FIG. 51. A DIALYSER.

in the form of a white curd.   Then shake this liquid with about the same volume of water.
Note that the precipitated albumin re-enters colloidal solution.

5. Make a 10 per cent solution of gelatine by warming in water and add a few drops
of phenolphthalein solution.   Fill a wide tube, stoppered at one end, with the warm solution
and allow to cool.   When it has set, invert the tube over a 10 per cent solution of caustic
soda.   The upward diffusion of the alkali through the colloidal gel will be indicated by a
reddening of the gelatine.

6. Cut three rectangles of exactly equal dimensions from sheet gelatine.   Let one remain
dry as a control.   Place a second in water ; note that its dimensions increase as a result
of the imbibition of water.   Measure its final area and compare with the control.   Place the
third rectangle in a very dilute aqueous solution of Methylene Blue.   Note that the swelling
is accompanied by a progressive staining of the gelatine.   Eventually the gelatine adsorbs
practically all the dye from the solution.

7. Half fill a test tube with an aqueous solution of Night Blue, add as much activated
carbon as would lie on a sixpence, shake vigorously for a short time, and then place the
tube in the rack for the carbon to settle.   Inspection of the supernatant fluid will then show
that the solution has been decolourised owing to the adsorption of the dye by the carbon.

8. Take a number of dry broad bean (or pea) seeds and measure their displacement
by placing them in a measuring glass partly filled with water.   Place them in moist sawdust
for two days and then again measure their displacement.   Estimate the percentage increase
in volume due to the imbibition of water.

9. Place a shoot of *Elodea* in a very dilute solution of Methylene Blue.   Leave for a few
hours and note that the *Elodea* becomes deeply coloured by adsorption of the dye.

CHAPTER X

INTERNAL ORGANISATION OF LIVING THINGS

All living organisms are composed of one or more living units called **cells** (see Chap. IX). The structure of all cells, whether of plant or animal, is *fundamentally* the same. Most plant cells are, however, surrounded by a cell-wall, while most animal cells are not. A general idea of the size of cells can be gleaned from the fact that it would require four to five million average-sized animal cells to cover one square inch.

THE CELL

The chief constituent of the cell is the protoplasm, which under normal conditions is in constant motion. Except in the case of the very simplest of microscopic organisms, this is divided into two main portions. The bulk of it is made up of a colourless substance, in the colloidal state, containing many non-living granules embedded in it. This part of the protoplasm is called the **cytoplasm** (Figs. 52, 53 and 54). The layer of cytoplasm in contact with the wall forms the **plasma membrane**. This appears to be a gel membrane formed by an orderly arrangement of protein and fat molecules.

Embedded in this cytoplasm is the other portion of the protoplasm, which is called the **nucleus**. The protoplasm composing the nucleus is often referred to as **nucleoplasm** in contradistinction to the cytoplasm. The nucleus is usually ovoid in shape, and is situated either near the centre of the cell or in a peripheral position. But, wherever it is, the nucleus is always completely surrounded by cytoplasm. The nucleus, like the cytoplasm, is composed of protoplasm, but the protoplasm of the two portions varies considerably, both chemically and physically. For example, the cytoplasm contains ribose nucleic acid as part of its protein-manufacturing system while the nucleus contains in addition desoxyribose nucleic

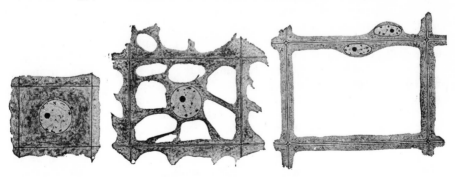

FIG. 52. A PLANT CELL.

Left, a very young cell without vacuole and surrounded by a cell-wall; middle and right, stages in growth and vacuole formation. Note that the nucleus is always surrounded by cytoplasm; note also the bridles. (Diagrammatic).

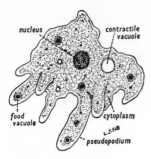

FIG. 53. AMOEBA, A UNICELLU-
LAR ANIMAL.
Note the absence of a cell-
wall (× 175)

acid which is vitally concerned in the process of nuclear division (p. 108). Only the nucleus contains nucleoproteins, which themselves contain the element phosphorus (see p. 73). The nucleus, too, is usually in a more viscous state than the cytoplasm. The nucleus is surrounded by a thin membrane, probably of the nature of a protein gel, referred to as the **nuclear membrane**.

The cytoplasm of many plant cells also contains small protoplasmic bodies known as **plastids**. These develop from minute bodies present in the cells of growing points which are termed **proplastids**. These enlarge in photosynthetic cells into green, ellipsoidal **chloroplasts** (p. 176) while in cells concerned with the storage of starch they develop into colourless **leucoplasts** (p. 68). In the cells of many red and yellow coloured flower petals and fruits they develop into irregularly shaped **chromoplasts**, where the pigment consists mainly of xantho-phyll and carotene. In many such examples, however, chloroplasts are present in the young petals or fruits and these later become converted into chromoplasts.

Mitochondria are also included in many cells. These are minute rod- or thread-like protoplasmic bodies which are difficult to see unless special techniques are used. Their role is uncertain but it is suggested that they are concerned with protein synthesis.

Nearly all plant cells are surrounded by the cell-wall. This consists of a thin membrane of **pectin** (a substance chemically related to carbohydrates) called the **middle lamella**, on which is deposited layers of a purer carbohydrate, cel-lulose. The protoplasm of each cell is, however, united to that of the adjoining cells by fine protoplasmic threads passing through extremely minute pores (normally invisible under the microscope) in the intervening cell-walls. This is referred to as the **continuity of protoplasm** (see p. 110).

The young cells at the growing points of plants are completely filled with protoplasm. Older cells, further back from the apex, increase in size and undergo a process of **maturation**. During this process the cytoplasm develops small spaces known as **vacuoles**, which are filled with an aqueous solution of various substances (Fig. 52). These spaces increase in size by the intake of more water and may finally coalesce to form one large central vacuole. The volume of the cell is naturally increased by this process and since little or no new protoplasm is manufactured, the mature cell often possess only a thin lining of cytoplasm. The nucleus may be embedded in this lining or it may be slung in the centre of the cell by strands or **bridles** of cytoplasm (Figs. 52 and 54). The cytoplasm forms a gel membrane, similar to the outer plasma membrane, where it is in contact with the vacuole fluid or **cell-sap**. The increase in size of the cell is naturally accompanied by an ex-pansion of the wall. As it is stretched by the expanding contents new cellulose is deposited on it by the cytoplasm so that it becomes thicker as the cell matures.

The cell-sap consists of water with numerous crystalloids dissolved in it and other substances may be present in colloidal solution. The substances present in the cell-sap vary, and in some cases there is a high percentage of certain crystalloids. For example, in the cells of the beet the sap contains a high percentage of sucrose while in those of peas and beans it contains a large amount of potassium nitrate.

Organic acids, proteins, alkaloids and a great variety of other substances may also be present. The vacuoles of the cells of many plants also contain pigments, particularly anthocyanins (p. 84).

In many living cells, other non-protoplasmic inclusions may occur, according to the function of the cell. For example, starch grains and aleurone grains may be present as a food store. Both of these inclusions are embedded in the cytoplasm. Other inclusions sometimes occurring are crystals of calcium oxalate (probably a waste product), which are usually present in the vacuoles.

Cells vary considerably in size and form. For example, the cells of some bacteria are spherical and about 0·001 mm. in diameter (Fig. 5), whereas the cells of some fleshy fruits are 1 mm. in diameter. Cells may be spherical, cubical, cylindrical, polyhedral or prismatic in shape. The cells which develop into the fibres of some plants are long and tapering, and may reach a length of 20 cm.

CELL DIVISION

In some of the lower plants any cell of the body may divide, but, in the higher plants, cell division usually takes place only in young cells, either those in young organs, such as developing fruits, or those in growing points (p. 130). Such cells are said to be **meristematic**. Each meristematic cell divides into two and each of the daughter cells may divide again. This process is continually repeated so that the cells multiply in number and add to the body of the organism. Mature cells do not normally divide ; but they may do so to form cambium (p. 149) or, following injury, to produce protective tissues and adventitious growths (p. 159).

The division of a cell involves the division of all the protoplasmic contents into two daughter cells. The nucleus plays a leading part in this process, and its own division is brought about by a series of complex but precise changes known as **mitosis**. Only rarely does a nucleus divide by simple constriction into two daughter nuclei by a process known as **amitosis.**

The nucleus of a cell which is not dividing is said to be in a **resting condition**, although it is very active in a physiological sense. The resting nucleus of a living cell often appears to be optically homogeneous apart from one or several highly refractive bodies, the **nucleoli**(Figs. 55a and 56n). It is, however, possible to show by special techniques that the nucleus is surrounded by a gel membrane, the nuclear membrane, and that it contains a definite number of very long and coiled double threads which are called **chromosomes**. Each chromosome consists of two **chromatids** which are held firmly together at one point by a structure called

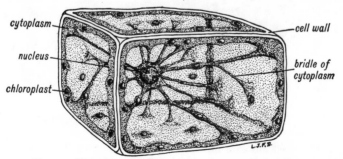

cytoplasm

cell wall

nucleus

bridle of cytoplasm

chloroplast

FIG. 54. DIAGRAMMATIC REPRESENTATION OF A PLANT CELL.

the **centromere.** At this stage the chromosomes are so highly hydrated that they are difficult to differentiate from the **nuclear sap**, a clear fluid in which they are immersed. It is also difficult to stain the chromosomes since they consist largely of proteins which do not take up basic dyes.

When cell division is about to occur, the chromosomes become more dense and so more easily seen, even in a living nucleus. They also become stainable with basic dyes since nucleic acid (derived partly from the nucleoli and partly from the cytoplasm) unites with the protein of the chromosomes to form a nucleoprotein, the deeply staining chromatin. Each chromosome now shortens and thickens as the result of each of its chromatids becoming spirally coiled. At this stage, therefore the nucleus consists of the membrane and, embedded in the sap, a number of deeply staining chromosomes. The changes leading to this condition constitute the **prophase** of mitosis (Figs. 55,*b-d* and Fig. 56, 1, 2).

At the end of the prophase the nuclear membrane and the nucleoli disappear and a stage known as the **pro-metaphase** (Fig. 55, *e* and *f*) is begun. This is characterised by the development of a **spindle**, a gelatinous body formed mainly from the nuclear sap by a change from the sol to a gel condition (Fig. 56, *s.f.*). This change is probably initiated by the centromeres of the chromosomes. The spindle often shows the appearance of fine striations or ' spindle fibres ' due to the fact that the long-chain molecules of the gel are all lying parallel to the longitudinal axis of the spindle. The formation of the spindle follows a rather different course in most animal cells and in those of some algae and fungi. In these, a body known as a **centrosome** (Fig. 56 *cs*) is constantly present. It plays an important part in the development of the motile organs of the male gametes, but it also plays a conspicuous part in the mitosis of these forms. During the prophase it divides into two parts, each of which is surrounded by a system of radiations known as an **aster.** The two asters separate from one another and move to the two poles of the cell where they take part in the organisation of at least part of the spindle.

Spindle formation is followed by the **metaphase** (Fig, 55, *g* and 56, 3). The chromosomes come to lie in the wider middle region of the spindle on the so-called

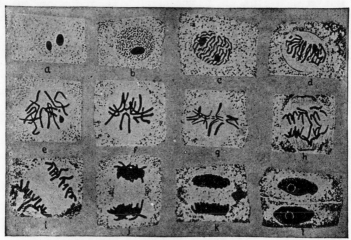

FIG. 55. SUCCESSIVE STAGES OF MITOSIS IN THE ROOT-TIP OF THE ONION (× about 1200)
(*After Bĕlăr : ' Die Cytologischen Grundlagen der Vererbung.'*)

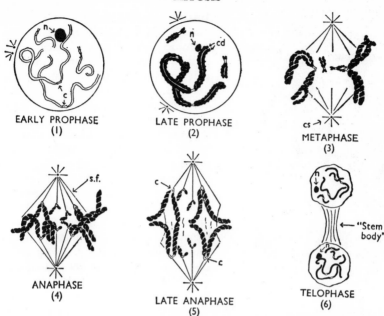

(*From Wilma George ' Elementary Genetics* ').

FIG. 56. MITOSIS.

A pair of large and a pair of small chromosomes are shown. *n*, nucleous ; *c*, centromere ;
cd, chromatids ; *s.f*, spindle fibres.

equatorial plate. Each of them is attached to the spindle by its centromere.
During this stage the chromosomes are at their shortest and thickest, and their
tightly coiled chromatids are so closely applied to each other that each chromosome
appears as a single structure.

The metaphase lasts only a short time and the **anaphase** then follows (Fig. 55,
h-j and 56, 4, 5). The centromere of each chromosome suddenly divides and the
two halves move up the spindle towards the opposite poles, each half pulling a
chromatid behind it. Fig. 57, 1–3, shows the separation of the two chromatids of
a chromosome which is attached to the spindle by a median centromere ; 4 and 5
show this process in a chromosome with its spindle attachment nearer one end.
(see also Fig. 58.). If the chromosome is a short one the two chromatids are
completely pulled apart by this process but longer ones may not be completely
separated. At this point, however, the middle region of the spindle elongates so
that the separation of the longer chromatids is also completed. The completely
separated chromatids are now called the daughter chromosomes.

The anaphase thus results in the presence of two groups of chromosomes near
to the poles of the spindle. In each group the chromosomes are identical in number,
size and shape. Each group is now reconstituted as a daughter nucleus by a series
of changes, which resemble the prophase in reverse which is called the **telophase**
(Fig. 55 *k*, *l* and 56, 6). A nuclear membrane is formed round each of the chromosome groups. The individual chromosomes begin to lose their nucleic acid and
as a result they begin to uncoil and lengthen and also to lose their staining properties. During this process the nucleoli are also reformed by definite regions of
one or more chromosomes and they become filled with proteins and nucleic acid.

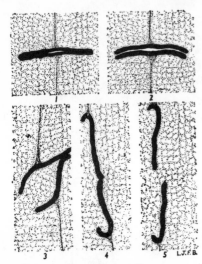

FIG. 57. STAGES IN THE SPLITTING OF A CHROMOSOME.

In this way the nucleus once more enters into the resting condition, where the chromosomes, now devoid of nucleic acid (except in certain nuclei where small patches of it persist giving deeply staining granules) are difficult to observe. It is probable that each of the chromosomes divides during the resting period into two chromatids thus giving them the double structure which would be seen in the prophase of the next division.

The changes described are continuous, and it is merely for convenience of description that they are divided into the various phases. The time taken for a complete sequence varies from a few minutes (ten minutes in the fruit fly, *Drosophila melanogaster*) to several hours (many plants). Prophase and telophase take relatively longer than metaphase and anaphase.

The division of the nucleus is followed by the formation of a wall between the daughter cells. This begins at the end of the telophase by the appearance of a liquid film across the equator of the parent cell (Fig. 55, *l*). This film soon becomes firmer by the deposition of pectin and is now known as the **middle lamella**. The cytoplasm then deposits a layer of cellulose on each side of the middle lamella and the cell division is now complete. During this formation of a new wall it is probable that very fine strands of cytoplasm remain, penetrating the wall substance and giving a living connexion between the two daughter cells (p. 106).

The necessity for this complex series of changes in the nucleus rests essentially upon the fact that the chromosomes carry many of the inherited characters of the organism. Each chromosome has a very large number of small granules or **chromomeres** distributed along the length of each of its constituent chromatids. These chromomeres are, in fact, the **genes** which determine the development and structure of the organism. It is suggested that many of these genes are enzymes of protein nature. Some genes may carry the factor determining the colour of the flowers, others the factors controlling leaf shape, and so on. It is thus essential that every cell in an organism shall have the same number of chromosomes and that each of these chromosomes shall carry exactly the same genes. Mitosis ensures that this shall be so (Chap. XXVIII).

It has already been seen that when a nucleus divides one chromatid of each chromosome passes to one pole while the other chromatid passes to the other pole. The result is that the two daughter nuclei each possess exactly the same number of chromosomes as the parent nucleus.

The number of chromosomes is constant in any one species but varies considerably in different plants and animals. Thus, in the body cells of man there are 48 ; fruit fly, 8 ; raspberry, 14 ; potato, 48 ; apple, 34 ; lilies, 24 ; *Dahlia*, 64. Chromosome numbers so far determined for various plants and animals range from two to five hundred (see also Chap. XXVIII).

The entire set of chromosomes in each of these examples is known as the

chromosome complement.
The individual chromosomes of
such sets usually show considerable variation both in size and
shape. The shape depends largely
on the position of the centromere,
since chromosomes usually bend
at this point. If the centromere is
median, the chromosome will
show the common U- or V-shaped
form. If, however, the centromere
is very nearly at one end, the chromosome will be rod–shaped since
one arm of the V or U will be very
short. All intermediate positions
of the centromere are possible so
that a wide range of chromosome
shape is given by the varying in-
equality of the two arms.

These variations in shape and
size of the individual chromosomes
often makes it possible to observe
another very important feature,
namely that each complement in
a body or vegetative cell consists
of pairs of similar chromosomes
known as **homologous pairs.**
One of each pair has been derived
from the female parent, the other

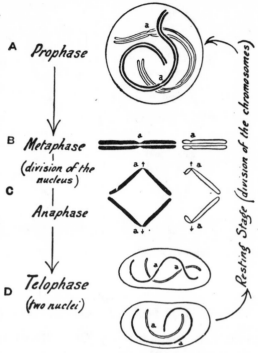

a, attachment constriction.
FIG. 58. DIAGRAM SHOWING THE CHANGES
UNDERGONE BY TWO CHROMOSOMES IN THE COURSE
OF MITOSIS.
(*From Crane and Lawrence after Darlington.*)

from the male parent. They became associated together when fertilisation
occurred. It follows that the number of chromosomes in the body cells of
animals and the higher plants is always an even number. It is in fact a double
number and such cells are said to be **diploid.** The sexual cells themselves are
haploid, that is, they have only one chromosome of each sort. There must obviously
be a halving of the diploid chromosome number during the formation of the
gametes and this is accomplished by a type of nuclear division known as **meiosis**
(p. 265).

Mitosis not only ensures that each daughter cell has the same number of
chromosomes but also that each chromosome in one cell carries exactly the same
genes as a corresponding chromosome in the other cell. This means that there must
be a duplication of each chromosome during mitosis. It is suggested that this
duplication takes place as a result of the attachment of nucleic acid during the
prophase, the nucleic acid acting as a pattern during the copying of the gene
molecule. By anaphase the separating chromatids or daughter chromosomes
(which originally possessed a single series of genes) have a double series, still, however, in intimate contact. It is only during the resting phase that the two series
become separated to give the two chromatids which are visible in the succeeding
prophase.

These general features of mitosis are obviously of great importance in all

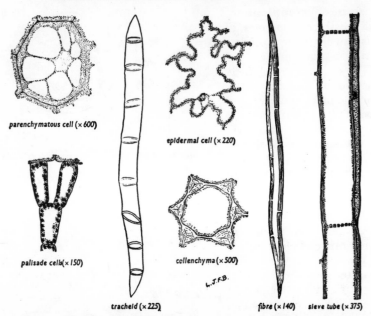

parenchymatous cell (×600)

epidermal cell (×220)

palisade cells(×150)

collenchyma (×500)

L.J.F.B.

tracheid (×225)   *fibre (×140)   sieve tube (×375)*

FIG. 59.   VARIOUS FORMS OF PLANT CELL ELEMENTS.

questions relating to heredity (Chap. XXVIII).   It should be realised, however, that not all the characters of an organism are determined by the genes of the chromosomes.   Certain characters relating to the development of chlorophyll seem to be determined by genes located in the plastids, which, like the chromosomes, are able to reproduce themselves by division.   Such genes are called **plastogenes.** The cytoplasm itself also carries hereditary factors and these are called **plasmagenes.**   Cell division results not merely in the precise division of the nuclear genes, but also in a less-precise division of the plastogenes and plasmagenes by the formation of the new wall between the two daughter cells.

DIVISION OF LABOUR IN THE PLANT

Many plants are composed of hundreds of cells all surrounded by their cell-walls.   In some big plants, such as trees there are many millions of cells.   Such plants are said to be **multicellular.**   On the other hand, there are plants which are each composed of one cell only—for example, bacteria (Fig. 5) and *Chlamydomonas* (Fig. 1).   Such plants are said to be **unicellular.**

The cells of a multicellular plant can be compared with human beings.   If a person were living alone, absolutely cut off from contact with any other human being, he would have to do everything for himself.   A unicellular plant is like that.   It performs all the processes of life within itself.

Now within a community of people nobody does *everything* for himself.   People split themselves up into groups, and each group does one or perhaps several kinds of work *only*.   Some grow food, such as farmers.   Others manufacture it, such as bakers ; others make clothes, and others are responsible for the health of the rest ; and so forth.   This splitting up of all necessary work and allocating it to different people or groups of people is called **division of labour.**

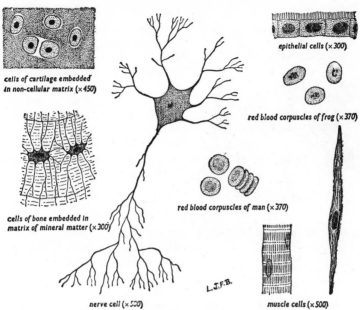

cells of cartilage embedded
in non-cellular matrix (×450)

epithelial cells (×300)

red blood corpuscles of frog (×370)

cells of bone embedded in
matrix of mineral matter (×300)

red blood corpuscles of man (×370)

L.J.F.B.

nerve cell (×500)   muscle cells (×500)

FIG. 60. VARIOUS FORMS OF ANIMAL CELL ELEMENTS.

In this respect, the plant represents the community, and each cell represents one human being within the community. Human beings become modified to do their special work ; farmers become proficient in tilling the soil, sowing and reaping ; doctors in treating the sick, and so on. So do cells become modified in order to carry out their special work successfully. Figs. 59 and 60 show different forms of plant and animal cell elements, which have become modified to perform various functions. Thus, in all living things, especially the more advanced, there is a *morphological differentiation of structure correlated with a physiological division of labour.*

TISSUES

Division of labour has another important characteristic. Men having the same kind of work to do often group themselves together. Coal miners, for example congregate around the coal mines, and farmers relegate themselves to the land. Thus, within a community of people, so far as is possible, there is a segregation according to work (or function). So also in living organisms many cells which have the same kind of work to perform are grouped together. Such groups of cells are called **tissues.** They may be **simple tissues** (for example, parenchyma) consisting of cells all having the same structure ; or **complex tissues** when cells of several types are grouped together to perform a particular function (for example, wood).

When they become modified, some cells still remain living, whereas others utilise non-living structures produced at the expense of the living protoplasm.

LIVING TISSUES

The young cell is composed of nucleus, cytoplasm and cell-wall. Then it grows in size and vacuoles appear in the cytoplasm. These unmodified cells when

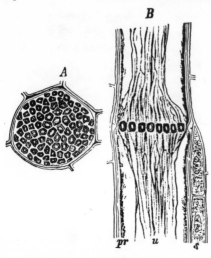

FIG. 61. SIEVE TUBES OF THE
PUMPKIN (*Cucurbita pepo*).

A, surface view of sieve plate ; *B*, longitudinal section of a sieve tube showing a
sieve plate between adjacent sieve elements ; *pr*, peripheral cytoplasm ; *u*,
mucilaginous contents ; *s*, companion cells.
(*After Strasburger*)

collected together form a tissue which is called **parenchyma**. Food-storage tissue is usually composed of parenchymatous cells, such as those in the beetroot and the potato tuber, and the fleshy parts of the fruit of the apple. Young parenchymatous tissue gives a close, honey-comb appearance, but as it gets older and the cells expand, so spaces appear where three or more cells meet. These are called **intercellular spaces**.

In all the more advanced plants there are tissues which are specialised for the transport of organic foodstuffs from the leaves, where they are mostly elaborated, to the living cells in other parts of the plant. Such foodstuffs are never transported in the solid state, but in true solution in water.

The cells used for this purpose are originally vertical rows of parenchymatous cells. The individual cells of each row elongate and often become broader so that they form a long chain of tubular cells. The cross-walls between adjacent cells of this chain become perforated by a large number of pores, so that foodstuffs can readily pass down the whole length of the tube. Each such perforated wall looks like a sieve and is, in fact, called a **sieve plate**. The entire tube is called a **sieve tube** (Fig. 61) and its component cells are called **sieve elements**. Each sieve element has a thin lining of cytoplasm and an abundant slimy cell sap and these are continuous through the pores with the corresponding contents of adjacent elements. The nuclei of the sieve elements degenerate. **Callus** (a carbohydrate) is often deposited on the sieve plates and as the sieve tube becomes older and ceases to carry food the pores are occluded by this substance. Each sieve element sometimes has a small cell in contact with it. This **companion cell** has a nucleus and dense cytoplasmic contents but little is known about its function. It is the same length as the sieve element, having been produced by longitudinal division of the potential sieve element. Such a companion cell may, however, be subdivided by transverse walls.

Sieve tubes such as those described occur only in the flowering plants where they are grouped together with other kinds of cells, especially parenchyma, in the complex tissue known as **phloem**.

The phloem of ferns and pines is less specialised and consists of **sieve cells**. These are long pointed cells with sieve plates scattered over their longitudinal walls. They do not form tubes and they have no companion cells.

As will be seen on p. 115 certain dead cell elements are produced by the deposition of thickening materials on their walls, and are used for strengthening purposes. Some living cells serving a similar purpose, however, are thickened in restricted areas, chiefly the corners, merely by the addition of more cellulose.

Such cells form a tissue known as **collenchyma** (Fig. 59), and are found chiefly in young growing organs.

Many plants grow in very dry regions, such as deserts and semi-deserts. Such plants are forced to take advantage of what little rain there is, collect more than they require at the moment and store the excess. For such storage of water, ordinary parenchymatous cells are used, but they become modified in that they grow to an abnormal size in order that their vacuoles can be very large and store the water. Such a tissue of cells is called **water-storage tissue** (Fig. 62).

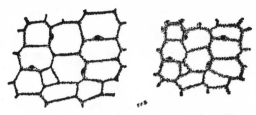

FIG. 62. WATER-STORAGE TISSUE.
On the left, tissue containing stored water ; on the right, the same tissue reduced in volume after some of its water has been removed for the use of the plant (× about 170).

DEAD TISSUES

Dead cell tissues are frequently found in the **skeletal** parts of the plant ; they are also used for conducting water from one part of the plant to another, because such transportation must be at a much quicker rate than the transportation of food-stuffs (see Chap. XVI).

Such cells are thickened and strengthened by the deposition of **lignin** (p. 70) on the cell-wall. Lignin is deposited on the primary cellulose wall by the activity of the cytoplasm. This results in the death and gradual disappearance of the cytoplasm, leaving a dead, thick-walled element. There are several forms of lignified elements, and their variety depends on the manner in which the lignin is deposited on the wall.

One type is that where the lignin is deposited almost completely over the whole surface of the cell-wall. A single cell thickened in this manner is called a **sclereide**. The tissue formed of such cells is called **sclerenchyma**. There are two main types of sclerenchymatous cells, namely, **stone cells** and **fibres**.

Stone cells scarcely ever form real tissues since they usually exist separately or in groups of just a few, and are generally embedded in a matrix of parenchymatous tissue. The stone cell is usually isodiametric in shape—that is, more or less spherical, though somewhat pushed out of shape by the close proximity of other cells. It is heavily thickened by the deposition of lignin on the wall, so that the **lumen**—that is, the cell cavity, is practically obliterated.

At certain spots on the surface of the cell wall, however, no lignin is deposited. As the deposit gets

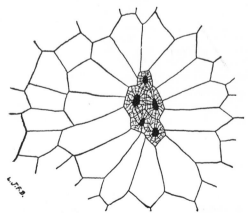

FIG. 63. A GROUP OF STONE CELLS EMBEDDED IN THE PARENCHYMATOUS CELLS OF THE PEAR FRUIT.

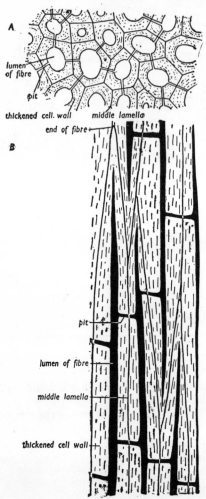

A

lumen of fibre

pit

thickened cell wall   middle lamella

end of fibre

B

pit

lumen of fibre

middle lamella

thickened cell wall

FIG. 64. *A*, TRANSVERSE SECTION OF WOOD FIBRES IN A STEM OF *Helianthus* ; *B*, WOOD FIBRES IN LONGITUDINAL SECTION.
(*From Murray: 'Biology'.*)

thicker and thicker, these unthickened spots form **pits** in the layer of lignin. Wherever a pit is formed in one cell, another is formed in the stone cell adjoining it. Therefore there is a communicating channel between the two stone cells ; but the channel is not complete, for the original cell-wall still exists across it and is called the **pit membrane**. That part of original cell-wall which persists and remains embedded in the middle of any thickened cell-wall is chiefly the middle lamella. However, despite this continuation of the middle lamella across the pits, such pits greatly facilitate the passage of water and dissolved substances from one stone cell to another. Some pits are branched. Stone cells are not very common in plants ; but they exist in great quantities embedded in the parenchymatous tissue of the flesh of the pear fruit (*Pyrus communis*). Their woody structure is responsible for the gritty nature of this fruit (Fig. 63).

Another type of sclerenchymatous cell is long and narrow and is called a **fibre**. It resembles a stone cell in all features except that, instead of being isodiametric, it is very elongated (Fig. 64). Sometimes the thickening in fibres is so great that the lumina are completely blocked up. Tissues made up of fibres are extremely hard and elastic and have great tensile strength. The mechanical advantages of such tissues, especially to the stem, can well be realised.

Another important function of lignified tissues besides that of giving mechanical strength through its fibres, is that of conducting water throughout the plant. Lignified elements are therefore found in most roots, stems and leaves.

The water-conducting elements are long pipe-like structures. These elements all begin their development as living parenchymatous cells. Then they elongate into cylindrical, pipe-like shapes. These pipes are prevented from collapsing by the deposition of lignin on their walls. But the lignin is never deposited all over the wall.

Certain water-conducting elements are called **tracheids**. The simplest type of tracheid is that in which the deposition of wood takes the form of parallel rings of lignin, deposited on the inside of the cell-wall at the expense of the protoplasm, which disappears. This gives an **annular** tracheid. Another type is formed by the deposition of wood in a spiral form, and such a tracheid is said to be **spiral**.

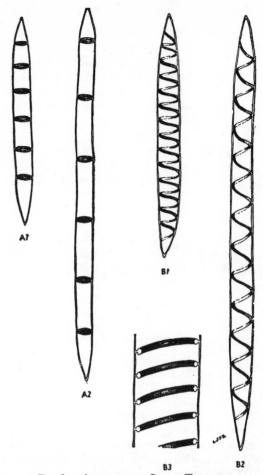

A1, annular ; B1, spiral ; A2 and B2, the corresponding tracheids elongated ; B3, part

FIG. 65. ANNULAR AND SPIRAL TRACHEIDS.
A1, annular ; B1, spiral ; A2 and B2, the corresponding tracheids elongated ; B3, part
of a longitudinal section through a spiral tracheid showing how the lignin is deposited
(× 300).

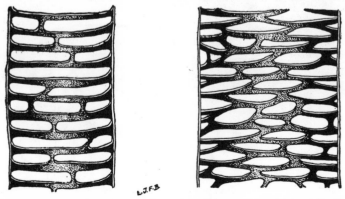

FIG. 66. PART OF A SCALARIFORM (LEFT) AND A RETICULATE (RIGHT) TRACHEID (× 200).

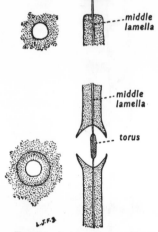

FIG. 67. ABOVE, SIMPLE PIT IN SURFACE VIEW (LEFT) AND LONGITUDINAL SECTION (RIGHT) ; BELOW, BORDERED PIT IN SURFACE VIEW AND LONGITUDINAL SECTION (× 375).

In the next form the wood is deposited all over the cell-wall except that unthickened parts are left in parallel lines, giving a ladder-like or scalariform effect. These are **scalariform** tracheids. The last form is similar except that the unthickened portions are left in a form of reticulum or network, thus forming the **reticulate** tracheid (Figs. 65 and 66).

Since water-conducting is an immediate necessity, even to a very young plant, such elements are formed in those parts of the root and stem which are still growing in length and, later on, those which are not growing in length. The tracheids themselves, once formed, cannot elongate by growth since they are dead ; but since cellulose is elastic and lignin is not, the unthickened portions of the tracheids can be stretched. This is possible, as may well be imagined, in the case of the annular and spiral tracheids, for in the former, the annular thickenings would merely be pulled farther apart, and in the latter, the spiral could be pulled out like a spring (Fig. 65) ; but not in the other two cases.

Therefore in the young growing parts the wood is formed of annular and spiral tracheids, whereas in the older non-growing parts, the scalariform and reticulate tracheids appear.

Tracheids, like sieve tubes, are superimposed on each other, forming chains of conducting elements. However, in this arrangement, there are cross walls where the tracheids meet, and these impede the progress of the passage of water. There are, however, certain other conducting elements which, in this respect, are even more efficient. They are very similar to the four types of tracheids, but differ in that the cross walls disappear during their formation. These are called **vessels**, and form good conducting channels for water. However, apart from this added facility, vessels are like tracheids. There are annular, spiral, scalariform and

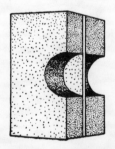
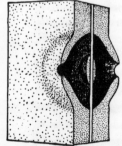

FIG. 68. DIAGRAMMATIC REPRESENTATION OF A SIMPLE PIT (LEFT) AND BORDERED PIT (RIGHT) IN LONGITUDINAL SECTION.

Note the middle lamella passing across the pit in each case ; in the bordered pit the torus is formed by the deposition of more lignin on part of the lamella passing across the pit.

reticulate vessels, and they differ from tracheids only in that, whereas tracheids are each formed from one cell, vessels are formed from several cells or **vessel segments**.

In some cases the vessel is thickened to a high degree, and the regions which remain unthickened are comparatively small in size though large in number. The circular unthickened areas are pits ; but they differ considerably in structure from the simple pits already considered. A glance at Fig. 67 will show that a simple pit in surface view takes the form of a circle, whereas that of the pit in question is bordered by another circle. Hence such a pit is called a **bordered pit.** When viewed in the solid the simple pit is of a cylindrical shape, with the pit membrane passing across it. In the case of the bordered pit, however, on each side of the middle lamella there is a dome formed by the thickening of lignin. The top of each dome is, of course, perforated, else there would be no real pit. Fig. 67 shows how this structure appears to be bordered in surface view ; for lignin, under the microscope, is rather transparent ; so the actual perforation of the dome appears as one circle and the base of the dome at the middle lamella appears as another circle, surrounding the first in a concentric position. A portion of the middle lamella, actually crossing the bordered pit may, particularly in the wood of conifers, become thickened by lignin to form a disk-shaped structure called the **torus.** Fig. 68 perhaps gives a better idea of the structure of simple and bordered pits. Vessels in the older parts of wood are usually covered with bordered pits, as shown in Fig. 69.

FIG. 69. PART OF WOOD VESSEL SHOWING THE REGION WHERE THE CROSS WALL HAS DISINTEGRATED.

Note the large number of bordered pits.

PRACTICAL WORK

SIMPLE AND COMPOUND MICROSCOPES

Whereas the compound microscope is not necessary for a great deal of the work involved in the study of plants, especially for beginners, the simple microscope (commonly called a 'lens') is practically indispensable. A folding type is the best, since it gives two or three lenses of different magnifying powers (Fig. 70). Where magnifications of not more than 20 diameters are required, a simple microscope will do.

On the other hand, when one comes to the study of the internal structure of the plant, and to examine cellular structure, a compound microscope is necessary, since much higher magnifications are required.

The structure of a compound microscope is seen in Fig. 71. The image of the object is obtained by the objective, and this image is then magnified by the eyepiece. Both these parts contain lenses. The coarse adjustment is used for raising and lowering the tube of the microscope, thus getting the image into focus. For more exact focusing, the fine adjustment is used. When, however, the object has been mounted on the stage, it should be got into the low-power field of the microscope by gently moving it about ; then it should invariably be focused by the coarse adjustment. The fine adjustment can be used finally, though the latter is scarcely ever necessary except during the course of high-power work. Two objectives are usually enough ; one to magnify about 50 diameters and another to magnify 300–400 diameters.

FIG. 70. A TRIPLE POCKET MAGNIFIER (SIMPLE MICROSCOPE).

Here are three lenses of different magnifying power.

B.I.B.

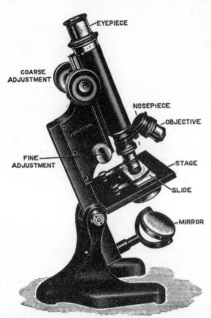

FIG. 71.

A TYPICAL COMPOUND MICROSCOPE.

In the course of all normal microscopic work, it is necessary that the object should be more or less transparent, for the light must pass through it. The light is directed up through the object on the stage by means of the mirror, the plane of which is adjustable. It is always best to work with daylight. A good light is necessary, but direct sunlight should be avoided. Artificial light is not desirable, unless it is possible to use ' daylight ' electric bulbs or a special microscope lamp fitted with a blue screen.

Many objects to be examined under the microscope are already sufficiently small and transparent to be placed on the stage complete ; but in other cases, such as a root or a stem, they are far too large. In such cases, microscope sections have to be prepared. The way to do this will be described as occasion arises.

In all cases, however, the object should be placed on a clean glass slide, and mounted in a mounting medium such as water or glycerol. There is, of course, the risk of some of the liquid medium getting on to the lens of the objective ; so this is prevented by covering the mounted object with a thin sheet of glass, which is either rectangular or square. Objects should seldom be mounted dry, and never without a cover glass (or slip).

It is useless to mount the object in water, then place the cover glass immediately on top, by means of the fingers, for air bubbles almost invariably appear in the medium. The cover glass should be held obliquely on the slide, near the object, then slowly lowered by means of a dissecting needle (See Fig. 72).

1. Cut a bulb of the onion and, from one of the cut fleshy leaves, take off a portion of the outer very thin, tissue-like layer. Mount this in water on a slide and examine the structure of the cells. Make a drawing of a collection of such cells. Then examine one cell in detail under the high power of the microscope. In order to do this, it is best to stain the cell contents. This can be done by placing one drop of iodine solution in contact with the edge of the cover slip, and drawing the solution under the cover slip by placing a little blotting paper on the opposite edge.

Note the cytoplasm enclosing several large vacuoles ; also the nucleus, which, normally colourless, is now coloured a pale yellow by the iodine solution. Note the thin cellulose cell-wall which is surrounding the whole cell.

2. Remove a very small portion of the pulpy tissue immediately beneath the skin of a tomato fruit (enough to cover a pin's head is sufficient). Mount this on a slide in water and then tease it out by means of dissecting needles. Cover it with a cover slip and examine under the high power of the microscope. Note the parenchymatous cells containing chromoplasts as well as cytoplasm, nucleus and vacuoles. Stain with iodine and examine the structure in detail.

3. Scrape off a little of the green encrustation from a piece of damp wood and examine the single cells of *Pleurococcus* (see Fig. 2). Stain with iodine solution. Note the cell surrounded by the cell-wall. Note also the cytoplasm, and nucleus, and the large irregularly-shaped chloroplast, packed within the cell. Look for examples showing a division into two or more cells. Such daughter cells would eventually round themselves off and separate from one another.

4. Examine a prepared slide showing cells in various stages of mitotic division. A stained preparation of a longitudinal section through the root-tip of the onion is a good example to obtain.

5. Plant an onion or shallot in moist peat in a warm place and so obtain a number of roots. Cut off 1 cm.-long pieces of root and fix them in 1 part glacial acetic acid to 3 parts absolute (or 95 per cent) alcohol. Leave for 24 hours. Then take a root and place in a drop of aceto-carmine on a slide. Cut off 3 mm. of the tip and discard the rest. Gently warm over a spirit lamp and then place a cover-slip over the drop of stain and apply gentle pressure. This will separate the cells and these, on examination, will show stages in mitosis.

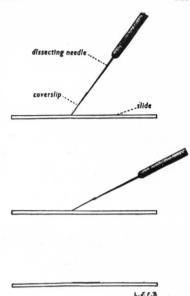

6. Take a very small portion of the gritty tissue of the pear fruit and tease it out on a slide. Examine the structure of the stone cells. Note the absence of cytoplasm, and look for simple unbranched and branched pits. Staining with iodine or aniline sulphate (or chloride) will show up lignified walls more clearly.

7. Obtain a prepared slide of a transverse section and a longitudinal section of a stem (a good example is that of *Cucurbita*). In these prepared sections examine, draw and describe as many conducting elements (sieve tubes and various types of tracheids and vessels) as is possible in transverse and longitudinal section.

FIG. 72. METHOD OF LOWERING A COVER SLIP OVER MATERIAL MOUNTED ON A MICROSCOPE SLIDE.

8. Prepare some slides of various woody elements by maceration. The material required is a small woody twig. Cut up a portion of this into very fine pieces, then place the pieces in a beaker or crucible. Cover with concentrated nitric acid, and add a few crystals of potassium nitrate. Warm this in a fume cupboard. When the reaction has finished, remove all the acid by repeated rinsing with water. Then mount some of the tissue and tease it out. Identify, draw and describe as many different examples of conducting and strengthening elements as possible.

9. Similarly macerate a small piece of wood of poplar. Look especially, in this case, for some examples of vessels with bordered pits. Examine the detailed structure of a bordered pit.

CHAPTER XI

ABSORPTION OF MATERIALS BY THE LIVING CELL

It may be said that the cell is a veritable chemical laboratory containing, besides a good supply of water, many chemical substances both organic and inorganic, namely, carbohydrates, proteins, fats, mineral salts, etc. This opens up the question : how do all the substances of which the cell is composed get into it? Some of the substances enter the cell from outside, being obtained from the environment or from other cells of the plant. This would apply to the mineral salts and perhaps to the sugar that is usually present. All substances entering the cell (or leaving it) do so in solution in water. This also applies to substances which normally exist in a gaseous form. Colloidal or insoluble constituents of the cell, such as proteins or starch grains, arise as the result of metabolic processes within the cell where they occur. These substances cannot, as such, leave the cell, but must first be converted to soluble products by enzymic action.

PRESSURE AND PERMEABILITY

Many substances will take up water, some very vigorously indeed. For example, concentrated sulphuric acid is used in desiccators because it will absorb a great deal of water, even when the latter is present as water vapour.

Whatever mechanism is used for forcing water into the absorbing substance, a pressure must be involved. This pressure, which, of course, varies considerably, is called **suction pressure.** Again, it is easy to conceive that if two absorbing substances are brought into contact with each other, then the one with the greater suction pressure will absorb water from the one with the less, until each contains the same relative amount of water : that is, until there is an *equilibrium* between them.

A system in which a solution of some crystalloidal substance is separated from water by a **membrane** (a thin film of material with pores of molecular dimensions) is of particular interest biologically. If the membrane is **permeable** to the solute (that is, if the molecules of the latter are able to pass through the pores of the membrane), then the main result will be that diffusion of solute will occur until an equilibrium is attained at which the concentration is the same on both sides of the membrane. Membranes of greatest interest in the present connexion are those which are **semi-permeable** to a given solution, that is, the water is able to pass through the membrane but the solute is not, probably because the pores of the membrane are too small. It is now found that water passes through the membrane from the pure-water side to the solution side. This movement of water is called **osmosis,** and it is due to what is termed the **osmotic pressure** of the solution.

This process is well illustrated by a simple experiment (Fig. 73). Through a bored rubber stopper fit tightly a piece of comparatively wide glass tubing. Then fix another stopper into one end of a cylinder of sausage paper (a fairly semi-permeable membrane towards sugar solutions), and ensure no leakage by binding

it with strong wire. Fill the sausage paper with
a strong sugar solution ; then fit the first rubber
stopper, to which the glass tube is attached, into
the other end of the sausage paper. Bind this
with wire also. Then immerse, as shown in the
diagram, in a vessel of water. Mark with a piece
of adhesive paper the position of the sugar solution
in the tube. After a few hours the solution will
have passed some distance up the tube, thus
showing that water has passed from the vessel
into the cane-sugar solution, a hydrostatic pressure being set up as a result. The passage of
water *into* the vessel by osmosis is referred to as
endosmosis. If the inner vessel be filled with
water, and the outer vessel with sugar solution,
then osmosis would cause water to pass out of the
vessel. This is known as **exosmosis.**

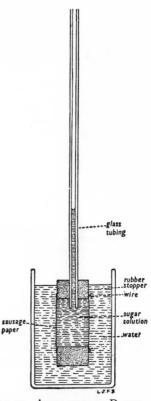

The process of osmosis was first observed by
the **Abbé Nollet** in 1748. If two cane-sugar
solutions be separated by a semi-permeable membrane, the water will pass from that of lower concentration to that of higher, and will continue to
do so, assuming that no hydrostatic pressures are
set up, until the concentrations are equal on both
sides of the membrane. That is, *osmosis aims at
establishing an equilibrium in concentration on both sides
of the membrane.*

FIG. 73. APPARATUS FOR DEMON-
STRATING OSMOSIS.

Just like other forms of pressure, osmotic pressure can be counteracted by another pressure
acting in the opposite direction. For example, if a sucrose solution be placed
inside a semi-permeable membrane and immersed in pure water, as shown in
Fig. 73, then a certain weight placed on the surface of the solution, the entry
of water by osmosis could be prevented and the smallest possible weight to
prevent any entry of water would be that which exerts a downward pressure
equal to the osmotic pressure tending to force water inwards. In the case of a
1 per cent solution of sucrose, this is equivalent to a weight of 10½ lb. per square
inch of the surface of the solution, as illustrated in Fig. 74.

That osmotic pressure varies directly as the concentration was proved by
Professor W. Pfeffer by his determination, shown in the following table, of the
osmotic pressures of sugar solutions of different concentrations :

OSMOTIC PRESSURE OF SUCROSE

Concentration Per Cent	Osmotic Pressure In Atmospheres
1 - - - -	0·686
2 - - - -	1·340
4 - - - -	2·750
6 - - - -	4·040

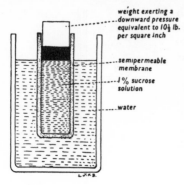

weight exerting a
downward pressure
equivalent to 10¼ lb.
per square inch

semipermeable
membrane

1% sucrose
solution

water

FIG. 74. DIAGRAMMATIC REPRESEN-
TATION OF THE PRESSURE DUE TO
OSMOSIS.

The weight is just sufficient to prevent water passing through the membrane into the inside vessel.

The skin of a pig's bladder also has semi-permeable properties. Experiments on osmosis, using this membrane, were first made by **Du-trochet** during 1827-1832 and **Vierodt** in 1848. If a pig's bladder were filled with a solution of sugar and then immersed in water, the water would pass into the bladder by osmosis. The result would be that the bladder would swell and, since it is elastic, it would become disten-ded. When elastic bodies are stretched, it is well known that, owing to the cohesion of the particles (that is, the attractive force tending to keep the particles of the elastic material to-gether), there is a tendency to resist further stretching, and the more such bodies are stretched the greater does this resistance be-come. Therefore, as the bladder becomes more and more distended, the more and more does the membrane of the bladder resist this stretching effect. It follows from this that the elastic tension, by resisting the stretching of the membrane, is also resisting the entry of water into the bladder, since it is the latter which is causing the stretching. In other words, while osmotic pressure is tending to force water *into* the bladder, the elastic tension of the membrane is tending to force water *out of* the bladder.

This can be well illustrated by the pumping up of a bicycle tyre. Here the water is represented by air. The pressure exerted to force air into the tyre by means of the pump represents the osmotic pressure. As the air passes into the tyre, it forces the tyre to swell, but the elastic properties of the rubber tyre make it resist this swelling. The more the tyre swells, the more does it resist swelling, and that is why the more a tyre is pumped up the harder it becomes to pump. The pump is forcing air in, but the swollen tyre is trying to force air out again. It does not succeed in this, of course, for there is no passage outwards for the air.

That such a pressure exists is clearly demonstrated if the tyre is punctured ; for as soon as the air is supplied with an outward passage, it is forced out. The puncture also shows that the more a tyre is swollen, the greater is its outward pressure, for the harder a tyre is pumped up the greater is the rush of air when there is a puncture. This is very similar to what is going on in the case of the pig's bladder. Osmotic pressure (OP) forces the water in, setting up the internal hydrostatic pressure which is termed the **turgor pressure**. The stretched wall of the bladder exerts a counter pressure which may be termed the **wall pressure** (WP). The capacity, at any stage, of the bladder to absorb further water (that is, the **suction pressure**, (SP) depends on the extent to which osmotic pressure exceeds this wall pressure, that is,

$$SP = OP - WP.$$

Now suppose that the bladder containing the solution of sugar is initially immersed not in pure water but in another sugar solution. If the concentration of sugar inside the bladder is the same as the concentration outside, there will be no passage of water, since osmosis tends towards an equilibrium of concentration, and that equilibrium is already established. If, however, the solution outside is stronger, then exosmosis will take place, with the result that the bladder will lose water. On the other hand, if the solution outside is weaker, then endosmosis will occur and the bladder will absorb water, but not to such an extent as it would if it were immersed in pure water.

In this case there is another force acting ; that is, the osmotic pressure of the sugar solution on the outside. If the osmotic pressure of the solution inside the bladder be represented by OP_1 and the osmotic pressure of the less-concentrated solution outside by OP_2, then OP_1 is greater than OP_2, and the suction pressure which forces water into the bladder at any moment can now be represented by the equation :

$$SP = OP_1 - (WP + OP_2).$$

In its relation to water, the living plant cell is very similar to the bladder containing the sugar solution. There is one great difference, however. The mechanism depends on elasticity and semi-permeability, and both characteristics are found in the one structure in the case of the bladder, that is, the bladder membrane. In the case of the cell, the cell-wall is elastic, but quite permeable to solutes. Therefore it supplies the necessary elasticity to give the wall pressure, but not the differential permeability to permit osmosis. The membrane necessary is provided by the cytoplasmic lining of the cell. This has semi-permeable properties, for while permitting free passage of water it restricts the leakage outwards of the cell-sap solutes, and also the entrance of some solutes (such as sucrose) from outside. The osmotic solution is provided by the cell-sap which is present in the vacuole.

Thus, if the cell be immersed in water it will, provided that it is not already replete with water, absorb the latter in a manner similar to that in which the bladder does, and will continue to do so until the wall pressure becomes equal to the osmotic pressure (that is, until $SP = 0$). If the cell is immersed in a solution of lower osmotic pressure than the cell-sap, then water will be absorbed until the wall pressure plus the external osmotic pressure together equal the osmotic pressure of the cell, as in the case of the bladder.

If we have two cells lying side by side, the movement of water between them will depend on the respective suction pressures. Water will pass from one cell to that having the higher suction pressure, that is, to the cell having the greater unsatisfied capacity for water absorption. It is important to realise that although the osmotic pressure of the cell places an ultimate limit on the water-absorbing power, *it is on the suction pressure that the absorbing power at any particular moment depends.*

TURGIDITY AND PLASMOLYSIS

When considering two solutions of the same substance but of different concentrations, in connexion with the phenomenon of osmosis, the solution of lower concentration is said to be **hypotonic** ; and that of the higher concentration, **hypertonic.** If both solutions are of equal concentration they are said to be **isotonic.**

We have seen the course of events when a cell is immersed in water or in a hypotonic solution. Now, what would happen if the cell were immersed in a solution hypertonic to the cell-sap? By the previous reasoning, it is quite clear

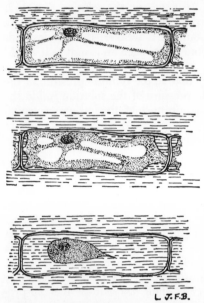

L. J. F. B.

FIG. 75. PLASMOLYSIS IN A PLANT CELL. The top figure represents the cell just as it has been immersed in a strong solution of cane-sugar. Note that the cell contents are pressed firmly against the surrounding cell-wall. The next figure represents the cell a few minutes afterwards. Plasmolysis has commenced, the cell contents are breaking away from the wall (the spaces between becoming filled with sugar solution), and the vacuoles are smaller. The lower figure represents complete plasmolysis. Note here the disappearance of the vacuoles.

that the water of the cell-sap would pass *out of* the cell (exosmosis), and instead of swelling the cell would shrink. This is exactly what does happen. At first the cell as a whole shrinks, but the shrinkage of the cell-walls soon ceases owing to their more rigid nature and to their permeability to solutes. The cell contents continue to shrink, and soon contract away from the cell-wall. The space between the shrunken cell contents and the wall is occupied by the solution of immersion. In this condition, the cell is said to be **plasmolysed,** and this process of shrinkage is called **plasmolysis.** Plasmolysis can easily be seen under the microscope if cells, such as those of the beetroot, which contain a coloured cell-sap, are used and immersed in a rather highly concentrated solution of salt or sugar (Fig. 75).

If such cells are not kept too long in the plasmolysed condition they will recover their normal condition if replaced in pure water. This process of recovery from plasmolysis is called **deplasmolysis.**

When the cells of a plant organ such as a leaf or a root contain as much water as they possibly can, they are each pressing firmly against their containing cell-walls, thus imparting rigidity to the organ as a whole. The cells in this condition are said to be **turgid.** If, owing to some change in circumstances, such as immersion in a hypertonic solution, the cells begin to lose water, they plasmolyse, lose their turgidity and firmness, and thus the organ wilts. If the cells are left in this condition too long, they will not recover in any circumstances, and they and perhaps the whole plant, finally die.

What has been said of cells with regard to their relations with water *applies only to living cells*. Dead cells will not absorb water by osmosis and will not plasmolyse, chiefly because the whole protoplasmic complex has broken down, and the semi-permeable membrane has disappeared.

ABSORPTION OF DISSOLVED SUBSTANCES BY THE CELL

It has been noted that the cytoplasm of the cell constitutes a membrane which is to a certain extent impermeable to the solutes of the cell-sap and to solutes such as sucrose which may be presented to the cell externally. The impermeability is not, however, complete. This could be deduced from the fact that some of the cell-sap solutes must have entered the cell from without in the first instance.

Experiments of various types also show that cells are permeable to many substances, though the degree of permeability varies very much towards different solutes. Substances towards which cells are very permeable include ethyl alcohol, and many weak acids and alkalies (such as ammonia). Compared with these, the penetration of sugars and many inorganic salts is very slow.

Water, as we have seen, enters plant cells by osmosis. Solutes enter by a process generally analogous to diffusion. Factors concerned in the process include the relative concentration of a particular solute outside and inside the cell, and the permeability of the cell towards the solute. But it is quite clear that ordinary diffusion alone cannot account fully for the observed facts about the absorption of dissolved substances by cells. Thus as previously noted (Chap. VI), absorption by a cell often continues *after* the internal concentration of a particular solute has become equal to the external concentration. The mechanism of this absorptive power of the cell is still under investigation. Adsorption may be involved, and it is also possible that in some way the energy of respiration (Chap. XVII) is applied to the absorptive process.

PRACTICAL WORK

1. Demonstrate the phenomenon of osmosis. This can be done in several ways ; but a very good method is that already described, and illustrated in Fig. 73.

When setting up this apparatus make sure that the sausage paper is tight around the two rubber stoppers. This can be done by means of copper wire or string. If the paper is not fixed tightly, the solution will leak out. Instead of the sausage-paper a parchment thimble fitted with a single stopper may be used.

Make a labelled drawing of the apparatus. Mark by means of a piece of adhesive paper the initial level of the solution in the tube, and in a similar way mark the level after successive twenty-four hour periods. Explain why the rate at which the solution rises in the tube gradually decreases with time.

2. The following experiment demonstrates an osmotic flow of water through plant tissue, as a result of setting up a gradient of suction pressure in the cells.

Prepare a ' thimble ' of plant tissue. That of a potato tuber forms excellent material. Choose a large tuber, and pare it roughly into the shape of a cone. Then with a sharp penknife hollow it out. Half fill this ' thimble ' with some 10 per cent solution of cane sugar, then, by means of a large needle, suspend it in a vessel of pure water (Fig. 76).

Note that after some time the solution inside the ' thimble ' shows signs of rising, and continues to do so slowly.

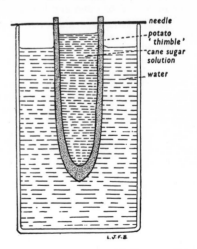

3. The pressure set up by osmosis is easily demonstrated in the case of a strong solution of cane sugar or common salt. Fill a large basin with some of the solution to be used, then immerse in the solution a glass cylinder or a wide glass tube sealed at one end. Across the mouth tie or wire on a piece of the membrane of a pig's bladder. All this must be done under the solution in the basin, to prevent any air bubbles from entering.

Take the cylinder containing the solution and thoroughly wash it under the tap, to remove all traces of solution from the outside. Then immerse it completely under water and leave for a few hours.

FIG. 76. DEMONSTRATION OF THE PASSAGE OF WATER THROUGH PLANT TISSUE BY MEANS OF SUCTION PRESSURE.

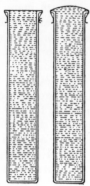

FIG. 77. METHOD OF DE-
MONSTRATING PRESSURE PROD-
UCED BY OSMOSIS.

The tubes contain sugar
solution and are then sealed
with pig's bladder membrane.
On the left, a tube before
immersion in water; on the
right, after immersion. Note,
in the latter, the distended
membrane.

By this time, the solution inside the cylinder will be exerting a pressure on the membrane, causing it to stretch outwards (see Fig. 77).

4. Cut some slices, about 3 mm. thick, from a potato tuber. By gently bending them between the fingers, notice their comparative firmness. This is due to the cells being full of water and therefore turgid. Then place the slices in a 3 per cent solution of common salt and leave for about half an hour. Note now that the slices have become very flabby. Replace in pure water and leave for half an hour, and then note that they have regained their firm texture.

5. Prepare about 30 disks from a potato tuber by piercing a large specimen with a wide cork borer (about $\frac{1}{2}$ inch in diameter), then cutting up the cylinder of tuber tissue thus obtained into circular disks about 3 mm. thick. Prepare a molecular solution of sucrose by dissolving 85·5 gm. of the sugar in 250 c.c. of distilled water. Put a little of this solution in a wide test tube, and from the rest prepare solutions of $M/2$, $M/3$, $M/4$ to $M/10$ sucrose by dilution and put each in a test tube. Dry the potato disks between sheets of blotting paper, ensuring that the tissue is not damaged, by pressing lightly. Weigh the disks in pairs. Then place one pair in each of the sucrose solutions. Leave for 3 to 4 hours, then remove the disks, dry between blotting paper and weigh again. In some cases there will be loss of weight; in others a gain. Explain this. That pair of disks which shows the least change in weight will reveal the sucrose solution nearest the suction pressure of potato tuber cells. Knowing that the osmotic pressure of a molecular solution of sucrose is equal to 26 atmospheres, state the approximate suction pressure of the potato tuber cells.

6. By means of a sharp razor, cut some thin sections of beetroot. Mount them in water and draw one or two of the cells, under the high power. Note especially the vacuoles containing a red cell-sap. Now, by means of a small piece of blotting paper placed at the edge of the cover slip, draw some 30 per cent sugar solution underneath it. This will cause the cells to plasmolyse. Examine and draw various stages of plasmolysis. Then irrigate the sections with water and look for deplasmolysis.

7. The same experiment can be performed with whole leaves of Canadian pondweed (*Elodea canadensis*). Mount a complete leaf in water on a slide and draw one or two of the cells, under the high power. Note especially the small green granules (chloroplasts) which contain the chlorophyll. Then irrigate the leaf with a strong solution of sugar or salt. The green chloroplasts help one to see plasmolysis taking place more easily.

Now place a small sprig of *Elodea* in boiling water for a few minutes. This kills the cells. Then mount one of the leaves and treat it as before. Note that plasmolysis does not occur in this case. This shows that only living cells possess semi-permeable membranes and are therefore able to absorb water by osmosis.

CHAPTER XII

THE ROOT

The root system of most plants serves to anchor the plant and to absorb water and dissolved mineral salts (Chap. IV). Many plants may, however, have their roots modified to perform other functions such as storage (Chap. V).

EXTENT OF THE ROOT SYSTEM

Both the successful anchorage of the plant and the absorption of adequate supplies of water and mineral salts depend on the development of an extensive root system. The extent of this system is often not realised since when a plant is pulled up the greater part of the root system is often left behind in the soil.

Tap roots, for example those of the parsnip and dandelion, may penetrate to a considerable depth and such roots are so firmly anchored that they are very difficult to pull up. Not only do such roots give a firm establishment of the plant in the soil but also they, along with the lateral roots they bear, tap a large volume of soil for supplies of water and mineral salts. Desert plants may have a tap root going down 40 ft. in order to reach the deeply placed water supply.

Fibrous root systems, for example, that of the wheat (Fig. 78) are also extensive and they penetrate to a depth of five feet or more. The total length of such a system is difficult to realise. If all the roots were placed end to end they would often be found to extend several miles. Such systems firmly anchor the plant and exploit a very large volume of soil. It has been estimated, for example, that a single cabbage plant may have roots ramifying through 200 cubic feet of soil.

The extent of the root system is much modified by external conditions such as the texture of the soil and the amount of water and mineral salts which it contains. For example, the roots of an oak tree may penetrate many feet into the soil if it is loose textured. If, on the other hand, the soil is shallow, with rocks near the surface, the roots spread horizontally

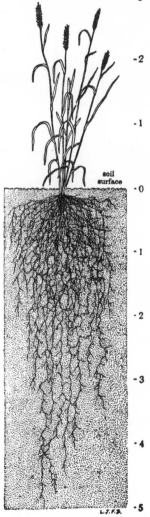

FIG. 78. A COMPLETE WHEAT PLANT, SHOWING THE LARGE FIBROUS ROOT SYSTEM.

The numbers indicate feet.

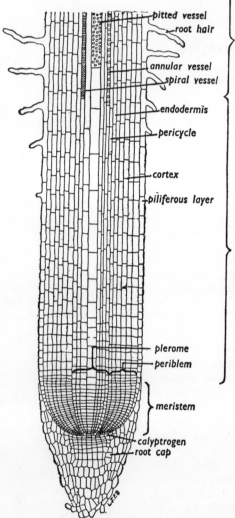

zone of differentiation

zone of elongation

- pitted vessel
- root hair
- annular vessel
- spiral vessel
- endodermis
- pericycle
- cortex
- piliferous layer
- plerome
- periblem
- meristem
- calyptrogen
- root cap

FIG. 79. LONGITUDINAL SECTION THROUGH THE
TIP OF A ROOT.

beneath the soil to an area much in excess of the area covered by the spreading branches.

GROWTH OF THE ROOT

The development of such root systems naturally involves the growth in length of each root. It also involves the production of many lateral roots (p. 30).

The growth and development of a root is best considered in relation to the structure of its terminal half inch or so. This is shown diagrammatically in Fig. 79 and the detailed structure of the actual growing point in Fig. 80.

The whole growth of the root depends upon the existence of a meristem, where all the cells are capable of repeated division. This meristem is protected by a hood-shaped **root cap** consisting of parenchymatous cells. The outer layers of the cap become mucilaginous and are rubbed off by contact with the soil particles as the root pushes its way downwards. As the cells are worn away, however, new cells are added on the inside by the activity of a layer of the meristem known as the **calyptrogen** (Fig. 80).

The apex of the root proper is dome shaped and its meristematic cells are arranged in definite zones which vary somewhat in different roots, that shown in Fig. 80 being characteristic of many Monocotyledons. The central mass of cells is known as the **plerome.** This is surrounded in turn by a zone of **periblem** and a single layer of cells, the **dermatogen.**

Just behind the actual meristem, where cell division takes place in all the zones, there is always a smooth region of the root, usually about 1 cm. long. This is the **region of elongation.** It is here that the individual cells undergo the changes known as maturation (p. 106) and become elongated. This elongation of the individual cells naturally results in the increase in length of the root itself, which is thus localised in a short region just behind the actual meristem.

Still further back from the apex, the now elongated cells differentiate to form the various tissues of the mature root. The cells of the outermost layer (dermatogen) remain thin walled and produce hair-like outgrowths, the root hairs. Most of the cells of the periblem undergo little further change and remain as parenchyma. The innermost layer, however, becomes specially modified as the **endodermis** (p. 133). The plerome, that is the central core of the root, becomes highly differentiated as the conducting tissue or **stele**. The outermost layer remains parenchymatous as the **pericycle**. Vertical series of cells at a number of places around the periphery of the stele form narrow annular and spiral vessels. These constitute the first-formed xylem or **protoxylem**. At a slightly higher level in the root, series of cells, progressively nearer the centre, form the pitted vessels of the later-formed or

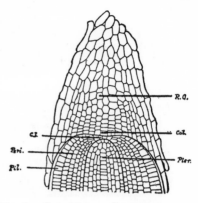

FIG. 80. LONGITUDINAL SECTION THROUGH ROOT TIP OF BARLEY. (Apex upwards.) *RC*, root cap ; *cal.*, calyptrogen ; *CI*, common initials for piliferous layer (*Pil.*) and periblem (*Peri.*) ; *Pler.*, central cylinder of plerome. (× 100). (*After Janczewski.*)

metaxylem. In this way, what is really a deeply fluted column of water-carrying xylem is formed. At about the same level as the protoxylem is formed, or even nearer the tip, cells situated between the protoxylem groups form sieve tubes and so columns of phloem are differentiated in the grooves of the xylem column. Cells of the plerome not differentiated either as xylem or phloem remain parenchymatous as **conjunctive parenchyma.**

It is important to realise that growth and differentiation are continuous processes. If one could trace the development of an individual cell it would be seen first as a cubical cell formed by division in the meristem ; it would then gradually mature and, according to its position in the root, differentiate as a cell of the piliferous layer, or as a cortical cell, or as a conducting element. It would be situated at a continuously increasing distance from the tip of the root since new cells are constantly formed by the meristem and these in turn go through the same sequence of development.

The general result of the entire process is the continuous downward growth of the root. The process is aided by the firm anchorage given by the root hairs. As the root grows, new root hairs are produced at about 1 cm. back from the tip while those on the old regions further back often die and disappear.

ABSORPTION OF WATER AND MINERAL SALTS

By far the most important function of the root is the absorption of certain raw food materials, namely, water and mineral salts from the soil.

A close examination of the soil in which the root grows shows it to be composed largely of fine particles of rock. The particles are of various shapes, and, therefore, though they are closely packed together, there are many spaces between them. In these spaces air is present. This is essential to the plant, for the root must have air as well as water (see Chap. XVII).

In a well-aerated soil, the water itself forms a film around each soil particle. To get at this water film, therefore, the root must be pressed very closely to the soil

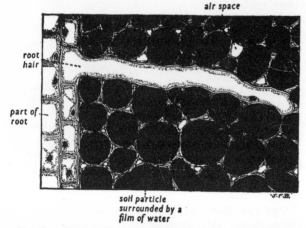

FIG. 81. PART OF A LONGITUDINAL SECTION THROUGH A
ROOT, PASSING THROUGH A ROOT HAIR, BENEATH THE SOIL.
Note how closely the hair adheres to the soil particles,

particles. The purely cylindrical portion of the root obviously cannot do this to any great extent owing to its shape. But the long thin root hairs which are found some distance behind the tip can do so by taking up a vermiform shape and the contact is made even more intimate by a gelatinisation of their walls (Fig. 81).

Since, therefore, the root hairs are in direct contact with the water films in the soil, it is chiefly through these hairs that the soil water, together with its dissolved mineral salts, passes into the root. Each root hair is composed of one long parenchymatous cell, with a very thin cellulose cell-wall, a thin lining layer of cytoplasm, a nucleus, and a very large vacuole. Thus, here is a cell surrounded by water and, therefore, by suction pressure, as explained in Chap. XI, the water passes into it. As there may be a thousand hairs on a 1 m.m. length of a root the absorbing surface is obviously very extensive.

INTERNAL STRUCTURE OF THE ROOT

The structures which are responsible for conveying the water together with its dissolved substances from the root up to the shoot are the pipe-like elements called vessels. These are found in the middle of the root, as shown in Fig. 82. Here the vessels are grouped together to form a tissue called **xylem.** The xylem, as is seen in Fig. 83, is embedded in a tissue of parenchymatous cells.

The root is a living part of the plant and therefore must have food, so it is clear that there must also be a downward passage of water containing dissolved foodstuffs to supply the root. The elements used in the conveyance of these organic food-stuffs are sieve tubes. The xylem is arranged in a more or less star-shaped pattern, when seen in transverse section. Small protoxylem vessels occur at the points of the star and the rest consists of the larger metaxylem vessels (p. 134). The protoxylem groups vary in number. Usually the number is small in dicotyledons. That shown in Fig. 82 has five protoxylem groups and is therefore said to be **pentarch.** That in Fig. 83 is **tetrarch** and other examples may be **triarch** or **diarch.** Monocotyledonous roots usually have a large number, and are therefore said to be **polyarch.**

Between the points of the xylem star may be seen groups of sieve tubes. Each group forms a tissue called **phloem,** and the phloem, like the xylem, is embedded in a tissue of parenchyma. Note also that the phloem groups never abut directly on the xylem. There is always a certain amount of parenchyma between and, in the case of Dicotyledonous roots, some of the cells are meristematic, constituting the **cambium** which is concerned with secondary thickening (see p. 149). This mass of tissues for conveying water and dissolved salts up to the shoot (xylem), and dissolved foods down to the root (phloem), together with its matrix of parenchyma and the layers of parenchymatous cells surrounding the whole (known as the **pericycle**), is called the **stele**.

Surrounding the whole of this collection of tissues is a cylinder of parenchyma called the **cortex**. The innermost layer of the cortex is specially modified to form the **endodermis**, which controls the passage of substances in solution into and out of the stele (p. 187). In a transverse section (Fig. 83), the endodermis is recognised by the presence of a slight suberised thickening of the middle region of each radial wall. This appearance is due to the fact that there is a suberised ribbon, the **Casparian strip**, running right round each cell on its radial and transverse walls. In older roots, particularly those of Monocotyledons, all the walls of the endodermal cells be-

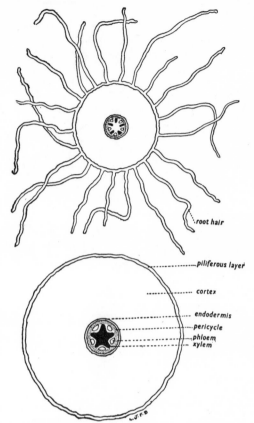

FIG. 82. DIAGRAMMATIC SKETCHES OF THE ROOT IN TRANSVERSE SECTION (LOW POWER).
Above, near the tip; below, in an older, but unthickened, part.

come strongly thickened with ligno-cellulose. Certain cells, opposite to the proto-xylem groups, remain thin walled and are called **passage cells**. The outermost layer of the cortex may also be modified, forming an **exodermis** of rather larger cells with suberised or lignified walls.

The outermost layer of the root is the **piliferous layer** characterised by the development of root hairs. Some roots, particularly those of plants growing in very wet soil, may not develop root hairs.

ORIGIN OF LATERAL ROOTS

Lateral (or branch) roots originate in a way quite different from the origin of lateral appendages of the stem (see p. 164). Most of the branches of the stem arise near a stem apex. Those of the root never arise near its apex, but in older parts just behind the region of differentiation. Moreover, branch roots arise from *internal* tissues, but stem appendages arise from the outer tissues. Therefore,

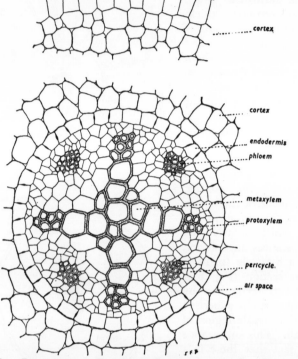

FIG. 83.   PORTIONS OF A TRANSVERSE SECTION OF A ROOT (HIGH POWER).
Above, the outer tissues ; below, the stele.

branch roots are said to be of **endogenous** origin, and branch stems and leaves of
exogenous origin (p. 162).

Lateral roots originate in the pericycle.   A small group of pericycle cells
becomes meristematic and the cells divide by walls parallel to the surface.   This
results in the formation of a hemispherical mass of actively dividing cells which
pushes into the cortex.   This mass soon begins to show the zonation of tissues
characteristic of a root apex.   Cell division continues and the young root penetrates
the cortex, sometimes aided by enzymes (secreted by the endodermis which still
surrounds it) which cause the breakdown of cells in the way of the root's outward
passage.   Eventually the lateral root bursts right through the cortex and piliferous
layer and emerges into the soil.   The split caused by this process can often be
observed with the naked eye and, in some examples, the ruptured outer tissues
form a little collar at the base of the lateral root (Fig. 84).

During the outward growth of the lateral root and throughout its further
development, the various tissues are differentiated as in the main root.   In
particular, it should be noted that the conducting tissues of the lateral root come
to be in functional continuity with those of the main root.

The lateral roots usually arise from patches of pericycle lying opposite to the
protoxylem groups so that, for example, a tetrarch root will bear four rows of

lateral roots. In other cases, however, the active pericycle cells lie between the protoxylem and the phloem so that a tetrarch root of this type would show eight rows of lateral roots.

ROOTS AND FUNGI

The roots of a considerable number of plants are regularly associated with the filaments or hyphae of soil fungi. Thus the roots of certain moorland plants, orchids, and many forest trees have fungal threads growing inside the cortical cells, or closely adpressed to the root surface. The term **mycorrhiza** is applied to such an association of fungus and root. The view has frequently been taken that this association is mutually beneficial to the two organisms concerned, and that it can therefore be classed as an example of **symbiosis.** As will be indicated later, this interpretation has not been universally accepted.

Mycorrhiza usually occurs on plants growing in soils which are rich in humus (that is, decaying organic matter) but poor in mineral salts, for example, the soils of woods and heaths. The majority of such plants possess chlorophyll and so are able to manufacture their own carbohydrates, but, since the soil is deficient in nitrates, they need an alternative source for their nitrogen supplies. This appears to be provided by the fungus which, unlike the roots of the higher plant, is able to break down the protein constituents of the humus into simpler, soluble organic compounds. The fungus absorbs these compounds, and some of them may be transferred to the root with which they are associated. It is possible that in return the fungus receives supplies of carbohydrates from the green plant. There are, however, some plants with mycorrhizal roots which possess little or no chlorophyll, such as the bird's-nest orchis (*Neottia nidus-avis*) mentioned below. Here the higher plant must derive not only its nitrogen but also its carbohydrates from the humus, presumably through the intermediary action of the fungus. Since *Neottia* manufactures no carbohydrates it is difficult to see what nutritional advantage the fungus gains, and it may be that the root merely provides the fungus with a suitable habitat.

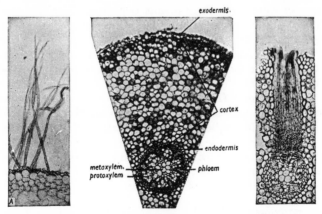

FIG. 84. PARTS OF TRANSVERSE SECTIONS OF YOUNG ROOTS OF *Ranunculus* (buttercup). *A*, showing root hairs ; *B*, just behind the region of root hairs ; *C*, the formation of a lateral root.

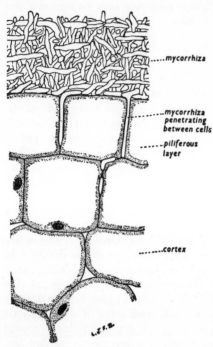

mycorrhiza

mycorrhiza
penetrating
between cells

piliferous
layer

cortex

FIG. 85. TRANSVERSE SECTION OF PART OF
A ROOT OF THE SCOTS PINE (*Pinus sylvestris*),
SHOWING THE PILIFEROUS LAYER, SURROUNDED
BY ECTOTROPHIC MYCORRHIZA. NOTE SOME
OF THE FUNGAL HYPHÆ PENETRATING *between*
THE CELLS.

Two main forms of mycorrhiza are recognized, namely (*a*) the ectotrophic type in which the fungus grows only on the surface of the root and *between* its cortical cells (Figs. 85 and 86), and (*b*) the endotrophic type, in which the fungus *penetrates* into and occupies the cells of the root. Intermediate conditions also occur.

(a) Ectotrophic Mycorrhiza

Pines (*Pinus* species), larch (*Larix decidua*) and spruce firs (*Picea* species) form symbiotic relationships with many toadstool fungi. These fungi, such as species of *Boletus, Scleroderma, Amanita* and *Russula*, consist of many threads which grow between the soil particles, like those of the mushroom (p. 340) ; the familiar aerial parts being the spore-bearing structures. These threads form an ectotrophic mycorrhiza around and in the young roots of such coniferous trees (Fig. 85). Experiments have shown that the presence of the fungus is beneficial to the trees. For example, conifer seedlings grown in soil containing the mycorrhizal fungus, resulting in the fungus-root association being set up, grow more strongly than similar plants grown in soil lacking the fungus. The reason for this benefit is not known for certain. One possibility is that, through the action of the fungus, the tree gains food materials derived from humus. Other theories have suggested that the absorption of water and mineral salts is promoted by the presence of the fungus. Beeches (*Fagus* species), oaks (*Quercus* species) and many other trees also possess ectotrophic mycorrhiza. The mycorrhizal roots of such trees are very different in external appearance from the normal roots. For example, in the pine, the roots remain short and fork repeatedly to give a coral-like mass of swollen rootlets (Fig. 86).

The yellow bird's nest (*Monotropa hypopithys*)—not to be confused with the bird's-nest orchis—also possesses an ectotrophic mycorrhiza. The aerial part of this plant, which grows in fir and beech woods, consists of a flowering shoot on which however, there are no green leaves. Below ground there is a much branched system of roots, each covered with a felt of fungal threads. Since the plant is unable to carry out photosynthesis, all its nourishment must be obtained from the humus, presumably with the aid of the fungus.

(b) Endotrophic mycorrhiza

Endotrophic mycorrhiza often occurs in the roots of heather (*Calluna vulgaris*), a plant which usually grows on poor, peaty soils of an acid nature. The new roots produced in the spring are infected with the threads of a fungus which grow not only on the root surface but also penetrate and grow inside the cells of the narrow cortex. Later, the cells of the cortex appear to digest the fungal threads and may thus obtain organic foodstuffs, some of which may have been absorbed from the humus by the fungus. There may be more than one type of fungus concerned but a species of *Phoma* is usually regarded as the important one. It has been stated that this fungus can build up nitrogenous compounds from atmospheric nitrogen (that is, carry out the process of nitrogen fixation) and that *Calluna* seedlings benefit from this additional supply of nitrogenous foodstuff. This experimental evidence is, however, not very sound.

Many orchids possess endotrophic mycorrhiza, the fungus concerned being a species of *Rhizoctonia*. The fungal threads form close coils inside the cortical cells of the roots and tubers and some threads pass out into the surrounding soil. In some of the cells (host cells) the

Dr. G. Bond

FIG. 86. PARTS OF ROOT SYSTEM OF SCOTS PINE, SHOWING REPEATEDLY FORKING MYCORRHIZAL ROOTS. (× 2.)

fungus appears to remain in a healthy condition; in other cells the fungus is digested by enzymes. Presumably the products of digestion are utilised by the orchid. The mycorrhizal fungus is important in the germination of the minute orchid seeds. The latter, if sown under sterile conditions and without a supply of organic food fail to germinate. If the fungus is introduced, it enters the cells of the seedling and germination proceeds. Germination can, however, be obtained in the absence of the fungus by sowing the seeds on a medium containing organic materials, particularly sugars. This method, known as asymbiotic germination, is now largely used in horticultural practice. Under natural conditions it seems that the fungus makes

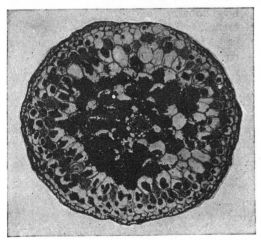

Dr. G. Bond

FIG. 87. TRANSVERSE SECTION OF ROOT OF *Neottia nidus-avis.*

The central small stele is surrounded by inner cortical cells containing dark-stained starch grains. The outer cortical cells contain fungal hyphae, those cells filled with greyish material being host cells, those containing dark-stained masses of disorganised hyphae being digestive cells. (× 30.)

FIG. 88. A CLOVER PLANT SHOWING
ROOT NODULES.
*(By courtesy of the U.S. Bureau of Plant
Industry.)*

available organic foodstuffs to the minute embryo of the orchid so that germination can proceed.

The bird's-nest orchis (*Neottia nidus-avis*) consists of a short rhizome bearing numerous fleshy roots and a brownish flowering shoot (p. 350). Brown scale leaves are developed but green leaves are entirely absent, so that the plant can be credited with little or no power of photosynthesis. The cortical cells of the fleshy roots contain coils of fungal threads (Fig. 87). As in the green orchids, some of the cells are host cells while in other cells the fungus is digested, presumably thus providing the foodstuffs needed by *Neottia*.

Very little is known about the way in which the exchange of foodstuffs between fungus and root takes place in either of the two types of mycorrhiza. It has, indeed, been suggested that the fungus is merely a parasite which the higher plant keeps in check. Any observed benefit to the host plant is, on this view, thought to be due to the fact that the mycorrhizal fungus, together with other fungi growing in the neighbourhood of the root, break down the organic matter in the humus into soluble forms which the root itself can then absorb directly. The balance of evidence, however, suggests that mycorrhiza is in fact a symbiotic relationship.

ROOT NODULES

For centuries it has been known that one way to enrich the soil is to grow on it plants which belong to the family Leguminosae (p. 455). Some of the best-known plants belonging to the family Leguminosae are peas, beans, clover, etc. Such plants actually enrich the soil with nitrogen compounds, rather than make them poorer.

For this reason, clover is often cultivated on farm land. It is a crop often used in crop rotation. Sometimes, especially if the clover turns out to be a poor crop, it is merely cultivated until the end of its growth and then just ploughed into the soil, where, during the winter, it decays, thus giving a splendid natural manure to the soil. More often, however, the crop of clover

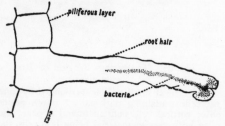

FIG. 89. *Rhizobium* PENETRATING A ROOT
HAIR OF A BROAD BEAN.

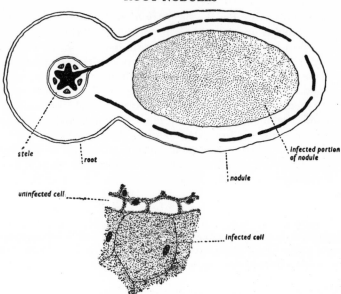

FIG. 90. ABOVE, A TRANSVERSE SECTION OF A BROAD BEAN ROOT, PASSING THROUGH A
ROOT NODULE. BELOW, A FEW CELLS FROM THE NODULE, ENLARGED.

is harvested ; but much addition of nitrogen compounds to the soil has by then
taken place.

This beneficial effect of leguminous crops on the soil was recognised by the
Greeks and Romans, but it was not explained until towards the end of the nineteenth century. In experiments then carried out, plants, such as wheat, were
grown in soil deficient in nitrates. They were found to become unhealthy,
although the atmosphere around them contained, of course, a large proportion of
nitrogen. On the other hand, clover and other leguminous plants were found to
flourish quite well on such soils. After much investigation, it was concluded that
leguminous plants were able to utilise atmospheric nitrogen in their growth,
provided that their roots were closely associated with a certain soil bacterium
now known as *Rhizobium* (formerly called *Bacillus radicicola*).

The *Rhizobium* organism inhabits the cells of the special structures called
nodules which are usually to be found on the roots of clover, broad bean, or
other leguminous plants (Fig. 88). The organism also exists in large numbers in
most soils, the exceedingly minute spherical or rod-shaped cells being able to move
slowly through the soil by means of flagella. They appear to be attracted to the
root hairs of leguminous plants and proceed to penetrate through the tips of the
hairs, forming inside an infection thread which grows down the root hair (Fig.
89) and enters the cortical cells of the root. The presence of the bacteria stimulates the cortical cells to commence meristematic division, with the result that a
swelling or nodule develops, the central region of which consists of enlarged cells
containing large numbers of the bacteria, while in the outer part of the nodule
vascular strands linking up with the stele of the root are present (Fig. 90). The
central infected region in a fresh nodule is usually red in colour, and this has been
shown to be due to the presence of haemoglobin, a pigment which, apart from this
instance, is confined to the animal kingdom.

The *Rhizobium* organism has the property of being able to utilise atmospheric nitrogen and to build it up into organic form, though this activity is only shown when the organism is associated with the leguminous plant in the nodule cells. This utilisation of atmospheric nitrogen is known as **nitrogen fixation.** A large proportion of the products of fixation passes from the nodule into the rest of the plant, with corresponding benefit to nutrition.   Indeed, as is suggested by the experiments already mentioned, the leguminous plant will grow vigorously in a rooting medium free of combined nitrogen, the plant deriving, under these conditions, all its nitrogenous materials from the nodules.   Like most other bacteria, *Rhizobium* requires to be supplied with organic carbon compounds such as sugars before it can grow.   These are obtained from the leguminous plant.   Thus a clear example of symbiosis is provided, both organisms deriving benefit from the association.

This fixation of nitrogen explains the beneficial effect of leguminous crops on the soil.   Besides the members of the family Leguminosae a few other plants possess nitrogen-fixing root nodules inhabited by a soil micro-organism.   Two examples are alder (*Alnus glutinosa*) and bog myrtle or sweet gale (*Myrica gale*, Fig. 173, p. 233).   Other instances of nitrogen fixation are considered below.

THE NITROGEN CYCLE

Unlike root nodule plants, the great majority of plants must be able to procure, from the soil, proper amounts of combined nitrogen in the form of nitrates or ammonium salts (Chap. VI).   A supply of these is maintained in the soil as a result of the **nitrogen cycle.**

We have seen in Chapter VI how the remains of dead plants become incorporated into the humus of the soil.   Through the action of soil bacteria and fungi the organic nitrogenous compounds at first present in the humus are broken down, much of the nitrogen being brought into the form of ammonia.   This tends to form ammonium carbonate by reaction with the carbon dioxide present in the soil.   Some of the ammonium salts may be absorbed by higher plants, but a greater part is promptly oxidised to the form of nitrate as a result of the activities of two soil bacteria known as the **nitrifying bacteria.**   These carry on rather unusual metabolic reactions, from which chemical energy is gained by the organisms.   The first, *Nitrosomonas*, oxidises ammonium salts to nitrites, while the second organism, *Nitrobacter*, oxidises the nitrites to nitrates.

Thus as the result of the activities of various soil micro-organisms the complex organic nitrogenous materials in humus are again simplified to a form suitable for absorption by the roots of higher plants. There are, however, processes at work in the soil which lead to a loss of available nitrogen and thus to a decrease in

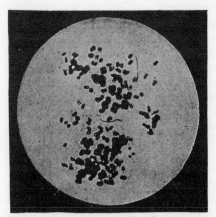

Fig. 91.   Bacteria (*Azotobacter*) found in all Fertile Soils.
(*Photomicrograph, Rothamsted Experimental Station.*)

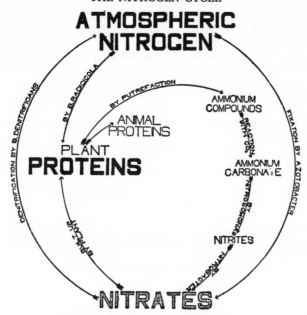

FIG. 92. THE NITROGEN CYCLE.

fertility. Thus there is loss by drainage, and through the action of certain bacteria which reduce nitrates with the evolution of gaseous nitrogen. These losses are compensated first as a result of the symbiotic nitrogen fixation considered in the preceding section, and secondly through the activities of certain free-living, nitrogen-fixing bacteria present in most soils, the chief of these being *Azotobacter* (Fig. 91). Under agricultural conditions, in which a large part of the plant growth is removed annually as a crop, it is often necessary to add to the soil additional nitrogenous fertilisers, such as ammonium sulphate.

The operation of the nitrogen cycle is shown in diagrammatic form in Fig. 92. From the above consideration we learn that although some bacteria are a menace to plant and animal life and responsible for serious diseases, others have a very beneficial role. The nitrogen-fixing bacteria, both symbiotic and free-living, are particularly beneficial, and in fact their activities are essential to *all* life on the earth, since they play such an important part in maintaining the fertility of the soil.

PRACTICAL WORK

1. Examine and draw an entire young root of the mustard (*Brassica nigra*), which has been cultivated on damp blotting paper (p. 41, No. 3). By means of a lens, or a low power of the microscope, examine and draw the younger regions enlarged, showing the cylindrical root, the root cap and the root hairs.

2. Examine the roots of garden cress (*Lepidium sativum*) obtained in the following way. Thoroughly soak a flower pot in water ; scatter the cress seeds thinly over the inner surface, where they will stick by means of their mucilaginous seed coat ; invert the pot over a dish containing enough water to cover the rim of the pot ; place in a warm place. The seeds will quickly germinate and provide suitable roots for examination. Cut off the terminal half-inch of one of these and mount in lactophenol-erythrosin. Note : the arrangement of tissues at the apex ; the region of elongation ; the development and structure of root hairs. After making these observations, crush the specimen and examine the older region for annular and spiral vessels of the protoxylem.

3. Allow some of the cress seedlings to grow on until the radicle shows lateral roots. Cut off the radicle just above the smallest visible laterals and mount this terminal length in lactophenol-erythrosin, coiling it round if necessary. Note and draw stages in the development of the lateral roots. The fine roots of chickweed or of a grass may be used for the same purpose.

4. Prepare a transverse section of a young bean or buttercup root for examination under the microscope.

The root should be placed between two pieces of pith or carrot tissue, in order that the whole may be held easily between the thumb and forefinger. Then dip the material in water in order to moisten it (never cut dry material). For cutting microscope sections, a very sharp razor is required. Dip the blade of the razor in water too. Then cut the sections holding the material vertically and drawing the razor quickly across it. Aim at cutting the sections as thin as possible, for it is impossible to see the structure under the microscope if they are thick. Cut a number of sections ; then wash them into a small dish of water. By means of a camel-hair brush, select the thinnest and mount it in water on a slide and cover with a cover-slip. Irrigate in the usual way with aniline sulphate, which will colour the xylem elements yellow.

Look for : (1) the piliferous layer, with its root hairs ; (2) the cortex, composed of (a) cortex proper and (b) the endodermis ; (3) the stele, composed of (a) xylem (protoxylem and metaxylem), (b) phloem and (c) pericycle, all of which are embedded in parenchyma.

Make a low-power drawing of the section, showing the relative positions of the various tissues. When making a low-power sketch like this, *never indicate cells*. Just indicate the various groups of tissue as in Fig. 82.

Then examine the tissues under the high power and make detailed drawings, showing the cellular structure. It is a waste of time to draw the whole section in detail. Select just enough to show a certain amount of all the tissues (see Fig. 83), making several separate drawings if necessary.

Cut transverse sections of a root of *Iris* (or of *Polygonatum*). Note that the general structure of this Monocotyledon root is similar to that of *Ranunculus* but that there are more protoxylem groups and that the metaxylem does not extend to the centre of the stele.

5. Make a large selection of different storage roots and, by the various methods described in Chap. VI, test them for food reserves.

6. Examine the mycorrhizal roots of birch or pine. These may be found growing in the surface layers of leaf mould below the trees. Note the characteristic branching and cut sections to see the ectotrophic mycorrhiza.

7. Collect young, thin roots of heather in the spring. Mount a length of one of them in lactophenol-erythrosin and look for fungal threads in the narrow cortex. Examine also a section of *Neottia* root, if available, as another example of endotrophic mycorrhiza.

8. Examine the root system of various leguminous plants, such as broad bean and clover, and look for root nodules. Make a drawing of them.

Prepare a transverse section of such a root to pass through a nodule. Note the red colour usually shown by the central part of the nodule and study the large infected cells present in this region. (Specially prepared slides are usually desirable for this, since the bacteria require special staining to be clearly visible.)

CHAPTER XIII

THE STEM

FUNCTIONS OF THE STEM

The stem is the axis upon which the other parts of the shoot, that is, the leaves and flowers, are borne. As described in Chap. V, the stem may be variously modified for special purposes, such as storage of food or water ; but the erect stems characteristic of the majority of flowering plants serve to display the leaves to the light needed for photosynthesis. The flowers are borne in suitable positions for the process of pollination and eventual fruit or seed dispersal. Mechanical tissues are developed in such stems in relation to their supporting function and give them rigidity and the ability to withstand bending strains.

The stem also carries water and mineral salts from the root system, not only into its own tissues, but also into the leaves and flowers. Substances manufactured

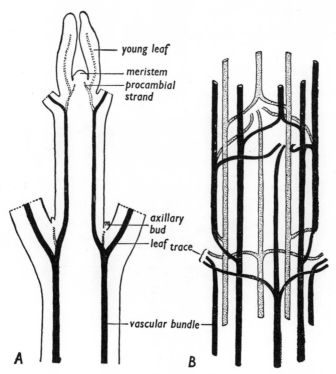

FIG. 93. *A*, LONGITUDINAL SECTION OF THE SHOOT APEX ; *B*, THE VASCULAR SYSTEM OF A DICOTYLEDONOUS STEM.

Both diagrams are based on the dog's mercury (*Mercurialis perennis*).

in the leaves are transported in the stem to the flowers and fruits and also down into the root system. Conducting tissues are therefore a conspicuous feature of stem structure.

STRUCTURE OF THE STEM

The development of a stem follows essentially the same course as that of a root, but with differences related to the fact that it is usually growing in air rather than in the soil. Growth depends on the activity of a meristem which gives rise to new cells which then elongate and finally differentiate to form the various tissues of the adult stem. The meristem is a dome-shaped mass of actively dividing cells. It is protected by young leaves folded over it and not by a cap as in the case of a root (Fig. 93, *A*). It is usually possible to distinguish a central core of cells from which the conducting tissues are ultimately formed, but the layering of the meristem is not so marked as in root apices.

The region of elongation behind the meristem is usually much longer than in the root and often extends over several inches. Plants showing the **rosette** habit,

Fig. 94. GREAT PLANTAIN (*Plantago major*).
Note rosette habit.

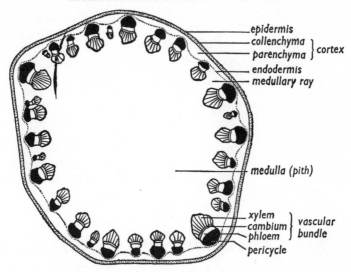

FIG. 95. DIAGRAM OF A TRANSVERSE SECTION OF A STEM OF SUNFLOWER
(*Helianthus annuus*).

for example, the plantain (*Plantago major*, Fig. 94) and dandelion (*Taraxacum officinale*), have, however, very short stems, showing little growth in length. Their large leaves radiate out from the stem and lie firmly pressed against the soil surface. On the other hand, **twining plants** have very long stems and the growing region extends over a considerable length behind the growing point.

Differentiation of the various tissues of the mature stem begins even in the growing region. This process will, however, be best considered after the adult tissues have been examined.

(a) Dicotyledon Stems

The tissues concerned with the translocation of water and foodstuffs consist of the same types of element as those already seen in the root, namely xylem and phloem. Their arrangement in the stem is, however, entirely different from that in the root. In place of the central column of vascular tissue seen in the root there is a number of separate conducting strands, or **vascular bundles,** each containing both xylem and phloem. If these vascular bundles, which are relatively firm, are dissected out from the surrounding softer tissues, it will be found that they form a cylindrical network with strands, known as **leaf traces,** going out to the leaves (Fig. 93, *B*). A transverse section of an internode (Fig. 95) will therefore show a circle of vascular bundles. Each bundle (Fig. 96) consists of a group of xylem vessels to the inside (with the protoxylem next to the pith) and a group of phloem to the outside. Between the xylem and the phloem there are several layers of thin-walled cells arranged in radial rows. One of these layers is the cambium, whose meristematic activity later leads to secondary thickening. Since it is difficult to say just which layer is the cambium, it is usual to call the whole zone the cambium, or more accurately, the cambial region.

The ring of vascular bundles surrounds the pith or medulla and the rays of tissue between the bundles are called medullary rays. This central set of tissues is the

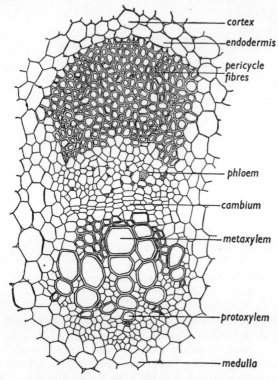

FIG. 96. TRANSVERSE SECTION OF A VASCULAR
BUNDLE OF SUNFLOWER (*Helianthus annuus*).
(*Dr. S. Williams.*)

stele of the stem. It is limited on the outside by a pericycle which may simply consist of parenchyma, but more frequently it develops as pericycle fibres. These fibres may form a continuous cylinder round the stele but they often form separate groups on the outside of each vascular bundle, the bundles then being called **fibro-vascular bundles.**

The tissues surrounding the stele form a usually narrow cortex. The innermost layer of this may be similar to the endodermis of the root, but usually the Casparian strip is absent and the cells are recognised only by the presence of abundant starch grains. The endodermis in this latter case is spoken of as the **starch sheath** which may be concerned with the perception of the stimulus of gravity (p. 230). The other layers of the cortex are largely parenchymatous and frequently contain chloroplasts. In most stems, however, rigidity is given to the stem by the development of collenchyma or sclerenchyma. These tissues may form an almost continuous ring at the outside of the cortex. This is so in the sunflower (*Helianthus annuus*) (Fig. 97), whereas in other cases, the mechanically strong tissue is present in discontinuous groups and in such examples as ridged and angled stems like those of the white deadnettle (*Lamium album*) such groups occupy the angles of the stem.

The outermost layer of the stem is the epidermis which serves to protect the inner tissues from desiccation and mechanical damage. Stomata are present and hairs of various types may be developed (see p. 167).

The origin of this system of tissues from the apical meristem may now be briefly traced. The epidermis arises from the outermost layer of the meristem without any marked changes apart from an increase in size and a deposition of cutin on the outer walls. A few layers of meristem to the inside of this differentiate into the parenchyma, collenchyma, and sclerenchyma of the cortex.

The vascular bundles develop in the following way. A short distance below the apex, a number of little groups of cells around the periphery of the central core begin to divide longitudinally and in this way strands of narrow, elongated cells are formed ; these are called **procambial strands.** Each strand is continuous with a similar one formed in a developing leaf. Cross connections are established in the stem so that the whole system of strands forms a cylindrical network. At

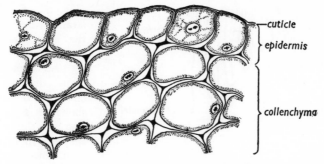

cuticle

epidermis

collenchyma

FIG. 97. EPIDERMAL AND YOUNG COLLENCHYMA CELLS IN STEM OF *Helianthus annuus*.

lower levels each procambial strand gradually differentiates into the tissues of a vascular bundle. In this way the vascular network of the adult stem (Fig. 93, *B*) is constantly added to from above as the stem grows in length, and conducting channels are thus established to supply both the stem apex and each developing leaf.

The development of a vascular bundle from a procambial strand results from the formation of xylem from the cells on the pith side of the strand (the protoxylem being next to the pith) and of phloem from the cells on the outside. The formation of xylem and phloem does not, however, extend right across the strand and a zone of meristematic cells, the cambium, is left between these tissues.

The above account has been confined to stems showing a network of separate vascular bundles. There are, however, many dicotyledonous stems which have a complete cylinder (a ring in transverse section) of xylem and phloem. In such examples the vascular tissue arises by the very early lateral extension of the procambial strands to give a complete procambial cylinder from which the xylem, phloem and cambium are differentiated as in a single vascular bundle.

There is also some variation in the structure of the individual vascular bundles. For example, in some climbing plants, such as, white bryony (*Bryonia dioica*) and vegetable marrow (*Cucurbita pepo*), each bundle has an internal group of phloem in addition to the outer one. Such bundles are said to be **bicollateral**, the normal type being **collateral**. This additional phloem, combined with the large size of the individual sieve tubes, presumably gives efficient food transport through the long stems of such climbing plants. It may be noted, too, that the metaxylem of such plants usually consists of very wide vessels which will permit the rapid movement of water.

(b) Monocotyledon stems

A few monocotyledon stems (for example, those of the black bryony, *Tamus communis*) resemble the dicotyledonous type in having a cylindrical network of vascular bundles. The majority, however, show a more complex arrangement of a large number of bundles. The occurrence of so many bundles is related to the fact that numerous vascular strands supply each leaf whereas in Dicotyledons the leaf supply consists of only one to three bundles. Moreover, the alternating leaves usually encircle the stem so that the numerous leaf traces enter from all sides. The complex course taken by the bundles in the stem is largely due to the fact that the meristem is in the form of a cone which widens very rapidly towards its base

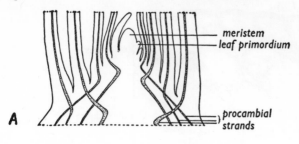

meristem
leaf primordium

A

procambial
strands

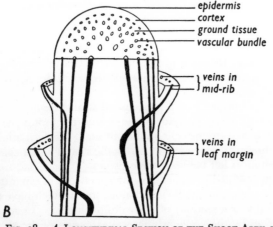

epidermis
cortex
ground tissue
vascular bundle

} veins in
} mid-rib

} veins in
} leaf margin

B

FIG. 98. *A*, LONGITUDINAL SECTION OF THE SHOOT APEX OF
A MONOCOTYLEDON ; *B*, THE VASCULAR BUNDLE SYSTEM OF A
MONOCOTYLEDONOUS STEM. (*Dr. S. Williams.*)

(Fig. 98, *A*). Procambial strands are formed near the apex and these are carried rapidly outwards by the expanding cone. Thus the main strands come to be arranged as shown in Fig. 98, *B*, passing into the centre, where they are more bulky, and then downwards and outwards, becoming thinner in the lower part of their course. Differentiation of the vascular tissues follows the same course as in dicotyledons. The strands frequently anastomose with one another in the outer region of the stem.

A transverse section (Fig. 99) shows an epidermis on the outside, a very narrow cortex, and a ground tissue, in which the bundles are embedded. The central bundles are larger but less numerous than the outside ones. The latter are very numerous owing to the fact that additional bundles are formed after the expansion of the apical cone has taken place, and these pass straight downwards in a peripheral position.

The individual bundles, for example, those of maize (Fig. 100), are very different from those of a dicotyledon, the most important differences being the absence of cambium and the presence of a sclerenchymatous bundle sheath. The xylem and phloem, however, occupy the same relative positions as in the dicotyledonous bundle. The metaxylem consists of two large pitted vessels with a number of small tracheids between them. The protoxylem consists of two or three annular and spiral vessels. These adjoin a large intercellular space, the protoxylem cavity. The phloem, which consists only of sieve tubes and companion cells, forms a clearly defined group on the outside of the bundle. A line of crushed small cells on the outer edge of the phloem is the protophloem. As mentioned before, no cambium is present, and the bundle is surrounded by a sheath of sclerenchyma.

Some monocotyledon bundles have the xylem developed more strongly than in those of the maize. The xylem elements then form a V- or Y-shaped mass, with the point of the V or the tail of the Y consisting of protoxylem, and the arms of a number of metaxylem elements. The phloem lies between the arms of metaxylem. Some monocotyledon rhizomes, for example, lily-of-the-valley (*Convallaria majalis*), have so-called **amphivasal bundles** in which the xylem completely surrounds the phloem.

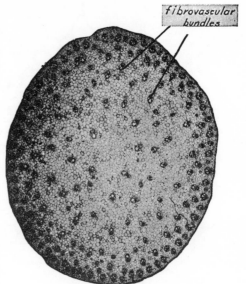

fibrovascular bundles

SECONDARY THICKENING

The main result of secondary thickening is the addition of more xylem and phloem to the conducting systems of stems and roots by the activity of the **cambium**. This addition is rendered necessary by the continued increase in size of the plant as growth proceeds. In annuals such as the sunflower, the amount of additional tissue formed is never large; but in woody perennials secondary thickening proceeds year after year from the seedling stage until the shrub or tree dies.

FIG. 99. TRANSVERSE SECTION OF A MONOCOTYLEDONOUS STEM—MAIZE (*Zea mays*).
Note the large number of vascular bundles.
(From Robbins and Rickett's "Botany", courtesy of D. Van Nostrand Co., Inc.)

The cambium, which is the formative layer for all the new conducting tissues, consists of flattened, prismatic cells (Fig. 101). Their most essential characteristic is an ability to divide repeatedly by longitudinal walls parallel to their flat surface. Since the latter is placed tangentially in the stem or root, each division of the cambial cell gives rise to an inner and an outer daughter cell.

Fig. 101 represents diagrammatically the result of such repeated division in one cell. Following a division, one of the two daughter cells remains meristematic and enlarges to the size of the original cell. If, as in the figure, the inner cell remains meristematic, the outer cell enlarges as an element of the secondary phloem. The cambium cell now divides again and perhaps this time the outer of the two cells remains meristematic. In this case the inner segment enlarges as an element of the secondary xylem. This process is repeated again and again. There is, however, no regular alternation in the production of xylem and phloem and many more secondary xylem elements than phloem elements are produced. It will be noted in Fig. 101 that the production of secondary xylem results in the pushing outwards of the cambium and all the tissues outside this. This leads to a constantly increasing circumference of the cambium cylinder, the increase being provided for by occasional radial divisions in the cambium cells.

(a) Secondary thickening in Dicotyledon stems

In many annuals such as the sunflower the primary stem has a ring of separate vascular bundles, each with fascicular cambium. When secondary thickening commences, cells of the medullary rays adjoining these fascicular cambia become meristematic. Sheets of interfascicular cambium are thus formed, extending laterally from each bundle, giving a complete cylinder of cambium (Fig. 102).

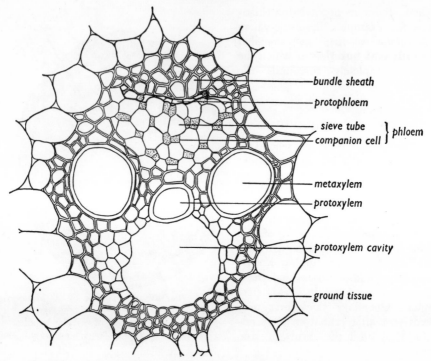

bundle sheath

protophloem

sieve tube
companion cell } phloem

metaxylem

protoxylem

protoxylem cavity

ground tissue

FIG. 100.   TRANSVERSE SECTION OF A VASCULAR BUNDLE OF MAIZE (*Zea mays*).
(*Mrs. Williams.*)

Secondary xylem is now formed to the inside and secondary phloem to the
outside of the cambium.   At intervals, cells of the cambium give rise only to
parenchyma both to the inside and to the outside so that ribbons of living cells
extend from the pith to the cortex.   These ribbons are called **medullary** or
vascular rays and serve for the radial transport of water and foodstuffs.   The
cylinder of secondary tissues thus produced never becomes very thick in the stems
of annuals.   Its development, however, results in the separation of the primary
xylem in each bundle (which remains *in situ*) from the primary phloem, which
is pushed outwards as the girth increases.   The cortex and the epidermis are also
pushed outwards and the cells of these tissues are stretched tangentially, and often
divide, so as to keep pace with the increasing circumference of the stem.

Secondary thickening in woody perennials is much more extensive than in the
annuals just described (Figs. 103 and 104).   The young twig already has a complete cylinder of primary vascular tissues and secondary thickening commences
early in the first year of its growth.   The cambium functions in the way already
described and secondary xylem is formed to the inside and secondary phloem to
the outside.   The secondary xylem elements are varied and consist of vessels,
tracheids, fibres and xylem parenchyma (Fig. 104, *C*).   The secondary phloem is
also a complex tissue with sieve tubes (with very oblique sieve plates), phloem
fibres and parenchyma (Fig. 104, *B*).   Special groups of cambium cells, known as
ray initials (Fig. 101), produce radial ribbons of living cells instead of xylem and
phloem.   These are the primary vascular (or medullary) rays.

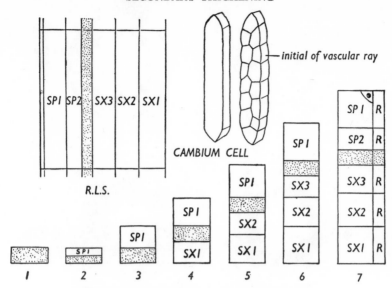

PRODUCTS OF A CAMBIUM CELL AS SEEN IN T.S.

FIG. 101. DIAGRAM ILLUSTRATING THE CAMBIUM AND CAMBIAL ACTIVITY.
Cambium stippled ; *SX* 1-3, secondary xylem ; *SP* 1 and 2, secondary phloem ; *R*,
medullary ray ; *R.L.S.*, radial longitudinal section ; *T.S.*, transverse section.
(Description in text.)
(*Mrs. Williams.*)

The cambium is inactive during the autumn and winter but renews its activity
in the spring. It then proceeds to form secondary xylem and secondary phloem as
before. The xylem produced in the spring, when the shoot is growing rapidly and
the leaves are expanding, consists mainly of large vessels. Later, as growth slows
down and eventually ceases, the xylem produced consists of smaller vessels with
a large admixture of tracheids and fibres. Then in the autumn the cambium
again becomes inactive only to renew its activity in the following spring and again
produce large vessels to the inside and phloem to the outside. Each annual
increment of secondary xylem is therefore a clearly defined zone known as an
annual ring (see Chap. IV and Fig, 105). The annual zonation of the phloem is,
however, not so clear (see also Fig. 106).

During this continued production of secondary tissues, the primary vascular
rays are maintained by the activity of the ray initials which add on new ray tissues
as the girth of the stem increases. Since, however, the mass of xylem between the
rays increases in width as the girth of the stem increases, further provision is made
for the radial transport of water and foodstuffs by the production of secondary
vascular (or medullary) rays. Certain cells of the cambium, which had previously
given rise to vascular tissues, divide up to form ray initials which then proceed to
form the ribbons of parenchyma constituting the secondary rays. The number of
secondary rays is small in a three-year-old twig ; but the number is increased as
each year's increment is added.

It is important to realise that the cambium of a tree is a continuous much-
branched cylinder extending from the topmost twig down through the branches
of increasing size into the trunk, and indeed, down into the roots. Each spring,

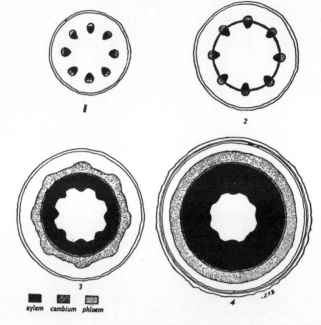

FIG. 102. EARLY STAGES IN SECONDARY THICKENING OF A DICOTYLEDONOUS STEM
(DIAGRAMMATIC).

this cylinder, probably activated by hormones liberated by the opening buds
(p. 224), begins to produce the secondary vascular tissues. There is thus a continuous cylinder of new xylem and new phloem right from the roots to the top of
the tree and it is in this new tissue that conduction of water and foodstuffs mainly
takes place (Chap. IV).

The number of annual rings increases from one in terminal twigs to greater
and greater numbers as one passes down through branches of increasing size into
the trunk. The top of the trunk has many fewer annual rings than the base but
the rings are thicker at the top, possibly because they are nearer the food supply
manufactured by the leaves, than they are at the bottom. This accounts for the
fact that the trunk remains approximately cylindrical and does not taper upwards
as one might expect.

Secondary thickening in climbing plants follows a similar course to that
described above but there are modifications to retain the necessary pliancy of the
stem. For example, in the stem of wild clematis or traveller's joy (*Clematis
vitalba*), the interfascicular cambium only produces parenchyma, so that the wide
medullary rays are maintained and the secondarily thickened vascular bundles
remain as separate strands, giving the stem a strong but pliant rope-like construction. The large tropical climbers or **lianes** show many and varied modifications of the mass of secondary tissues. In many of these, the secondary tissues are
split up into separate strands so that again a rope-like construction is attained.
In others, the cambium is only active at two opposite points of the stem so that a
pliant, ribbon-shaped stem is produced.

(b) Secondary Thickening of Dicotyledon Roots

Secondary thickening takes place in most dicotyledon roots (Fig. 107). The first sign of secondary activity is the development of an arc of cambium on the inside of each phloem group, that is, between the metaxylem and the phloem. Parenchyma cells adjoining these strips then become meristematic and form cambium cells which eventually link up the original strips to form a corrugated cylinder. In a transverse section, this appears as an undulating ring, extending round the xylem rays and inside the phloem groups (Fig. 107). The activity of this cambium leads to the production of secondary tissue as in the stem. More secondary xylem is produced by the cambium lying inside the phloem groups than by that part of it extending over the xylem rays. The result is that the corrugations of the cambial cylinder are thrust outwards and it becomes circular in section. Continued activity giving a mass of secondary xylem to the inside and secondary phloem to the outside then follows. Vascular rays are formed as in

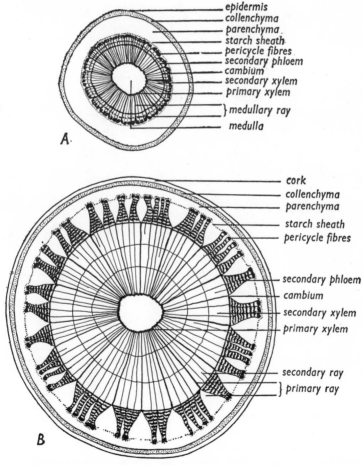

FIG. 103. *A*, TRANSVERSE SECTION OF A LIME TWIG AT THE END OF THE FIRST YEAR'S GROWTH ; *B*, TWIG AFTER THREE YEAR'S GROWTH.
(*Mrs. Williams.*)

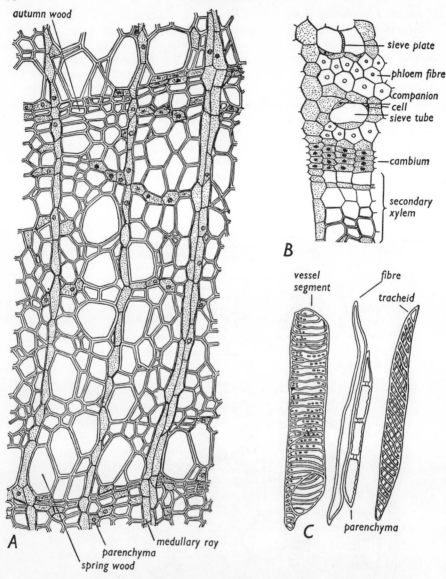

FIG. 104. *A*, TRANSVERSE SECTION OF A PORTION OF SECONDARY WOOD SHOWING THE SPRING AND AUTUMN WOOD IN ONE ANNUAL RING ; *B*, TRANSVERSE SECTION IN THE REGION OF THE CAMBIUM ; *C*, ELEMENTS OF SECONDARY XYLEM IN MACERATED MATERIAL. (*Mrs. Williams.*)

the stem and particularly wide ones usually radiate from the protoxylem groups. Annual rings are not so obvious as in stems, a fact possibly related to the more uniform growth conditions in the soil.

In storage roots, the secondary tissues produced are largely parenchymatous. Thus, in the turnip and radish, food is stored in the parenchymatous secondary xylem. In the carrot, on the other hand, storage takes place in the abundant secondary phloem, the amount of secondary xylem remaining small (as the yellow core).

(c) Secondary Thickening in Monocotyledons

The vascular bundles in the stems of monocotyledons are closed, that is, they have no cambium (Fig. 100). This fact, together with the complex arrangement of the bundles, makes impossible the type of secondary thickening characteristic of the dicotyledons. A few arborescent (tree-like) members of the Liliaceae are, however, secondarily thickened. The most common are *Dracaena*, *Cordyline*, *Aloe* and *Yucca*. All are tropical plants, though certain examples are cultivated in warm temperate regions, such as *Yucca* (Adam's needle), which is a familiar ornamental plant in parks and gardens, especially on the south coast of England.

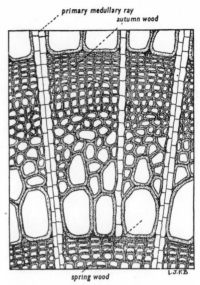

FIG. 105. TRANSVERSE SECTION OF A PORTION OF SECONDARY WOOD, SHOWING SPRING AND AUTUMN WOOD IN ONE ANNUAL RING.

The anomalous secondary thickening (Fig. 108) arises from a secondary meristem of cambium which arises in the innermost layers of the cortex, just outside the ground tissue in which the normal vascular bundles are embedded. By the formation of tangential walls, the cambial cells cut off a few secondary cortical cells to the outside and a secondary ground tissue with embedded secondary vascular bundles towards the inside. In the formation of the secondary bundles some of the cells formed from the cambium divide repeatedly by longitudinal walls so that strands of elongated cells are formed. The peripheral cells of each strand differentiate as scalariform tracheids and the central ones as phloem, thus giving an amphivasal vascular bundle. A transverse section of an old stem will show the primary ground tissue, with scattered vascular bundles, surrounded by radial rows of secondary bundles embedded in secondary ground tissue. The secondary bundles anastomose with one another so as to form a complicated network.

Anomalous secondary thickening in a monocotyledonous root is exemplified by *Dracaena*. Here, the cambial ring arises in the cortex just outside the endodermis.

The most outstanding example of anomalous secondary thickening is the dragon's blood tree (*Dracaena draco*) of Crotava, Teneriffe. This huge tree was destroyed by a storm in 1869. It was 70 ft. in height, and near the ground its stem measured 45 ft. in diameter. It was estimated as being six thousand years old. The name, dragon's blood, is derived from a resin which this and other species exude. The resin is used as a polish, chiefly for violins.

CORK, BARK AND LENTICELS

As secondary thickening proceeds in a stem the mass of wood and phloem continually increases in diameter and thus pushes the cortex and epidermis outwards. This results in a tangential stretching of these outer tissues and such

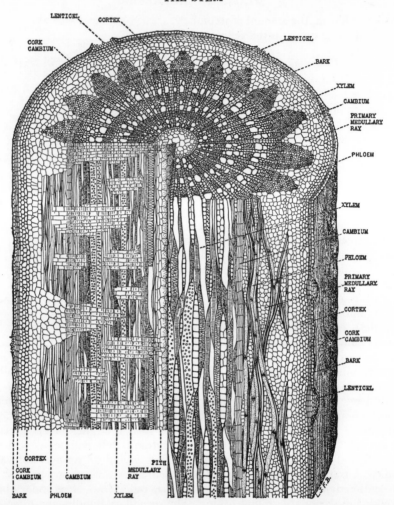

FIG. 106. DIAGRAMMATIC REPRESENTATION OF A DICOTYLEDONOUS STEM IN ITS THIRD YEAR, SHOWING THE STRUCTURES IN TRANSVERSE, RADIAL LONGITUDINAL AND TANGENTIAL LONGITUDINAL SECTIONS.

stretching is also accompanied by radial divisions in the cells so that, for a time, the cortex and epidermis are able to keep pace with the expanding girth. Presumably there is a limit to the ability of the outer tissues to increase their circumference and when this is reached the danger of rupture ensues. Before such rupture occurs, however, a new protective tissue, the cork, is formed (Fig. 109).

The cork is formed by the meristematic activity of a cork cambium or **phellogen.** This arises as a result of tangential divisions either in the epidermis (for example, apple), or, more usually, in the sub-epidermal layer of the cortex (for example, lime, poplar). A cylinder of cork cambium is thus formed and its cells divide in precisely the same way as those of the vascular cambium. The cells cut off to the outside have their walls suberised as they mature and the living contents disappear. These radial rows of closely fitting, suberised cells form the cork, which is impervious to both gases and liquids. Segments cut off to the inside of the cork cambium remain parenchymatous and constitute secondary cortex or **phelloderm** ; frequently this is not well developed. The cork cambium and its derivatives are often spoken of collectively as the **periderm**.

In trees with a smooth bark, like the birch (*Betula* species), the same cork cambium functions year after year. In the birch, each annual zone of cork consists of a layer of large thin-walled cells (formed in the spring) and a layer of flattened, thick-walled cells (produced later in the year). The rupture of the former results in the cork peeling off in the characteristic papery sheets. Commercial cork is derived from the cork oak (*Quercus suber*) and in this example too, the same cork cambium is functional year after year. The cork is stripped from the tree every nine years or so.

In most trees, however, the original cork cambium only functions for a few

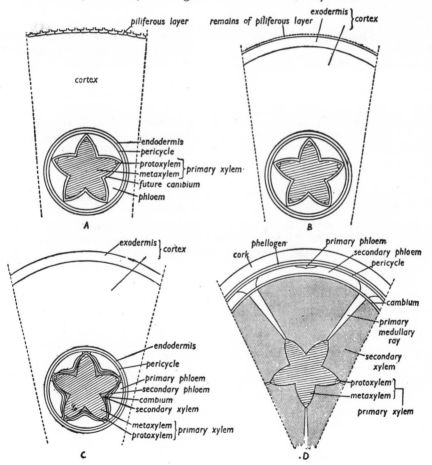

FIG. 107. TRANSVERSE SECTIONS (DIAGRAMMATIC) OF THE MOST IMPORTANT TISSUES OF THE DICOTYLEDON ROOT.

A, through the region of the root hairs ; *B*, just behind the root-hair region ; *C*, early secondary thickening ; *D*, advanced secondary thickening. (The magnification decreases from *A* to *D*.)

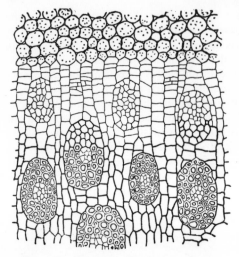

FIG. 108. PART OF A TRANSVERSE SECTION
THROUGH THE TRUNK OF *Dracaena marginata.*
Note the formation of secondary vascular
bundles. The cortical parenchyma near the
top of the diagram contains air spaces, and immediately beneath this is the cambium from which
the new bundles are being formed.

years. When it ceases to be active, a
second cambium arises deeper in the
cortex. This is followed at intervals of
a few years by a succession of cambia
arising at lower levels in the outer
tissues, the innermost ones being
formed in the outer regions of the
phloem. Each successive cambium
produces a layer of cork and, since
this tissue is impervious, the tissues
outside the innermost layer of cork
die. Thus the bark (or **rhytidome**)
of most trees is a composite mass of
cork, dead phloem and dead cortex.

The stems of various currants
(*Ribes* species) are interesting since
the cork is here formed in the pericycle region. This results in the
death and subsequent shedding of the
cortex and epidermis. Examination
of a shoot of black currant (*Ribes
nigrum*) will show that the present-
year portion is thicker than the
older parts which have lost their cortex as a result of cork formation in the
pericycle.

The roots of many herbaceous forms do not form cork. The roots of trees and
shrubs, however, usually develop cork from a deeply seated cork cambium (usually
in the pericycle). As in the stem of *Ribes*, this results in the death and eventual
loss of the cortex and piliferous layer. Young tree roots may therefore be
considerably thicker than older ones which have undergone secondary thickening.

The impervious cork does not form an uninterrupted covering on stems and
roots since breathing pores, or lenticels, are always present. In stems these are
usually formed in positions formerly occupied by the stomata of the epidermis and
originate even before the sheet of cork has developed.

In the region of a lenticel (Fig. 109, *A*) the cork cambium does not produce
cork cells but a tissue known as complementary tissue, the essential feature of
which is that it has intercellular air spaces. It may be a fairly compact tissue, as
in the elder (*Sambucus nigra*), or it may consist of loose, powdery tissue held in position by firmer diaphragms or closing layers, as in the plum (*Prunus domestica*).
In either case the complementary tissue allows the inward passage of the oxygen
needed for respiration.

In trees such as the birch, where there is a persistent cork cambium, the original
lenticels persist and become transversely drawn out, forming a conspicuous feature
of the bark. Persistent lenticels are also clearly shown by bottle corks. In trees
producing a succession of cork cambia, new lenticels are produced by each successive cambium and these open into cracks in the bark.

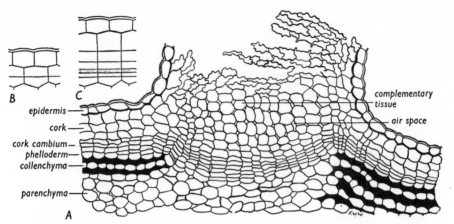

FIG. 109. *A*, Transverse Section of the Outer Region of the Stem of Elder (*Sambucus nigra*), showing Cork and Lenticel ; *B* and *C*, the Origin of Sub-Epidermal Cork (Cork Cambium stippled).

WOUNDS

The healing of wounds in plant organs is made possible by the fact that living cells in the vicinity of a wound are stimulated to meristematic activity.

If the damage is confined to parenchymatous cells, the wound is quickly protected by the production of a layer of cork. Thus, if a potato tuber is cut across, the layer of exposed cells collapses and dies. Cells lying a little deeper become meristematic and form a cork cambium by the activity of which a protective layer of cork is formed over the cut surface. When the tissues of a leaf are torn, the wounded surfaces are protected in a similar way. The wound cork in all such cases gives protection against desiccation of the exposed tissues and against the entrance of disease-causing organisms.

If, however, the wound penetrates as far as the vascular tissues, or if a stem or root be completely severed, then the healing of the wound depends on the production of a mass of undifferentiated parenchyma known as **callus**. This healing tissue can be formed by any living cells in the neighbourhood of the wound but the most active tissues are the cambium and the very young vascular tissues in the cambial region. Such callus often differentiates later into corky or woody tissue, as in the case of damaged trees.

Sometimes the complete branch of a tree is torn off, thus leaving an extensive wound. Such a wound exposes the living tissues of the cambium and phloem to the danger of desiccation and attack by disease-causing bacteria and fungi. These tissues, however, produce a mass of callus which later becomes corky. If the wound is not too large, the callus eventually covers it completely and thus gives adequate protection (Fig. 110). If, however, the wound is very large, callus formation may only cover the living outer tissues, the wood nearer the centre being left unprotected. This seldom has any serious effect since such wood is dead and often contains preservative substances. Water does, however, sometimes cause the wood to rot, with the result that eventually a deep pit is formed in the trunk from which the branch was torn. In horticultural and forestry practice, such rotting is prevented by covering the wound surfaces with a layer of tar, cement or lead paint.

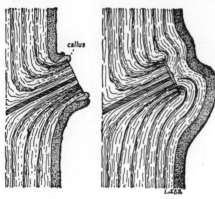

FIG. 110.   HEALING OF A WOUND CAUSED
BY CUTTING OFF A BRANCH.

In this case, note that the callus took three
years to cover the wound completely.

In some very old forest trees, rotting
of the wood exposed by wounds proceeds
to such an extent that the trunk becomes
completely hollow.   Such trees remain
alive since the loss of the heart wood
does not prevent the normal transport
of water which always takes place in the
outer sap wood.

GRAFTING

As has already been seen in Chap.
V, stems are often modified as organs of
vegetative propagation.   Another type of
propagation involving the stem is that
artificially brought about by the horticultural practice of **grafting**.   This process
is also important in surgery and zoology
and great strides have been made in recent years in the transplantation and grafting of tissues, such as skin and bone, in the higher animals, including man.   In
all cases, the process involves fixing together cut surfaces of similar living tissues,
so that further growth will lead to their union.   Its success in plants depends
upon the wound reactions described in the previous section.

Grafting is used in horticultural practice as a means of rapidly propagating
varieties of fruit trees and flowering shrubs.   This method is used since plants
derived from cuttings often form weak root systems and take a long time to reach
the stage of flowering and fruiting.   Propagation by seeds is also unsatisfactory
since the seedlings grow slowly and the plants obtained are often not true to type.

Shoots of the variety which it is desired to propagate are grafted on to rooted
plants of a related wild species or common variety.   Thus a twig of a choice variety
of apple may be grafted on to a crab apple.   The twig is known as the **scion**
and the rooted plant upon which it is grafted is the **stock**.

The top of the stock is cut off and the cut surface at the base of the scion is
brought into close contact with the cut surface of the stock.   It is essential that the
exposed growing tissues (mainly those in the cambial region) of the scion and the
stock should be brought into close contact with one another over as large an area
as possible.   Many ways of achieving this have been devised.

One of these, called **whip** or **tongue grafting** (Fig. 111) is used when the stem
of the stock and that of the scion are about equal in diameter.   The top of the
stock and the base of the scion are both cut across obliquely.   A vertical notch is
then made in the stock and the scion is slit vertically to form a tongue.   The
tongue of the scion is then dovetailed into the notch of the stock and the junction
bound firmly with raphia.   The region of the graft is covered with a thick layer
of grafting wax to prevent evaporation from the cut surfaces and the entry of
disease-causing organisms.   A simpler method is **splice grafting**.   Both stock
and scion are cut across obliquely at the same angle and the two cut surfaces are
then firmly fixed together as in the previous example.   When the stem of the stock
is of much greater diameter than that of the scion other methods of grafting have to
be used.   **Crown grafting** is one of the methods frequently used.   The stock is

cut across transversely and then several vertical slits are made through the bark from the cut surface downwards and the flaps of bark loosened from the underlying tissues. A scion, with its end cut very obliquely, is then inserted in each slit and the whole region of junction is bound and waxed.

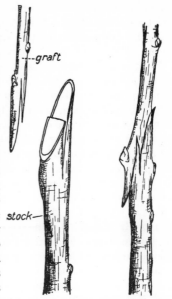

Whichever method of grafting is used, the success of the operation depends upon the fact that the cut surfaces of both scion and stock produce callus, mainly from their cambial regions. The callus of the stock intermingles with that of the scion and so a living connexion is established. Xylem elements are differentiated in the callus thus linking the xylem of the stock with that of the scion ; in a similar way continuity of the phloem and cambium is also established. The scion is thus able to receive a supply of water and mineral salts from the roots of the stock and the latter to receive foodstuffs manufactured by the scion. The establishment of this close connexion between stock and scion involves active growth so that grafting operations are usually performed in April or May when the conditions for growth are most favourable.

Ready for grafting   Ready for binding

FIG. 111.   WHIP GRAFTING.

The stock and the scion must belong to related plants. Thus apples can be grafted on the crab apple (*Malus sylvestris*) or, as is now usual, on an Asiatic type known as the Paradise variety. Pears (*Pyrus communis*) can be grafted on other pear varieties, on quince (*Cydonia oblonga*) and even on mountain ash (*Sorbus aucuparia*). Choice varieties of rhododendron and lilacs are grafted on to common varieties.

The effect of the stock on the scion and *vice versa* is not very great and the scion retains all the characteristic features of the variety from which it was taken. The stock does, however, influence the size and vigour of the tree produced. Thus, in the case of apples, the East Malling Research Station has developed a range of stocks suitable for the production of dwarf bushes or of standard trees of varying size. The age at which the tree will bear fruit and its period of productivity are also influenced by the type of stock.

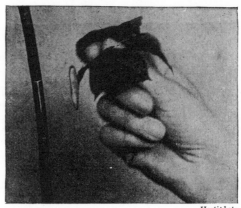

Hortiphoto

FIG. 112.   BUDDING A ROSE BUSH.
On the left is the stock with its T-shaped incision.
The bud is about to be fixed into it.

BUDDING

The horticultural practice of **budding** (Fig. 112) is a special type of grafting in which the scion is a single bud of the desired variety and the stock a rooted plant of a related species or variety.   Roses, peaches and plums are frequently propagated in this way, the operation being carried out in July or August.

In the case of roses, briars of the wild rose (*Rosa canina*) are often used for stocks and the following procedure followed.   A bud is removed from a not quite mature branch of the desired variety by a tangential cut so as to obtain a shield-shaped piece of bark and wood with the bud about in the middle of its length.   The wood is removed so as to leave the shield with a lining of its cambial region.   A vertical cut, about an inch long and just penetrating through the bark, is then made in the stock, followed by a cross cut at the top so as to give a long T-shaped incision. The corners of the incision are then lifted so as to raise the bark from the wood, and the shield inserted under the flaps of bark.   The edges of the bark are folded over the shield and finally raphia is bound round, leaving the bud free.   The cambial regions of the stock and scion are now in close contact and if the bud ' takes ', as indicated by a swelling of the region in about a month's time, the raphia is loosened. The bud will remain dormant over the winter and grow out into a shoot of the desired variety in the following spring.   The stock is then cut back, and care must be taken to remove all the buds belonging to the stock itself.

Bushes are obtained by budding basal branches of the stock.   On the other hand, if standards are required, then the buds are inserted on branches borne at the top of a stout briar, three or four feet high.

ORIGIN OF LATERAL APPENDAGES

The leaves arise at the stem apex as small protuberances formed by the active cell division of a few of the outer meristem layers.   Their origin is thus superficial or **exogenous**.   Some of the cells in the axil of each leaf remain meristematic and develop as the axillary buds which may later grow out as axillary branches. Such axillary, or lateral, branches are also exogenous and are thus very different in origin from the endogenous lateral roots (p. 134).

PRACTICAL WORK

1. Prepare a transverse section of a dicotyledon stem for detailed examination under the microscope.   Almost any stem of an annual will serve the purpose ; the sunflower is a good example.

Mount the section and irrigate with an aniline salt, in order to stain the woody tissues. Examine and draw a diagram of the whole section under the low power of the microscope. Make no attempt, at this stage, to show cellular structure.   Note the epidermis, cortex and central cylinder.   Look for any layers of the cortex which may be thickened to give additional strength.   Note the vascular bundles composed of xylem, phloem and cambium.   Note the position of the protoxylem and, from this, deduce the direction of development of the meta-xylem.   Compare this with that in the root.   Note the pith, and see if the stem is solid or hollow.

Then examine the stem under the high power and note the cellular structure in detail. Make two drawings ; one of the epidermal and some of the cortical tissues, and the other of a complete vascular bundle.

Then examine a similar stem by means of radial longitudinal sections.   Note the appearance of the collenchyma ; the spiral and annular protoxylem vessels ; and the pitted metaxylem vessels.

2. Cut transverse and longitudinal sections of a vegetable marrow stem. Note the general arrangement of the bicollateral bundles and then examine the phloem in detail, paying particular attention to the sieve tubes.

3. Examine the structure of a monocotyledon stem. That of the maize is a good example. Under the low power note the small size, large number, and irregular arrangement of the vascular bundles.

Under the high power, note especially the complete absence of cambium.

4. Make a detailed examination, by means of transverse sections, of a secondarily thickened stem. A three or four years old twig of the lime or horse-chestnut is good material upon which to work.

The various stages of secondary thickening can be examined very well by taking several transverse sections through the stem of various ages. The following will act as a guide: (1) near the apex of the shoot; (2) the middle of the first year's growth; (3) near the bottom of the first year's growth; (4) about the middle of a later year's growth.

Make low-power sketches of the various stages, then examine one of the older ones in detail, and prepare drawings of the various tissues. Note the secondary wood and examine the vessels of the spring and autumn wood. Look for medullary rays. On the outside of the cambium, note the secondary phloem, containing thickened cells known as phloem fibres. Note how the medullary rays widen out in the phloem. Near the periphery, look for cork cambium, and note the layers of cork cells produced on the outside of this.

Then examine a secondarily thickened stem by means of radial longitudinal and tangential longitudinal sections, using Fig. 106 to help in the recognition of the various tissues.

5. In a similar way, examine the structure of a secondarily thickened root. The roots of broad bean, nettle or sycamore may be used for this purpose. Note the origin of **pericyclic cork.**

CHAPTER XIV

THE LEAF

Foliage leaves are the chief photosynthetic organs of the plant, manufacturing carbohydrates from carbon dioxide and water in the presence of light. Proteins are also synthesised in the leaves. Apart from this elaboration of foodstuffs, the leaves also give off excess water in the form of water vapour and their tissues are continually respiring. The structure of leaves is clearly related to the performance of these physiological processes.

The varying external form of leaves has already been described (Chap. IV). The essential part is the flattened expanse of green tissue known as the **lamina**. In this, the chlorophyll-containing tissues are supported by a framework or skeleton of veins (Chap. IV and Fig. 113). These veins, which consist of vascular tissue, bring water and mineral salts to the leaf tissues and carry back manufactured foodstuffs into the stem. The lamina may or may not be supported by a leaf stalk or **petiole**. Where present, the petiole serves to adjust the position of the lamina in relation to the incident light and to prevent tearing of the lamina by wind, by allowing it to move freely.

ORIGIN OF LEAVES

Foliage leaves first appear just below the tip of the stem. They arise as small exogenous prominences (**leaf primordia**) as the result of repeated divisions in groups of cells on the outside of the stem. In Dicotyledons (Fig. 93*A*), the primordia are hemispherical or conical protuberances, in which one to three procambial strands (the young veins) develop. In Monocotyledons (Fig. 98*A*), the primordia are in the form of ridges which extend almost completely around the stem apex and which are supplied by a relatively large number of young veins.

Leaf primordia always arise in a definite sequence, the latest formed being

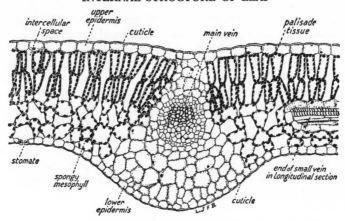

FIG. 114. TRANSVERSE SECTION OF PART OF A LEAF, PASSING
THROUGH THE MIDRIB (× 100).

nearest to the apex. They also arise in definite positions round the periphery of
the stem, their arrangement varying in different plants (p. 171).

The primordia increase rapidly in size and soon become flattened tangentially
to the stem. The side facing the stem is the upper or adaxial surface ; that facing
outwards is the lower or abaxial surface. Such a flattened primordium will become a simple type of leaf if further growth is general. If growth is localised at
certain points round its periphery, it will develop into a lobed or compound type.
In Monocotyledons, growth is localised at the base of the primordium and this
intercalary growth leads to the common ribbon-shaped leaf, such as is seen in the
grasses.

INTERNAL STRUCTURE OF THE LEAF

The petiole, when present, has a stem-like structure. It is usually cylindrical or
more or less deeply grooved on its adaxial face, so that in transverse section it
appears circular, or U- or V-shaped. Embedded in its ground tissue (which is
often collenchymatous), there is, in the case of Dicotyledons, a ring or crescent of
vascular bundles. In other examples, there is a single large U-shaped vascular
strand. The Monocotyledon type of petiole has scattered vascular bundles. The
petiolar vascular strands are those which have already been described (p. 147) as
being connected to the vascular system of the stem and their structure is similar to
that of a stem vascular bundle. The xylem is always on the adaxial side and the
phloem on the abaxial side. In Dicotyledons, the petiolar strands (where more
than one is present) unite at the base of the lamina to form the mid-rib ; in Monocotyledons, the strands remain separate and traverse the length of the lamina as
the system of parallel veins.

The structure of the lamina of a Dicotyledon leaf may now be considered
(Fig. 114). The upper surface is covered by a single layer of tabular cells forming
the **adaxial epidermis**, while a similar layer, the **abaxial epidermis**, covers
the lower surface. The tissue between these two sheets of cells is the **mesophyll**,
in which the veins are embedded.

The mesophyll consists mainly of parenchymatous cells containing chlorophyll.

It is usually differentiated into two types of tissues. The first type, known as **palisade tissue**, lies immediately below the adaxial epidermis and this is the main photosynthetic tissue of the leaf. The individual cells are usually cylindrical in form and are placed with their long axes at right angles to the leaf surface. They contain numerous chloroplasts which line the vertical elongated walls, where they are embedded in the cytoplasmic lining. The cells are more or less closely packed together but, since the cells are cylindrical, narrow air spaces, extending the full length of the cells, are present where the curved walls are not in contact. The number of layers of palisade cells varies in the leaves of different species from one to three. Variation in this respect also occurs between different leaves of the same plant. For instance, leaves on the outside of a beech tree exposed to full sunlight (**sun leaves**) have more layers of palisade cells than do leaves which are constantly shaded (**shade leaves**).

The tissue lying between the palisade tissue and the lower epidermis is the **spongy mesophyll**. The cells are irregular in shape and are often only in contact with one another at the tips of short radiating arms, so that the whole tissue is spongy in texture with large intercommunicating air spaces. The cells contain chloroplasts, which, however, are not so numerous as in the palisade cells.

All the cells of the mesophyll are in contact with air spaces, so that the gases involved in the various physiological processes are able to enter or leave the cells in solution by way of the moist cell walls. Moreover, the internal air spaces are in contact with the external atmosphere through pores in the epidermis, so that diffusion of gases into and out of the leaf can take place (see pp. 168, 177).

The system of veins is embedded in the mesophyll and in a transverse section they are seen cut across in all directions. In the midrib itself and in the region of the stronger veins, the palisade and spongy mesophyll tissues are replaced by collenchyma or sclerenchyma, thus helping to give rigidity to the leaf. Each vein (except the smallest ones) has the structure of a vascular bundle, with the xylem facing the adaxial and the phloem the abaxial surface. The network of finer veins, the smallest of which may consist of only a single tracheid, is embedded in the spongy mesophyll, each vein being surrounded by a sheath of closely fitting parenchymatous cells. The vascular tissues of the veins are continuous with those of the petiole, stem and root. The xylem of the veins is thus able to supply the leaf tissues with water and mineral salts (originally absorbed by the roots), while foodstuffs elaborated in the leaf can be transported by the phloem to other parts of the plant.

As already mentioned, the mesophyll is bounded above and below by an epidermis. The epidermal cells, since they do not contain any chloroplasts, are not photosynthetic. The epidermis gives mechanical protection to the mesophyll and prevents excess evaporation of water from the leaves. This latter function is made possible by the deposition of what is called a **cuticle** on the outer walls of the epidermal cells. This cuticle is formed of a waxy substance called **cutin** which allows very little water to permeate through it. It is usually thicker on the upper surface than on the lower surface of the leaf, since it is from the upper surface of a normal leaf that more evaporation would tend to take place. In many plants the epidermis also serves for the storage of water and, in some examples, the cells are greatly enlarged for this purpose.

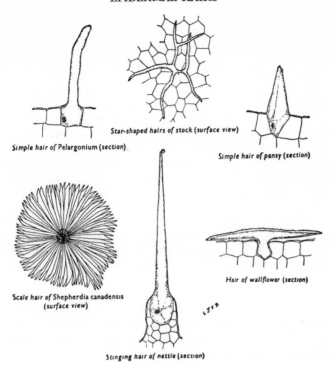

Simple hair of Pelargonium (section)

Star-shaped hairs of stock (surface view)

Simple hair of pansy (section)

Scale hair of Shepherdia canadensis
(surface view)

Hair of wallflower (section)

Stinging hair of nettle (section)

FIG. 115. TYPES OF EPIDERMAL HAIRS FOUND ON LEAVES.

Sometimes some of the epidermal cells grow out to a considerable length, thus giving rise to hairs, the presence of which may serve to restrict loss of water or to give protection against animals. These may be simple, branched or star-shaped (Fig. 115). In many cases, such hairs are still living cells ; whereas in others they are dead and the protoplasm is replaced by air spaces. The latter is well seen in the hairs of the great mullein (*Verbascum thapsus*) which give the leaf a whitish woolly appearance. On the leaves of many thistles the hairs are dead, hardened spines.

An interesting type of hair produced by the epidermis of the leaf is that, unfortunately all too familiar, of the stinging nettle (*Urtica dioica*) (Fig. 115). It is composed of a single long and tapering cell with a swollen base. The tip of the hair contains silica, and is therefore brittle. When this is touched, for example, by the hand, it breaks off, the sharp point of the hair enters the skin and, by the pressure exerted on the bulbous portion of the hair, the contents of the hair are injected into the puncture. Ejection of the contents is rendered sudden because in the intact hair the contents are under a high turgor pressure which is suddenly released when the tip is broken. The poisonous content of the hair is a protein, the composition of which is, at present, not known.

The hairs of many leaves, for example those of lavender (*Lavendula* species), mint (*Mentha* species) and *Pelargonium*, are glandular, and secrete essential oils, which give those leaves their characteristic fragrance.

An essential and constant feature of the epidermis is the presence of pores or stomata leading into the air spaces of the mesophyll.

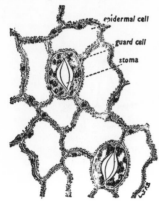

Each stoma arises as the result of the division of a single young epidermal cell to form two sausage-shaped cells, attached at their ends and having a slit or pore between them (Fig. 116 and 117). The two cells are termed **guard cells** since, by changes in their shape and size, they control the width of the **stomatal pore** between them and thus the rate of gaseous interchange between the air spaces of the leaf and the external atmosphere. Usually guard cells, unlike the other epidermal cells, contain a number of chloroplasts. When turgid, the two cells are strongly curved, thus leaving a wide stomatal pore between them ; when flaccid they lie together and the pore is then closed. The conditions under which the opening and closing of stomata take place are discussed in Chapters XV and XVI. The detailed mechanism of the movement is complex and variable. In part it is due to the fact that the elongation of the guard cells as they become turgid is accommodated by their curvature away from one another. Another factor operating in the stomata of some plants is the fact that the various walls of the guard cells are differentially thickened. For example, the guard cells shown in Fig. 124 have thick inner and outer tangential walls while the side walls are relatively thin. As the figure indicates, this leads to the opening of the pore when the cells are turgid and to its closing when they are flaccid. When fully open, the

FIG. 116. PART OF THE LOWER EPIDERMIS OF A POTATO LEAF SHOWING THE STOMATA (× 100).

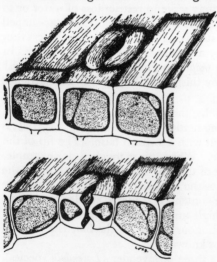

FIG. 117. ABOVE, A DIAGRAMMATIC VIEW OF PART OF THE EPIDERMIS OF A LEAF, SHOWING A STOMA ; BELOW, THE SAME CUT IN SECTION. Note the shape of the guard cells which contain chloroplasts.

pore of a stoma of average size measures about 18μ long by 6μ wide.

The distribution of the stomata varies in the leaves of different plants. In the dorsiventral types of most Dicotyledons, there are many more stomata per unit area on the lower epidermis (where they open into the air spaces of the spongy mesophyll) than on the upper epidermis, and, indeed, they are commonly absent from the latter. For example, in the lilac, there are 160,000 stomata per square inch on the lower surface of the leaf, and only 3-10 on the upper surface. In the holly, there are 63,000 per square inch on the lower surface and none on the upper. Other types of leaf may, however, show a very different distribution of the stomata.

For example, in the case of

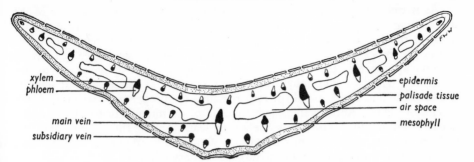

FIG. 118. TRANSVERSE SECTION OF LEAF OF *Narcissus*.
(*Mrs. Williams*.)

the upright growing leaves of many monocotyledons (for example *Narcissus, Iris*), unlike horizontally placed leaves, both surfaces are equally exposed to the atmosphere and light. In relation to this, palisade tissue is developed on both sides of the leaf, the mesophyll occupying the central region (Fig. 118). The stomata, too, are present in equal proportions on both surfaces, namely, about 11,600 per square inch. The veins of such **isobilateral** leaves are arranged in two or more series, thus giving an adequate conducting system to and from both sides of the leaf.

In the case of an aquatic plant with floating leaves, for example, the water-lily, stomata in the normal position, that is, on the lower surface, would be worse than useless, for they would not only allow of no access to air, but would also cause the internal parts of the leaf to be flooded with water. In this case all the stomata are on the upper surface.

Stomata occur on the surface of other parts of a plant where there is a living covering epidermis, such as young stems, scale-leaves, etc.

AUTUMN LEAF-FALL

The majority of long-living perennials are shrubs, bushes or trees. In the tropics, these are mostly evergreen ; but in temperate regions, such as Great Britain, the majority are deciduous. The reason for that is not far to seek. Temperate regions experience two definite seasons during the year, namely, winter and summer. Winter is not conducive to the good growth of a plant, since a low temperature and lack of sunshine have a bad effect on nearly all life processes. On the other hand, summer weather is favourable.

Leaf-fall is obviously related to seasonal changes. The loss of the leaves prevents loss of water during the winter period when the low temperature of the soil makes absorption of water difficult. The loss of the leaves naturally results in the cessation of photosynthesis as well, but, as respiration and growth are also slowed down, no foodstuffs beyond those already stored are required.

Such facts indicate that winter is a period of physiological rest. All perennial plants show an internal rhythm, an active period alternating with a dormant one.

The onset of leaf-fall in deciduous trees is due to this internal rhythm but it is influenced by factors which are little understood. Autumn brings lower temperatures ; the light becomes less intense ; and the length of day becomes shorter. These factors, perhaps particularly the last named one, may set in train the changes leading to leaf-fall.

As is well known, leaf-fall is usually preceded by a change in colour of the leaf leading to various autumn tints. During the spring the amount of chlorophyll in the leaf increases. The young leaves are yellowish-green, but, as the season progresses, the deep green of the mature leaves is developed. Then, as autumn approaches, the leaf begins to age and the chlorophyll starts to disintegrate. The two chlorophylls (*a* and *b* ; see p. 175) disappear, but some of the carotene and xanthophyll (p. 176) remain—the proportion varying with the species of plant. Anthocyanins sometimes also are manufactured at this stage, chiefly from any foods still left in the ageing leaf. Among the autumn tints, the yellows are due to varying mixtures of carotene and xanthophyll, and the reds, browns and purples to different mixtures of anthocyanins.

Few herbaceous plants display autumn tints ; but among the exceptions are certain species of dock (*Rumex* species), spurge (*Euphorbia* species), herb Robert (*Geranium robertianum*) and the dove's-foot crane's bill (*G. molle*). Among shrubs and climbers are the blackberry (*Rubus fruticosus*) with its yellows and reds, certain cultivated grapes (*Vitis* species) displaying yellow, red and purple foliage, and of course, the two Virginian creepers (*Parthenocissus tricuspidata* and *P. quinquefolia*).

But the most familiar autumn tints are presented by deciduous trees. Only two coniferous trees are deciduous, namely, larch (*Larix decidua*) the leaves of which turn pale green or yellow, and the rare, exotic deciduous cypress (*Taxodium distichum*) whose foliage turns red during autumn.

Among the broad-leaved, angiospermous trees, the following may be mentioned : Ash (*Fraxinus excelsior*)—little change ; sycamore (*Acer pseudoplatanus*)—little change ; maples (*Acer* species)—yellow and red ; wych elm (*Ulmus glabra*)—little change ; tree of heaven (*Ailanthus glandulosa*)—little change ; lime (*Tilia vulgaris*, T. *cordata* and T. *platyphylla*)—golden yellow ; birch (*Betula verrucosa* and B. *pubescens*)—bright yellow ; white poplar (*Populus alba*) and blank poplar (*P. nigra*)—bright yellow ; walnut (*Juglans regia*)—dull green or yellow ; willows (*Salix* species)—yellow ; hornbeam (*Carpinus betulus*)—yellowish gold ; hazel (*Corylus avellana*)—yellow ; common elm (*Ulmus procera*)—yellow ; sweet chestnut (*Castanea sativa*)—yellowish gold ; horse-chestnut (*Aesculus hippocastanum*)—brown or dull gold ; London plane (*Platanus acerifolia*)—greenish yellow ; white beam (*Sorbus aria*)—dull red, spotted with brighter red ; hawthorn (*Crataegus monogyna* and *C. oxyacanthoides*)—red and brown ; common oaks (*Quercus robur* and *Q. petraca*)—dull brown ; cultivated exotic oaks (for example, *Quercus coccinea*)—fiery red ; spindle (*Euonymus europaeus*)—yellow, pink and red ; dogwood (*Cornus sanguinea*)

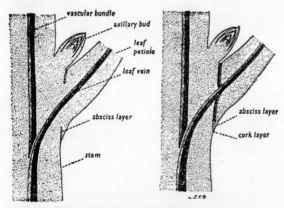

FIG. 119. PARTS OF LONGITUDINAL SECTIONS THROUGH A STEM, PASSING THROUGH A LEAF-BASE, SHOWING STAGES IN LEAF-FALL. Left, the absciss layer beginning to form ; right, absciss and cork layers almost completely formed.

—orange and bright red ; wayfaring tree (*Viburnum lantana*)—yellow and pink ; alder (*Alnus glutinosa*)—dull yellow.

Apart from the disappearance of chlorophyll and the withdrawal of foodstuffs, anatomical changes take place in preparation for the separation of the leaf from the stem.

Several layers of cells, forming a diaphragm right across the leaf-base become spherical (Fig. 119). The layer which does this is several cells in thickness. The rounding-off process, accompanied by a gelatinisation of the cell walls begins in the outer tissues of the leaf-base and gradually works its way across it. Naturally by the time all the cells in a complete cross-section of the leaf-base have done this, that area is no longer firmly knit together and the leaf is then held in position only by the veins. This layer of rounded-off cells is called the **absciss layer** or **separation layer**. The leaf then falls by virtue of its own weight, though the process is frequently accelerated by wind and the action of frost.

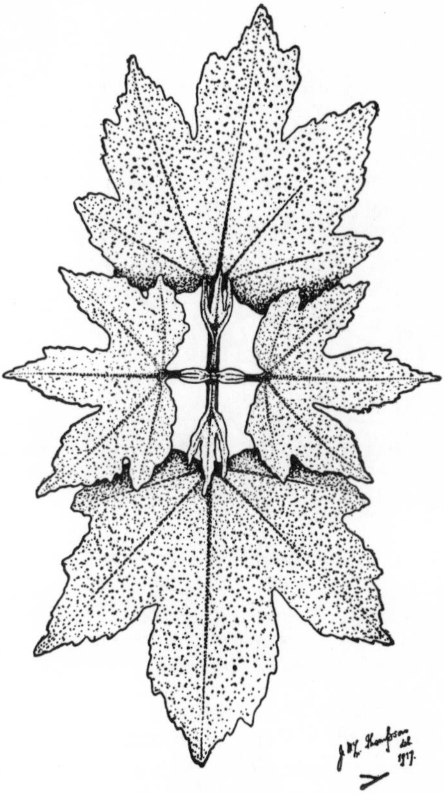

However, this is not the only process which takes place. If it were, it

FIG. 120. YOUNG LEAFY STEM OF SYCAMORE, SEEN FROM ABOVE, ILLUSTRATING DECUSSATE LEAF-MOSAIC.

would mean that there would be an open wound of thin parenchymatous living cells left exposed to rain and disease-causing bacteria and fungi. This is prevented by a process which takes place simultaneously with the formation of the absciss layer.

While the cells gradually become exposed by this rounding-off process, the layers of cells immediately beneath deposit cork on their walls. This goes on at such a rate that, by the time these cells are completely exposed, they are no longer living, and are thick-walled cork cells. This is called the **primary protective layer**. In some trees it is supplemented by the development just below it of a cork cambium. This forms a layer of cork known as the **secondary protective layer**. The result is that, when leaf-fall is complete, a layer of cork is left covering the wound on the stem. The leaf-scar seen on a winter twig is really an area of cork, and the marks left by the vascular bundles, which originally passed up into the living leaf, may clearly be seen.

PHYLLOTAXIS

The arrangement of the leaves on the stems of many plants is such as to ensure their maximum illumination. The leaf arrangement or **phyllotaxis** follows

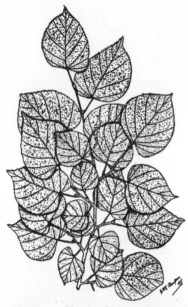

FIG. 121. YOUNG LEAFY STEM OF LIME, SEEN FROM ABOVE, ILLUSTRATING COMPACT, DORSI-VENTRAL LEAF-MOSAIC.

various characteristic patterns in different plants. In many plants, for example, two leaves are borne opposite each other at a node, for example, the privet (*Ligustrum vulgare*). Such leaf arrangement is known simply as **opposite**. If more than two leaves occur as a node, the arrangement is said to be **whorled**. A good example of this is cleavers or goosegrass (*Galium aparine*) which bears its leaves in whorls of six to eight. When successive pairs are placed at right angles, the arrangement is known as **decussate**, and affords adequate illumination for all the leaves which, when looked at from above present a space-saving **leaf-mosaic**. This is well seen on a shoot of sycamore (*Acer pseudoplatanus*) (Fig. 120).

An even more common phyllotaxis is the **spiral**, when only one leaf is borne at each node and in a series of nodes they are inserted spirally. The simplest case is that where the third leaf is immediately above the first ; in other words, the leaves come off alternately right and left. (At one time such an arrangement was called **alternate** ; but now it is considered to be the simplest form of spiral.) An example of this can be seen in the lime (common lime, *Tilia vulgaris*). This gives a shoot of such leaves a very flattened form. Such a shoot is said to be **dorsi-ventral**, and it presents a good leaf-mosaic (Fig. 121).

Other spiral arrangements may occur. In some plants, the fourth leaf is above the first ; but perhaps most common of all is that when the sixth leaf comes above number one, as seen in oaks (*Quercus* species), apple (*Malus sylvestris*), etc. The spiral phyllotaxy of a shoot can be represented by a fraction representing the proportion of the stem circumference separating successive leaves in the spiral. Starting from one leaf and drawing a line through the following leaves above until one comes to the next leaf immediately above the first leaf, one draws a spiral, going around the stem a definite number of times and passing through a definite number of nodes. The numerator of the fraction is the number of times one goes around the stem ; the denominator the number of nodes passed through, not counting the first. So the simplest spiral (alternate) would be $\frac{1}{2}$ and the next $\frac{1}{3}$. In Nature the next is usually $\frac{2}{5}$, the next $\frac{3}{8}$, the next $\frac{5}{13}$. In other words, adding two consecutive numerators and their corresponding denominators we get the next spiral phyllotaxy to be found in Nature. The phyllotaxy $\frac{3}{8}$ is seen in the radial branch of a *Rhododendron* which also gives a beautiful radial leaf-mosaic (Fig. 122). These various leaf arrangements naturally give a varying number of vertical rows of leaves (or **orthostichies**) on the stem. Thus the $\frac{1}{2}$ arrangement gives two rows, the $\frac{1}{3}$ arrangement 3 rows and the $\frac{2}{5}$ arrangement 5 rows.

This series of fractions ($\frac{1}{2}, \frac{1}{3}, \frac{2}{5}, \frac{3}{8}, \frac{5}{13}$ and so on) is known by mathematicians as a Fibonacci series. Its interesting feature from the point of view of leaf arrangement is that the value of all the members of the series lies between $\frac{1}{2}$ and $\frac{1}{3}$. This means that the succession of young leaves being produced at the apex are separated from one another by at least one third of the circumference of the young stem. This spacing of the leaves may be due to the fact that a growing leaf rudiment is drawing on the available food supplies from the tissues in its vicinity. The next primordium will therefore be likely to arise some distance away, perhaps separated by $\frac{1}{2}$, $\frac{1}{3}$ or $\frac{2}{5}$ of the circumference as the case may be. It is also possible that a young leaf produces a growth substance (p. 224) which diffuses out from the

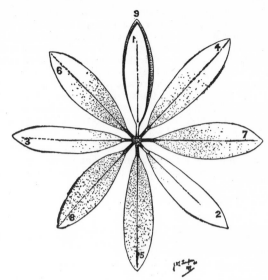

FIG. 122. LEAFY STEM OF *Rhododendron* SEEN FROM ABOVE, ILLUSTRATING RADIAL LEAF-MOSAIC DUE TO 3/8 SPIRAL PHYLLOTAXY.

leaf and actually inhibits the development of another primordium in its immediate vicinity. These are just possibilities and, in spite of much research, very little is definitely known about the factors which determine the very precise leaf arrangements described above.

PRACTICAL WORK

1. Choose a simple form of leaf, such as that of the elm, beech, apple, or *Hydrangea*, and examine its external appearance in detail. Make a clear drawing of the leaf, showing the swollen leaf-base, the petiole and the lamina. Examine the type of venation, and note how the veins gradually diminish in size, until the ultimate branches are scarcely visible.

2. Make a detailed examination of the internal structure of a typical leaf. This is done by preparing a microscope section of the leaf, cut transversely. Prepared sections may be used ; but if time permits, it is better to make such sections. Naturally only a portion of a leaf can be cut successfully. Choose a small portion of the leaf (such as lilac or *Hydrangea*), which contains some of the mid-rib, and fix it between two pieces of elder pith or carrot root, in such a position that the mid-rib may be cut transversely. Mount the section and stain with an aniline salt.

Examine, first of all, the whole structure under the low power and make an orientation sketch of the tissues, noting the upper epidermis covered with a layer of cuticle, palisade tissue, spongy tissue, and lower epidermis covered with a slightly thinner layer of cuticle. Note also the vascular bundle forming the mid-rib, composed chiefly of xylem and phloem.

Then make two detailed sketches, (*a*) of the mid-rib embedded in a sheath of parenchymatous cells and (*b*) the thin portion of the leaf. Here, note the absence of chloroplasts in the upper and lower epidermis, a large number in the palisade mesophyll cells and a relatively smaller number in the spongy mesophyll cells. Note the shape and position of the chloroplasts. Examine the shape of the cells of the palisade tissue and note the number of layers here. The cells are separated by small air spaces. Note the more or less irregular shape of the cells of the spongy mesophyll, and the presence of large air spaces in this tissue.

3. Detailed examination of the stomata is best made on material like that of the *Iris* or *Narcissus* leaf.

In surface view. Take a small portion of such a leaf, and carefully tear away the epidermis. This comes away fairly easy as a thin, tissue-like layer. Mount this flat in water and examine it under the high power of the microscope. Look for a stoma and draw it, with several of the surrounding cells. The stoma is just a small pore, surrounded by two kidney-shaped guard cells. Note that whereas the guard cells of most species contain chloroplasts, the normal epidermal cells do not. Irrigate the mount with salt solution and note that the guard cells, which become plasmolysed and flaccid, now close the pore.

In section. Cut a transverse section of the leaf and look for a stoma cut in section. Make a drawing of this structure under the high power. Notice the shape of the guard cells and of the pore itself. Note also the large air space in the mesophyll. immediately adjoining the stomatal pore.

4. In the section of the *Iris* or *Narcissus* leaf already prepared, note the general arrangement of the tissues as an example of an isobilateral monocotyledon leaf. Note in particular the development of palisade tissue on both surfaces and the arrangement of the veins.

CHAPTER XV

MANUFACTURE OF PLANT FOOD

In Chapter VI it was seen that the most important food-stuffs for plants (as indeed for all living things) are water, mineral salts, carbohydrates, proteins and fats. The last three of these are complex organic materials, and it is the origin of these that we are concerned with in the present chapter. Green plants have the distinctive property of being able to manufacture these complex food-stuffs from simple inorganic raw-materials, and the key process in these food-synthesising operations is that in which carbohydrates are built up from carbon dioxide and water. This particular process is known as **photosynthesis** since the energy of sunlight is used in it, this energy being absorbed in the first place by the green pigment (**chlorophyll**) present chiefly in the chloroplasts of the leaf. The green leaf in fact constitutes the food factory of the plant, but in addition has a much wider significance since, directly or indirectly, all life on the earth is dependent for food on the activities of the leaf.

CHLOROPHYLL

A great deal of our knowledge of chlorophyll has been obtained by extracting the pigment from the leaves. This can be achieved by taking a few thin green leaves such as those of *Tropæolum*, *Fuchsia* or *Hydrangea*, immersing them in boiling water for a few moments, and then placing them in a flask containing warm alcohol or acetone. The green pigment is quickly extracted.

The chlorophyll extract so obtained is bright green in transmitted light, but red in reflected light ; that is, if the solution is held up to the light, with the light shining through it, it is green. If, on the other hand, it is held up against a dark background, with the light shining on it, it appears red. Chlorophyll is therefore said to be green, with a red **fluorescence.**

The name chlorophyll was given to the green pigment in 1817 by **Pelletier** and **Caventou** ; but it is certain that its importance was realised long before this, since it is on record that Nehemiah Grew extracted it from green leaves in 1682 by means of oil. But it was not until the earlier part of the present century that the chemical nature of chlorophyll was elucidated, largely as the result of the work of **Willstätter** followed by that of **Fischer.** It has been shown that two slightly different forms of the green pigment are present in the leaf, namely **chlorophyll a** and **chlorophyll b**, the empirical formulae of these being as follows :

$$C_{55}H_{72}O_5N_4Mg \text{ (chlorophyll } a) \text{ and } C_{55}H_{70}O_6N_4Mg \text{ (chlorophyll } b).$$

The magnesium atom is considered to occupy a central position in the molecule, with the four nitrogen atoms attached to it and forming parts of complex ring structures. It is of great interest that there is a close resemblance between the chemical structure of chlorophyll and that of the substance haem, which, joined to a protein, forms the hæmoglobin of animals. In the animal pigment iron replaces magnesium.

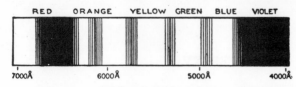

FIG. 123.ABSORPTION SPECTRUM OF AN ETHER SOLUTION
OF CHLOROPHYLL *a*.

Since, as has been mentioned, an important function of chlorophyll is to absorb light-energy for use in photosynthesis, it is natural that the light-absorbing properties of the pigment should have been carefully studied. If an extract of chlorophyll is examined through a spectroscope it will be seen that of the various colours that make up white light, certain red rays and also blue-violet rays are strongly absorbed, while the intervening green and yellow rays mostly pass through unabsorbed (see Fig. 123).

The red fluorescence noted above as being shown by chlorophyll is due to a re-radiation of some of the light-energy absorbed.

Two yellow pigments are regularly present in the chloroplasts of the leaf along with the chlorophylls. These are **carotene** and **xanthophyll**. Carotene is a hydrocarbon ($C_{40}H_{56}$), several forms being known. The commonest of these is β carotene, and this is sometimes termed provitamin A since in the animal body it is converted to that vitamin (p. 89). Xanthophyll ($C_{40}H_{56}O_2$) also exists in a number of slightly different forms. The function of these carotenoid pigments in the leaf remains uncertain. Their absorption spectra show absorption of blue-violet rays.

Carotene and xanthophyll also occur separately from chlorophyll in the plant. Thus they are responsible for the colour of many yellow flowers and fruits, while carotene occurs in the root of the carrot and in this way gained its name. **Lycopene**, the red pigment of the tomato fruit, is closely related to carotene and has the same empirical formula.

The percentage by weight of these pigments present in plants varies with the plants. On the average, however, fresh green leaves contain 0·2 per cent. chlorophyll *a*, 0·075 per cent chlorophyll *b*, 0·17 per cent. carotene, and 0·03 per cent xanthophyll. Largely by use of the electron microscope it has been shown that the pigments are not dispersed uniformly through the chloroplast, but are concentrated in particular regions of it, these being termed **grana.**

THE PHOTOSYNTHETIC PROCESS

As the term implies, photosynthesis is a constructive process occurring in the presence of light, the latter (usually in the form of sunlight) providing the energy needed for the synthesis of sugars from carbon dioxide and water. The amount of energy utilised, expressed in terms of heat energy, is indicated in the following equation which sums up the process of photosynthesis :

$$6CO_2 + 6H_2O + 674 \text{ kgm. cals.} = C_6H_{12}O_6 + 6O_2$$

A feature shown by the equation but not previously mentioned is that **oxygen** is produced in photosynthesis.

Observations and experiments of various kinds show that only those parts of the plant that contain chlorophyll can carry on photosynthesis. Thus a strong indication that this is so is provided in simple tests with variegated leaves in which certain areas only are green, the rest being yellow in colour. After being exposed to sunlight for several hours, such leaves are tested for photosynthetic products. Sugars themselves are not so easily detected as the polysaccharide starch, which in most leaves is quickly formed from the initially produced sugars, and which can be recognised by means of the iodine test. The application of this test to the variegated leaves shows that starch has formed in the green areas but not in the yellow. Further details of this experiment are given in practical work at the end of the chapter.

Before passing to consider several aspects of photosynthesis in more detail, it may be noted that the rate of photosynthesis can be measured by observing the amount of carbon dioxide absorbed or of oxygen evolved in a given time, or that of carbohydrates formed in the leaves. The results obtained in such determinations will depend on the particular plant employed and on external conditions such as light intensity and temperature; but, to quote a single example, the leaf of the potato in full sunlight at a temperature of 20° C. was found to utilise 19 mgm. of carbon dioxide per square decimetre per hour, corresponding to a formation of 13 mgm. of glucose.

THE CARBON DIOXIDE FOR PHOTOSYNTHESIS

The plant growing on land absorbs the carbon dioxide utilised in photosynthesis from the atmosphere, as is indicated by an experiment described later (see practical work) which demonstrates that if a leaf is denied access to this atmospheric carbon dioxide, then no photosynthesis occurs. Alternatively normal air can be passed over illuminated leaves enclosed in a wide glass tube and the issuing air shown to have a diminished content of carbon dioxide.

Carbon dioxide occurs in the atmosphere to an average extent of 3 volumes per 10,000 volumes of air, though there may be local variations from this figure. Thus in the vicinity of dense vegetation the carbon dioxide content of the atmosphere may, owing to photosynthesis, fall considerably below that figure during the middle part of the day. Again, carbon dioxide is constantly escaping into the atmosphere from the soil, as a result of the respiration of soil organisms and plant roots (Chap. XVII), so that the atmosphere immediately above soil-level may be above the average in its carbon dioxide-content.

In order to reach the chloroplasts (the actual site of photosynthesis) the molecules of carbon dioxide must first diffuse through the **stomata** of the leaf. The general structure of these and their frequency of occurrence were considered in the previous chapter, where it was also noted that the guard cells, by their movements, control the size of the stomatal pore. This stomatal mechanism is closely co-

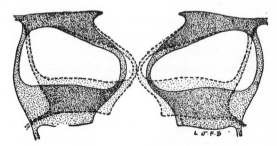

FIG. 124. DIAGRAMMATIC REPRESENTATION OF THE ACTION OF GUARD CELLS.

The deeper shading indicates the position of the guard cells when the stoma is open, and the lighter shading their position when the stoma is closed.

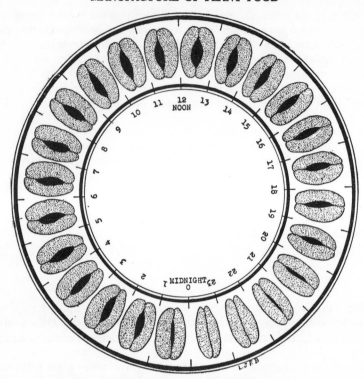

FIG. 125.   DIAGRAMMATIC REPRESENTATION OF THE SIZE OF STOMATAL PORES, REGULATED
BY THE MOVEMENT OF THE GUARD CELLS, DURING THE TWENTY-FOUR HOURS OF THE DAY.

ordinated with the requirements of photosynthesis since it normally ensures that
the stomata are open during the hours of daylight.   The immediate cause of the
opening of the stomata in the presence of light is an increase in the turgor of the
guard cells, this resulting in the latter curving apart as seen in surface view (p.
188) so that the pore opens.   When at night-time the turgor falls the guard cells
relax into a straighter position, and the pore closes.   Fig. 124 illustrates the movements of the guard cells as seen in section.   The changes in turgor are due to
alterations in the amount of osmotically active sugars in the sap of the guard-
cells.   As already noted, the latter, unlike the normal epidermal cells,
contain chloroplasts, and in the presence of light sugars will be formed by photosynthesis, while additional sugar is probably formed as a result of a hydrolysis
of starch which, during hours of darkness, is present in the chloroplasts of the guard
cells.   The increased osmotic pressure produced in this way leads to an intake of
water from adjacent cells, with resulting greater turgor.   Fig. 125 shows diagrammatically the diurnal changes in stomatal aperture in a particular plant.

Very rapid ingress of carbon dioxide can occur through the open stomata.
This is partly due to their huge numbers, but also to the less obvious reason that,
under similar conditions, far more gaseous molecules can diffuse through a unit
area of a minute pore such as a stoma, than through the same area of a larger
pore, as was first shown by **Brown** and **Escombe**.

Afterwards the carbon dioxide diffuses through the continuous system of air-passages that permeates the leaf, and becomes dissolved in the moisture present in the cell-walls bounding the air-spaces. Finally it diffuses in solution through the cell-walls into the interior of the mesophyll cells and so reaches the chloroplasts. A steady utilisation of carbon dioxide in photosynthesis will lead to a constant inflow of further carbon dioxide from the external atmosphere into the leaf, because of the concentration gradient that is thereby set up.

Experiments have shown that when other factors affecting photosynthesis (such as light intensity and temperature) are at a favourable level, the rate of photosynthesis can often be increased by supplying the plants with extra carbon dioxide, over and above that normally present in the atmosphere. This indicates that under natural conditions photosynthetic activity in plants is frequently limited by a shortage of carbon dioxide, and in view of this some attention has been given to the possibility of increasing the supply of this gas to crop-plants, with the expectation that if such a measure proved practicable the yield would be improved.

UTILISATION OF LIGHT-ENERGY

In the practical work at the end of this chapter experiments are described which show that photosynthesis only proceeds when light is falling on the leaf, and that the effect of light is localised, so that if only part of a leaf is illuminated, photosynthesis is confined to that part.

A maximum incidence of light on the leaf for purposes of photosynthesis is ensured by the various types of leaf arrangement on the stem (phyllotaxy, see pp. 171–3), and also by the property of transverse phototropism shown by many leaves (Chap. XX), whereby the leaf places itself during growth in a position at right angles to the direction from which light falls upon it.

With most plants the rate of photosynthesis increases during the morning and reaches a maximum round about midday, when the light is usually strongest ; a plant which for any reason is partly shaded will be prevented from photosynthesising at a normal rate and its growth will suffer accordingly. There is of course a number of plants which habitually grow in shady situations, for example many woodland species. Such plants are specially adapted for securing normal growth in the rather low light intensities prevailing.

Earlier in this chapter we have seen that chlorophyll shows strong absorption of red and blue-violet rays. Various experiments have shown that if plants are exposed to light of different colours, photosynthesis is most rapid in the red and (at least according to some investigators) blue-violet rays, that is, the same rays that are absorbed by chlorophyll. From this it may be concluded that chlorophyll absorbs the light-energy utilised in photosynthesis.

EVOLUTION OF OXYGEN

The oxygen produced in photosynthesis passes out of the leaf by a route that is the reverse of that followed by the entering carbon dioxide. Methods of gas analysis are employed to prove that the leaves of land plants do evolve oxygen during photosynthesis, but this is more readily demonstrated in the case of water plants. Thus if a cut shoot of Canadian pondweed (*Elodea canadensis*) lying in water is exposed to bright light, a continuous stream of bubbles is seen to escape from the cut end of the shoot, especially if the water has been enriched with extra carbon dioxide. An arrangement by which the gas may be collected and shown to be chiefly oxygen is described in the practical work. The production of the bubbles is due to the oxygen accumulating under pressure in the intercellular spaces of the shoot, so that it escapes from any cut surface.

The equation for photosynthesis given on p. 176 shows the volume of oxygen produced as being equal to that of carbon dioxide absorbed. This is in agreement with numerous actual measurements that have been made on leaves during photosynthesis. The gaseous exchange associated with photosynthesis is of great significance to the animal kingdom, since by virtue of it the atmosphere is ' purified ', its oxygen content being restored and the carbon dioxide kept in check. Observations on this complementary relation between the gaseous exchanges of plants and animals were made as long ago as 1771 by **Joseph Priestley,** who showed experimentally the ability of green plants to improve the air and to maintain it in a condition favourable to the life of animals.

It should be carefully noted that it is only in the presence of light that green plant organs exhibit the above type of gaseous interchange with the atmosphere. In darkness oxygen is absorbed and carbon dioxide evolved, owing to the occurrence of respiration (Chap. XVII).

MECHANISM OF PHOTOSYNTHESIS

It has already been indicated that the photosynthetic process can be summed up by the following equation:

$$6CO_2 + 6H_2O \xrightarrow{\textit{light energy}} C_6H_{12}O_6 + 6O_2$$

This only shows the beginning and the end of the synthesis, but there is no doubt that many intermediate steps are involved. A theory which for a time gained much popularity was advanced in 1870 by the chemist **Baeyer,** and suggested that by reaction between water and carbon dioxide, formaldehyde (H.CHO) was formed, and by combination of six molecules of this latter substance, hexose sugar arose. There was never much experimental evidence in favour of this formaldehyde theory, and in view of the results of more modern investigations the theory has been to a large extent discarded.

In recent years important information regarding photosynthesis has been gained by the use of isotopes of oxygen and carbon. In experiments first carried out by **Ruben** and his collaborators, **heavy oxygen** (^{18}O) has been employed to determine the source of the oxygen that is evolved during photosynthesis. It will be noted that the equation given above implies that at least half the oxygen must come from carbon dioxide; but in order to put this to the test, plants actively photosynthesising were supplied with special carbon dioxide, or in other cases, special water, containing heavy oxygen. Only when the isotope was supplied in the water did ^{18}O appear in the oxygen evolved during the photosynthesis, suggesting that all the oxygen of photosynthesis comes from water, and none from carbon dioxide.

If this conclusion is finally proved to be correct, then the over-all equation for photosynthesis becomes more informative if it is amended in the following way, the oxygen in or derived from water being distinguished by heavy type:

$$6CO_2 + 12H_2\mathbf{O} = C_6H_{12}O_6 + 6\mathbf{O}_2 + 6H_2O$$

This provides for all the oxygen originating from water. Molecules of water appear in both sides of the equation, but actually those on the right are not identical with any of those on the left, being newly formed during the reaction.

The above result, together with evidence obtained by the study of the activities of chloroplasts isolated from mesophyll cells and of special types of photosynthesis carried on by certain bacteria, has led to the view that the first main stage in photosynthesis is a splitting of water to produce hydrogen and oxygen. This is a photochemical process in which light energy absorbed by the chlorophyll molecules is utilised. The oxygen is evolved, while the hydrogen is used to reduce carbon dioxide with the eventual formation of sugar.

This reduction of carbon dioxide constitutes the second main stage in photosynthesis, and there is evidence that this stage does not involve a direct utilisation of light energy. In experiments carried out especially by **Benson** and **Calvin**, some progress in the elucidation of the steps by which the reduction is effected has been made by using the **radio-active isotope of carbon** (^{14}C). Plants have been allowed to photosynthesise for short periods in the presence of special carbon dioxide containing ^{14}C, and the plant cells then killed and analysed for compounds containing the isotope, in order to discover the nature of the intermediate products into which the carbon dioxide had passed in course of being built up to sugar. The radio-active nature of the isotope facilitates its detection, and methods of *paper chromatography* have been widely used in these studies. After only five seconds of photosynthesis some of the ^{14}C had already been built up to sugar, but there was evidence that the carbon passed first into the form of **phosphoglyceric acid**, a substance which, as will be seen on p. 198, is also formed in the breakdown of sugar in respiration.

The above very brief statement only touches on a few of the important investigations that are being carried on by plant physiologists, chemists and physicists in many different countries into the mechanism of photosynthesis, from which conclusions of great significance are emerging.

ORGANIC PRODUCTS OF PHOTOSYNTHESIS

It is theoretically probable that, as we have so far assumed, the hexose sugars **glucose** and **fructose** are the first to arise in photosynthesis, and there is some experimental evidence supporting this. These initially formed sugars are quickly converted by appropriate enzyme action to the disaccharide **sucrose** and (in many plants) also to **starch,** the latter being deposited in the chloroplasts of the leaf. These products of photosynthesis do not remain permanently in the leaf, but are conveyed to other parts of the plant by the process of translocation (Chap. XVI). Thus if a plant is allowed to carry on photosynthesis until the leaves give a strong starch reaction and is then transferred to darkness, by further use of the iodine test it will be found that the starch content of the leaves gradually falls, so that after twenty-four hours they will probably be free of starch.

The other organic foodstuffs mentioned at the beginning of this chapter (**proteins** and **fats**) are not direct products of photosynthesis; but the sugars formed in photosynthesis undoubtedly form the starting point for the formation of these further foodstuffs.

As already indicated (Chap. VI), the protein molecule is built up from **amino-acids**. There is some evidence that these acids arise in the tissues of the plant by interaction between ammonia, derived from the nitrates absorbed from the soil, and **organic acids** commonly present in the tissues. Examples of such plant acids are **malic, citric, oxalic** and **succinic acids**. These acids are usually considered to arise in the plant from sugars, and may be intermediate products in the respiratory breakdown of the sugars (Chap. XVII). From the amino-acids, proteins are thought to be formed by a process of condensation. Protein synthesis is probably essentially independent of the presence of light, but the leaf may again be the most important site for the synthesis, since it is there that the sugars are most abundant, while the nitrogenous salts absorbed from the soil are largely conveyed to the leaf in the transpiration stream.

As was stated in Chap. VI, fats are formed by the interaction of **fatty acids** and **glycerol**. There is good evidence that in the plant fats arise from sugars. Presumably the fatty acids and the glycerol are separately formed from sugar, followed by the production of the fat ; but the precise nature of the chemical changes by which these transformations are achieved remains rather obscure.

PRACTICAL WORK

1. Make a rough extraction of chlorophyll from green leaves. Thin leaves should be chosen, such as those of *Tropæolum* or *Fuchsia*. First of all, kill the leaves by immersing for a few minutes in boiling water. Then place a large test-tube, about half full with ethyl alcohol, in a vessel containing water which has just been boiled, but be sure that the burner has been turned off. Immerse the killed leaves in the alcohol. Note how the leaves gradually lose their colour owing to the alcohol dissolving the chlorophyll.

FIG. 126. EXPERIMENT TO PROVE THAT CHLOROPHYLL IS NECESSARY TO PHOTOSYNTHESIS.

Left, a variegated leaf, showing the green parts (shaded) and the white parts (unshaded) ; right, the same leaf after the iodine test for starch, showing the area which has reacted positively (black) and the area which has reacted negatively (white).

FIG. 127. EXPERIMENT TO PROVE THAT A CERTAIN CONSTITUENT OF THE AIR IS NECESSARY TO PHOTOSYNTHESIS.

Above, a leaf with part of its surface smeared with 'Vaseline' (shaded) ; below, the same leaf after the iodine test for starch, showing part which has reacted positively (black) and part which has reacted negatively (white)

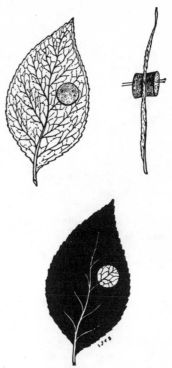

FIG. 128. EXPERIMENT TO PROVE THAT
CARBON DIOXIDE IS NECESSARY TO PHOTO-
SYNTHESIS.

FIG. 129. EXPERIMENT TO SHOW
THAT LIGHT IS NECESSARY TO PHOTO-
SYNTHESIS.

Top left, a leaf, a portion of which
is covered by two corks fixed with
pins ; top right, side view ; below,
the same leaf, tested for starch
twenty-four hours afterwards.

2. Make an extract of chlorophyll in 85 per cent acetone from dried nettle-leaf powder.
Place some of the powder in a funnel fitted with a filter paper, slowly add the acetone and
collect the filtrate in a beaker or test-tube. The filtrate is an acetone solution of almost pure
chlorophyll. Note the fluorescence as seen by looking through the solution held against
a dark background.

3. Show that some green leaves are capable of manufacturing the carbohydrate starch,
by means of photosynthesis. Choose a leaf of the potato, *Fuchsia*, or lilac and treat it as in
Experiment 1, until it becomes whitish through loss of colour. Then subject it to the iodine
test for starch. Perform a similar experiment, using onion or hyacinth leaves.

4. Prove that chlorophyll is necessary in order that photosynthesis may take place.
Perform an experiment similar to Experiment 3, choosing some type of thin variegated leaf.
In this type of leaf, there are certain areas green with chlorophyll and others devoid of
chlorophyll (Fig. 126). Note that the green areas give a positive starch reaction, whereas
the white areas react negatively.

5. Prove that a certain constituent in the air is necessary for photosynthesis. Choose a
green leaf (do not remove it from the plant) and smear some 'Vaseline' over a portion of its
area (Fig. 127). Do this on both sides of the leaf, making the areas coincide with each other.
Allow the leaf to remain in such a condition for about two days, then take it off the plant
and remove the 'Vaseline'. Subject the leaf to the iodine test for starch.

6. Prove that the constituent of the air necessary to photosynthesis is carbon dioxide

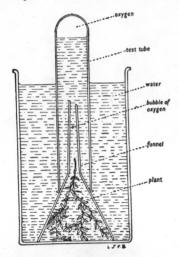

FIG. 130. EXPERIMENT FOR DEMONSTRATING PHOTOSYNTHESIS IN AN AQUATIC PLANT.

N.B.—This apparatus is more easily assembled completely under water.

Fix up the apparatus as shown in Fig. 128. Make sure that the split rubber stopper is fixed firmly, and smear 'Vaseline' over the surface of the stopper to prevent any air entering the vessel either between the stopper and the neck or the stopper and the plant stem. Place the apparatus in a warm room near the window, or under any conditions such that photosynthesis can take place freely. Leave for two or three days, then perform the iodine test for starch on a leaf from that part of the twig inside the vessel and also a leaf outside the vessel.

Realising that caustic potash absorbs carbon dioxide strongly, fully explain the results obtained.

7. Show that a good supply of water is necessary for photosynthesis. Place a potted plant in a dark room for forty-eight hours, so that at the end of that time there is no starch present in the leaves. Then remove two leaves. Stand one in water and place in the light ; put the other in the light too, but do not supply it with water. After about eight hours, test both leaves for starch.

8. Show that light is necessary for photosynthesis. Choose two small corks equal in size, and by means of a couple of pins, fix them opposite each other on the surfaces of a leaf as shown in Fig. 129, taking care not to crush the tissues of the leaf. Do this without removing the leaf from the plant. After about twenty-four hours and towards the end of the day, remove the leaf and test it for starch by means of the iodine test.

9. Show that green aquatic plants carry out photosynthesis. There is normally sufficient carbon dioxide dissolved in tap-water to supply such submerged plants. Choose some fresh green Canadian pondweed (*Elodea canadensis*) and place it in the apparatus as shown in Fig. 130. The test-tube is initially full of water. Place the apparatus in bright sunlight or near a 100-watt lamp. Notice that small bubbles are given off periodically at the cut ends of the pondweed and collect in the top of the tube, gradually displacing the water. When a sufficient supply of the gas has been collected, remove the test-tube and test the gas with a glowing splinter. The gas proves to be oxygen. If a little potassium bicarbonate, sufficient to give a 1 per cent concentration, is dissolved in the water in which the *Elodea* shoots are lying, the evolution of oxygen will be favoured, since in this way additional carbon dioxide is supplied. If the light falling on the shoots is reduced by shading, the rate of bubbling will immediately slow down, providing further evidence of the importance of light for photosynthesis.

CHAPTER XVI

TRANSPIRATION AND THE INTERNAL TRANSPORT OF WATER AND SOLUTES

It has been seen in previous chapters that water is one of the essential foods for plants. Besides being actually used in certain chemical reactions in the plant, such as photosynthesis, water is the medium in which the protoplasm works and is also essential if the cells are to attain their proper size and turgor. But the root system of the plant absorbs far more water than would be required for these purposes alone. This is because of the necessity of replacing water given off from the aerial parts of the plant, especially from the leaves, in the form of water vapour, a process which is called **transpiration**. It is because of transpiration that the inside of the glass of a greenhouse is often covered with condensed moisture, especially in the early morning when the ventilators have been closed for some time and the outside temperature is low. Under favourable conditions transpiration may be very rapid, so that an oak tree may transpire as much as 150 gallons of water in one day. Methods for demonstrating and measuring transpiration are described in the practical work.

SIGNIFICANCE OF TRANSPIRATION

From earlier chapters we know that the tissues of the leaf are permeated by a continuous system of air-spaces communicating with the external atmosphere through the stomata. The cellulose cell-walls bounding the air-spaces contain imbibed moisture. It is obvious that when the stomata are open moisture will under normal conditions evaporate from these cell-walls, this being followed by the diffusion of the vapour through the air-spaces and out through the stomata, as a result of the gradient of concentration that will exist. Experiment shows that besides the loss of water in this way, additional evaporation occurs direct from the outer surface of the epidermal cells, because in many leaves the cuticle is not fully waterproof. This is termed **cuticular transpiration**. But by far the greater part of the total water loss is due to **stomatal transpiration**, particularly in the day-time when the stomata are normally open.

Some plant physiologists have taken the view that transpiration, the occurrence of which as we have seen follows automatically from the structure of the leaf, is a process of little value to the plant and, in times of water-shortage, detrimental to it. This is an extreme view, and to a plant well supplied with water it is probable that transpiration has definite advantages. In the first place the evaporation has a cooling effect which helps to prevent harmful overheating of the leaf, especially when it is exposed to strong sunshine, just as evaporation of moisture from the human skin has a cooling effect. Secondly the flow of water from the soil up through the plant to the leaves which, as will be seen later, is initiated by transpiration, probably assists in the absorption of mineral salts by the roots from the soil and certainly accelerates their transport up the plant. The fact that in most plants there is a subsidiary mechanism bringing about an upward flow of water in the plant when transpiration has ceased suggests that such a flow is of importance to the plant (see root pressure, p. 188).

FACTORS AFFECTING TRANSPIRATION

In the first place transpiration is affected by the same factors that control the rate of ordinary evaporation. Thus one important factor is the **humidity** of the atmosphere, for water evaporates at a much greater rate in a dry atmosphere than in a very humid one. **Wind** has a very potent effect in increasing evaporation and transpiration. Very often gardeners protect their plants from harmful excessive transpiration due to wind by erecting ' wind-breaks ' constructed from reeds, branches or similar material. Such breaks are a common sight in that part of France where the well-known piercing and bitter wind, the mistral, sweeps down to the Gulf of Lions.

Temperature is also an important factor. A rise in temperature increases not only the rate of transformation of liquid water into water vapour, but also the amount of water vapour that the atmosphere can absorb. Unlike many animals, plants tend to take on the temperature of the environment, so that a change in the temperature of the atmosphere leads to a corresponding change in leaf temperature and in the rate of transpiration. **Atmospheric pressure** has some effect. The rarer the atmosphere the greater is evaporation, and it is probable that to be partly correlated with this is the finding that alpine plants, growing at high altitude where atmospheric pressure is low, frequently show modifications likely to prevent excessive transpiration.

The factors so far considered affect ordinary evaporation and transpiration similarly. In the case of the plant process, the necessity for most of the water vapour to pass through the stomata before reaching the external atmosphere results in further control, since any factor affecting **stomatal aperture** is liable to exert some control over transpiration. For this reason **light intensity** has an important effect on transpiration, since as was seen in the previous chapter stomatal aperture is chiefly controlled by this factor. Thus transpiration takes place much more rapidly during the day than at night-time, when the stomata are typically closed. The significance of this night closure of the stomata is probably that unnecessary loss of water by transpiration is avoided over a period when there is little need for gaseous exchange with the atmosphere.

Stomatal aperture is also affected to some extent by the **water content** of the leaf tissues, which in turn depends on the balance between the rate of transpiration and that at which additional water is procured from the roots. A fall in water content and turgor of the guard cells will tend to cause stomatal closure, and it was formerly thought that in this way the stomata exercised a close control over transpiration and adjusted it in accordance with the available supply of water. Experiment has indicated that often a very considerable reduction in stomatal aperture must occur before transpiration is itself curtailed, because when the stomata are wide open their diffusive capacity may not be fully utilised. If transpiration is so excessive as to cause visible loss of leaf turgor or **wilting**, the stomata usually (though not always) tend to close completely, so that leaf moisture is conserved. Irrespective of this stomatal regulation, transpiration does in some degree keep in step with the supply of water, for if the latter becomes inadequate

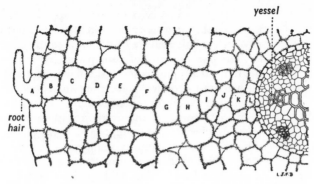

FIG. 131.   PART OF A TRANSVERSE SECTION OF A ROOT TO SHOW HOW WATER PASSES FROM
THE ROOT HAIR TO THE STELE.

the cell-walls in the leaf become somewhat drier, with consequent reduction in the
rate of evaporation from them.

THE TRANSPIRATION STREAM

The water that is lost in transpiration from the leaf is normally made good as the
result of an upward stream of additional water absorbed by the root from the soil.
This current of water (containing small amounts of mineral salts in solution)
flowing through the plant from the root hairs to the leaves is known as the **transpiration stream.**

If cut shoots are stood in dye solutions until the dye has reached the upper
regions of the shoots, sections then prepared from the stem will show that only the
xylem tissues are stained (see practical work). It may therefore be concluded
that the transpiration stream ascends the plant through the vessels and tracheides
of the xylem, a tissue which forms a continuous system extending through root,
stem and leaf.

It is obvious that in order to reach the xylem of the root, the water absorbed
must first pass across the living cells lying between the root hair and the xylem.
If as the result of some mechanism (to be considered below) water constantly
passes from the inner living cells, such as those of the pericycle, into the xylem
then a gradient of **suction pressure**, decreasing as we move outwards to cell A
(Fig. 131), will be automatically set up and will lead to a continued passage of water
across the cortex. To make this a little clearer it can be supposed that for a short
time the movement of water into the xylem ceases, and also that an ample supply
of pure water is available to the root hairs. The result will be that all the cells in
the series A to L (together with the endodermal and pericyclic cells) will become
fully turgid. If now a little water passes from the pericycle into the xylem, the
pericycle cell will develop a certain suction pressure which leads it to absorb water
from the adjacent endodermal cell, which will in turn absorb from cell L. By a
repetition of these events water is eventually withdrawn from cell A, with the result
that the root hair takes water from outside. Actually these events occur as a
continuous process, so that there is continual passage of water across the cortex.

We have next to consider the nature of the mechanism that effects the movement of water into the xylem vessels of the root and then up the root and stem to
the leaves. Clearly very considerable forces must be involved in raising water, in

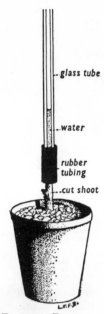

FIG. 132. EXPERIMENT
FOR   DEMONSTRATING
ROOT PRESSURE.

the quantities observed to be transpired, to the leaves of taller plants such as trees, and there has been much discussion as to the nature of the mechanism responsible.

The phenomenon of **root pressure** must first be considered. If a plant is severed a few inches above soil-level an exudation of water (containing small amounts of material in solution, chiefly of mineral nature) is often found to occur from the stump. If the exudate is allowed to collect in a tube attached to the stump, as in Fig. 132, or, still better, if some form of mercury manometer is attached, considerable hydrostatic pressures are often set up. Thus the stump of a *Fuchsia* plant will often set up a pressure in the region of 1 atmosphere (15 lb. per square inch).

Closely related to root pressure is the exudation (or **guttation**) of drops of water from the leaves of intact plants which is shown by many species. The phenomenon is best seen in the early morning in summer-time. Frequently the drops form round the margin of the leaf, as in *Fuchsia*, lady's mantle (*Alchemilla vulgaris*) and *Tropaeolum*, but in grasses they appear at the leaf-tips and are often mistaken for dew. Guttation appears to be due to the root pressure mechanism continuing to force water up the plant during the night when transpiration, owing to the closure of the stomata, is slow. The plant becomes gorged with water, and the latter is forced out in liquid form from special glandular structures (**hydathodes**) present in the epidermis (Fig. 133). Sometimes these take the form of permanently open stomata situated over the ends of veins, as in *Tropaeolum*, or of projecting, club-shaped structures scattered over the leaf surface, as in *Phaseolus*. Guttation is very common in tropical rain-forests, where transpiration is curtailed by the high humidity.

It is clear that there exists in the root system a mechanism capable of forcing water up the plant, but though it is generally agreed that this is osmotic in nature the elucidation of the mechanism is incomplete. One theory suggests that sufficient solutes accumulate in the water occupying the xylem of the root to set up an osmotic pressure high enough to lead to active absorption of water from the living cells in contact with the xylem. According to another, an osmotic mechanism exists in these living cells which results in water being forced out into the xylem.

The question naturally arises as to how far root pressure can account for the upward flow of the transpiration stream. Consideration of the magnitude of root pressure in different plants and of the rate of exudation from severed plants compared with that of transpiration from similar intact plants, has led most investigators to the conclusion that root action alone is quite inadequate to account for the upward flow of the transpiration stream, especially in shrubs and trees. Some much more powerful mechanism must accordingly be present, and it is fairly generally accepted that the **cohesion theory** originally propounded by **Professor H. H. Dixon** provides the best explanation yet available as to the nature of this mechanism.

According to this theory the upward movement of water is chiefly due to a pulling force exerted by the leaves. The loss of water in transpiration tends to

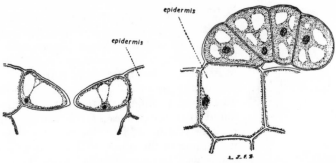

Fig. 133. Left, Section through a Hydathode of *Tropæolum majus* ; Right, Hydathode of *Phaseolus multiflorus* (Scarlet Runner).

increase the suction pressure of the mesophyll cells, with the result that they
absorb additional water from the xylem elements of the veins permeating all
through the leaf. The water in these elements is visualised as forming the upper
ends of thread-like columns of water stretching right down through the vessels
of the stem and root ; the withdrawal of water from the top of one of these leads to
the column as a whole being drawn up the plant, and to a corresponding amount of
water being sucked into the xylem from the adjacent living cells in the root.

·Now the possibility of pulling up a column of water enclosed in a tube such as
a xylem vessel depends upon the existence of strong mutual attraction (**cohesion**)
between the molecules of the water, and of a corresponding attraction (distinguished
as **adhesion**) between the molecules of water and those of the materials forming
the wall of the vessel. Cohesion is a property of substances in general, though it is
stronger in solids than in liquids. But Dixon and other investigators have shown
by direct and indirect methods of experiment that columns of water enclosed in
perfectly clean glass tubes or otherwise supported can withstand very great ten-
sions without breaking down, that is, without the molecules being torn apart.
Thus in one experiment a column of water remained intact although subjected to
a tension exceeding 100 atmospheres.

The so-called **Askenasy** experiment is considered to present a close analogy
to the system operating in the plant as conceived by the cohesion theory (Fig. 134).
The porous pot and the long glass tube are initially filled with water, with complete exclusion of air bubbles, while at the base the tube dips into mercury.
Evaporation of water occurs from the surface of the porous pot, with the result
that the wall of the pot absorbs further water from the interior. By cohesion,
water and subsequently mercury are drawn up the glass tube. If proper precautions
have been taken in setting up the apparatus, the mercury may rise to a considerable
height, for example, exceeding 100 cm.

In addition to the evidence showing that by virtue of cohesion it is possible to
raise a column of water by a pull applied at the top, other observations support the
cohesion theory. Thus there is evidence that during active transpiration the
contents of the xylem vessels are actually in a state of tension, and also that the
cells of the leaf have an adequate osmotic capacity for exerting the required pull.
The commonest criticism of the theory is that the stipulated columns of water,
extending unbroken right through the plant may not actually exist, since tests
suggest that air bubbles are frequently present in the vessels. It is possible,
however, that a sufficient proportion of the vessels are entirely filled with water.

The cohesion theory provides a satisfactory explanation of an internal redistribution of water in the plant of which evidence is sometimes obtained. For example, in an experiment carried out by **Maximov** two leafy branches were cut from an orange tree, one of the branches bearing fruit in addition. No water was supplied to the cut branches. It was found that the leaves of the fruit-bearing branch remained fresh and free of wilting for a longer period than those of the other, and it was concluded that water was being transferred from the fruits to the leaves, as a result of the greater increase in suction pressure occurring in the more vigorously transpiring leaves.

XEROPHYTES

Some plants show a special ability for growing in places where the water supply would be inadequate for ordinary plants, either on account of an actual deficiency in the soil, or due to atmospheric conditions encouraging excessive transpiration, or to both. Such plants are termed **xerophytes** and commonly exhibit special features in their form and internal structure which seem likely to have the effect of curtailing transpiration, though it is usually not possible to determine by direct experiment the extent of the curtailment.

Externally xerophytes often show relatively small

Fig. 134. Apparatus for Askenasy Experiment.

leaves, as for examples gorse (*Ulex europaeus*) and broom (*Sarothamnus scoparius*), and, as in those two plants, the stem may add to its normal functions that of considerable activity in photosynthesis. Many xerophytes show provision for the storage of water in leaf or stem, these organs being in consequence swollen and fleshy, as in the desert cacti (Fig. 29), the houseleek (*Sempervivum tectorum*, Fig. 32), which grows on old cottage roofs and walls, and in several native species of *Sedum*. The development of spines or thorns is a common feature and gives protection against herbivorous animals. In some instances, as in the mullein (*Verbascum thapsus*), the leaves are covered with a dense growth of hairs which seems likely to impede transpiration (p. 167).

In other xerophytes the leaf is rolled lengthwise in such a way that the stomata lie on the inner surface. The rolling may be permanent, as in the crowberry (*Empetrum nigrum*, Fig. 135) and the cross-leaved heath (*Erica tetralix*), both of which grow on exposed moors, or may be shown only during dry weather as in various grasses such as marram (*Ammophila arenaria*, Fig. 135) which grows on sand dunes, and a number of moorland types (Fig. 136).

In these grasses the stomata are confined to the upper surface, which is deeply grooved and lies to the inside when the leaf is rolled. The illustrations (Fig. 136) show the leaves in the rolled-up condition, but in moist weather special ' aqueous cells ' present at the base of each groove swell and tend to open-up the grooves, with the result that the leaf unrolls and assumes a flattened form.

Many xerophytic leaves show unusually thick epidermal walls and cuticle, resulting in diminished cuticular transpiration, and this together with a free development of mechanical tissue often gives a leathery texture to the leaf. The stomata are frequently sunk in deep pits below the level of the epidermis, as for example in the pine and in an Australian desert plant, *Hakea* (Fig. 137), the pocket of moist air which collects in the pit being likely to depress transpiration. It seems probable that these various types of stomatal protection will be attended by the disadvantage that the entrance of carbon dioxide for use in photosynthesis will also be impeded.

The leaf-fall shown in autumn by ordinary mesophytic trees and shrubs of deciduous type is undoubtedly in part related to what has been termed the 'physiological drought' liable to be experienced by plants during the winter months. By this leaf-fall the transpiring surface is greatly reduced at a season when, owing to low soil temperatures, root activity is curtailed. The leaves of those evergreens which retain their foliage over the winter usually show some degree of xerophytic structure.

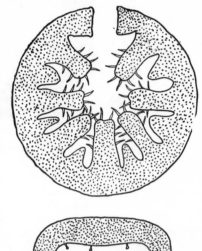

FIG. 135. OUTLINES OF TRANSVERSE SECTIONS OF LEAVES OF MARRAM GRASS (ABOVE) AND CROWBERRY (BELOW) SHOWING THEIR ROLLED-UP FORM.

Hairs are present on the ridges in the marram grass and at the entrance to the leaf chamber in the crowberry. The stalked structures in the leaf chamber of the latter are glands.

INTERNAL TRANSPORT OF DISSOLVED SUBSTANCES

The organisation of the plant necessitates a constant transference of material in solution from point to point within the plant. The upward transport of mineral salts within the transpiration stream has already been noted. In addition there is the very important transport of manufactured food materials, and to this the term **translocation** is applied.

We have seen that the leaf is the chief food 'factory' for the plant. From the leaf the carbohydrates and proteins manufactured there are distributed to other parts of the plant. All the living cells of the plant require such materials, but in particular the foods pass to actively growing parts of the plant, such as the growing points of root and shoot, or regions of cambial activity. Another important line of flow is from the leaves to regions where food is stored for future use. Thus in many herbaceous plants storage is effected in parts specially distended for the purpose, as in tubers, bulbs, corms and rhizomes, while in trees food is stored in the medullary rays, wood parenchyma and cortex of the trunk and branches. In addition seeds usually contain a considerable food reserve which meets the needs of the seedling in the early stages of growth.

Although the translocation of manufactured materials out of the leaf is a

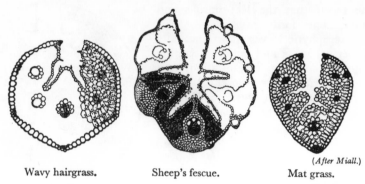

(*After Miall.*)

Wavy hairgrass.   Sheep's fescue.   Mat grass.

FIG. 136. TRANSVERSE SECTIONS OF LEAVES OF MOORLAND GRASSES.
The upper surfaces bear the stomata, and are infolded.

continuous process, it is most readily demonstrated when the leaf is darkened so
that further food synthesis is arrested. Thus, as already noted, by application of
the iodine test to leaves kept in darkness it can be shown that the starch content
steadily diminishes, this being due chiefly to translocation. It is only in the soluble
form that foods can be translocated. Insoluble material, such as starch, must
first be converted to soluble products by appropriate enzyme action. There is
evidence that sucrose is the form in which translocation of carbohydrate material
occurs, while nitrogenous food may be moved as amino-acids.

There is good evidence that translocation of these organic materials takes
place chiefly through the sieve tubes of the phloem which form a continuous
system extending through most parts of the plant. As noted in Chap. X, the
structure of these elements appears to be adapted to facilitating a lengthwise
conduction, while again their contents are often relatively rich in carbohydrate and
nitrogenous substances. Further evidence has been obtained from ' **ringing** '
experiments, in which all the tissues external to the cambium are removed over a
short zone of a stem. Some of the earliest experiments of this kind were carried
out many years ago by **Stephen Hales**, a curate of Teddington, Middlesex, who
developed a great interest in plant and animal physiology and carried out important pioneer investigations in these fields.

It is found that when the outer tissues are removed in this way, the leaves on the
ringed branch remain fresh, confirming that the flow of water is not interrupted.
But various observations indicate that translocation is impeded. Thus an accumulation of food materials can often be demonstrated to have occurred on one side of
the ring, while differences in growth activity may develop. As is well-known, a
tree frequently dies if the main trunk is ringed. These observations show that
some tissue external to the cambium is concerned with translocation, and taking
into account structural considerations it is reasonable to conclude that the sieve
tubes are responsible. This view has been supported by more critical experiments
carried out by **Mason and Maskell**, which revealed a close correspondence
between the daily fluctuations in the concentration of carbohydrate and nitrogenous materials in the leaf tissues and those in the phloem of the stem below the
leafy zone.

The mechanism which actually brings about the movement of materials
through the sieve tubes is not fully understood. Although there are superficial

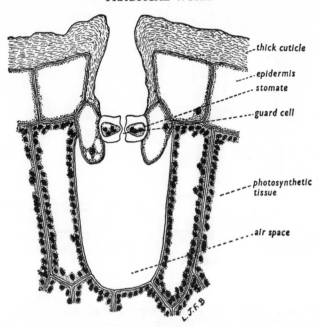

thick cuticle

epidermis

stomate

guard cell

photosynthetic tissue

air space

FIG. 137. PORTION OF A SECTION THROUGH A *Hakea* LEAF, PASSING THROUGH A STOMA.
Note the sunken stoma and the very thick, waxy cuticle.

resemblances between translocation and ordinary diffusion, consideration of the rate at which the plant process occurs indicates that diffusion alone cannot provide a satisfactory explanation. An osmotic mechanism is visualised in the **mass-flow theory** advanced by **Münch**. According to this, the accumulation of sugars and other foods in the sieve tubes of the leaf is accompanied by an osmotic absorption of water from adjacent xylem elements, so that a hydrostatic pressure is set up. This results in the contents of the sieve tubes flowing out of the leaf to other parts of the plant, such as storage organs, where, owing to the rapid utilisation of soluble food materials, pressure in the sieve tubes is relatively low. In such regions water is returned from the sieve tubes into the xylem.

PRACTICAL WORK

1. Show that water vapour is given off from leafy shoots.

Take several leafy shoots of a plant and enclose them in a bell-jar sealed to a glass plate, or in a suitable stoppered jar. Set up a similar but empty container as a control. After a few hours, moisture will be seen to have condensed on the inside of the container with shoots.

2. Demonstrate the occurrence of transpiration by means of a weighing method.

A well-watered plant growing in a pot is required, such as a *Pelargonium* or *Fuchsia*. The surface of the soil and the sides and base of the pot must be covered with water-proof material. The pot can be placed in one of the special metal containers that are available for this type of experiment, or in a metal food-can of suitable size. The soil surface may be covered by a piece of sheet rubber which is slit at one point to admit the stem and then re-sealed with rubber solution. The rubber is secured to the metal container by means of a special clamp or with string. The whole apparatus is then weighed on a suitable scale or balance, preferably with an accuracy of not less than 0·1 gm., after which the apparatus is stood in a greenhouse or on a window-sill for several hours. Re-weighing will reveal an

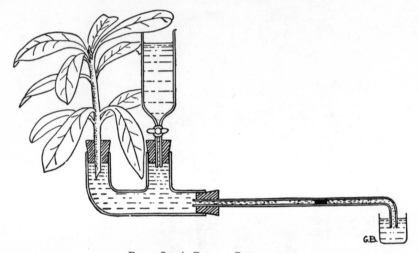

FIG. 138. A GANONG POTOMETER.
The air-bubble (black) can be forced back to the right-hand end of the tube by admitting
water from the funnel.

appreciable loss in weight, this being due to transpiration. A direct measure of the rate
of transpiration is thus obtained, and the way in which the process is affected by different
factors can be studied.

3. Fit up a potometer.
With this apparatus (Fig. 138) it is really the absorption of water by a cut shoot that is
measured, but since under a fairly wide range of conditions absorption runs parallel to
transpiration the apparatus also gives information about the latter process.

Before the shoot is inserted the wider part of the apparatus, to the left in the figure, is
to be filled with water. To do this, the end of the capillary tube is closed temporarily with
'Plasticine', and then by admission of water from the reservoir and by tilting the apparatus
to dislodge air bubbles it can be filled as required. A shoot of some shrubby plant such as
one of the garden laurels or *Rhododendron*, cut from the plant some hours previously and
stood in water in the interim, is now fitted into a split rubber stopper or into a short length
of thick-walled rubber tubing, and this inserted into the appropriate arm of the potometer.
This joint must be made air-tight by means of 'Vaseline' or 'Vaseline'-beeswax mixture. The
plasticine is then removed and water admitted from the reservoir to displace the air from the
capillary tube. By means of filter paper a little water is absorbed from the end of the capillary and the latter then immersed in the small beaker of water.

The rate of movement of the bubble of air thus trapped in the capillary serves as an
index of the absorption of water by the shoot and indirectly of transpiration. A ruler or
other scale is fixed in position behind the capillary tube, and the time taken by the bubble
in traversing a definite number of divisions is noted. The effect of different conditions can
be studied, for example, light and darkness, still air and moving air.

4. Show that transpiration occurs chiefly through the stomata.
Method (*a*). Choose four similar leaves of some thin-leaved plant known to have
stomata on the lower surface only, and treat as follows :

1. Smear both surfaces completely with 'Vaseline'.
2. Smear the upper surface only with 'Vaseline'.
3. Smear the lower surface only with 'Vaseline'.
4. Leave untouched.

Then pin the leaves up and allow them to hang freely exposed to the air. After a day,
inspect the leaves and compare the extent to which they have wilted.

Method (*b*). Dip some half-inch squares of filter paper into a 5 per cent solution of
cobalt chloride and then dry them in an oven. Notice that when dry they are blue in

colour, but if one piece is exposed to a humid atmosphere or dipped in water it will turn pink. The dried paper can be stored in a specimen tube over anhydrous calcium chloride.

Now choose a leaf known to have stomata on the lower surface only, place a dry piece of cobalt chloride paper in contact with this surface and quickly cover with a small piece of glass. Note the time required for the blue colour to fade. Repeat for the upper surface, where a considerably longer period will be required for decoloration.

5. Show that the path of the transpiration stream is through the xylem.

Dilute some red ink, or prepare a solution of eosin, and stand some white flowers such as *Narcissus*, and also some cut leafy shoots, in it. Within a short time the dye will reach the petals or leaves, and particularly in the former it will be obvious that the staining is confined to the veins. If the stems are cut across it will be seen that it is the xylem of the bundles that is stained.

6. Demonstrate root pressure in a plant.

Obtain a single-stemmed potted plant (*Fuchsia* is suitable) and by means of a sharp knife or secateurs sever the stem about two inches above soil level. Fit a short piece of rubber tubing to the stump (Fig. 132) and fill the bore of this tubing with water. Then insert the end of a piece of glass tubing about two feet long and approximately one-eighth inch bore. Each day measure the height attained by the exudate which will soon commence to collect in the tube. Alternatively a mercury manometer can be attached instead of the simple glass tube.

7. Show that transpiring leaves exert a considerable suction.

Set up the apparatus shown in Fig. 139, taking care to make all joints tight and to

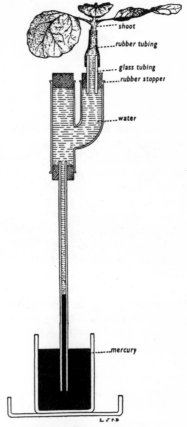

FIG. 139. EXPERIMENT FOR DEMONSTRATING SUCTION DUE TO TRANSPIRATION.

avoid the inclusion of any air bubbles. If the experiment is properly set up the mercury may rise to a considerable height in the narrow glass tube, which should be three feet in length.

CHAPTER XVII

RESPIRATION AND FERMENTATION

All living things must perform a certain amount of work in order to continue living, for many processes in the living organism involve **work**, and this in turn necessitates the utilisation of **energy** in some form. Thus in animals energy is obviously used in running, swimming or flying, breathing, maintenance of body temperature, etc., and although the need of the plant for energy is less patent, there is no doubt that food-building processes other than photosynthesis (the special source of energy for which has already been considered) require energy, while in addition it is possible that internally produced energy is used by the plant during the absorption and transport of mineral salts and other solutes.

Energy exists in several definite forms. Energy due to motion is said to be **kinetic** ; for example, water-wheels are turned by the kinetic energy of rivers. Energy which is stored is called **potential** energy, such, for example, as water at the top of a cliff, or the wound-up spring or weights of a clock. The potential energy of a wound-up spring is liberated as kinetic energy when the spring is released. A charged Leyden jar possesses energy in the potential state, but, on discharging, the energy is released as **electrical** energy. This electrical energy can again be transformed into another form of energy : that of **heat**. Thus there are several forms of energy—potential, kinetic, chemical, electrical, heat, light, radiant and so forth. Not only do these forms of energy exist, but also it is possible in certain circumstances, to convert one form into another.

SOURCE OF ENERGY FOR VITAL ACTIVITIES

The energy which living things, plants and animals, utilise in vital activities of the kind mentioned in the first paragraph is really that which is fixed in the first place by photosynthesis in plants. We saw in Chap. XV that a very considerable amount of radiant energy derived from certain visible light rays is utilised during that process and embodied in the sugar molecules, and it will be useful if we recapitulate to the extent of repeating the over-all equation for photosynthesis :

$$6CO_2 + 6H_2O + \text{radiant energy} \rightarrow (C_6H_{12}O_6 + \text{potential energy}) + 6O_2$$
$$\text{\textit{from the sun}}$$

A large proportion of the sugar formed in this process becomes incorporated into more complex substances such as cellulose required in the formation of new parts of the plant, or is deposited in various forms in storage organs ; but some of the sugar is utilised in a further process to be considered below by means of which the potential energy of the sugar molecules is released for use in energy-needing activities of the plant. The animal likewise obtains energy from organic foods such as sugars, though in this case the foods have not been built up by the animal itself but rather obtained ready-made from its environment. The ultimate origin of the energy obtained in this way by animals is, as already stated, that it too was fixed in the first instance by plant photosynthesis from sunlight.

RELEASE OF ENERGY IN LIVING ORGANISMS

Since energy is stored in complex foods such as sugars while they are being built up, it is clear that the energy would be released if the foods could be broken down again into their raw materials. This is what plants and animals do in the process of **respiration**, a term which actually refers to the act of breathing which, more visibly in animals, is the external sign that respiration is proceeding.

Thus it comes about that in many ways respiration in the plant is the opposite of photosynthesis, since it normally involves the absorption of oxygen, the breaking down of foods (usually sugars), the formation of carbon dioxide and water, and the release of energy. There is good evidence that the respiration process can in fact be summed-up by reversing the equation for photosynthesis, in the following way :

$$(C_6H_{12}O_6 + \text{potential energy}) + 6O_2 \rightarrow 6CO_2 + 6H_2O + \text{released energy}$$
$$(674 \text{ kgm. cals.})$$

By means of this oxidatory breakdown energy originally fixed in the leaf can be released and used for vital activities in any part of the plant and at any stage in plant development. Thus the food-store in the seed, laid down by the activity of the parent plant, provides for the energy requirements of the developing seedling.

RESPIRATORY GASEOUS EXCHANGE

It is generally considered that in the presence of light the green parts of the plant, in addition to photosynthesising, also carry on respiration ; but since photosynthesis goes on much more rapidly than respiration it is the gaseous exchange of the first process that is exhibited, that is, carbon dioxide is absorbed and oxygen is liberated. In darkness the same green organs show the gas exchange of respiration—oxygen is now absorbed from the atmosphere and carbon dioxide is released. Non-green organs, for example, roots, flowers, storage organs such as potato tubers, fruits such as the orange or the apple (where chlorophyll is present only in the skin), show the respiratory type of gas exchange all the time irrespective of the light factor, as also applies to animals.

In consequence of these activities the plant can be regarded as ' contaminating ' the atmosphere during the hours of darkness, since all parts of it then absorb oxygen and liberate carbon dioxide, like the animal. In daytime, however, the photosynthetic activity of the green organs more than compensates for the respiratory events in the non-green parts, the net result being, as already noted in Chap. XV, that the plant now exerts a ' purifying ' effect on the atmosphere.

As in photosynthesis the plant relies on ordinary gaseous diffusion to effect the movement of the required oxygen from the external atmosphere to the interior of the plant organs and of the carbon dioxide in the opposite direction. Stomata and lenticels facilitate the passage of the gases through epidermis or bark, while in the case of young roots the gaseous exchange with the soil atmosphere takes place at the root surface. Thus the plant exhibits a simpler arrangement than is found in most animals, since in these oxygen is taken up in the lungs by the special oxygen-carrying pigment, haemoglobin, and transported by this substance in the blood stream to all parts of the body, the carbon dioxide being returned to the lungs in the same blood stream.

SUBSTRATE FOR PLANT RESPIRATION

In the equation for respiration given above, hexose sugar is shown as forming the initial substrate. Evidence of various kinds indicates that this is the case. For example there is that provided by the value of the **respiratory quotient**, which is the term applied to the ratio of the volume of carbon dioxide evolved to that of oxygen absorbed during the respiration of a given sample of plant tissues over a certain time. According to the equation already given this ratio should be unity, and when the gas volumes are actually measured most plant tissues give a quotient of 1 or very near to it. While this does not point specifically to hexose being the substrate, since the same quotient would be expected with other carbohydrates such as sucrose, it does exclude non-carbohydrate substances such as fats or proteins, the oxidation of which would be attended by quite different quotients. There is also analytical evidence showing that the respiration of organs detached from the plant, such as fruits or darkened leaves, is accompanied by a fall in their carbohydrate content. Other considerations indicate that hexose sugars are the particular carbohydrates that are used.

Fats do, however, sometimes form the substrate for respiration, especially during the germination of seeds rich in fat, such as sunflower or flax.

CHEMICAL MECHANISM OF RESPIRATION

As in the building-up of sugars in photosynthesis, so in their respiratory breakdown many intermediate stages are involved. The identity of all these stages has not been finally established, but there is good evidence that most, if not all, of them are enzymic in nature and that the first part of the sugar breakdown is due to the action of the group of enzymes known collectively as **zymase**, already mentioned in Chap. VIII.

This enzyme complex, zymase, is best known in the fungus **yeast** (see Chap. XXIV) and is responsible for the process of **alcoholic fermentation** characteristically associated with that organism, in which sugar is broken down to ethyl alcohol and carbon dioxide. The chemistry of this process has been very carefully studied by **Meyerhof** and others, and there is good evidence that the breakdown of the sugar in the yeast cell follows the course shown in the following scheme :

Hexose sugar
↓
Hexosediphosphate
↓
Phosphoglyceraldehyde
↓
Phosphoglyceric acid
↓
Phosphopyruvic acid
↓
Pyruvic acid (+ phosphate)
↓
Acetaldehyde + CO_2
↓
Ethyl alcohol

The successive steps in this sequence are promoted by different individual enzymes of the zymase complex. It will be noted that phosphate compounds figure prominently in the earlier stages, and this is of great importance in connection with the transference of respiratory energy as will be explained later.

The zymase complex is also present in higher plants, and it is a commonly held theory that the initial stages in the respiratory breakdown of sugars in these plants are due to zymase action and that they are identical or similar to those shown above. The later stages of the breakdown proceed differently. In the higher plant (provided oxygen is present) the zymase action is thought to be arrested at some intermediate stage, probably at that in which pyruvic acid is produced, and now a new group of **oxidising enzymes** begins to operate, with the final result that the pyruvic acid is oxidised to carbon dioxide and water.

Of the various types of oxidising enzymes concerned with this second phase of respiration, great importance is attached to the **dehydrogenases**, investigated by **Thunberg** and others. As already noted in Chap. VIII, these enzymes oxidise organic compounds, particularly organic acids, by catalysing the removal of hydrogen from them, as exemplified by the action of malic acid dehydrogenase in promoting the oxidation of malic acid to oxalacetic acid:

$$
\begin{array}{ccc}
\text{COOH} & & \text{COOH} \\
| & & | \\
\text{CHOH} & \xrightarrow{\text{enzyme}} & \text{C}=\text{O} \quad +\text{H}_2 \\
| & & | \\
\text{CH}_2 & & \text{CH}_2 \\
| & & | \\
\text{COOH} & & \text{COOH}
\end{array}
$$

Starting with pyruvic acid as the initial substrate, a series of oxidations promoted by different dehydrogenase enzymes is believed to ensue, involving the temporary formation of various organic acids, including the two shown in the above equation. But of course hydrogen is not evolved during respiration. There is evidence that it is immediately taken up by special **carrier substances** shown to be present in all cells, and that these are in turn re-oxidised by **oxidase** enzymes, with the result that the hydrogen is combined with atmospheric oxygen to form water. In fact according to this view the main function of the oxygen absorbed in respiration is to combine with hydrogen in this way.

Accompanying the action of the dehydrogenase enzymes is that of the **carboxylases**, enzymes which split off carbon dioxide from keto-acids, such as pyruvic and keto-glutaric acids, in the following way:

$$
\text{R.CO.COOH} \xrightarrow{\text{enzyme}} \text{R.CHO} + \text{CO}_2
$$
$$
\text{keto-acid} \qquad\qquad \text{aldehyde}
$$

It is thought to be mainly through the action of dehydrogenases and carboxylases that the oxidatory phase of respiration, resulting in the formation of carbon dioxide and water, is completed.

The above views as to the course of higher plant respiration are to a certain extent based on work on yeast, as noted, and also on animal material, such as muscle tissue, and although there is considerable evidence of a strong resemblance

(*From the drawing by Dr. Ulric Dahlgren, by courtesy of the Franklin Institute.*)

FIG. 140. A LUMINOUS DEEP-SEA FISH.

between respiratory events in higher plants and those in the other cases mentioned, further proof of this is desirable.

It is important to notice that the respiratory process, besides releasing energy, may also be the means by which substances required by the plant for food-building purposes are produced. Thus some of the organic acids mentioned above may be withdrawn from the respiratory sequence, and by reaction with ammonia produce amino-acids for use in protein synthesis (Chap. XV).

UTILISATION OF RESPIRATORY ENERGY

As observed at the commencement of this chapter, living organisms carry on a variety of energy-requiring activities, and it is assumed that the energy for these is obtained by respiration. Thus we may note again the necessity of chemical energy for food-building processes in plants and the possible use of energy in the transport of materials within the plant, while in the animal there is the energy required for muscular action, for the functioning of the brain and nervous system, together with some unusual examples such as the generation of electricity by the ray fish and certain eels, or the emission of light energy by glow-worms, fire-flies and certain deep-sea fishes (Fig. 140).

Insight into the way in which the energy of respiration is applied to these various purposes has been gained through the recent work of **Lipmann** and others on the significance of the participation of phosphate in the respiratory process. The phosphate shown in the scheme on p. 198 as released at the pyruvic acid stage is actually in combination with a complex cell-constituent called **adenosine**, which functions as a donator and acceptor of phosphate in respiration. More phosphates of this substance may be formed in the oxidative phase of respiration. They are distinguished first by having a very high energy content, this energy being that released from the sugar broken down in respiration, and secondly by being chemically very active, so that they can take part in a wide range of cell re-actions.   This is probably the way in which the energy of respiration is applied to other processes, such as the synthesis of sucrose and starch from hexose sugars, the synthesis of proteins, and, in animals, the contraction of the fibrils in muscular action.

A proportion of the energy of respiration emerges as heat.   This is not nor-mally a noticeable feature in plants because the heat production is relatively small and is dissipated by radiation.   But by insulating a quantity of plant material, as for example by enclosing germinating seeds in a thermos flask (see practical work), a considerable accumulation of heat can be obtained.   It seems likely that in the case of plants this heat represents wasted, surplus respiratory energy.   In warm-blooded animals the body is attuned to function properly only if it is maintained at a certain temperature, and the heat of respiration is employed for this purpose.

RATE OF RESPIRATION

Respiration is usually measured by observing the rate of evolution of carbon dioxide or that of absorption of oxygen, and many types of apparatus have been devised which permit of these being determined with great exactitude.

A close relation is usually found between the rate of respiration and that of growth.   Thus in the life-cycle of a plant the most rapid respiration (per unit weight of plant) is shown during the germination stage, which is one of intense growth activity.   In the later development the average respiratory activity of the plant as a whole diminishes ; but there is now considerable variation from one part of the plant to another.   Thus rapidly growing parts such as opening buds respire at a greater rate than mature parts of the same plant.   Dormant organs such as dry seeds show exceedingly low rates of respiration.

As in other physiological processes, the rate of respiration is also affected by various external conditions, two of these that we may note being temperature and oxygen supply.

COMPARISON OF RESPIRATION AND PHOTOSYNTHESIS

Some of the differences between these two processes were noted on p. 197. The following table provides a more detailed comparison.

PHOTOSYNTHESIS	RESPIRATION
1. Takes place only in chlorophyll-con-taining cells.	1. Takes place in all living cells.
2. Raw materials are carbon dioxide and water.	2. Raw materials are sugars and oxygen.
3. Takes place only in the presence of light.	3. Takes place at all times, day and night.
4. Is an energy-absorbing process.	4. Is an energy-releasing process.
5. Products are sugars and oxygen.	5. Products are carbon dioxide and water.

It may also be noted that not only is photosynthesis confined to chlorophyll-containing cells, but also (as indicated in Chap. XV) within those cells the process is almost certainly restricted to the chloroplasts.   No such recognisable centres exist in the cell for respiration, so that this process may occur in all parts of the proto-plasm, including perhaps the chloroplasts as well.

Since at most stages in its development the plant accumulates carbohydrates in the process of photosynthesis more rapidly than it oxidises them in respiration,

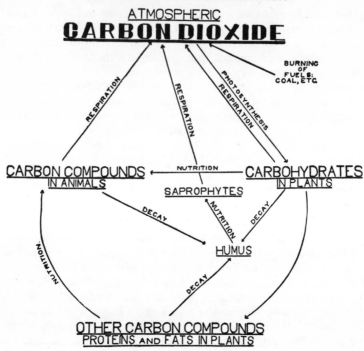

FIG. 141. THE CARBON CYCLE.

it follows that on balance the plant takes much more carbon dioxide from the atmosphere than it returns as the result of respiration. Eventually, however, this carbon accumulated by the plant finds its way back into the atmosphere, either as the result of natural processes of decay after the death of the plant, or through the activities of the animal kingdom. There is thus a carbon cycle in Nature, as is shown in Fig. 141.

COMBUSTION OF PLANT MATERIALS

Respiration is very similar to combustion. The latter process involves the chemical absorption of oxygen with the giving off of carbon dioxide, together with the splitting up of some complicated chemical compound. At the same time, energy is released in the form of heat. Respiration, therefore, may be looked upon as a slow form of combustion.

The heat of combustion of an element is measured by causing it to combine with oxygen in an enclosed chamber and measuring the heat evolved calorimetrically. For example:

$$C + O_2 \rightarrow CO_2 + 94 \cdot 8 \text{ kgm. calories}$$
$$H_2 + O \rightarrow H_2O + 69 \text{ kgm. calories}$$

Thus, 12 grams of carbon yield 94·8 kgm. calories on complete oxidation and 2 grams of hydrogen, 69 kgm. calories. The combustion of 1 gram-molecule of

hexose sugar (180 gm.) yields 674 kgm. calories, or 3·7 per gram. This same amount of
energy will be released when the same amount
of sugar is broken down in respiration in the
plant. We have noted that fat is sometimes
used in plant respiration, and the same may
be true of protein. The heats of combustion
of these substances, per gram, are 9·5 and
5·7 kgm. calories respectively. Thus those
seeds that have fatty food reserves stand to
gain more energy from unit weight of the
reserve than do seeds with carbohydrate
reserves.

Fig. 142 illustrates the calorific values of
some common constituents of the human
diet.

Many of the built-up substances in the
plant body, though they are not foodstuffs,
naturally contain some potential energy
since they have been primarily manufactured by photosynthesis. Yet they are not
usually made use of in respiration. For
example, the wood vessels, the cellulose cell-
walls, etc., form the plant skeleton, and
seldom enter into the process of respiration ;
yet they are indirect products of photosynthesis. Their potential energy is therefore
stored throughout the life of the plant.

Yet in certain everyday activities of man,
the presence of such potential energy is made
clear. For example, timber is a combustible body. Given certain well-known
conditions, it will burn, and in doing so absorb oxygen and give off carbon
dioxide, among other gases, together with the freed energy in the form of heat.
So also will cotton, straw, etc.

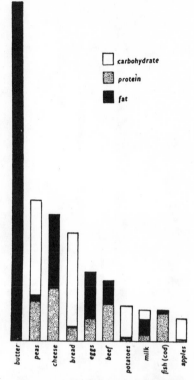

FIG. 142. DIAGRAMMATIC REPRESEN-
TATION OF THE COMPARATIVE RESPIR-
ATORY VALUES OF COMMON FOODS.

COAL

A still better example is **coal**. This is a well-known combustible body which
gives off carbon dioxide and other gases, together with heat energy, at the same
time absorbing oxygen during its combustion.

From the geologist's point of view, coal is a carbonaceous rock. It was formed
from great masses of partially decayed plant remains. Traces of plant fragments
may sometimes be seen on a dull piece of coal. Special methods, such as the
examination under special lighting of highly polished coal surfaces and the cutting
of sections thin enough to be translucent, have also demonstrated the presence of
cellular tissues, spores and pieces of cuticle.

In Great Britain, coal-beds are associated with the Carboniferous rocks (see
p. 400), which were formed some three hundred million years ago. At that
time, great forests covered vast areas of low-lying swamps and deltas. The forest

FIG. 143. RESTORATION OF FERNS, ETC., GROWING IN THE CARBONIFEROUS AGE.
(Scott: ' Extinct Plants and Problems of Evolution '. After Grand 'Eury.)

trees belonged to types of plant now extinct or represented at the present time only by small herbaceous forms. The majority of them were Pteridophytes. Clubmosses, such as *Lepidodendron*, were the most abundant of the trees ; they were related to the small club-mosses which belong to the living genera *Lycopodium*, *Selaginella*, and *Isoetes* (pp. 372). Horsetails (*Calamites*), 100 ft. high, were also abundant in the coal forests ; they were the gigantic ancestors of the present-day small species of *Equisetum*. Ferns, of types now extinct and often very different in appearance from modern ferns, as well as many plants with fern-like foliage but bearing seeds (not spores), added to the rich flora (Fig. 143). Gymnosperms of extinct types (for example, *Cordaites*) were also important trees in some of the Carboniferous forests.

As these low-lying forests persisted for long periods of time, the bodies of successive generations of the plants accumulated in a water-logged condition or actually immersed in water. Complete decay did not take place under these anaerobic and probably acid conditions. Nevertheless, a partial breakdown took place through the activity of anaerobic bacteria (p. 206). The result of this would be the evolution of carbon dioxide, methane and sulphuretted hydrogen while the remaining carbohydrate material of the plants would be converted into a compound of the coal type with a much higher percentage of carbon.

Eventually, the whole delta or swamp was submerged and the forest trees killed. Vast quantities of sand and mud (eventually becoming sandstones and shales) were then deposited over the accumulated mass of plant debris. The great pressure exerted by these and later-formed overlying rocks completed the change of the plant material into coal, the latter now having up to 94 per cent of carbon in its composition. It has been estimated that a depth of about 50 ft. of forest debris would yield, on compression, only a 4-ft. seam of coal.

Below such coal-seams there is nearly always a bed of unstratified clay, known variously as the underclay, seat-earth or rootlet bed. This clay was in fact the soil in which the forest trees were rooted and it is characterised by the occurrence of the root systems of *Lepidodendron* and related club-mosses. These root-bearing parts of the club-mosses are preserved as the fossils of the genus *Stigmaria* (Fig. 144).

House coal, known also as bituminous coal, and anthracite were probably formed in the way described on the actual site of the forest. There are, however, types of coal formed from very different material. Cannel coal and boghead coal, for example, were probably formed in small ponds and lakes within the general area of the swamp. The former often consists largely of the spores of the club-mosses growing in the neighbouring forest, the latter of algae with thick gelatinous walls containing much oil. Anaerobic partial decay followed by intense pressure were again the probable agents in the conversion of the plant material into coal.

Evidence that coal was in fact formed as just described is provided by observations on peat. This substance, of more recent origin and still being formed today in moorland and fen districts, consists of a mass of readily recognisable plant remains. These remains are only partially decayed since the aerobic bacteria, which normally break down organic matter completely, are absent due to water logging, low temperature and deficiency of calcium. It has been shown experimentally that if peat is subjected to great pressure it gives rise to a black, glistening, coal-like substance.

(*Crown Copyright. Reproduced by permission of H.M. Stationery Office.*)

FIG. 144. NATURAL CASTS OF LOWER ROOTED PORTIONS OF *Lepidodendron* TREES OF LOWER CARBONIFEROUS AGE.
'The Fossil Grove', Victoria Park, Glasgow.

It is clear that the heat energy given off when coal is burnt is the energy first fixed by the photosynthetic activity of the plants of the coal forests. The heat is, in fact, the radiant energy from the sun, fixed by photosynthesis some three hundred million years ago.

Apart from its use as fuel, coal yields, on distillation, coal-gas, ammonia and coal-tar, leaving the coke which is used in blast furnaces. Redistillation of the tar gives toluene, benzene and a host of other substances which are the raw materials of the drug, dye and plastics industries.

RESPIRATION IN ABSENCE OF OXYGEN

The normal respiratory process of plants, that which we considered in the earlier part of this chapter, necessitates the presence of air, particularly of the oxygen component of the air, and is termed **aerobic respiration**. If they are deprived of oxygen, most plants or plant organs, though they cease to make any appreciable growth, remain alive for a period and exhibit a modified type of respiration known as **anaerobic respiration**. Carbon dioxide is again evolved in this, as is shown by an experiment described in the practical work ; but a new end-product now accumulates in the tissues, namely ethyl alcohol.

It is generally considered that under these conditions the sugar-breakdown is again initiated according to the scheme on p. 198. The absence of oxygen makes it impossible for the oxidising enzymes to function, so that the zymase action goes on uninterrupted and results in the formation of alcohol. It is probably fairly accurate to represent the anaerobic respiration of higher plants by the over-all equation for alcoholic fermentation as carried on by yeast, namely :

$$\overset{zymase}{C_6H_{12}O_6 \ \rightarrow \ 2C_2H_5OH + 2CO_2 + 50 \text{ kgm. cals.}}$$

Comparison of this with the equation for aerobic respiration (p. 197) shows that very much less energy is released now, because the sugar is less completely disrupted. This is no doubt one reason why plants cease to grow and usually die quite soon if deprived of oxygen.

It should, however, be noted that a few exceptional plant organisms thrive best under anaerobic conditions, and in fact some of these actually die if they are exposed to oxygen. Most of these **anaerobes** are bacteria. For example, various bacteria of the genus *Clostridium*, responsible for tetanus and for some forms of gangrene and food poisoning, are anaerobes.

PRACTICAL WORK

1. Demonstrate the process of aerobic respiration in plants.

There are several methods of doing this, the best depending upon the production of carbon dioxide during the process. The following are suggested methods.

(a) Collect a large number of heads of living flowers and push them down into the spherical part of a flask. Then invert the flask over a dish of mercury, so that the mouth is below the surface. Now, by means of a bent tube, allow some strong caustic potash solution to rise to the surface of the mercury inside the neck of the flask (Fig. 145).

After a few hours, the mercury will rise inside the flask. This can be explained by the respiratory activity of the flowers. During the process, oxygen is absorbed from the enclosed atmosphere of the flask, and an almost equal volume of carbon dioxide is given off. Therefore, there should be no change in the volume of the atmosphere in the flask. But

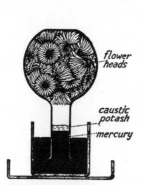

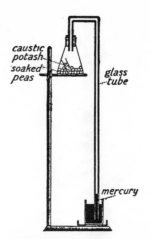

FIG. 145. EXPERIMENT FOR
DEMONSTRATING AEROBIC RES-
PIRATION.

FIG. 146. EXPERIMENT FOR
DEMONSTRATING AEROBIC RES-
PIRATION.

caustic potash absorbs carbon dioxide strongly. Hence the reduction in volume of the atmosphere of the flask.

(*b*) Set up the experiment as shown in Fig. 146. By means of gummed paper, mark the level of the mercury in the tube at hourly intervals.

(*c*) Set up the apparatus shown in Fig. 147 (one of the bottles containing lime water to the left of the peas can be omitted). Care must be taken that all the connexions are air-tight. Then draw air slowly through the apparatus by fixing the end indicated to a filter pump.

After a time, notice that although the lime-water, through which the air current bubbles *before* passing through the soaked peas, remains clear, that through which the current bubbles *after* having passed through soaked peas becomes milky.

Realising that soda-lime and caustic potash absorb carbon dioxide, fully explain these results.

2. Show that respiration takes place both in the light and in the dark.

Perform any of the experiments suggested in Experiment 1 in the light and in a dark room, and compare the results.

3. Show that normal plants can respire for a time in the absence of oxygen (anaerobic respiration).

Fill a test-tube with mercury and, sealing the end with the finger, invert it over a dish of mercury, with the mouth of the tube below the surface. Then, by means of the fingers,

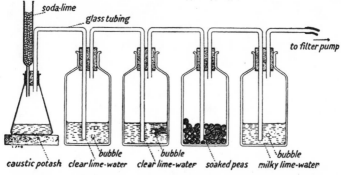

FIG. 147. EXPERIMENT FOR DEMONSTRATING AEROBIC RESPIRATION.

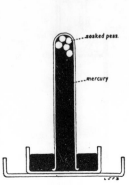

FIG. 148. APPARATUS FOR DEMONSTRATING ANAEROBIC RESPIRATION.

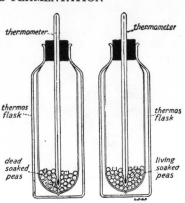

FIG. 149. EXPERIMENT FOR SHOWING THAT HEAT IS LIBERATED DURING RESPIRATION.

insert some soaked peas, one at a time into the mouth of the tube. These peas will rise to the top of the tube (Fig. 148). Support the tube in this position by means of a clamp.

After about twenty-four hours, the peas will have given off a gas which has forced the mercury some distance down the tube. The gas can be shown to be carbon dioxide by allowing a little strong caustic potash solution to rise into the test-tube from a bent tube inserted at the mouth. The potash, now in contact with the carbon dioxide, absorbs it, and the mercury rises again in the tube.

4. Show that energy, in the form of heat, is liberated during the process of respiration. Take some soaked peas and divide them into two groups of equal numbers. Place one lot in a thermos flask fitted with a thermometer (preferably with a long stem), making sure that the stopper is replaced tightly (Fig. 149). Put the other lot of peas in boiling water and boil them for about ten minutes. Remove them and allow to cool down to the normal temperature. Then treat them as the first lot. By boiling the peas, they have been killed.

After a day record the temperature indicated by each thermometer, and explain the results obtained.

CHAPTER XVIII

CARNIVOROUS PLANTS

Although plants are consumed whole or in part by animals, there are very few cases where plants make direct use of animals as a source of food. Normally they only utilise animal bodies and excretions after these have been converted by bacteria into simple nitrogenous substances (Chap. VI).

Nevertheless, there are examples of plants preying on living animals for the purpose of nutrition. They actually catch the animals, kill them, and then digest them. Animal-eating plants have caught the imagination of many a person, and are especially common in stories of travellers of days gone by. For example, for many years travellers in the unknown parts of certain tropical regions used to come home with awe-inspiring tales of a 'man-eating tree', that is, a tree which was able to trap a man if he got too near it and then digest him. Though this myth held sway for a long time, no such tree exists. In fact, comparatively few animal-trapping plants exist, and these catch only insects and other very small animals.

CARNIVOROUS PLANTS

Some plants, since they consume insects and other small animals, are called **insectivorous** or **carnivorous** plants. Although comparatively rare, they are of extreme interest in that they have all developed most wonderful methods for catching their prey.

Much work on the insectivorous plants was done, chiefly by the best method, that of field observation, by Charles Darwin, who published an account of them in 1875. Before considering any examples of carnivorous plants, it is necessary to reflect on them in general, chiefly to get rid of any exaggerated ideas that we may have of them. Some of them may be found in Great Britain. It is quite true that they

Ernest G. Neal

FIG. 150. COMMON BUTTERWORT.

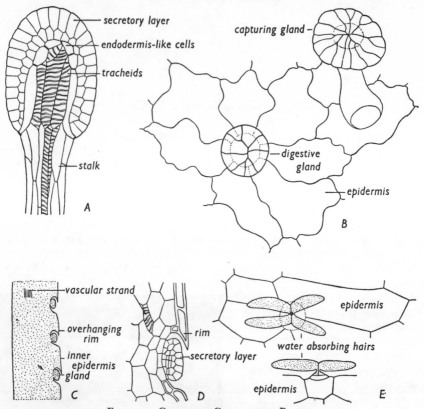

FIG. 151. GLANDS OF CARNIVOROUS PLANTS.
A, head of tentacle of *Drosera* ; *B*, surface view of part of a leaf of *Pinguicula* ; *C* and *D*,
wall of the lower part of a pitcher of *Nepenthes* ; *E*, the quadrifid hairs on the inside of the
bladder of *Utricularia*, in surface view (above) and in section (below).
(*F. W. Williams*.)

trap insects and other small animals, and absorb certain digestible parts of
the body of their victims. But not one carnivorous plant depends *completely* on
insects for its nutrition. It is possible to go even further than that : *insects
never form more than a very small fraction of the food of carnivorous plants.* There is no
need to watch such a plant and see how many insects it catches in a certain length
of time to learn this, for, when it is realised that all carnivorous plants possess
green leaves, it can be concluded immediately that their main source of food
supply is the normal one, that is, photosynthesis.

Since this is the case, the question arises : why do such plants trap insects?
The majority of carnivorous plants grow in very wet situations, such as swamps and
wet moors. In such regions the waterlogged condition interferes with the bacterial
breakdown of organic matter and so leads to the accumulation of peat. This is
deficient in mineral salts and particularly in nitrates. A few carnivorous plants
grow as epiphytes in wet tropical forests and these, too, will tend to suffer from
nitrogen deficiency. Now insects and other small animals which these plants trap
have a high protein content in their bodies. Thus, by digesting them, carnivorous
plants obtain an extra supply of nitrogen, to make up for the deficiency of that
element in such habitats.

Though carnivorous plants actually can do without such animal food, there is little doubt that the nutrition obtained by them is valuable.

CARNIVOROUS PLANTS OF TEMPERATE REGIONS

A few species of plants of insectivorous habit are to be found growing wild in Great Britain.

One rather common carnivorous plant is the **butterwort** (*Pinguicula vulgaris*).

This is a perennial herb with leaves 1 to 3 inches long, arranged in rosette fashion (Fig. 150). The upper surface of the leaves is covered with a pale yellow, sticky substance which looks like a thin layer of butter ; hence the common name and the generic name, the latter being derived from the Latin *pinguis*, oily. An unwary insect alighting on this sticky surface gets caught like a fly on a fly-paper. The margins of the leaf are slightly incurved (Fig. 150), and when an insect is caught, they curve over still further, thus gripping the insect.

Ernest G. Neal

FIG. 152. ROUND-LEAVED SUNDEW.

On the surface of the leaves are certain microscopic structures called **glands**. These are of two kinds (Fig. 151, *B*). Stalked glands, looking like miniature mushrooms, secrete a sticky mucilage and are the **capturing glands**. Sessile glands, borne at the level of the leaf surface, are **digestive glands**. When an insect has been captured, these produce an abundant acid secretion which contains proteolytic enzymes. The proteins of the insect body are thus broken down into simpler nitrogenous substances which are then absorbed by the surface of the leaf. After absorption the leaf unrolls, and in due course the insect remains are either blown or washed away.

This plant is quite common in Great Britain, especially on the Somersetshire Plain, and in Yorkshire and Scotland. It is very widely distributed in the north temperate zone, and is common in certain parts of Canada and the United States.

There are a few rarer species of butterwort, for example, the Irish butterwort (*P. grandiflora*), Alpine butterwort (*P. alpina*) and Western butterwort (*P. lusitanica*).

Butterwort flowers appear during May to July. They are borne on erect, thick stalks and are purple in colour, two-lipped in form with a backwardly projecting spur.

Another British insectivorous plant, but one with a more complicated trapping mechanism, is the **round-leaved sundew** (*Drosera rotundifolia*) (Fig. 152). The

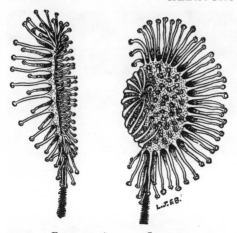

FIG. 153. LEAF OF SUNDEW.

Left, side view ; right, front view after being touched with a pencil.

leaves, all of which are radical, again supply the trapping mechanism. They are green with patches of red on them, and possess long petioles. The laminae are round or oval according to the species. The leaves are again arranged in rosette fashion. The flowers are white, and are arranged in rather long, upright inflorescences.

The edges and upper surface of the leaf of the sundew are covered with long, hair-like outgrowths called **tentacles** (Fig. 151, *A*). There are about 200 tentacles on one lamina. Each tentacle ends in a club-shaped glandular head which secretes a sticky substance looking like dew ; hence the common name and the generic name which is derived from the Greek *drosos*, dew. When the insect alights on the leaf, it is thus caught by the sticky tentacles. Then there is a general bending of all the tentacles towards the centre of the lamina, with the result that the insect is caught and pressed firmly against the leaf surface by the tentacles (Fig. 153). The ends of the tentacles then secrete a protein-digesting enzyme, and thus the insect is digested and the products are finally absorbed. After the process is finished, the leaf attains its normal form again and the indigestible parts of the insect are exposed and blown away by the wind. The characteristic nastic movements (p. 234) of the tentacles are stimulated both by contact and by chemical substances such as ammonium salts and proteins. They are brought about by increased growth of the lower side of the tentacles near their base. Subsequent unfolding results from growth of the upper surface of the tentacles.

The round-leaved sundew (*Drosera rotundifolia*) is common in the swampy regions of Great Britain, favouring acid bogs and the moist hollows of sandy heaths. There are two other species, the long-leaved sundew (*D. anglica*) which is rare in the south of England, and the lesser long-leaved or the narrow-leaved sundew (*D. intermedia*, formerly *longifolia*) which is fairly well distributed but not so common as the round-leaved. Sundews are also common in North America, and are abundant in Australia.

Sundew flowers are borne on erect, thin, stalks varying from two to eight inches in height. The flowers appear during June to August in circinate cymes and are white in colour.

An even more specialised method for trapping small animals is shown in another well-known insectivorous perennial plant called the **bladderwort** (*Utricularia vulgaris*), a brackish-water plant which grows in ditches and ponds in certain parts of Great Britain.

The leaves of the bladderwort are very finely divided, as is the case in many examples of submerged leaves. Certain of these leaf segments become much modified into bladder-like structures, thus giving the plant its familiar name (Fig. 154). Each bladder has one opening which is protected by a valve. The valve will open inwards, but will not open outwards. At one time it was considered that small aquatic animals forced their way into the bladder (probably seeking sanctuary from enemies). This, however, is not true. They are actually captured by the bladder. The organism enters the vestibule of the bladder and may stimulate the valve to open by touching sensitive hairs growing on the outer surface of the valve itself. The walls of the bladder, hitherto compressed, now expand and draw water in, thus sucking in the prey. The valve now closes again so that egress of the prey is prevented. The whole of the ' gulping ' action resulting in the capture of the prey takes only about $\frac{1}{30}$ of a second. Water-fleas and minute insect larvae are the usual prey. These swim about in the bladder for a time and then die. There is no evidence that digestive enzymes are secreted by the bladder but the body of

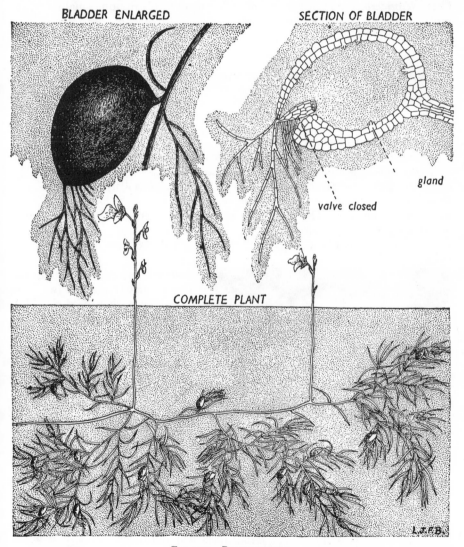

BLADDER ENLARGED   SECTION OF BLADDER

gland

valve closed

COMPLETE PLANT

L.J.F.B.

FIG. 154. BLADDERWORT.

the prey is broken down by bacterial action. Some of the soluble nitrogenous substances so produced are absorbed by the wall of the bladder.

The trap is then set again, a process taking about twenty minutes. Curious four-rayed hairs clothe the inner surface of the bladder, and these absorb water Fig. 151, *E*). Since the bladder, sealed with its valve, is impervious to water the volume of the bladder is decreased and the walls sucked in. A considerable tension is thus set up, and the bladder is again in the ' set ' condition.

The flowers of this plant appear during June and July borne on long, thin stalks which project above the surface of the water. They are rich yellow in colour with a pronounced spur.

There is a smaller species of bladderwort known as the lesser or small bladderwort (*Utricularia minor*) which is more localised in distribution, mainly in the north. Its leaves have not so many forked lobes. Two other species are even more localised.

CARNIVOROUS PLANTS OF THE TROPICAL AND SUB-TROPICAL REGIONS

The carnivorous plants just described are the only ones which grow wild in Great Britain. This shows how comparatively rare carnivorous plants are. But there are a few very interesting plants which demonstrate this habit, native to tropical and sub-tropical regions.

One such plant is called **Venus's fly-trap** (*Dionaea muscipula*). This plant is very common in the peat bogs of North and South Carolina. The first full description of the insectivorous habit of this plant was given by Charles Darwin, though it had actually been noticed before. The insectivorous habit is again connected with the leaf. The leaves are arranged in rosette fashion, as in British terrestrial insectivorous plants.

Each leaf consists of a winged petiole which is joined by a short cylindrical region to the lamina, which is modified as a trap for insects. The lamina consists of two lobes united by a thick mid-rib. The margins of the lobes bear long, firm spikes (Fig. 155), while the surface of the lobes bears glands of two kinds. A narrow marginal zone is occupied by glands which secrete nectar, a sugary solution attractive to insects, while the rest of the surface bears digestive glands having a deep red colour. Each lobe also bears three stout bristles, each jointed at the base and sensitive to touch.

When an insect alights on the surface of the lamina and moves about in search of nectar it cannot help but touch one or more of the sensitive bristles. The result is

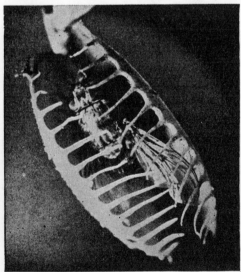

L. *Keinigsberg*

FIG. 155. VENUS'S FLY-TRAP CLOSING ON A FLY.

that in less than a second the lobes snap together and the marginal spikes interlock with one another so that the lobes cannot be pushed apart. This first rapid movement effectively traps any large insect (Fig. 155) but allows the escape of small ones : it is followed by slower movements which bring the lobes closer together, and the lobes themselves become flatter so that the prey is held firmly between them.

The digestive glands secrete proteolytic enzymes and the products of digestion of the insect body are then absorbed. The lobes open again after some days.

The movement is 'triggered' by the bristles, which are able to fold over at their joints as the lobes close together. The closing movement is caused by a rapid increase in size of the lower leaf surface and the slow reopening by increased growth of the upper surface. Changes in turgidity and growth are both involved.

W. Anderson

FIG. 156.   *Nepenthes rafflesiana* IN THE GLASGOW BOTANIC GARDEN.

It is interesting to note the instructive experimental observation that Darwin made on this trapping mechanism. His experiment showed what a great deal of liquid containing the protein-digesting enzymes is secreted by the glands of this leaf. He waited until an insect was caught in a trap, then immediately made a small perforation at the base of the lobes, without removing the crushed insect. The digestive fluid poured out of this hole and flowed down the leaf-stalk, and it continued to do this for nine days.

Some carnivorous plants are known as **pitcher plants** since their leaves are modified to form receptacles or pitchers which act as passive traps for insects and other animals.

The genus *Nepenthes* is an outstanding example. It has about sixty species which are distributed throughout the old world tropics, extending from Madagascar to North Australia and being especially abundant in North Borneo (Fig. 156).

The species are of varying size and habit ; but the majority of them are plants of the wet jungle, growing rooted in the swampy soil or as epiphytes on the forest trees. They usually have a rhizome from which climbing shoots grow up through the jungle vegetation, sometimes to the tops of trees 50 ft. high. The leaves are highly modified both for climbing and as traps for insects. Each leaf has a basal region clasping the stem. Above this is a flattened region which resembles the lamina of a normal leaf both in appearance and function. Beyond this again is a cylindrical tendril of varying length bearing a pitcher at its tip. This pitcher is the

W.Anderson

FIG. 157. *Sarracenia mitchelliana* IN THE GLASGOW
BOTANIC GARDEN.

last part to develop, and before it does so the tendril has often twined round a support, thus aiding the climbing habit and also serving to support the pitcher. The latter always assumes a vertical position in response to the stimulus of gravity, suitable growth taking place just below the pitcher. It has proved impossible to relate the various parts of this highly modified leaf to the parts of a normal leaf and many different interpretations have been advanced.

The pitchers themselves are usually tubular, with the upper third somewhat constricted, and bear two wing-like structures down their length. The mouth of the pitcher, which has a conspicuous rim, is usually placed obliquely and is overhung by a lid. When the pitcher is young the lid tightly seals the opening; but, as the pitcher grows, it is raised and, remaining stationary, it serves to keep rain from entering the now open pitcher. The size of the pitchers varies in different species, from the size of a thimble to that of a quart mug. They are also variously coloured. The general colour is usually yellowish or bright green and there are blotches or stripes of bright colours, red, purple or yellow. Variations in the intensity of the colour are governed by factors such as light intensity and soil conditions. These bright colorations, like those of flowers, serve to attract insects.

The detailed structure of the pitchers is related to the attraction, capture and digestion of the prey. The lid, apart from keeping out rain, attracts insects by means of nectar secreted on its under surface. The pitcher itself has a conspicuous rim which forms a sort of parapet with a general tendency to point downwards into the pitcher. It is hard and glossy, with prominent ridges which end in sharply pointed teeth. Large nectar glands are present between the teeth. The inner surface of the pitcher is differentiated into two zones. In the upper zone, starting just below the rim, the epidermal cells secrete minute flakes of wax on their outer surface. The lower zone bears numerous glands, each being seated in a small depression, the upper rim of which overhangs the gland (Fig. 151, *C* and *D*). These glands secrete a slightly acid liquid during the development of the pitcher so that young pitchers, sealed with their lids, are already filled with a sterile fluid.

Insects are attracted to the pitchers by their bright colours and by the copious secretion of nectar by the glands on the under surface of the lid and on the rim. Small insects, such as ants, are able to walk on the glossy rim; but when they attempt to walk on the wall just below the rim they immediately slip down into the pitcher liquid. This is because the small flakes of wax produced there stick to their adhesive foot-pads and render them useless. Larger insects, such as cockroaches, beetles and flies, find the rim itself treacherous and, in attempting to reach the nectar glands on its inner margin, they too fall into the pitcher. These slipping zones are in fact extremely effective. The escape of any insect so trapped is prevented by the zone of wax and by the downwardly pointing teeth of the rim.

The insect soon dies in the pitcher liquid, possibly because the liquid contains a wetting agent. The presence of the insect results in active secretion by the digestive glands. This secretion is acid and contains an enzyme which breaks down the proteins of the insect body. The products of digestion are then absorbed by the glands. The enzyme is very active and insect bodies are disintegrated within forty-eight hours and only the chitinous skeletons remain.

A different type of pitcher is formed by the genus *Sarracenia* (side-saddle flowers). The nine or so species are all native to marshy habitats in the eastern regions of North America, extending from Labrador to Florida (Fig. 157).

The rhizome bears a rosette of radical leaves. These have a broad sheathing base, above which the whole leaf forms a long, tubular pitcher, with a single photosynthetic wing down its length and a conspicuously coloured upright flap at its mouth.

The flap attracts insects by the secretion of nectar, and such insects tend to move towards the pitcher cavity since downwardly pointing hairs make movement in other directions difficult. The rim round the mouth of the pitcher secretes nectar, and nectar glands are also present on the inner wall below the rim. This upper zone of the wall is extremely slippery although no wax is present and insects fail to maintain their foothold. Below the slipping zone are numerous digestive glands. The amount of liquid secreted is small but contains enzymes similar to those of *Nepenthes*. Digestion and absorption follow as in the latter.

The cobra plant (*Chrysamphora*, formerly *Darlingtonia*) belongs to the same family as *Sarracenia*. The pitchers are bent sharply forward at the top and the small mouth faces downwards. The forked lid curves forwards and acts as a platform leading to the opening of the pitcher. The whole rosette of pitchers looks like " a number of hooded yellow snakes with heads erect, in the act of making a final spring ". Insects landing on the flap are lured inside by nectar and fail to escape mainly because they are attracted to translucent areas, or windows, in the wall of the pitcher rather than to the actual opening which is on the shaded lower side. The trapped insects eventually die and, although no enzymes are secreted by the pitcher, their bodies are broken down by the action of bacteria and some of the simple nitrogenous substances produced are absorbed by the lower part of the pitcher wall.

The pitchers of the carnivorous plants just described are obviously efficient in the trapping and digestion of small animals. It is interesting to note, however, that a large variety of animals utilise the pitchers and their contents as homes and as a source of their own food supplies. For examples, certain species of mosquitoes and gnats breed exclusively in the pitchers of *Nepenthes*, and other species of the same insects only in those of *Sarracenia*. The larvae of these insects live in the fluid of the pitchers. Protozoa, small crustaceans, diatoms and desmids also live in this fluid, and it is suggested that all these organisms are able to survive in a fluid which digests other animals by reason of their possessing substances which inhibit the action of the enzymes.

Pitchers frequently become filled with a decaying mass of insect debris. Flies of various species lay their eggs in this debris so that the developing larvae are provided with a plentiful food supply. Birds are attracted to the pitchers and feed on the larvae after slitting the pitcher walls with their beaks.

Other animals utilise the attractiveness of the trapping mechanisms. Thus, there are several species of spiders which are found living only in the mouths of the pitchers of *Nepenthes* and *Sarracenia*. Small toads and lizards may also live just inside the pitchers, and these, like the spiders, take their toll of the visiting insects.

It is clear, therefore, that a carnivorous plant is the centre of a whole living community, each member of which is adapted in relation to the very special conditions arising from the carnivorous habit.

PRACTICAL AND FIELD WORK

British carnivorous plants are studied best in their native habitats.

Butterwort and sundew are to be found on the acid soils of moors, where there is a lack of nitrogenous compounds. Both are usually found in the wetter parts of moors, and sundew may be found even in boggy soils.

Both plants can be kept in a room for some considerable time, provided there is plenty of the original soil left around the roots. This should be kept damp.

The commonest British butterwort (*Pinguicula vulgaris*) has rosette leaves. These should be examined and a specimen drawn, to show the incurved margins. Note the sticky nature of the upper surface. Mount a small piece of leaf with. the upper surface uppermost. Examine under the low power of the microscope and note the stalked capturing glands and the sessile digestive glands. Another species, *Pinguicula lusitanica*, is fairly common near the shores of the west of England.

The commonest British sundew is *Drosera rotundifolia*, which may be found in bogs. It has creeping rhizomes. Note the rosette arrangement of the leaves. This species has fairly short petioles. Lightly touch the leaf with a pencil and try to get the tentacles to respond as if they were trying to trap an insect. This experiment is usually more successful in the field, rather than on a disturbed plant. Remove and mount a few tentacles. Examine under the microscope and note (*a*) the stalk and (*b*) the glandular head.

The commonest bladderwort (*Utricularia vulgaris*) is to be found in ditches and pools of brackish water. Obtain a specimen and make drawings of it. There are no roots to this plant. The leaves are very finely divided. The flowers project above the water. Note especially the shape of the bladder and the presence of hairs at the orifice. Open several bladders and look for the remains of animal prey.

The tropical examples of carnivorous plants are not, of course, so easily available. However, specimens of *Nepenthes* and *Sarracenia* are usually to be seen in the greenhouses of botanic gardens. When visiting such gardens take the opportunity of examining any specimens there and, if possible, make drawings, especially of the pitchers, noting and explaining their modified nature.

CHAPTER XIX

GROWTH AND DEVELOPMENT

In previous chapters it has been shown that the plant carries on a wide range of chemical activities, which collectively are said to constitute **metabolism**. These metabolic processes fall into two general categories. Some are of constructive or **anabolic** type, leading to increase in the substance of a plant, and of these photosynthesis is the chief. Others, distinguished as **katabolic**, are reactions of breakdown leading to loss of plant substance, as in respiration.

Anabolic processes proceed more rapidly than those of katabolic type at most stages in plant development, so that the substance of the plant tends to increase. This provides the basis for **growth**. But there is much more to plant growth than mere accumulation of anabolic products or even increase in size of existing parts of the plant. New organs—leaves, shoot and root branches, and in due course flowers and fruits—are developed. Thus growth is accompanied by increase in size and in complexity of form and internal structure. These events obviously require a co-ordination of various processes already considered, such as food manufacture, translocation and water absorption, together with the operation of factors controlling the formation of new cells, the differentiation of tissues, and the relative development of different organs of the plant so that a proper balance is maintained. As we shall see later, the understanding of these latter aspects of growth has been helped by recent studies of hormone substances in plants.

Although the growth processes of all plants are essentially alike, the external form that is assumed varies very greatly from species to species, even though the plants that are being compared have grown side by side in a similar environment. This is because of differences in the hereditary factors of the plants, carried in the nuclei.

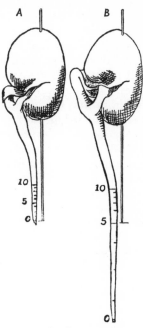

Fig. 158. Localisation of Growth near the Root-tip of the Broad Bean.

A, the root-tip marked in ten zones by lines 1 mm. apart; *B*, the same root after 2 hours.

LOCATION OF GROWTH IN THE PLANT

Though most organs and tissues play some part in facilitating the continued growth of the plant, the actual growth processes, namely the formation of new cells, the increase in size of cells and the differentiation of tissues, are restricted to certain regions. The leaf is of course the organ of chemical synthesis which provides the organic foods needed for growth; but the leaf itself is typically an epheme-

Dr. E. L. Nixon

FIG. 159. REGION OF GROWTH IN A LEAF OF POTATO.
Left, young leaf marked into ⅛ in. squares ; right, the same leaf ten days later.

ral structure, not forming a permanent part of the plant. It is accordingly in the
stem and the root that the capacity for growth chiefly resides.

As already described, at the tips of stems and roots groups of permanently
meristematic cells are found. These cut off new cells, chiefly behind themselves. The new cells soon undergo a marked increase in size, and especially in
length, so that the height of the shoot or the length of the root is incremented.
Fig. 158 shows how the region of the root in which this cell elongation is occurring
can be located. Inspection of the spacing of the lines marked (originally equi-
distantly) on the root indicates that practically all the elongation has occurred in a
zone a few millimetres in length lying just behind the tip. Corresponding experiments with shoots show that the elongative process is less sharply localised and is
often spread out over several of the younger internodes. Fig. 159 shows a similar
technique in use for the study of the growth of a potato leaf, and it is clear that
greatest growth occurred in the lower region of the leaf.

The older parts of stem and root, although they show no further elongation,
often grow markedly in thickness as a result of the formation of new tissues by the
cambium. Lateral appendages are produced, namely leaves and axillary branches
in the case of the stem, formed as a result of the special growth potentialities of
groups of superficial cells recognisable shortly behind the apex and arranged in
a definite pattern for each species of plant. In the root, lateral roots arise as the
result of localised meristematic activity in the pericycle.

In some plants, especially trees, the life-span may extend over a great many
years, sometimes more than 1,000 years, as in the yew or the giant tree of California (Chap. III). Each year the branches of the stem and root increase in
length, and new branches and leaves are formed. Because of this constant
production of new organs in its growth the plant is said to show **continuous**

embryology. In most animals growth is much more limited than in the plant, and is largely confined to the further growth of organs already present at a very early stage in development. The relatively unlimited capacity for growth and organ formation of the plant endows it with much greater powers of regeneration and recovery from loss of parts than are possessed by the animal.

MEASUREMENT OF GROWTH

Growth is clearly a complex process involving changes in many attributes of the plant, so that methods for measuring growth will differ depending on which particular aspect of growth is being considered. In a stem or root measurement of increase in length or diameter may be necessary, or that of area in a leaf. For the study of growth in length various types of **auxanometer** have been devised, a simple form being shown in Fig. 160. For more accurate work a special microscope with a horizontal tube is used. The tube of this microscope can be

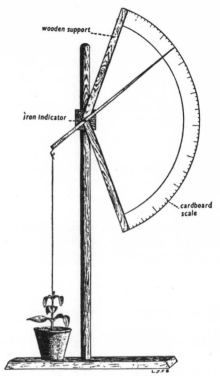

FIG. 160. APPARATUS FOR DEMONSTRATING GROWTH IN A SHOOT.

moved in a vertical plane in conjunction with a vernier scale, and in this way very small increments of growth can be accurately measured.

Since growth is usually accompanied by a proportionate increase in the weight of the plant this is often used as an index of growth in experimental work, just as the farmer uses weight as a measure of the success of his crops. Of course a high proportion of the fresh weight of plant material is due to water, and the actual water content is liable to vary considerably between different species of plant and between individual plants of one species or variety if they are growing under different conditions. To avoid this complication the dry weight of the plant can be determined (p. 58). It may be noted that the fresh and dry weights of a plant sometimes change in opposite directions. Thus during the early stages of germination, fresh weight increases as growth occurs but the dry weight of the seedling decreases because at this stage respiration is unaccompanied by photosynthesis.

Rates of growth vary very much, depending on the species of plant and on the external conditions (see later). Very rapid rates are shown in some tropical regions where the high temperature and conditions generally are very favourable to plant growth. Thus shoots of bamboo plants in Ceylon have been known to grow in length by 16 inches in a day. In temperate climates growth is usually slower. Fig 161 presents the results of measurements of the daily increase in length of a single internode of a young runner bean plant, beginning as soon as the internode emerged from the terminal bud and continuing until the mature length

Plant organ	Percentage elongation per minute
Bamboo shoot	1·27
Bryony shoot	0·58
Bean root	0·45
Grass stamens	60·00
Fungal hyphae	80 to 120
Pollen tubes	100 to 220

had been attained.   The figure shows that the fastest rate of growth reached was 14 millimetres per day.   After this the growth of the internode slowed down as the function of elongation was transferred to younger internodes that were now present.   To such a growth-cycle Sachs applied the term **grand period of growth**.

The above table, based on data compiled by **Buchner,** compares the growth-rates shown by a variety of plant organs.

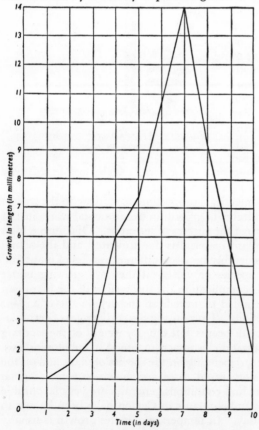

FIG. 161. GRAPHICAL REPRESENTATION OF THE GRAND PERIOD OF GROWTH OF THE INTERNODE OF THE RUNNER BEAN.

(Data from Sachs.)

PHASES IN GROWTH AND DEVELOPMENT

Three main phases can be recognised in the complete growth-cycle of a flowering plant, firstly the germination of the seed, secondly the period of vegetative development in which the plant body is built up, and thirdly the reproductive phase (often overlapping with the second) in which the flowers and fruits are formed.

These three periods are mostly clearly presented in annual plants.   In perennials there is typically a production of reproductive structures each year as well as an addition to the vegetative plant body, although there may be a period extending over many years of purely vegetative growth before the first production of flowers occurs, as is the case in many trees.

Germination is a stage usually characterised by rapid development of the embryo plant, intensive metabolism, and transition from the dependence of the young plant on the food stored in the seed to a state of independent nutrition. The course of germination and the effect on it of various environmental factors will be described in Chapter XXII.

(*From Robbins and Rickett's 'Botany'. Courtesy of the D. Van Nostrand Co., Inc.*)

FIG. 162. ETIOLATION IN THE RUNNER BEAN.

Left, grown for ten days in the light ; right, grown for ten days in the dark.

VEGETATIVE GROWTH

The processes of growth and development just outlined, though determined in the first instance by the inherent growth-potentialities of the plant itself, are also dependent on the presence of suitable external conditions. Thus growth will be hindered if adequate supplies of water and mineral salts are not available, or if light intensity or oxygen supply is defective.

Temperature, as we have seen already, is a factor exercising a potent effect on all physiological processes, so that naturally it is equally important for growth. Growth is only possible within a certain range of temperature, though the limits of this range vary for different plants. For tropical or sub-tropical plants such as maize the range is approximately 10° to 45° C., the optimum temperature being in the region of 30° C. For plants native to temperate climates the corresponding figures are about 5° lower, and for arctic and alpine plants are lower again. The rise in temperature in spring is a leading factor in the general resumption of growth at that season, while the beneficial effect of artificial heating is well-illustrated in the forcing of plants in horticultural practice.

A plant does not necessarily die if it is exposed to temperatures outside its growth-range, and in countries of temperate climate like Britain it is the effect of subjection of plants to low temperatures that is of greater interest to gardeners and farmers. Many plants are killed or severely injured by temperatures in the range 0° to − 5° C., which usually suffice to cause the water of the plant cells to freeze. Thus such temperatures are usually fatal to potato, tomato and *Dahlia*. But it is clear that many of our native plants are able to withstand still lower temperatures without injury. It has been shown that this frost resistance is at least partly due to seasonal changes in the tissues which help to delay ice formation, including an accumulation of sugars in the sap, a relatively low water content, and a binding of a proportion of the water to colloidal material in the cells.

Because of its importance for photosynthesis, it is obvious that if a green plant is denied access to proper **light** there is liable to be serious interference with growth. However, if food reserves are available then plants will grow in complete darkness. Thus quite large plants can be grown from food-containing structures such as seeds, tubers and bulbs in darkness. But the appearance of plants so grown is very abnormal. The shoot is yellowish-white in colour since no chlorophyll forms, which suggests that there is a photochemical stage in the synthesis of that substance in the plant. Again in many species the shoot is unusually tall and straggly and the leaves under-developed (Fig. 162). Such a plant is said to be **etiolated.** Clearly, in addition to its importance for photosynthesis, light has also

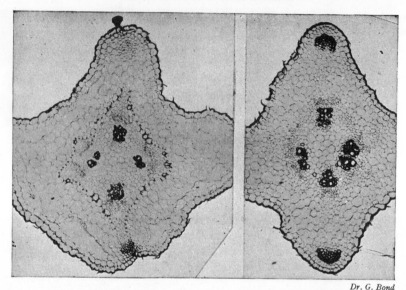

Dr. G. Bond

Fig. 163. Transverse Sections of Stem of Sweet Pea (*Lathyrus odoratus*).
Right, normal stem ; left, stem grown in continuous darkness, showing the presence of an
endodermis. Both sections are from the fourth internode (× 42).
(*Originally published in the Trans. Roy. Soc. Edinburgh*, **58** (1935).)

a developmental effect on shoot growth in that it permits chlorophyll formation, controls stem elongation, and promotes leaf development. There is evidence that the shorter wave-lengths of light, namely blue and violet rays are more effective in these respects than is red light, whereas in photosynthesis the opposite is true, as already noted. Differences in internal structure have also been observed, one notable feature being that in some plants an endodermis develops in the etiolated stem whereas it is absent from the light-grown stems (Fig. 163).

Partial etiolation is induced when plants, though receiving some light, do not get enough, as for example when they are placed too far from the source of light or are too crowded. Such plants become unduly tall and ' drawn '. The greater rate of elongation of the shoot in darkness is advantageous to the plant during germination, since the young shoot in consequence grows more rapidly through the soil and so attains conditions suitable for the initiation of photosynthesis.

GROWTH SUBSTANCES

In the growth and functioning of animals an important part is played by **hormones**, which are chemical substances produced by glands in certain parts of the body and carried in the blood stream to other parts, where effects on development and behaviour are exerted. It had for a long time been realised that in the plant also there are indications of a mutual influence or **correlation** in the development of different parts. Such indications are often provided when one part of a plant is removed. Thus if the terminal bud is removed from the shoot of a young broad bean plant, one or more of the lateral buds which had previously remained dormant will now grow out and gradually take the place of the injured main shoot (Fig. 166). In the same way, by trimming a hedge the development of buds further down the shoot is encouraged. It seems then that in the intact shoot the presence of the terminal bud in some way inhibits the development of the lateral buds. Other instances could be quoted.

It is only in modern times that clear evidence has been obtained of the participation of special growth-regulating substances of hormone type in such developmental processes in plants. In this field the work of **Boysen-Jensen**, **F. A.** and **F. W. Went**, **Kögl** and **Snow** has been particularly important.

W. Anderson and Dr. G. Bond

FIG. 164. EFFECT OF INDOLE-ACETIC ACID ON GROWTH OF SHOOTS OF BROAD BEAN ($\times \frac{1}{10}$).

The instance that has been most closely investigated refers to the relation between the tip of a shoot and the region of maximum elongative growth situated some little distance behind the tip, and in this work the seedling shoot of the oat or other grasses has been much used. Except in very early stages of development the growth of this organ is due to elongation of existing cells rather than to the production of new cells. If the extreme tip of the shoot is amputated the growth of the shoot is quickly arrested. Further study has shown that this is because in the intact shoot the tip secretes a hormone which diffuses down the shoot and stimulates the growth of the cells in the lower region. Thus in one experiment tips of oat shoots were cut off and allowed to stand for a time on small blocks of agar jelly, and the effect of placing these blocks on decapitated shoots then studied. They were found to promote growth strongly, evidently as the result of the diffusion of some substance from the agar into the shoot. Similar results have been obtained in corresponding tests with other plants.

From various plant and other material it has been possible to isolate chemical substances which exert a strong promoting effect on stem growth. The best-known of these is **indole-acetic acid**, which has the following structural formula :

$$\text{indole-acetic acid structural formula}$$

A solution containing only o·o1 mgm. of this substance per litre has been found to exert a definite growth effect. There is evidence that this substance is present and functions as a hormone in many plants. Fig. 164 illustrates the effect of indole-acetic acid on the growth of shoots of broad bean. Initially all the shoots were standing erect. Then in the case of the plants on the left a little of a preparation of indole-acetic acid in lanoline was applied to the right-hand side of each shoot. The photograph shows the appearance after twenty-four hours. Each

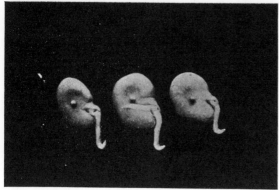

Dr. G. Bond

FIG. 165. EFFECT OF INDOLE-ACETIC ACID ON GROWTH
OF ROOTS OF BROAD BEAN (× ½).

treated shoot has bent over strongly, away from the side to which the hormone was applied, because that side has consequently grown more rapidly.

Some synthesised substances, most of them closely related to indole-acetic acid, have been found to have a similar growth-promoting effect on stems, and often the term ' **growth substance** ' is applied to all these compounds, both to those occurring naturally in plants (hormones proper) and to those synthesised in the laboratory. The promotion of growth seems to be due to the growth substance, directly or indirectly, stimulating the entry of water into the cells, but the mechanism is incompletely understood.

The reaction of the cells of the growth zone of roots to these same growth substances appears to be to a large extent opposite to that of the cells of the stem, since growth is retarded. Thus Fig. 165 shows the effect of applying the same indole-acetic acid preparation as was used above to broad bean roots, originally straight. After twenty-four hours the roots had curved *towards* the side to which the hormone was applied. Thus in this respect there seems to be a fundamental difference between stem and root cells.

The effect of growth substances is by no means confined to the control of cell elongation. Thus there is good evidence that the control exerted by the terminal bud of a shoot over the lateral buds (p. 225) is due to the secretion by the terminal bud of growth substance which arrests lateral bud development. By applying indole-acetic acid to a decapitated shoot the development of lateral buds is inhibited just as though the terminal bud was still in position (Fig. 166).

A somewhat comparable result was obtained in experiments with *Selaginella*, a member of the Pteridophyta (Chap. XXVI). In the species under consideration meristems are present in the angles of the branches of the shoot and normally give rise to special structures known as rhizophores which

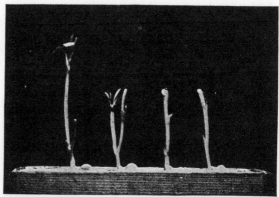

W. Anderson and Dr. G. Bond

FIG. 166. LATERAL SHOOT DEVELOPMENT IN BROAD BEAN

In the two plants on the left decapitation of the original shoot has been followed by the outgrowth of a lateral shoot. In the other two plants indole-acetic acid paste was applied to the cut surface after decapitation, and in these no lateral shoots have developed (× ¼).

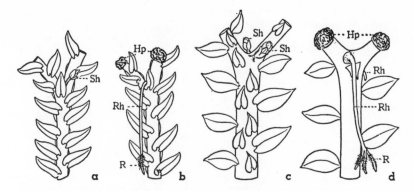

FIG. 167. EFFECT OF HETEROAUXIN PASTE ON *Selaginella.*
a, untreated length of shoot of *S. martensii* ; *b*, similar shoot with heteroauxin paste ; *c* and
d, comparable lengths of *S. lobbii*. *Sh*, shoot ; *Rh*, rhizophore ; *R*, root ; *Hp*, heteroauxin
paste. *N.B.* Heteroauxin is an alternative name for indole-acetic acid.
(*From Dr. S. Williams* ; '*Nature*', **139**, 966 (1937).)

descend to the soil and there produce roots. If, however, the shoot apices are cut
off before the meristem has become active, then a leafy shoot is produced instead
of a rhizophore (Fig. 167, *a* and *c*). But a rhizophore is still produced by the angle
meristem if indole-acetic acid paste is smeared on the cut surface after removal of
the shoot apices (Fig. 167, *b* and *d*). Thus here again the effect of applying the
growth substance is much the same as if the shoot apices had been left in position.

Another effect of growth substances is to stimulate the formation of adventitious roots on stem cuttings. The cuttings are stood for a few hours in a solution
of a growth substance after being taken from the parent plant, and are then
inserted in moist sand or similar medium in the usual way. Fig. 168 illustrates
the stimulation of root formation observed in many species of plants, and there
has been some practical application of this
finding in horticulture. The growth substances have also been shown to be instru-
mental in promoting cambial activity in
connection with secondary thickening.

In recent years growth substances have
assumed greater practical importance with
the discovery that they can be used for
the control of weeds on farm-land or on
playing fields and lawns. If solutions of
a growth substance of relatively high concentration are sprayed on to the shoots of
plant, gross malformation and stunting may
be caused, often resulting in the death of
the plant. But these harmful effects are
much more prone to occur with broad-
leaved dicotyledonous plants than with the
narrow and upright-leaved monocotyledonous type, so that a spray applied to a field
of wheat will suppress the weeds, most of
which are dicotyledons, and leave the crop

(*Photo. supplied by Dr. M. A. H. Tincker.*)
FIG. 168. EFFECT OF A GROWTH-SUB-
STANCE ON CUTTINGS OF *Buddleja fallowiana.*
Above, control cuttings (untreated);
below, treated cuttings after four weeks.

By courtesy of H. W. Gardener

FIG. 169.   USE OF PHENOXYACETIC ACID DERIVATIVES FOR SUPPRESSION
OF BUTTERCUPS IN PASTURE.
The plots on the front right and middle left had been sprayed with the growth-substance
earlier in the same year.
(*From S. J. Willis: 'Agriculture', November* 1950.)

almost undamaged.   In the same way weeds such as buttercup can be eliminated
from pasture, as shown in Fig. 169.   Derivatives of phenoxyacetic acid are widely
used for these weed control purposes.

FACTORS AFFECTING FLOWER AND FRUIT FORMATION

Sooner or later in the growth and development of the higher plant, flowers and,
from these, the fruits and seeds are formed. The structure of these will be dealt
with in later chapters, and here we shall merely discuss some of the factors that
affect the time and the extent of reproductive development.

Illumination is again important.   In many species flowering tends to be suppressed in individuals subject to inferior light, as for example is true of honeysuckle growing in a wood or ivy in a hedgerow.   The explanation of this is probably
provided by the investigations of **Klebs**, extended by **Kraus** and **Kraybill**, on
the effect of the balance of food substances in the tissues on flowering.   They
showed that a high proportion of carbohydrates relative to nitrogenous substances
tends to promote reproductive, at the expense of vegetative, growth. Poor illumination, by its effect on photosynthesis, will tend to lower this ratio, as will also excessive nitrogenous manuring in the case of cultivated plants.

The length of daily illumination is an important factor in determining the time of flowering, as was clearly demonstrated by **Garner** and **Allard**. In many species flower formation only occurs when the length of day (sunrise to sunset) attains a certain value. Some plants require a relatively short day, and in our latitudes such plants mostly produce their flowers in the late summer or autumn, as for example many varieties of *Chrysanthemum* and *Dahlia*. If these plants are exposed to artificially shortened days during their earlier development they can be brought into flower unusually early. Long-day, summer-flowering plants such as wheat, evening primrose and *Iris*, can be prevented from flowering by subjection to shortened days. These responses of the plant to length of day constitute what is termed **photoperiodism**. While the mechanism is not fully understood, there is some evidence that a special hormone is formed in the leaves under certain day-lengths, and that this is transported to the growing points where it induces the initiation of flowers. Besides these effects on flowering, day-length also influences some aspects of vegetative development, such as tuber and bulb formation, and leaf-fall.

Temperature also exerts an effect on the initiation of flowering. Thus it is frequently observed that biennials such as onion and turnip tend to flower in the first year (or ' bolt ') if sown too early so that the young plant is exposed to frost. The exposure of soaked seeds to low temperature has been found to hasten the time of flowering, and this has been put to practical advantage in the process of **vernalisation** by means of which crops such as wheat can be induced to come into ear more quickly than would be the case without the seed treatment.

PRACTICAL WORK

1. Demonstrate and obtain a relative measure of the growth of a young shoot by means of the apparatus shown in Fig. 160. This apparatus can be purchased, but it is comparatively easy to construct. The materials suggested are indicated on the diagram.

2. Germinate a bean seed, and when its radicle is about an inch long mark it by means of Indian ink from the tip upwards at intervals of exactly a millimetre, for about 10 millimetres (as in Fig. 158).

Take a wide-necked jar fitted with a cork and half-full of water. By means of a long pin pushed through the cork, suspend the seedling in the jar with the root pointing downwards. Leave the jar in a fairly warm place for 24 hours, and then examine the original millimetre markings, and note how they have moved apart due to the growth of the root.

3. Perform a similar experiment with a young shoot. Sunflower or castor-oil seedlings are excellent material for this. The seedlings may be grown in damp sawdust.

4. Grow some runner-bean seedlings and, when they are sufficiently developed, take daily measurements of the increase in length of the first internode.

Record these measurements up to about the end of a fortnight and then plot the results in the form of a graph (see Fig. 161).

5. Sow a few soaked seeds of pea or other plant in two pots of sand or soil, choosing seeds of uniform size. Place one pot in a greenhouse or other well-lighted place, and the other pot in the dark. After the seedlings produced in the light have attained a height of about two inches, examine and draw typical seedlings from both pots.

Obtain the dry weights of the plants from both pots and note that although the etiolated plants are much taller, the dry weight per plant is less than that of the normal plants.

6. If a small quantity of indole-acetic acid in the form of a paste in lanoline is purchased, experiments of the type illustrated in Figs. 164-166 can be carried out.

CHAPTER XX

ORIENTATION AND MOVEMENT

It is a matter of common observation that the organs of the plant show a definite and characteristic orientation in space. Under normal conditions most main stems point vertically upwards and main roots vertically downwards, while leaves occupy a horizontal or oblique position. These are the positions in which the respective functions of the organs are most effectively discharged, but the immediate reason for their assumption is that plant organs are sensitive or **irritable** towards certain influences which bear on the plant throughout its development. The chief of these is the **force of gravity.** Another, the action of which is normally confined to the shoot, is the **direction of light.**

It is such factors, acting as **stimuli,** that cause plant organs to assume definite positions during their growth. When the organs become displaced from their usual position or when the direction of one of these orientating influences is changed, then because of curvatures and twistings which soon develop as a result of changes in growth-rates, the organs are to some extent restored to their normal position relative to the influence. A type of slow **movement** is thus exhibited by the plant organs.

This sensitiveness on the part of the plant to a directive stimulus constitutes a **tropism.** The response to the force of gravity is termed **geotropism,** and that to the direction of light, **phototropism.** Some other tropisms will be mentioned later.

GEOTROPISM

That the upward growth of the main stem (shown even in continuous darkness) and the downward growth of the main root are indeed due to the influence of gravity might be deduced from the fact that gravity is the only directional factor of universal operation that could be concerned. Confirmation is provided by the Knight's wheel experiment to be described later.

The root, growing towards the source of the gravity stimulus, is said to be **positively geotropic,** while the main stem, growing away from the centre of the earth, is **negatively geotropic.** Organs, such as

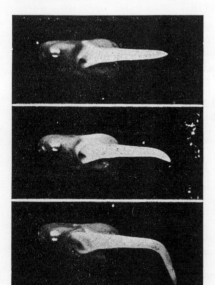

Dr.G. Bond

FIG. 170. GEOTROPIC CURVATURE IN A
BROAD BEAN ROOT.

Top, seedling fixed with root horizontal; centre, appearance of same seedling after four hours; bottom, after twenty-four hours ($\times \frac{3}{4}$).

230

many rhizomes and stolons and some leaves, which assume a horizontal position are said to be **diageotropic,** while those placing themselves in oblique positions are **plagiotropic.**

The root and the shoot normally assume their characteristic orientations as soon as the germination of the seed commences. If a seedling is fixed so that the root and shoot are in an unusual position then in the course of a few hours curvatures will develop and gradually bring the apical parts back into their normal positions. This is illustrated for a root in Fig. 170. In this particular experiment the first signs of curvature were seen after 90 minutes. The curvature occurs in the zone of cell-elongation (p. 220) and is due to the growth becoming greater, under the influence of the gravity stimulus, on the upper side of the root than on the lower. The older part of the root has no power of growth and does not regain its normal orientation.

W. Anderson

FIG. 171. GEOTROPIC RESPONSE IN A *Pelargonium* SHOOT.
Photographs taken three days after the plant had been placed so that the shoot was horizontal. An upward curvature has developed in the stem, while petiolar bendings have restored three of the four larger leaves to a horizontal plane ($\times \frac{1}{3}$).

Geotropic responses in a shoot are illustrated in Fig. 171. Curvature in the stem, again in the region of cell-elongation, has restored the apical part of the organ to the upward-pointing position, the first signs of this curvature having been visible four hours from the commencement. The stem curvature automatically restores the youngest leaves to their normal position, but somewhat older leaves have independent powers of adjustment of their position by means of petiolar bendings and twistings, as shown in the figure.

The development of geotropic curvatures in roots or shoots placed in a horizontal position is prevented if the organ is slowly turned about its long axis, an apparatus called a **klinostat** being used for this purpose (see practical work). Thus if a *Pelargonium* shoot is so fixed on a klinostat and rotated at the rate of one revolution per half hour, the stem will grow horizontally. Under these conditions the stem is not in any one position long enough for a geotropic response to arise, so that the organ continues to grow in whatever direction it is fixed.

One of the earliest experiments on geotropism was carried out at the beginning of last century by **T. A. Knight,** well-known for his pioneer investigations on plant physiology and horticulture. This experiment is still of great significance.

FIG. 172. GEOTROPIC
CURVATURE IN A GRASS
STEM, OCCURRING AT THE
NODE.

(After Noll.)

Knight fastened germinating seeds to the rim of a wheel which was rotated at high speed in a vertical plane, so that the seedlings were exposed to a considerable **centrifugal force.** He found that this latter force now controlled the direction of growth of the plant organs. The roots grew outwards, with the centrifugal force, while the shoots grew towards the centre of the wheel, that is against the force, just as they normally grow against the force of gravity. Thus in this Knight's wheel experiment gravity had been replaced by another similar but stronger force.

We now have to consider why plant organs should react to gravity. Since, as was seen in the previous chapter, the elongation of cells in shoot and root is controlled by growth substances diffusing back from the tip of the organ to the region of cell elongation, it is an obvious possibility that the unequal growth productive of a geotropic curvature is due to some effect of gravity on the growth substance. There is in fact some experimental evidence that in a shoot or root fixed horizontally, instead of the growth substance arriving at the zone of cell elongation in a state of equal distribution all round the organ, for some unexplained reason most of it accumulates on the lower side. Because of the opposite effects of growth substances on shoot and root (growth-promoting in the former, growth-retarding in the latter), this new distribution will result in an upward curvature in the shoot and a downward one in the root.

Though in general the development of geotropic curvatures is confined to young still-growing parts of the plant, an apparent exception to this is provided by the stem of grasses and cereals, and also carnations, and some related plants. In these geotropic curvatures are shown in older parts of the shoot, as for example in wheat and grass shoots flattened by wind and rain (Fig. 172). This is because the nodal regions of these shoots retain the power of intercalary growth.

Certain plant organs show unusual geotropic responses. Thus the so-called breathing roots of mangroves (Fig. 26) grow vertically upwards and project from the swampy substratum, experi-

W. Anderson and Dr. G. Bond

FIG. 173. ROOT ORIENTATION IN BOG MYRTLE (*Myrica gale*).

The upward-growing roots originate from numerous root nodules. It may be noted that these nodules fix atmospheric nitrogen, as suggested by the fact that this plant grew in water culture from the original seed without any combined nitrogen being supplied (Chapter XII) ($\times \frac{3}{10}$).

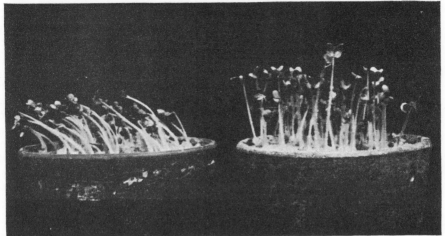

Dr. G. Bond

FIG. 174. PHOTOTROPIC CURVATURE IN SHOOTS OF WHITE MUSTARD (*Sinapis alba*)
SEEDLINGS.
Right, seedlings grown with normal, overhead illumination ; left, seedlings illuminated
from the left for five hours, now showing curvature in the hypocatyl ($\times \frac{1}{2}$).

ment suggesting that this is due to negative geotropism. In the native bog
myrtle (*Myrica gale*) certain roots originating from the root nodules present on
this plant grow vertically upwards, and this again appears to be due to negative
geotropism (Fig. 173).

PHOTOTROPISM

Most stems show **positive phototropism**, that is, they grow towards a source
of light. With overhead illumination, as usually prevails under normal conditions
of growth, the stem points vertically upwards, but if for some reason the illumination becomes lateral then curvature develops (Fig. 174). This is again due to
differential growth, the cells on the shaded side growing more rapidly than those
nearer the light, with the result that the apical part of the shoot is moved over
until it points to the source of light. During these events the tendency of the
shoot to remain upright as the result of geotropism is overcome, suggesting that
the sensitiveness to the direction of light is greater than that to gravity. There is
evidence that these phototropic curvatures of stems are occasioned by a redistribution of growth substance induced by the one-sided illumination, more of
the growth substance accumulating on the shaded side.

Most leaves are **diaphototropic**, placing themselves so that the upper surface
is presented at right angles to the direction from which most light comes. Adjust-
ments of leaf position are effected by means of growth-curvatures and torsions in
the petiole. In some plants growing in regions where sunlight is very intense the
leaves lie in a vertical rather than a horizontal plane, with the result that the
incidence of the sun's rays on the leaf is reduced.

Roots in general show no sensitiveness to the direction of light, but those of
some plants exhibit **negative phototropism**, as in mustard (Fig. 175). The
climbing roots of ivy (Fig. 26) are also negatively phototropic.

OTHER TROPISMS

The growth of plant organs is sometimes subject to control by other directive influences in addition to gravity and direction of light. Thus roots are often sensitive to the distribution of moisture in the soil, and tend to grow towards regions of greater moisture content. In so doing they exhibit **hydrotropism**. The roots of some plants (for example, lupin) have been shown to be sensitive to the distribution of other substances besides water, such as salts and acids, the roots exhibiting both positive and negative curvatures depending on the identity of the substance and the concentration. This is distinguished as **chemotropism**. The downward growth of the pollen tube through the tissues of the stigma and style (Chap. XXI) is in part directed by the chemotropic effect of substances such as sugars present in the tissues.

FIG. 175. SEEDLING OF WHITE MUSTARD WHICH HAS BEEN FIRST ILLUMINATED FROM ALL DIRECTIONS, AND THEN FROM ONE DIRECTION ONLY (INDICATED BY ARROW).
(*After Noll.*)

Haptotropism, or curvature in response to contact, is best known in the tendrils of climbing plants (pp. 51, 53). Contact of the tendril with a solid object causes growth on the side away from the point of contact to become faster, so that the tendril tends to coil around the object. Under favourable conditions a tendril may begin to curve within a few minutes of being lightly touched on one side ; but there is at present an incomplete understanding of the reason for these effects.

NASTIC MOVEMENTS

In addition to responses to stimuli coming from one direction only, many plant organs show changes in orientation in response to diffuse stimuli present equally all round the plant, such as changes in temperature or in light intensity. Such responses are said to be of **nastic** type. Some nastic movements are due to differential growth and therefore resemble tropic responses in being confined to still-growing parts and in taking place rather slowly. But many of them are occasioned by differential changes in turgor, and as such can occur in mature parts and often with considerable rapidity.

The closing movements of many flowers and of some foliage leaves in the evening are examples of nastic responses, and are often termed ' **sleep** ' **movements.** Reverse movements occur in the morning. In some flowers, such as crocus, tulip and lesser celandine, changes of temperature are mainly responsible for the closing and opening movements. These induce differential growth-changes on the inner and outer surfaces of the perianth or petals, and it is a matter of common experience that if closed flowers of the plants mentioned are brought into a warm room, opening soon commences. In other plants floral movements are dependent more on changes in light intensity than in temperature. This is the case in dandelion, daisy, and other composites (Fig. 176). The closure of flowers at night or in wet weather probably serves to protect the inner floral organs against damage by low temperature or rain. However in a number of plants the flowers open in the evening and close during the day, as is the case in night-scented stock, some evening primroses, and white campion. Most such flowers are pollinated by night moths. In some cases the actual factor inducing the evening opening has been shown to be the increasing atmospheric humidity.

Sleep movements, as noted, are also shown by the leaves of a number of plants, as for example wood sorrel and many members of the family Leguminosae (clover, runner bean, *Mimosa*, etc.). These plants have compound leaves, and in the daytime the leaves are open and the leaflets displayed in a horizontal

FIG. 176. FLOWER-HEAD OF ROUGH HAWKBIT, CLOSED WHEN KEPT IN THE DARK; OPENED WHEN ILLUMINATED. (*After Detmer.*)

position, as shown on the left of Fig. 177 for *Amicia* (a South American leguminous plant) and of Fig. 178 for *Mimosa pudica*. As night approaches the leaves close up and assume the appearance shown on the right of the two figures just mentioned. In *Amicia*, as in wood sorrel, the leaflets droop : in *Mimosa* pairs of leaflets fold their upper surfaces together and the petiole droops. These movements are facilitated by the presence of **pulvini**, somewhat swollen joint-like structures present at the base of each leaflet and, in *Mimosa*, at the base of the petiole. Differential changes in turgor in the parenchymatous cortical tissue on the upper and lower sides of the pulvini occur in response to changes in light intensity, and lead to movement of the parts of the leaf. The value to the plant of these nocturnal leaf closures is not perfectly clear, but it may be said that they are likely to lead to some curtailment of transpiration and of loss of heat by radiation.

As is well-known, in the particular case of *Mimosa* (native to Assam, Malaya and other Asiatic regions) the leaves also close up if they are touched or if the plant is shaken, hence the common reference to this as the sensitive plant. The response is very rapid. If the stimulus is applied to the terminal leaflets the reaction can be seen to spread progressively to lower parts of the leaf, and then perhaps to other leaves. There is some evidence that a chemical substance is produced at the point of stimulation and is rapidly conducted through the leaf, affecting successive pulvini. The value of this sensitiveness to the plant is not very clear.

Some other examples of nastic movements are provided by the carnivorous plants (Chap. XVIII), as, for example, the movements of the tentacles of the leaf of *Drosera*, and the closure of the leaf of *Dionaea*. Again, the stamens of some flowers show nastic movements when touched. Such responses are seen in *Berberis* and cornflower, and form part of the pollination mechanism.

FIG. 177. *Amicia zygomeris*, SHOWING DIURNAL (LEFT) AND NOCTURNAL (RIGHT) POSITIONS OF LEAVES.

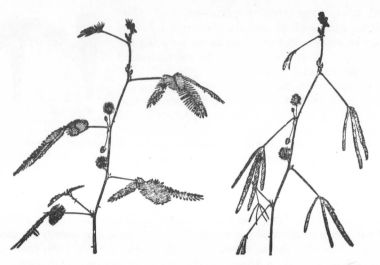

FIG. 178. *Mimosa pudica*, SHOWING LEAVES IN NORMAL POSITION (LEFT) AND AFTER
TRAUMATIC STIMULATION (RIGHT).

PHOTOTAXIS AND CHEMOTAXIS

The movements so far considered have been of individual organs of plants
which are rooted in the soil.   Some lower plants are of minute size, consisting in
many cases of a single cell, and by means of cilia are able to move through the
water in which they live.   An example is provided by *Chlamydomonas* (Fig. 1).   In
other cases, as in *Fucus* and the ferns, motile unicellular sperms are produced
during reproduction (Chaps. XXIII and XXVI).   The locomotions of such
motile structures is termed **taxis**.   The direction of movement is subject to control
by external factors of which light and the presence of chemical substances will be
mentioned here.

If a jar of water containing a good growth of *Chlamydomonas* is placed near a
bright window, the organisms will tend to swim towards the more brightly
illuminated side of the jar, so that the green colour due to the presence of the
organisms will become deeper on that side.   The mechanism of this **phototaxis**
is obscure.

An example of **chemotaxis** is provided by the sperms of the fern.   Pfeffer
showed that the sperms were strongly attracted by malic acid, so that if a capillary
tube containing a solution of that substance is introduced into a drop of water
containing the sperms, the latter may be seen under the microscope to swim
towards the entrance of the capillary tube.   There is evidence that it is this way
that the sperms are attracted to the archegonia containing the eggs, and fertilisation thus brought about (Chap. XXVI).

MOVEMENTS IN PLANTS AND ANIMALS

In plants locomotion is confined to certain lower forms of the type indicated
in the previous section.   Many other lower plants, and most higher plants, are
fixed in the soil or other substratum.   This difference between plants and animals
is related to divergences in the method of securing nutrition, as noted in Chapter I.

The animal has to search for the complex foods which it requires, while the simple raw materials utilised by the green plant are of almost universal occurrence.

Some of the movements exhibited by the individual organs of the plant bear a superficial resemblance to those of parts of the animal body, particularly in the case of the more rapid nastic responses. Actually they are brought about by quite different mechanisms. Movements in plant organs, as we have seen, are due to differences in growth-rate on two sides of an organ, or to reversible changes in the turgor of cells : no system of contractile muscle fibres exists in the plant. Neither is there any recognisable nervous system in the plant, the conduction of stimuli, so far as it has been elucidated, appearing to be brought about by movement of chemical substances.

PRACTICAL WORK

1. Soak seeds of broad bean in water overnight and then sow with the hilum downwards in sand or sawdust. Take a wide-necked glass jar or a rectangular museum jar, half-fill with water, and close the top by means of a cork or a piece of stout cardboard. Insert several long blanket pins through the latter. When the radicles have attained a length of about half an inch fix several seedlings to the pins inside the jar, with the radicles pointing in various directions. Seedlings with straight radicles should be selected, and they should be fixed clear of the water. Place the jar at a temperature of 15°—25° C. and inspect from time to time for positive geotropic curvatures in the radicles.

2. For the study of geotropic responses in shoots, somewhat older seedlings of broad bean may be used or potted plants from the greenhouse, such as *Pelargonium*, or again shoots may be cut from herbaceous plants in the field or garden and fixed in a large test tube of water by means of a rubber stopper, the hole in the latter being sealed up with ' Plasticine '. Fix the specimen so that the shoot is horizontal, preferably in a fairly warm place where uniform light is provided. Examine periodically for negative geotropic curvature in the stem and diageotropic responses in the leaves.

3. If a klinostat is available confirm that geotropic responses are no longer shown if roots and shoots are placed horizontally and slowly revolved about their long axes.

4. For the demonstration of phototropic responses seedlings of mustard, wheat, barley, etc., may be grown in pots, under conditions of uniform illumination. When a height of 2—3 inches has been attained, place the pots under a hood made from a box or a piece of black cloth arranged so that light now reaches the shoots from one side. Alternatively place the plants in a darkroom about 3 feet to the side of a 40-watt lamp. Note the time taken for positive phototropic curvature to appear.

Instances of diaphototropic responses in leaves can be found in plants grown in living-rooms or against a wall. The actual occurrence of the responses can be demonstrated by taking a *Pelargonium* plant which has been grown in uniform light and partly screening it by means of a black cloth so that light now reaches it from the side. In a few hours time it will be observed that the leaves have begun to adjust their position. A positive response will also develop in the stem.

5. Thermonastic responses in flowers can be demonstrated with crocus or tulip. If flowers of these in the closed condition are brought into the warm laboratory, opening usually commences promptly, even if the flowers are placed in a dark cupboard. The movements are reversed if the specimens are put into a refrigerator.

Experiments with the daisy or dandelion may be carried out to illustrate the photonastic properties of these flowers. They may be performed out-of-doors or daisy plants can be transplanted into pots in the greenhouse. If certain plants are covered over with a box in the afternoon, examination next day will show that at a time when exposed plants have opened their flowers, the flowers in darkness are still closed.

The sleep movements of leaves should be studied in suitable examples in the garden and field.

CHAPTER XXI

SEXUAL REPRODUCTION IN THE FLOWERING PLANT

The flower can be regarded as a shoot specially modified for the production of seeds. Inside each seed the essential structure is an **embryo**, a rudimentary plant already possessing a root and a shoot. Unlike vegetative propagation, where new plants arise simply from buds of the parent plant, the production of seeds involves a sexual process.

SEXUAL REPRODUCTION

Sexual reproduction always involves the union of two special reproductive cells known as **gametes**. In some primitive Algae, such as *Chlamydomonas*, both gametes are alike (see p. 299), but in the majority of plants there is a distinction between small male gametes (**spermatozoids**) and larger female gametes (**eggs or ova**). The union of a male with a female gamete is the process known as **fertilisation**, a process which is fundamentally the same in the sexual reproduction of all plants and animals. The fertilised female gamete or ovum gives rise to an embryo plant or animal by repeated cell divisions. It would appear that the male gamete serves to stimulate the ovum to divide although in some few examples an ovum may divide and develop into an embryo without fertilisation having taken place. This latter phenomenon is known as **parthenogenesis**. Another important aspect of fertilisation is that it leads to a mingling of the characters of the male and female parents with results which form the subject-matter of the science of genetics (Chap. XXVIII).

HERMAPHRODITE AND UNISEXUAL FLOWERS

In flowering plants, the structure which produces the eggs and the male gametes is the flower.

Many plants and many animals are capable of producing both eggs and male gametes in the one organism. For example, in the case of the buttercup, both eggs and male gametes are produced in the same flower ; the same applies to the majority of familiar flowers, such as the wallflower, tulip, pea, violet and a host of others. In animals, this production of both male and female gametes by the same individual is not so common. It is present in some animals, however ; for example, the earthworm. In such cases as these, where the same individual produces both kinds of gametes, the plant or animal in question is said to be **hermaphrodite**.

In other cases the male and female gametes are produced by different individuals. This is more common in animals (for example, mammals, birds, reptiles, fishes and so on) than in plants. Such types where an individual produces only male or only female gametes are said to be **unisexual**.

In Angiosperms, unisexuality is not by any means so common. Nevertheless

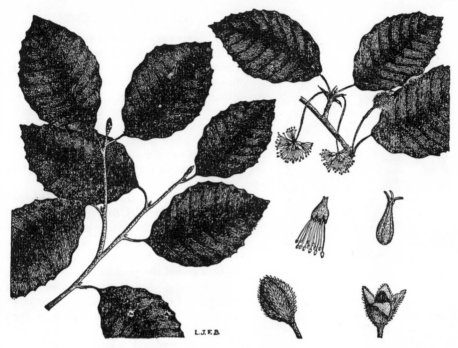

FIG. 179. BEECH (*Fagus sylvatica*).
Left, twig bearing foliage ; top right, clusters of male flowers and two female flowers
enclosed in a cupule ; middle right, single male and female flowers ; bottom right, ripe
cupule closed and one opening exposing seeds.

there are unisexual flowers.   The unisexual flower is that type of flower which can
produce either eggs or male gametes, but not both.   Unisexual flowers are again
subdivided, because in certain plants both male unisexual and female unisexual
flowers grow on one and the same plant, whereas in other cases the female flower
grows on one plant but the male flower grows on another plant.   Those plants
which bear both types of unisexual flowers are said to be **monœcious**.   Examples
of monœcious plants are the beech (*Fagus sylvatica*) (Fig. 179), oaks (*Quercus* species),
hazel (*Corylus avellana*), and sycamore (*Acer pseudoplatanus*).   Plants in which the
two types of flowers are borne on different plants are said to be **diœcious**.   Examples of these are the willows (*Salix* species), poplars (*Populus* species) (Fig. 180) and
hop (*Humulus lupulus*).

    There is a great number of advantages in sexual reproduction which vegetative
reproduction cannot give.   In the case of vegetative reproduction (for example,
the potato tuber or the bramble stolon) the young new plants produced cannot be
very far removed from their parents unless an artificial agency, like man, steps in
and helps.   On the other hand, seeds which contain the sexually produced embryos
are capable of being carried far away from the parent plant.   This is of great advantage, for whereas those plants which reproduce themselves vegetatively soon
begin to overcrowd themselves, the sexually reproduced plants can, by means of
their seeds, be distributed far and wide.   Further, and more important, advantages of sexual reproduction will be considered in Chap. XXVIII.

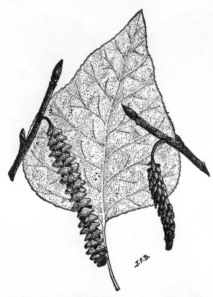

FIG. 180. BLACK POPLAR (*Populus nigra*).
Left, catkin of male flowers ; right, catkin
of female flowers.

THE INFLORESCENCE

Flowers may be borne singly as in the buttercup, tulip, wood anemone (*Anemone nemorosa*) and many other plants. Such flowers are said to be **solitary**. On the other hand, a number of flowers may be borne together on a branch system, as in the wild hyacinth or bluebell described below. Such flower bearing branch systems are usually clearly marked off from the vegetative region of the plant. They are described as **inflorescences**.

For example, the bluebell (*Endymion nonscriptus*) bears its flowers in inflorescences, each of which may be composed of anything from five to fifteen flowers. The main stalk bearing the complete inflorescence is called the **peduncle**. In this case it is a strong upright stem, just as it is in the case of the delphinium, foxglove, cowslip, etc. Each separate flower is connected to the peduncle by a short stem called the **pedicel**. This pedicel usually stands in the axil of a small tissue-like scale-leaf, known as a **bract** (Fig. 181). This is understandable when one considers that the flower can be regarded as a modified lateral shoot and that such shoots usually arise in an axillary position. In fact, one frequently finds inflorescences of bluebells in which some of the bracts, especially the lowermost, are grown out into green leaves, 2-4 inches long.

In the bluebell the inflorescence is not very complicated, being composed of a single, straight peduncle bearing a series of bracts, from the axils of which pedicels arise, each bearing a still smaller scale-leaf or **bracteole** and terminating in a flower. The oldest flowers, that is, the flowers which open first, are at the bottom of the inflorescence, and then the flowers are younger and younger going towards the top. This type of inflorescence is called a **raceme**, though in the bluebell it tends to become one-sided as the flowers open.

A type of inflorescence similar to the raceme is the **spike**. This differs only in that the flowers are borne directly on the peduncle, that is, they have no pedicels. This type is represented in the plantains (*Plantago* species) (Fig. 197).

In another type of inflorescence, the axils of the bracts, instead of giving off pedicels, give off branches of the peduncle. These branch peduncles themselves bear bracts, in the axils of which pedicels are produced. The final structure therefore is a peduncle bearing branch peduncles, each of which is a raceme. This ' raceme of racemes ' is called a **panicle**. The panicle is characteristic of the Adam's needle (*Yucca filamentosa*), a native of Central America and Mexico, frequently seen cultivated in parks and gardens in Britain (Figs. 182 and 183).

In a type of inflorescence closely related to the raceme it will be noticed that the length of the pedicel gets longer and longer from the top downwards. The

result is that the inflorescence, instead of being pyramidal in shape, is circular and flat, viewed from
above, since, owing to the different lengths of the
pedicels, all the flowers, in spite of the fact that they
are borne at different levels on the peduncle, are
themselves on a level with each other. This is well
seen in the candytuft (*Iberis amara*). This type of
inflorescence is called a **corymb.**

The same effect is obtained by another form of
inflorescence in that all the flowers are on the same
horizontal plane, but this type differs from the
corymb since all the pedicels are given off from the
same level on the peduncle, that is, the top, instead
of at different levels. This type of inflorescence is
called an **umbel**, and is characteristic of the cherry
(*Prunus* species) (Fig. 184).

In many plants the umbel is more complicated.
It resembles the simple umbel in that all the
branches are given off at the same level, but here
the branches are not single pedicels, but branches
of the peduncle, and each one of these branches
in its turn bears a collection of pedicels at its end.
Therefore, the whole inflorescence may be looked
upon as being an umbel of umbels or, better still,
a **compound umbel.** But the final effect is the
same in that all the flowers are on the same level.
It is clear that the compound umbel bears the same
relation to the simple umbel as the panicle does to
the raceme. The compound umbel is very common
in Nature, being represented in the parsley (*Petroselinum crispum*), parsnip (*Pastinaca sativa*), carrot
(*Daucus carota*), fool's parsley (*Aethusa cynapium*),
hemlock (*Conium maculatum*), chervil (*Anthriscus cerefolium*) (Fig. 185) and many other plants—members
of the family Umbelliferae (p. 464).

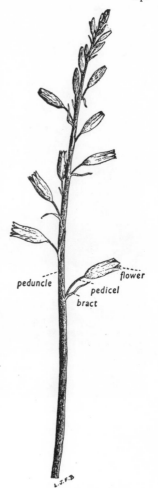

FIG. 181. INFLORESCENCE OF
BLUEBELL (RACEME).

In all members of the flowering-plant family called Compositae, there is a very
special kind of inflorescence and the flowers themselves are very peculiar in
structure. Many British plants belong to this family, such as the dandelion
(*Taraxacum officinale*), sunflower (*Helianthus annuus*), daisy (*Bellis perennis*) (Fig. 186),
etc. In a single daisy head, for example, what appear to be single white petals
given off in ray-like form from the circumference are, in fact, all single, separate
flowers. Each yellow structure, too, many of which form the yellow disk is a single
yellow flower. So here we have a large number of flowers all borne on the same
level on the enlarged and flattened summit of the peduncle. This type of inflorescence is called a **capitulum.** The whole inflorescence in the case of the
capitulum is surrounded and supported by a collection of bracts known as an
involucre.

Other types of inflorescence are classified as **cymes.** They are more definite
than racemose inflorescences because they each end in a flower, and the pro-

FIG. 182. ADAM'S NEEDLE (*Yucca filamentosa*).

duction of a flower at the end of a shoot prevents further growth in that direction. On the other hand, a raceme does not end in a flower.

There are several types of cymes, but in each case, since the inflorescence terminates in a flower, any further development must take place through a branch. Cymes may be divided into two groups, namely, a one-branched type, commonly called a **monochasium**, and a two-branched one, or **dichasium**. The former may be divided into two types: (*a*) in which the new branches invariably come off on the same side of the parent stem, as seen in the forget-me-not (*Myosotis palustris*); (*b*) in which the new branches come off on alternate sides of the parent branches, for example, buttercup (*Ranunculus* species).

A typical example of a dichasium is the mouse-ear chickweed (*Cerastium vulgatum*) (Fig. 187). These various types of inflorescences are understood better by reference to their diagrammatic representation in Fig. 188.

THE FLOWER

One of the more simple types of flower is that of the buttercup (*Ranunculus* species). In it there are four sets of different organs, all borne upon a swollen structure. This structure is really the swollen end of the stem, and is called the **receptacle** (Fig. 189).

Passing from the outside of the flower towards the centre, the four sets of organs may be clearly distinguished.

The outermost whorl of organs is composed of three to five green, boat-shaped organs, each of which is called a **sepal**. The complete whorl of sepals is collectively known as the **calyx**. The main function of the sepals is that of protection of the more delicate and much more important floral organs nearer the centre of the flower. In the buttercup, for example, when the flower is young and unopened, the sepals, being on the outside, surround the inner structures and protect them from rain, cold, etc. When the flower finally opens, the sepals help to hold the rest of the floral whorls together. In some flowers, however, the function of the sepals comes to an end when the flower-bud opens. This is demonstrated, for example, in the common poppy (*Papaver rhoeas*). Here there are two large, hairy sepals. They form a splendid protective covering when the poppy flower is still in bud ; but when the flower is opened the sepals fall off.

In the buttercup, the next inner whorl to the calyx is the **corolla**. This is composed of usually five bright yellow, heart-shaped **petals**, though this number varies from five to thirteen. At the base of each petal is a small sac which contains a sweet juice called **nectar**. Therefore the sac is called a **nectary**. Insects visit flowers chiefly to collect this nectar, and from it some species make honey. The petals alternate in their position with the sepals, when there are equal numbers. That is, between the five sepals there must be five spaces ; the petals are opposite these spaces and not opposite the sepals themselves.

Next in order to the petal whorl, passing inwards, are numerous **stamens**, yellow in colour and shaped like Indian clubs. These are arranged spirally on the receptacle and constitute what is collectively known as the **androecium**. The number of stamens in any one buttercup

Fig. 183. Inflorescence of *Yucca filamentosa* (Panicle).

flower is large, but the number varies considerably in different buttercup flowers. This is quite different from the case of, say, the bluebell, where the number of stamens is constantly six.

Numerous spirally arranged **carpels**, constituting the **gynoecium**, occupy the centre of the buttercup flower. Each carpel is more or less kidney-shaped and contains one **ovule**, or potential seed. Again, in many other flowers, the number of carpels is small and constant for the species.

The main function of the flower, as has already been seen, is to produce male and female gametes, and to allow male and female gametes to fuse together, thus producing the young embryo. Neither the sepals nor the petals have anything to do with the production of gametes. Therefore, as floral organs, they are looked upon as being only of secondary importance. In fact, in many flowers they are absent altogether. For example, in the unisexual willow the female flowers have no sepals, petals or stamens and the male flower has only stamens, and no other whorls.

Fig. 184. Inflorescence of Cherry (Simple Umbel).

Although sepals and petals are relatively unimportant, stamens and carpels are of the utmost importance, for it is these organs which produce the gametes. The stamens produce pollen grains, whose contents

FIG. 185.   INFLORESCENCE OF CHERVIL
(COMPOUND UMBEL).

ultimately give rise to male gametes, whereas the carpels bear ovules, inside which the female gametes are produced. In hermaphrodite flowers, like the buttercup, stamens and carpels are present in each flower. On the other hand, in unisexual flowers, stamens only are present in the male flowers and carpels only in the female flowers.

Nehemiah Grew, in 1676, was the first to suggest that the stamens and carpels produced the male and female organs, respectively, of plants. Nevertheless, it was not until 1694 that another botanist, **R. J. Camerarius**, proved the truth of this suggestion by carefully devised experiments. Although the details were not worked out until about 150 years after this, as will be seen later in the chapter, **Carl Linnæus** accepted the work of Camerarius and concluded that the stamens and carpels were so important that he used them as the main basis for classifying the flowering plants (see Chap. XXX).

The stamen of the buttercup is composed of a fine stalk which swells at the top into a lobed, ovoid structure. The stalk is called the **filament**, and the swollen head the **anther**. The latter is the more important part, and the stalk serves to convey food materials to it from the plant itself. To get a clear idea of how the stamen performs its important function of the production of male gametes, it is necessary to examine the anther under the microscope. Then it is seen to be, not a solid mass of tissue (at any rate when ripe), but to be composed of four cavities which run practically throughout its length. When ripe, these cavities or **pollen sacs** are filled with hundreds of spherical bodies called **pollen grains**. Pollen grains are not the male gametes, but the male gametes are produced from them, as will be seen when the process of fertilisation is considered, later. Until then, it will be best, therefore, to leave the structure of the stamen and consider that of the carpel.

Externally, the carpel of the buttercup looks like a tiny green kidney with a small hooked projection at the upper end. The main part is called the **ovary**, and the hooked projection the **style**. At the very tip of the style, the surface is sticky. This portion of the style is called the **stigma**.

FIG. 186.   INFLORESCENCE OF DAISY
(CAPITULUM).

The carpel can be thought of as a little leaf folded and fused along its margins so as to enclose a cavity. In the buttercup, at the base of this ovarian cavity there is borne a single, egg-shaped **ovule**, which, when it is ready for fertilisation, contains the female gamete or **ovum.**

DIVERSITY OF FLORAL STRUCTURE

From the above it will be clear that the stamens and the carpels are the parts of the flower which are concerned in the process of fertilisation. Since, however, these parts are separate from one another in the flower, a necessary preliminary to the fusion of the male and female gametes is the transportation of the pollen grains to the carpels.

FIG. 187. INFLORESCENCE OF MOUSE-EAR CHICKWEED (DICH-ASIUM).
(After Duchartre.)

This process is called **pollination.** The methods of pollination in flowering plants are manifold, and this is where the secondary organs, the petals, are helpful. So, before considering pollination and what happens after pollination takes place, it would be best to examine some different types of flowers.

The various whorls of flowers vary in almost every conceivable way. It is naturally impossible to consider them all, but a few examples will give some idea of this diversity of structure.

CALYX

The calyx shows comparatively little diversity. The number of sepals composing it varies ; there are two in the poppy (*Papaver* species), three in the lesser celandine (*Ranunculus ficaria*), four in the wallflower (*Cheiranthus cheiri*) and five in the buttercup (*Ranunculus* species). In a few plants, the sepals are more numerous, as in the globe-flower (*Trollius europaeus*), where there may be as many as fifteen.

Sometimes the sepals of the calyx are all joined to each other, forming a tube. This is well seen in the case of the primrose (*Primula vulgaris*). There, the calyx forms a definite tube, but it is easy to see that it is really composed of five sepals joined together, by the five long teeth at the top of the calyx tube (Fig. 190).

Sometimes the calyx, instead of being its normal colour, that is, green, becomes brightly coloured and enlarged, and assumes the function of the petals of the corolla. The calyx in such a plant is said to be **petaloid.** In the marsh marigold (*Caltha palustris*) for example, what look like five large yellow petals are really sepals. In this flower and also in the clematis or traveller's joy (*Clematis vitalba*), the petals are absent. In the Christmas rose (*Helleborus niger*), the sepals are petaloid, being large and white ; yet the petals are not absent but are reduced to small tubular nectaries in their correct position in the flower, that is, between the calyx and the andrœcium.

COROLLA

It is in the corolla where the greatest diversity of shape, colour and arrangement occurs.

In some flowers the petals are entirely absent, for example, the willow (*Salix*

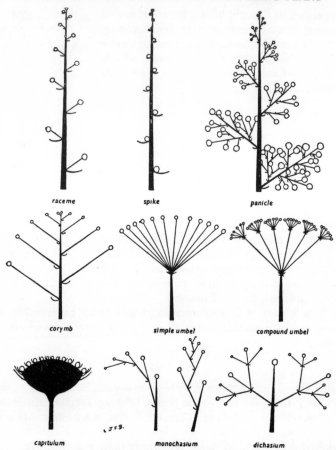

FIG. 188. DIAGRAMMATIC REPRESENTATION OF VARIOUS TYPES OF INFLORESCENCE.

species). In many cases the petals join to form a tube, as in the primrose (Fig. 190).
The number of petals, too, varies considerably, even more than in the case of the
sepals. For example, the wallflower has four, the pink (*Dianthus* species) and the
buttercup, five, and some, for example, the white water-lily (*Nymphaea alba*), have
an indefinite number.

Some flowers are regular in the arrangement of their petals. For example,
in the case of the buttercup or the wallflower, the petals and, indeed, all the floral
organs are symmetrical about any axis. That is, it does not matter through what
vertical plane the flower is cut, the two halves produced are the mirror images
of each other. Such regular flowers are said to be **actinomorphic** (Fig. 190).

On the other hand, many flowers are irregular. In the sweet pea (*Lathyrus
odoratus*), for example, there are five petals, but they are not all of the same shape.
Looking straight towards the inside of the flower, there is one large petal standing
up at the back. It is larger and more spreading than any of the others, and is
called the standard. Then, there are two wing-like petals, one on each side of the
standard. Each is called a wing. At the bottom are two still smaller petals,
facing each other and appearing similar to a ship's keel. The two together are

therefore called the keel. In this flower it is quite obvious that there is only one vertical plane through which the flower could be cut in order to produce two symmetrical halves. The plane would pass down the middle of the standard and between the two wings and the two petals forming the keel. Such irregular flowers as these are said to be **zygomorphic** (Fig. 190). Violets (*Viola* species) and the white deadnettle (*Lamium album*) are also examples of zygomorphic flowers.

In a large number of flowers, it is impossible to distinguish a separate calyx and corolla. Good examples of this are the tulip, bluebell, crocus, lily, etc. (Fig. 191). There are, however, two whorls of brightly coloured segments, an outer one of three and an inner one of three, which alternate with the three outer ones. In such examples, where there is no differentiation into sepals and petals, the two whorls are known collectively as the **perianth**. This characteristic is well exemplified in the natural order Liliaceae (p. 480).

STAMENS

The androecium of the flower also shows a great diversity, especially in number of stamens. The number is high and indefinite in the buttercup (Fig. 189); also in the poppy and the rose. But in many plants the number is definite. In the deadnettle (*Lamium* species) there are four; wallflower, six; hyacinth, six; and pea, ten. Sometimes the filaments of the stamens differ in length in the same flower; for example, the wallflower has four long stamens and two short ones (Fig. 192), and the white dead-nettle, two long and two short.

In the buttercup, wallflower, etc., the stamens are joined directly to the receptacle (Figs. 189 and 192). In other plants, however, instead of being fixed to

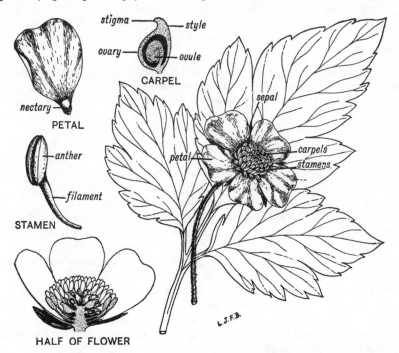

Fig. 189. The Buttercup Flower.

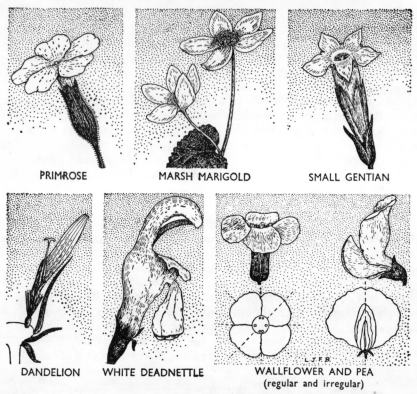

PRIMROSE   MARSH MARIGOLD   SMALL GENTIAN

DANDELION   WHITE DEADNETTLE   WALLFLOWER AND PEA
                                        (regular and irregular)

FIG. 190. TYPES OF FLORAL FORM.

the receptacle, the stamens are joined to the members of the perianth or to the petals. In the bluebell, for example, the stamens are joined to the perianth segments. There is one stamen on each perianth segment, thus giving six stamens in all (Fig. 192). Another example, which is even more curious, is that of the primrose. Here the stamens, of which there are five, are joined to the corolla tube some distance up. But the distance varies, in that in some flowers the stamens are about half-way up the tube and the style of the ovary is long, thus placing the stigma above the stamens (**pin-eyed** flower); whereas in other flowers, the stamens are fixed at the top of the corolla tube and the style is short, thus placing the stigma below the stamens (**thrum-eyed** flower) (Fig. 192).

In some flowers the stamens are joined together, as in the dandelion (*Taraxacum officinale*) where the anthers of the five stamens are joined, thus forming a tube around the gynœcium (Fig. 190). In others only some of the stamens are joined, as in the bird's-foot trefoil (*Lotus corniculatus*), where there are ten stamens, nine of which are joined by their filaments and the tenth is free (Fig. 192).

There is also much variation in the form of the individual stamens. The filament may be joined to the base of the anther (the **basifixed** condition); it may be joined to the back of the anther for some distance (**dorsifixed**); or it may be joined to a point about half-way along the length of the anther, so that the latter can swing freely (**versatile**).

CARPELS

The gynœcium shows many interesting forms. One of the simplest is that of the buttercup, where all the carpels are separate. The number, too, is high and indefinite (Fig. 189). In many flowers, on the other hand, the carpels are few and definite in number, and are very often joined to each other.

The methods of joining of the carpels are interesting and varied. The best way to examine the various types would be first of all to imagine the carpel as an open leaf, bearing its ovules on the margins. It must be remembered, however, that ovules never are borne naked and exposed in this way in Angiosperms ; but, starting with this *hypothetical* case, it is easy to see the various ways in which the ovules could conceivably become enclosed by the carpel.

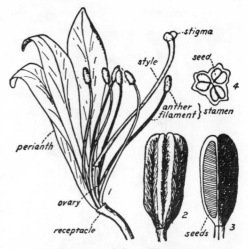

FIG. 191. A LILY FLOWER.
1. flower with part of the perianth removed to show the inner whorls ; 2, fruit ; 3, longitudinal section of the fruit ; 4, transverse section of the fruit.
(From Robbins and Rickett's ' Botany '. Courtesy of the D. Van Nostrand Co., Inc)

The simplest method for the carpel to enclose its ovules would be to fold in half, thus bringing together its two margins, bearing the ovules. Then, imagine these margins to become fused. Thus one carpel enclosing a vertical line of ovules would be produced. In the bean, for example, there is a series of about eight ovules arranged as one would expect them in such a case (Fig. 193).

The next method of enclosure would be for two carpels, facing each other, to join at their margins. This would give two carpels forming an ovary with one common cavity into which the ovules would project in two longitudinal rows. This is the case in the gooseberry (*Ribes uva-crispa*) (Fig. 193). A variation of this is seen in the tomato (Fig. 217). Here, the margins, four in all, two from each carpel, meet at a common centre. Thus there are two carpels, with two cavities, and a row of ovules projecting into each cavity. In many other cases the same method as that of the gooseberry applies, except that there are three carpels, with a common cavity into which three longitudinal rows of ovules project. This is seen in the violet (Fig. 193).

Then a type similar to that of the tomato (*Solanum lycopersicum*) except that it has three carpels with three cavities, into each of which one row of ovules projects, is also possible ; indeed, it is very common among flowers. It is seen, for example, in the tulip (Fig. 193). A completely different type is seen in the primrose (Fig. 193). Here, there are five carpels joined together ; but the ovules, instead of forming rows at the fused carpellary margins, are borne on a projection from the base of the ovary cavity.

The simple type of gynœcium where the carpels are all free is called **apocarpous**, whereas the ovary which is composed of joined carpels is said to be **syncarpous**.

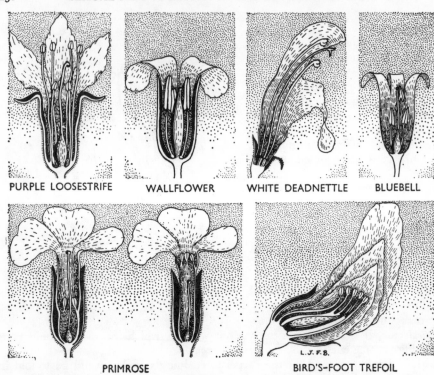

PURPLE LOOSESTRIFE   WALLFLOWER   WHITE DEADNETTLE   BLUEBELL

PRIMROSE   BIRD'S-FOOT TREFOIL

FIG. 192. DIVERSITY IN ARRANGEMENT OF STAMENS.

The arrangement of the ovules within the syncarpous ovary is referred to as **placentation**, because that part of the carpel on which the ovules are borne is called the **placenta**. In the case of the gooseberry and violet, where by the method of carpellary fusion the ovules lines the ovary wall in longitudinal rows, the placentation is said to be **parietal**. Where the fusion of carpellary margins takes place all together at the centre, as in the tomato and tulip, the placentation is **axile**. The primrose, however, is an example of the ovules being free from the ovary wall. Such placentation is therefore called **free central** (Fig. 194).

In many flowers, the floral receptacle changes its shape so much that the relative position of the various whorls is entirely altered. In the simplest case, such as the buttercup, the gynœcium is at a higher level than all the other whorls ; in other words, the gynœcium is **superior** to them, whereas they are **inferior** to the gynœcium. Since the other whorls are below the gynœcium, this flower is said to be **hypogynous**. On the other hand, the receptacle may become cup-shaped and fuse with the ovary wall, so that the gynœcium is placed below the other floral parts. Such flowers, for example, that of the pear, are said to be **epigynous** and their gynœcium is **inferior**. An intermediate condition is seen in examples where the receptacle widens out as a disk, or forms a cup, but does not fuse with the ovary wall. The sepals, petals and stamens are usually borne on the edge of the disk or cup ; the gynœcium is superior. This intermediate condition, seen for example in the cherry, is said to be **perigynous**.

Perigynous and epigynous types occur in the rose family (Rosaceae) (Fig. 195).

REPRESENTATION OF FLORAL STRUCTURE

There are three ways in which the structure of a flower is usually recorded.

One is by means of what is called a **floral formula**. By reference to a floral formula it is possible to tell (a) whether a flower is actinomorphic or zygomorphic ; (b) whether it has an inferior or a superior ovary ; (c) the number of sepals, petals, stamens and carpels ; (d) whether any or all of these are free or joined ; (e) whether the parts of any whorl are united to those of another whorl.

(a) Actinomorphy is represented by the sign $\oplus$, and zygomorphy by $\uparrow$; (b) an inferior ovary is represented by a line over the number of carpels in the gynœcium, and a superior ovary by a line beneath ; (c) sepals by K, then the number, petals by C, then the number, andrœcium by A, then number, and gynœcium by G, then number. If any of the parts are joined, then brackets are put around the numbers concerned. If one whorl is united to another, for example, stamens to petals, this is indicated by bracketing the symbols of the two whorls concerned.

The following floral formulæ are examples :

common buttercup (*Ranunculus acris*) : $\oplus K5\ C5\ A \infty\ G\underline{\infty}$
wallflower (*Cheiranthus cheiri*) : $\oplus K4\ C4\ A4 + 2\ G(\underline{2})$
garden pea (*Pisum sativum*) : $\uparrow K(5)\ C5\ A(9) + 1\ \overline{G}\mathrm{I}$.

primrose (*Primula vulgaris*) : $\oplus K(5)\ \overset{\frown}{C(5)\ A}5\ G(\underline{5})$

Where there is a perianth and not two separate whorls of sepals and petals, the perianth is represented by P. Therefore the floral formula for the tulip is :

$$\oplus\ P3 + 3\ A3 + 3\ G(\underline{3}).$$

The second method of representing floral structure is by means of the **longitudinal section** (Fig. 196). This gives an elevation of the flower. It is important as showing the vertical placing of the various parts on the receptacle and so will

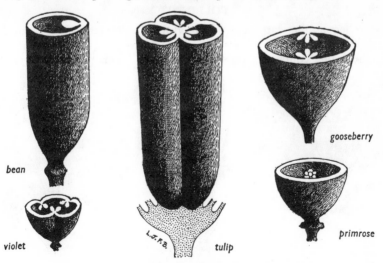

FIG. 193. TYPES OF FLORAL OVARIES.
Each cut across transversely, showing the lower half.

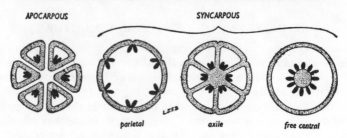

Fig. 194. Diagrammatic Representation of Placentation.

show clearly whether the flower is hypo-, epi-, or peri-gynous. The flower should be halved along the median plane and only the cut surfaces drawn. A drawing of the whole of the half flower may replace the longitudinal section. It will give fuller information about the shape of the various floral parts.

The third method of representation is the **floral diagram**. This is really a plan of the flower, the various parts being represented as simplified transverse sections. If the sepals, petals or stamens are joined, this is indicated by small brackets on the diagram. This is not necessary in the case of the carpels, since they are usually drawn more or less as they really appear in transverse section (Fig. 196). If the stamens are epipetalous, each is joined by a line to the appropriate petal.

When constructing the floral diagram, the flower should always be held so that the bract faces the observer and the axis bearing the flower is furthest away. The bract is said to be **anterior** and the axis **posterior**. A plane joining the two is the **antero-posterior** or **median** plane. It is along this plane that most zygomorphic flowers can be divided into equal halves (for example, pea, Fig. 196). The position of the various parts of the flower can be described as being anterior, posterior or lateral.

POLLINATION

Before fertilisation can occur, pollen grains (which later give rise to the male gametes) must be transferred from the stamens to the ovaries, inside which are borne the ovules containing the female gametes. This process is termed **polli-**

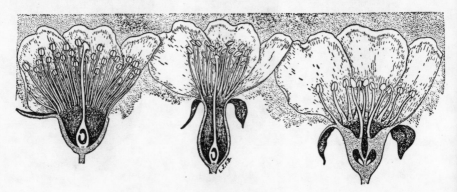

Fig. 195. Longitudinal Sections through Flowers of Plum (*left*), Cherry (*centre*) and Apple (*right*).

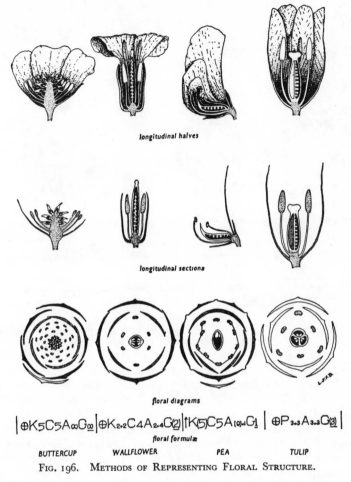

longitudinal halves

longitudinal sections

floral diagrams

$$| \oplus K_5 C_5 A_\infty G_{\underline{\infty}} | \oplus K_{2+2} C_4 A_{2+4} G_{\underline{(2)}} | \uparrow K_{(5)} C_5 A_{(9)+1} G_{\underline{1}} | \oplus P_{3+3} A_{3+3} G_{\overline{(3)}} |$$

floral formulæ

BUTTERCUP   WALLFLOWER   PEA   TULIP

FIG. 196.   METHODS OF REPRESENTING FLORAL STRUCTURE.

nation. It involves the transference of pollen grains from the stamens of a flower to the receptive stigma at the top of the ovary of a flower belonging to the same species. The pollen from a flower may be deposited on the stigma of the same flower, or on that of another flower on the same plant. In either of these two events, **self-pollination** is said to have occurred. On the other hand, the pollen from a flower may be carried to the stigma of a flower borne on a *different* plant, though the latter must of course belong to the same species. This is **cross-pollination**, which usually gives more seeds and stronger offspring than self-pollination, as well as other advantages which will be dealt with in Chap. XXVIII.

It is not surprising therefore that the majority of flowering plants are normally cross-pollinated. In fact, many flowers are so constructed that they may not only increase their chances of cross-pollination, but also prevent self-pollination.

SELF-POLLINATION

In self-pollinated flowers, the stigmas and stamens ripen at about the same time and the stamens are often placed so that the anthers are at a higher level than the

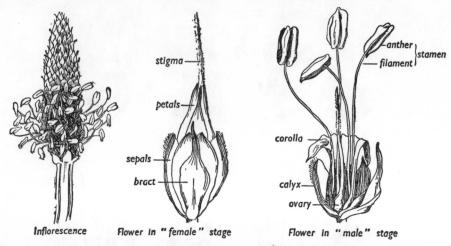

stigma

petals

sepals

bract

corolla

calyx

ovary

anther ⎫
filament ⎬ *stamen*

Inflorescence   *Flower in "female" stage*   *Flower in "male" stage*

FIG. 197. RIBWORT PLANTAIN FLOWER.

Showing the inflorescence of flowers, those above being in the ' female ' stage, those below having passed through that stage and now in the ' male ' stage. The enlarged diagram of a flower in the ' female ' stage (*centre*) shows a conspicuous stigma ready to catch pollen but the stamens of that flower have not yet developed. The diagram of the flower in the ' male stage ' shows the four stamens fully developed.

stigmas.   Thus, when the pollen is liberated it simply falls on to the stigma below. Many annuals, as, for example, the small chickweed (*Stellaria media*) are self-pollinated.   Although self-pollination is not so advantageous to a plant as cross-pollination, many flowers with modifications to favour cross-pollination have further modifications to bring about self-pollination in the event of the former not taking place.   This is the case in the harebell or Scottish bluebell (*Campanula rotundifolia*) and in many members of the Compositae (see p. 476).   Another example is seen in the sweet violet (*Viola odorata*) and dog violet (*V. canina*).   In the normal flowers of these plants, cross-pollination usually takes place.   But very often certain flowers are produced which always remain in bud.   They never open, and are very inconspicuous.   Such flowers are said to be **cleistogamic.**   Although they never open, stamens and carpels develop inside them.   Thus, when the pollen is ripe and the anther becomes ruptured, the pollen necessarily falls on the stigma of the same flower while yet remaining in bud.

CROSS-POLLINATION

In many plants there are most wonderful mechanisms which ensure cross-pollination ; but, on the other hand, there are some which, by various methods actually make self-pollination difficult or impossible.   In some examples, although the pollen from a flower reaches its own stigma, this self-pollination is not followed by fertilisation and the flowers are said to be self-sterile.   A more common instance is where the stamens of a flower and the stigmas of the same flower ripen at *different* times ; for self-fertilisation is impossible unless both male and female organs are ripe simultaneously.   The ripening of stamens and carpels at different times is known as **dichogamy.**   If the carpels ripen first, the flower is said to be **protogynous.**   A good example of this is seen in the ribwort plantain (*Plantago* *lanceolata*). The flowers of this plant are borne in a spike-like inflorescence and open in succession from below upwards. Each flower is first in a 'female' stage with the stigmas receptive; later, in the 'male' stage, the stamens are ripe and liberate their pollen. A zone of receptive stigmas will thus be exposed at successively higher levels on the spike, followed by a zone where the stamens are shedding their pollen. The ripe stamens will therefore always be below the receptive stigmas so that self-pollination will be unlikely (Fig. 197). Protogyny is common among wind-pollinated flowers such as grasses, rushes and sedges. A few insect-pollinated flowers, such as that of the horse-chestnut, are also protogynous. If the stamens ripen first, then the flower is said to be **protandrous**. Protandry is common among insect-pollinated flowers, such as the crane's bills (*Geranium* species), ivy (*Hedera helix*), and harebell (*Campanula rotundifolia*).

WIND POLLINATION

The simplest method of cross-pollination is by wind. This is very common among plants with unisexual flowers, such as poplar and hazel. In the latter, the male flowers, which, of course, produce the pollen, are borne in inflorescences familiarly known as catkins. The female flowers are far less conspicuous. When the pollen is ripe and exposed on the ruptured anthers, it is easily blown off by the wind, since the pollen is so dry. Naturally, the chances of any one pollen grain reaching the stigma of a female flower are very remote. Therefore, much more pollen is produced than is actually required. This is, therefore, a very wasteful method. Wind pollination is usual among most British trees and grasses, and for this they are well adapted. In most grasses, for example, meadow fescue (*Festuca pratensis*), the anther-heads are well exposed and pendulous on long, hair-like filaments and the stigmas are feathery, thus offering a large area for catching the pollen grains (Fig. 198).

INSECT POLLINATION

A more efficient method of cross-pollination involves the use of insects, chiefly bees, wasps, butterflies, moths, flies and beetles, as the agents of distribution of the

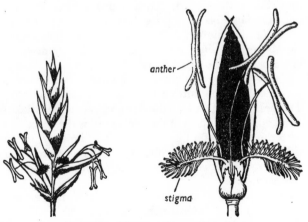

FIG. 198. MEADOW FESCUE FLOWER.
Left, a flowering shoot; right, a single flower.

T. *Edmondson.*

FIG. 199. BEE POLLINATING A FLOWER OF *Helenium.*

pollen. They collect the usually sticky or papillate pollen on their hairy backs and legs, when visiting a flower for its nectar or for the pollen itself. Then, as the insects pass on to another flower, the pollen collected from the first is rubbed on to the stigma of the next. This is what happens in its simplest form in the case of the buttercup and a host of other flowers (Fig. 199).

Some flowers, however, are specially modified, and various devices in the flowers make pollination, by means of the insect, certain.

First of all, of course, flowers are able to attract insects to them. This is done by the bright, attractive colours of some petals. This is one of the functions of the petals of most flowers. The delicate perfume so familiar in many flowers also performs a similar function. Insects are attracted by this perfume ; in fact, scent attracts insects more surely than bright colours do.

Flies and beetles, sometimes used in cross-pollination, have only short tongues, which are, therefore, useful only to widely open flowers like the buttercup. Bees and butterflies, on the other hand, have long tongues. The flowers that attract these, therefore, usually have their nectar deeply seated, so that the insect has to push its way right into the flower. Thus, the structure of the flower ensures that the insect touches both anther and stigma. For example, the colour of the yellow toadflax (*Linaria vulgaris*) attract bees. The petals are joined to form a long tube projecting backwards into a long, tapering spur which contains nectar. At the opening, the corolla is two-lipped. The two lips are pressed together so that only strong insects, such as large bees can prise them apart in order to get at the deeply seated nectar. An orange spot on the corolla acts as a ' honey-guide ' to the bee. Cross-pollination is ensured since the stigma of the flower meets the head of the bee before the stamens of the same flower do (Fig. 200). The primrose flower is pollinated by bees. The bee, alighting on a thrum-eyed flower (p. 248), pushes its long proboscis down to the bottom to obtain nectar. In doing so, the tip of the bee's head comes into contact with the ripe anthers and the sticky pollen adheres to it (Fig. 201 and 202). This part of the bee's head is the exact part which comes into contact with the stigma of a pin-eyed flower. Similarly, the pollen of a pin-eyed flower will adhere to the bee's proboscis at a level where, on the bee visiting a thrum-eyed flower, it will come into contact with the stigma. Thus cross-pollination is ensured.

Those flowers, such as the primrose, which exist in two forms are said to be **dimorphic**. Then there are **trimorphic** flowers, such as those of the purple loosestrife (*Lythrum salicaria*), in which there are three forms, each form growing on a different plant. Since, as will be seen, the pollen of the flower of a certain form

cannot reach the stigmas of a
flower of the same form, self-pollination is rendered impossible.
The purple loosestrife flower has
twelve stamens—six long and six
short. The two carpels are joined
and have a common style topped
by a stigma. In one form of flower
the style stretches beyond the
longer stamens (long-styled) ; in
the second form the style is shorter
than the shorter stamens (short-
styled) ; in the third form the style
is longer than the shorter stamens,
but shorter than the longer stamens
(intermediate-styled). Since the
visiting insect enters all flowers in
the same way and penetrates to the
same depth, it follows that: (*a*)
long-styled flowers can be pollinated by short-styled and intermediate-styled form only ; (*b*) intermediate-styled flowers by long-
styled and short-styled ones only ;
(*c*) short-styled flowers by long-
and intermediate-styled only, as
illustrated in Fig. 203. This gives
eighteen different methods of
cross-pollination.

A very pretty special adaptation to insect pollination is seen
in the highly zygomorphic flower

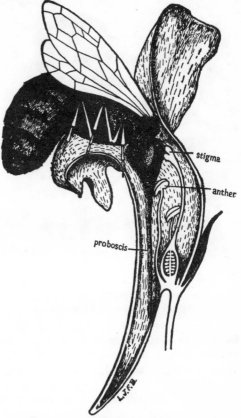

FIG. 200. BEE VISITING FLOWER OF YELLOW
TOADFLAX AND SO POLLINATING IT.

of the sage plant (*Salvia*). In this flower, the stamens ripen before the carpels.
These are two stamens. Each stamen is T-shaped, the upright line being the
filament and the cross stroke, the anther. Each anther consists of a drawn-out
connective bearing at the upper end a fertile lobe (that is, with pollen) and a
smaller, often sterile lobe at the lower end. The style of the ovary is long
and reaches out above the stamens ; so that it is not in the way of the insect.
The petals are joined to form a tube, at the base of which is the nectar.
The lips of the petals, however, protrude ; and the lower one grows outwards to form a landing stage for the insect. When the insect has landed, it
forces its head down the corolla tube to get at the nectar. In doing so,
it pushes against the lower lobes of the anthers, which then act as a lever,
and the upper lobes with their exposed pollen are forced over and touch the
insect's back, thus brushing the pollen on to it. During this time, the style is
above the insect, and is not touched. This does not matter, for the time being,
since the carpels are not ripe. Later, however, the stamens die, and then the
stigmas ripen. Elongation and downward curvature of the style brings the
stigmas into the path of the next insect visitor. The stigmas brush the back of

Dr. D. Lewis.

FIG. 201. PIN-EYED (LEFT) AND THRUM-EYED
(RIGHT) FLOWERS OF PRIMROSE.

the insect, and thus collect some of the pollen (gathered from a younger flower) from its back (Fig. 204).

Other methods of pollination are mentioned in Chap. XXX.

It will be clear from the examples described above that the structure of flowers is closely correlated with their polliatinon mechanisms.

This correlation could not be recognised until the true nature of stamens and carpels was demonstrated by Camerarius (1694). It was not, in fact, until nearly a century later that **Sprengel** clearly established the prevalence of cross-pollination and the importance of floral structure in relation to this process.

Although the importance of pollination was thus recognised by the end of the eighteenth century, the facts concerning the actual process of fertilisation were still unknown. The history of the solution of this difficult problem will be outlined when the facts, as we now know them, have been described.

STRUCTURE AND DEVELOPMENT OF A STAMEN

A very young stamen in a flower bud is a small, four-angled, upgrowth from the receptacle. This upgrowth will form the anther, the filament being delayed in development. In each of the four angles a **pollen sac** containing pollen grains will be elaborated as development proceeds. Fig. 205 illustrates these developments as seen in transverse sections of anthers of various ages, but it must be realised that longitudinal series of cells are involved, running the whole length of the anther.

The left-hand figures represent early stages. The anther consists at first of undifferentiated parenchym: but soon cells in the four corners, lying just beneath the epidermis, become active. Divisions take place giving a group of inner cells (shaded) surrounded by flattened, actively dividing cells. The inner cells ultimately give rise to the pollen grains while the outer cells form the walls of the pollen sacs. Divisions continue in both the inner and outer cell groups and the condition seen in the top right hand figure is attained, where each corner of the anther is occupied by a young pollen sac.

Each pollen sac has a wall several layers thick, the innermost being a layer of cells known as the **tapetum**, which later provides food for the developing pollen grains. The cells in the centre of the sac are **pollen mother cells** with dense cytoplasm and large nuclei. These, at first joined closely together, soon round themselves off and float freely in a nutritive liquid derived from the tapetum.

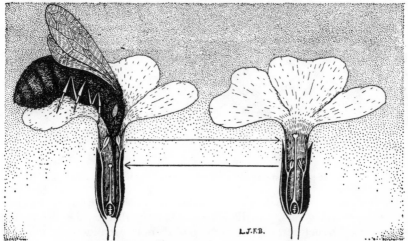

FIG. 202.   CROSS-POLLINATION OF PRIMROSE BY A BEE.
Left, thrum-eyed ; right, pin-eyed.

Each pollen mother cell now gives rise to four pollen grains by two successive divisions.   The nucleus divides into two and then into four.   Cytoplasm collects round each of the four daughter nuclei and partition walls are formed so as to give a tetrad of four pollen grains.   The type of nuclear division involved is reduction division or **meiosis** (p. 265).

As the anther reaches maturity, the tetrads of pollen grains break up and the separate pollen grains now lie freely in the cylindrical pollen sacs (Figs 205 and 206). Each young grain has at first one nucleus but this soon divides into two.   One of these, surrounded by denser cytoplasm, forms the **generative cell** (which ultimately gives the male gametes) and the other is the **tube nucleus** (Fig. 207).

The wall of the ripe pollen grain is a double one.   There is a delicate inner wall of cellulose (the **intine**) and a cuticularised outer one (the **exine**) of variable thickness.   Thin places, or pores, are present in the exine.   The grains vary in size from 3-300μ.   Their shape may be ellipsoidal, spherical, crystal-like with a varying number of facets, or many other shapes.   Further variation is given by characteristic sculpturing of the outer wall which may have spines, papillae, a network of ridges and so on.   Many plants can, indeed, be identified merely on the basis of their characteristic pollen.   It has proved possible to learn something about the changes which have occurred in our vegetation since the Ice Age by the identification of pollen grains preserved at various depths in peat deposits.   It may also be mentioned that the earliest known trace of Angiosperms is in the form of pollen grains found in coal of Jurassic age in Scotland.

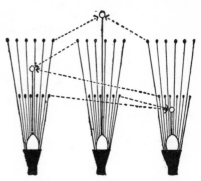

FIG. 203.   DIAGRAMMATIC REPRESENTATION OF CROSS-POLLINATION IN THE TRIMORPHIC PURPLE LOOSESTRIFE.

Left, intermediate-styled ; centre, long-styled ; right, short-styled.

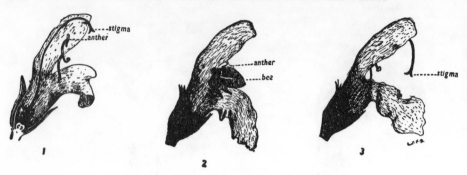

FIG. 204.   CROSS-POLLINATION OF A SAGE FLOWER BY A BEE.

As a result of the various developments described above the anther now has the structure shown in Fig. 206.   As the left-hand figure indicates, the pollen sacs are long, cylindrical structures.   There are two of them in each lobe, the lobes themselves being joined by a **connective** containing a vascular bundle.   The wall of each sac has now a highly differentiated structure.   The cells lying just below the epidermis have enlarged and their walls have been modified by the deposition of lignin.   This characteristic fibrous thickening is laid down on all the walls except the outer ones.   In some anthers several layers of the wall may show this type of thickening.

The thin outer walls of the cells of the **fibrous layer** shrink on drying and so cause a longitudinal rupture of each anther lobe along a line of specialised cells called the **stomium**.   This rupture is accompanied by a breaking down of the wall

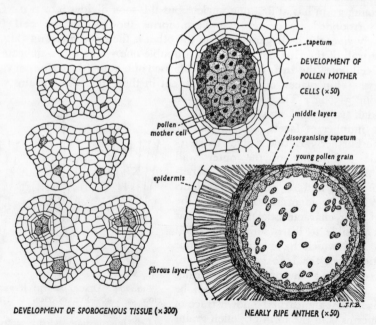

FIG. 205.   STAGES IN THE DEVELOPMENT OF A STAMEN, SEEN IN TRANSVERSE SECTIONS OF ANTHER HEADS OF DIFFERENT AGES.

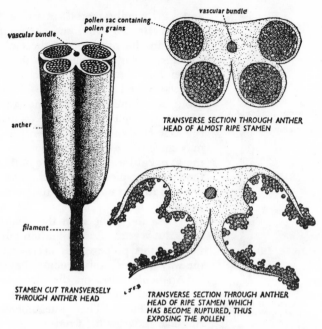

vascular bundle

pollen sac containing
pollen grains

vascular bundle

anther

TRANSVERSE SECTION THROUGH ANTHER
HEAD OF ALMOST RIPE STAMEN

filament

STAMEN CUT TRANSVERSELY
THROUGH ANTHER HEAD

TRANSVERSE SECTION THROUGH ANTHER
HEAD OF RIPE STAMEN WHICH
HAS BECOME RUPTURED, THUS
EXPOSING THE POLLEN

FIG. 206. A TYPICAL RIPE STAMEN.

between each pair of sacs. The pollen grains are thus liberated (as is indicated in the bottom right-hand drawing of Fig. 206) and pollination can now take place.

STRUCTURE AND DEVELOPMENT OF AN OVULE

The ovule when it is ready for fertilisation is more or less egg-shaped. It consists essentially of an ovoid mass of parenchymatous cells called the **nucellus** which is borne on a short stalk or **funicle** by means of which it is attached to the placenta in the ovary. The nucellus is invested by one, or more frequently two, **integuments** which are protective coats growing up from its base. These cover over the nucellus except at the extreme tip, where a narrow channel is left open as the **micropyle** leading down to the nucellus. Embedded in the nucellus there is a large ovoid sac bounded by a thin cell-wall. This is called the **embryo-sac**, since, after fertilisation, it will contain the embryonic plant.

In some plants, for example, rhubarb, the ovule is straight and the micropyle is at the end opposite to the stalk. Such ovules are said to be **orthotropous**. In other examples, for example, shepherd's purse, the nucellus and embryo-sac are curved so that the micropyle is placed near to the stalk. This is the **campylotropous** type of ovule. In the majority of ovules, however, the nucellus is straight but the entire ovule is inverted by a sharp curvature at the top of the stalk so that the micropyle lies near to the base of the stalk and also near to the carpellary wall. This is the common **anatropous** type of ovule (Fig. 208).

The embryo-sac at the time of fertilisation contains vacuolated cytoplasm in which a number of prominent nuclei are embedded. At the end furthest from the micropyle there are three large nuclei, each surrounded by granular cytoplasm

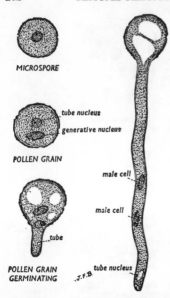

MICROSPORE

tube nucleus

generative nucleus

POLLEN GRAIN

male cell

male cell

...tube

POLLEN GRAIN   tube nucleus
GERMINATING   .J.F.B.

POLLEN TUBE GROWING
(MALE GAMETOPHYTE)

FIG. 207.   STAGES IN THE DEVELOP-
MENT OF A POLLEN GRAIN.

and a gel membrane ; these are called the **antipodal cells** (Fig. 209) and often appear to play a part in the nutrition of the embryo-sac. At the micropylar end there are three other large nuclei and these too are each surrounded by cytoplasm and a membrane. This micropylar group constitutes the important **egg apparatus**. The central cell is the female gamete called the egg cell or **ovum**. The two other cells, which may correspond to non-functional ova, are called the **synergids**.   In the centre of the embryo-sac there is a conspicuous nucleus which is called the **central fusion nucleus** or the **primary endosperm nucleus**.

When the ovule is in the condition just described it is ready for fertilisation but before considering this process it is necessary to examine the way in which this structure has developed.

An ovule arises on the placenta as a small upgrowth of parenchymatous tissue which is the young nucellus. Then integuments appear as ring-like upgrowths from its base and these gradually invest the growing nucellus until only the micropyle is left at its tip.   Meanwhile, if the ovule is of the anatropous type, the stalk has elongated and curved sharply at its top so as to bring the micropyle near the insertion of the stalk on the placenta.

It is during these developments that the embryo-sac is formed.   A single cell of the nucellus, situated just below its tip, becomes conspicuous by reason of its larger size and dense protoplasmic contents.   This is the **embryo-sac mother cell** which then divides first into two cells and then into a row of four cells.   As in the formation of four pollen grains from a pollen mother cell, the nucleus of the embryo-sac mother cell divides by the process of meiosis to give the four potential embryo-sacs.   Usually, however, it is only the innermost of these which develops further as the functional embryo-sac.   The other three collapse and are crushed as the surviving embryo-sac enlarges.

When first formed, the embryo-sac has a single nucleus (Fig. 209).   This nucleus divides by mitosis and the daughter nuclei move to the two poles of the sac. Each of the two nuclei divides again twice so the four nuclei occupy the micropylar end and four the opposite (**chalazal**) end of the sac.   One nucleus from each group of four now moves to a central position in the sac and these two fuse to give the central fusion nucleus.   The three at the micropyle end form the egg-apparatus already described and the other three form the antipodal cells.   The embryo-sac thus reaches the condition which is found in practically all Angiosperms at the time of fertilisation.

FERTILISATION

It is now necessary to trace the sequence of events following upon the deposition of pollen grains on the receptive stigma.

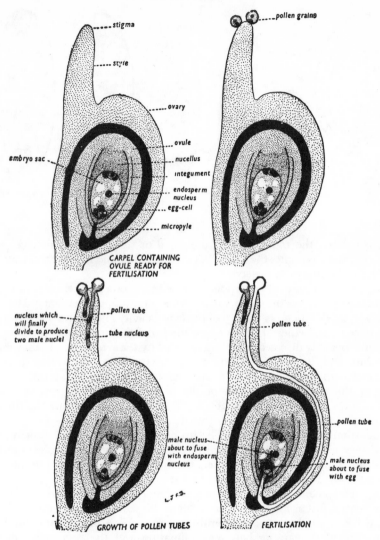

FIG. 208. STAGES IN THE PROCESS OF FERTILISATION.

The pollen grains germinate on the stigma where there is usually a liquid secretion, often containing sugar, but sometimes other substances are also present, as, for example, malic acid in the case of members of the Ericaceae. Germination results in the growing out of the thin inner wall of the pollen grain through one of the pores in the thick exine. This results in the production of a cylindrical **pollen tube** (Fig. 207), into which the tube nucleus and generative nucleus pass. The pollen tube, tending to grow towards chemical substances produced by the stigma, towards moisture and away from oxygen, then enters the stigma and begins to grow down the style. In some examples, for example, *Viola* species, there is an open canal in the centre of the style leading right down to the ovary; in others, for example, *Rhododendron* species, there is a canal filled with mucilage. The

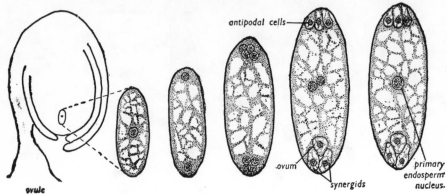

FIG. 209.   DEVELOPMENT OF THE EMBRYO SAC.

tube grows down the moist sides of the canal or through the mucilage, as the case may be, and so reaches the ovary.   It obtains water and food for this continued growth from the tissues of the style.

Having reached the ovary, the pollen tube either grows in the tissues of the ovary wall until the tip reaches the vicinity of an ovule (Fig. 208) when it enters the cavity of the ovary, or else it enters the top of the ovary cavity and grows down the moist inner surface to an ovule.   In either case the tip of the tube enters the micropyle of the ovule, possibly directed by a fluid secreted by the synergids.   It thus reaches the nucellus, and, by penetrating the few overlying layers of this tissue, it comes into contact with the embryo-sac in the vicinity of the egg-apparatus.

During the downward growth of the pollen tube the generative cell has divided to form the two **male gametes** (Fig. 207).   These occupy the tip of the tube, the tube nucleus having disintegrated.

Fertilisation now takes place.   The tip of the pollen tube becomes ruptured and the two male gametes, now spirally coiled, thread-like bodies, enter the embryo-sac.   One of the gametes enters the egg and fuses with its nucleus.   The egg, so fertilised, is called the **zygote** and it is from this that the embryo plant develops.   The second male gamete moves deeper into the embryo-sac and fuses with the central fusion nucleus.   The result of this second fusion is the production of a nutritive tissue, known as the **endosperm** ; this is used by the developing embryo.   The time elapsing between pollination and the occurrence of these fusions varies considerably.   In rye it is seven hours ; in maize, twenty-four hours ; in some trees it may be a year or more.

The elucidation of the sequence of the events just outlined occupied botanists for many years.   Although it was known from the time of Camerarius onwards that a fertilisation process did in fact take place, the details proved difficult to observe.   In 1823, **Amici,** an Italian naturalist, observed the production of pollen tubes by pollen adhering to a stigma and seven years later he traced the course of the pollen tube down into the ovary and into the micropyle of an ovule.   Later still (1846), the same observer showed that the ovule had an egg cell before fertilisation and that this, excited by the pollen tube, gave rise to the embryo, **Hofmeister** confirmed these observations in 1849 but it was not until 1884 that **Strasburger** observed the fusion of one of the male gametes with the nucleus of the ovum. It was still later (1898) before **Nawaschin** observed the fusion of the second male gamete with the central fusion nucleus in the centre of the embryo-sac.

We have now seen that fertilisation consists essentially of the fusion of a male gamete with a female gamete, the ovum, to form a zygote and ultimately an embryo. The second fusion is a subsidiary one and results only in the formation of nutritive tissue.

In the account of fertilisation given above the events have been described in relation to a single pollen grain and a single ovule. Normally, however, many pollen grains will be deposited on a stigma and they will all produce pollen tubes. If the ovary contains only one ovule, naturally only one of these tubes penetrates the micropyle and liberates its male gametes into the embryo-sac. If, however, as in the majority of ovaries, numerous ovules are present, then the production of numerous pollen tubes is a necessity if all the ovules are to be fertilised. In a flower like that of *Rhododendron*, if a transverse section is cut through a style some time after pollination has occurred, the section will show a large number of pollen tube sections embedded in the mucilage of the stylar canal.

MEIOSIS

An essential feature of fertilisation is the fusion of a male gamete with the nucleus of the female gamete. This results in the zygote or fertilised egg the nucleus of which contains the nuclear material of both the male and female gametes. If the male gamete has n chromosomes (a number varying with the species) and the female has n chromosomes the zygote will obviously possess $2n$ chromosomes. It was, in fact, noted in Chap. X that the chromosome complements in the vegetative cells of flowering plants consist of homologous pairs, one of each pair being maternal and the other paternal in origin. It is clear that these pairs came together in the act of fertilisation which gave rise to the plant.

As the zygote develops into the embryo, and then into the new plant, all the cell divisions which take place are mitotic and thus the $2n$ number of chromosomes is maintained. If, however, this type of division occurred throughout the life-history, then the gametes would each have $2n$ chromosomes and, when fertilisation occurred, the zygote would have $4n$ chromosomes. There is, however, a compensating process in the form of another type of nuclear division known as **meiosis** or **reduction division**, by which the number of chromosomes is halved. Meiosis always occurs in the formation of the gametes or in some cell division prior to such gamete formation. The result is that the gametes themselves always have the haploid or n number of chromosomes and the zygote formed as a result of fertilisation will have the diploid or $2n$ number.

In flowering plants, meiosis occurs in the formation of the four pollen grains from each pollen mother cell and also in the formation of the four potential embryo-sacs from an embryo-sac mother cell. As the subsequent nuclear divisions which take place both in the pollen grain and in the functional embryo-sac are all mitotic, it follows that the male gametes and the female gamete will have the reduced or haploid number of chromosomes. All plants which reproduce sexually show the occurrence of meiosis at some point in their life-history between successive acts of fertilisation but this point varies in the different big groups of plants (see table on p. 395).

Meiosis consists of two very much modified mitotic divisions, normally giving rise to four cells, but it is essential that it should be considered as a single process with two stages. Cells which are about to undergo this type of division are usually filled with dense cytoplasm and have a large nucleus.

The first indication that meiosis is in progress is the appearance of the chromosome threads just as in the prophase of mitosis, except that they are single and show the chromomeres more obviously along their length. There are, however, elaborations of the simple prophase of mitosis and the significance of these will be dealt with in Chap. XXVIII.

The stage where the strands are obvious as separate entities is called **leptotene** and this is followed by **zygotene** where the homologous chromosome strands come to lie together, so precisely paired that corresponding segments and even chromomeres are in juxtaposition. These paired chromosome structures are called **bivalents.** In each bivalent both of the chromosomes now split longitudinally (except in the region of the centromere) giving a four-stranded structure which is characteristic of the **pachytene** stage. Pachytene is a very short stage and is

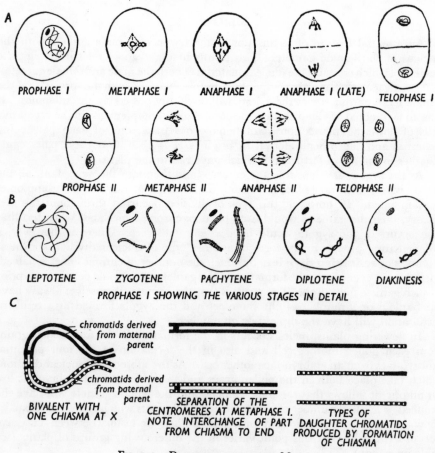

FIG. 210. DIAGRAM ILLUSTRATING MEIOSIS.
(A. M. M. Berris.)

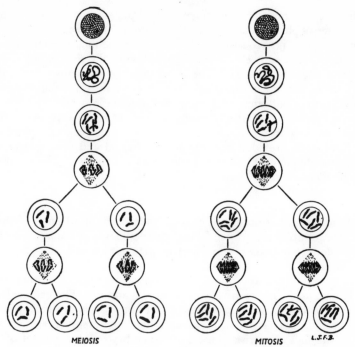

FIG. 211. DIAGRAMMATIC REPRESENTATION OF MEIOSIS AND MITOSIS.

seldom seen in preparations showing meiosis; but it is nevertheless important. In each bivalent the chromatids now contract and twist around one another but the members of each pair repulse the opposite members except at points which are called **chiasmata** (sing., chiasma) and it will be seen in Chap. XXVIII that these chiasmata explain cytologically the genetic problem of 'crossing over'. This stage is **diplotene** and it is followed by **diakinesis**, which is characterised simply by a greater shortening and thickening of the bivalents. The latter usually arrange themselves around the periphery of the nucleus and at this stage also the nuclear membrane and the nucleolus disappear. The above stages which constitute the prophase are shown diagrammatically in Figs. 210 and 211.

Metaphase I follows and is similar to the metaphase of a mitosis in that a spindle has now been formed and that the bivalents arrange themselves on the equatorial plate in exactly the same way as do the chromosomes in mitosis. The bivalents are placed with their two unsplit centromeres above and below the equatorial plate. Anaphase I now follows and the two centromeres of each bivalent move to the opposite poles of the cell, each pulling its pair of chromatids with it. Telophase I can be a very transient stage, and can frequently be regarded as non-existent since the daughter nuclei pass straight into the prophase of the second division. Metaphase II follows, each centromere (of which there are only half as many as in the original cell) moves on to the equatorial plate, taking with it the two chromatids which are attached to it. The centromeres then divide and at anaphase II the two chromatids separate. Four daughter cells are thus produced, each with half the number of chromosomes which were present in the parent cell.

One feature requires fuller treatment. At diplotene there is actual breakage and interchange of parts of chromatids, the position of the breakages usually being seen as chiasmata. Fig. 210 shows a chromosome which has one chiasma and the result of this chiasma as it effects the constitution of the segregating chromatids at anaphase I and anaphase II. The interchange of segments extends from the chiasma to the ends of the chromatids. This means that two of the chromatids are actually made up of both maternal and paternal material while the other two are either wholly maternal or wholly paternal in their make-up. More than one chiasma can be present in a bivalent and the interchange can take place between any two chromatids provided that one is of maternal and the other of paternal origin. Extremely complicated segregations can thus occur. The significance of these will be mentioned in Chap. XXVIII.

THE SEED

After fertilisation has occurred, changes take place in all parts of the ovule. The structure resulting from these further developments in the ovule is the characteristic reproductive body of the flowering plants, the **seed**.

Changes in the embryo-sac itself may be considered first. The fertilised egg forms a wall around itself and proceeds to divide so as to form a short chain of cells which is known as the **pro-embryo** (Fig. 212, 1 and 2). The cell nearest to the micropyle grows to a large size, while the cell at the tip of pro-embryo begins to divide in various planes, these divisions following a regular sequence. This terminal cell in fact gives rise to most of the embryo plant, the rest of the pro-embryo forming the **suspensor**. The large basal cell of the suspensor serves as an organ of attachment to the wall of the embryo-sac ; the filamentous region

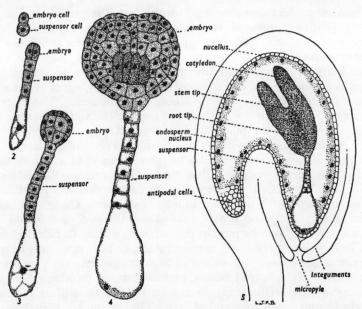

FIG. 212. STAGES IN THE DEVELOPMENT OF AN EMBRYONIC PLANT FROM THE FERTILISED EGG.

1-4, Young developing embryo only ; 5, embryo within the ovule (now seed).

continues to elongate and pushes the embryo into the nutritive tissue, or **endo-sperm**, which is developing in the embryo-sac at the same time. Continued cell division in the embryo soon results in the appearance of two lobes which are the two cotyledons (or of one lobe if the plant is a Monocotyledon). The stem apex is formed between the two cotyledons (or laterally at the base of the single cotyledon in a Monocotyledon). The lower part of the embryo in both Dicotyledons and Monocotyledons forms the bulk of the first root or radicle. The tip of the latter is, however, formed by divisions in the top cell of the suspensor. The tip of the root is thus attached to the suspensor and points towards the micropyle (Fig. 212, 5).

Simultaneously with the development of the embryo, or even preceding this, endosperm is formed in the embryo-sac. This results from the active division of the central nucleus (which, it will be remembered, is the product of the fusion of the second male gamete with the central fusion nucleus). Repeated divisions of this nucleus give rise to numerous daughter nuclei which are not, at first, separated by cell walls (a type of division known as **free nuclear division**). These nuclei take up a peripheral position in the embryo-sac (Fig. 212, 5). Continued nuclear division takes place and then wall formation commences. This results in the forma-tion of a cellular tissue which gradually encroaches on the cavity of the embryo-sac. This tissue, which is rich in food materials, is the endosperm.

As the embryo and endosperm develop inside the embryo-sac, the latter increases rapidly in size and, as a result of this enlargement, the tissues of the nucellus are crushed and practically obliterated. In a few examples, however, some of the nucellus persists as a nutritive tissue known as **perisperm**. The anti-podal cells may soon disappear or, in other examples, they may persist and either enlarge or divide to form a tissue which helps in the nutrition of the enlarging embryo-sac.

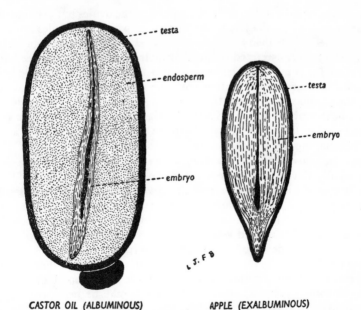

CASTOR OIL (ALBUMINOUS)   APPLE (EXALBUMINOUS)

FIG. 213. LONGITUDINAL SECTION THROUGH CASTOR OIL AND APPLE SEEDS.

ovary containing ovules

FIG. 214. OVARY AND FRUIT OF THE PEA.

Changes also take place in the integuments. These usually become fibrous or woody and constitute the protective seed coat or testa.

The total result of these post-fertilisation changes in the ovule is the structure known as a seed. The testa is, as just described, derived from the integuments of the ovule. Inside the testa is the embryo. In addition there may be endosperm present, or this tissue may be absent from the ripe seed. If the embryo grows rapidly during the ripening of the seed it will use up all the endosperm to obtain the requisite food material for its growth. Such a seed is said to be **non-endospermic** or **exalbuminous** (for example, the seeds of apple (Fig. 213), bean, pea, etc.). On the other hand, the embryo may cease to grow before all the endosperm has been used and then the ripe seed will be **endospermic** or **albuminous** (as in castor oil (*Ricinus communis*) (Fig. 213), wheat (*Triticum vulgare*), and coco-nut (*Cocos nucifera*).

THE FRUIT

The effects of fertilisation are not confined to the ovules but extend to the other floral parts and, in particular, to the carpels in which the ovules are borne.

The carpellary or ovary wall is stimulated to further development, possibly by growth substances secreted by the developing ovules. It may become dry and membranous, or it may thicken considerably and become either woody or fleshy. This modified ovary wall, together with the contained seed or seeds, constitutes a fruit, the modified wall being the fruit coat or **pericarp.** The various forms of fruit will be described in Chap. XXII, but Fig. 214 will serve to emphasise the fact that the ovary wall becomes the pericarp (in this case, the pea pod), and that the ovules become the seeds.

It may be noted that in a few cases, for example, banana (*Musa sapientum*), the fruit normally develops without fertilisation having taken place. This phenomenon is known as **parthenocarpy** and naturally gives rise to a seedless fruit. The development of seedless tomato fruits may be induced by spraying the flowers with a growth substance such as β-naphthoxyacetic acid. In this case the growth substance artificially supplied seems to act in a similar way to the growth substance produced by ovules developing after normal fertilisation.

After fertilisation the sepals, petals and stamens usually wither, although their dead remains may often still be seen, for example at the top of an apple or gooseberry. In some examples, however, the sepals may be persistent (for example, pea, tomato). In the mulberry (*Morus* species) the perianth becomes pulpy and embeds the true fruits.

The post-fertilisation changes described above are summarised in the following table :

FLOWER.	FRUIT.
Calyx.	Sometimes remains ; usually deciduous.
Corolla.	Usually deciduous.
Stamens.	Usually deciduous.
Stigma.	Usually withers away, though sometimes withered portions remain.
Style.	Usually withers away, though sometimes withered portions remain.
Ovary.	Fruit.
Ovary wall.	Pericarp.
Ovule :	Seed.
Integument.	Testa.
Micropyle.	Micropyle.
Nucellus.	Perisperm (rare).
Endosperm nucleus.	Endosperm (if still present).
Egg cell.	Embryonic plant.

PRACTICAL WORK

1. Examine the inflorescence of the buttercup. Note how comparatively simple it is. Compare this inflorescence with that of the cowslip. Note that in the latter, each flower is borne on a short stalk (pedicel), each of which is joined to a main flower-stalk (peduncle). Make drawings of these inflorescences.

2. Collect a large number of plants with varying types of inflorescences. Make detailed, labelled drawings of these, and group them according to the type of inflorescence they represent.

The chief types of inflorescence to be examined are : spike, raceme, panicle, corymb, umbel, compound umbel, capitulum, monochasial cyme and dichasial cyme.

When examining these inflorescences, note especially the peduncle, pedicels, and bracts, and their relative positions ; do not, at this stage, spend too much time on the details of floral structure.

From the specimens examined and drawings made, describe the various types of inflorescences in flowering plants.

3. Make a detailed examination of a relatively simple flower, for example, the buttercup. The structure of the flower is best appreciated by making careful drawings as directed. The drawings should be large and in sufficient detail to show the shape and structure of the various organs and their relation to each other.

First make a drawing of the complete flower, noting especially the number, arrangement, shape and colour of each set of organs, namely, sepals, petals, stamens and carpels. Then dissect the flower and make enlarged drawings of a specimen of each organ. Note that the sepal is pale green, boat-shaped, and covered, on the outside with hairs. The petals are yellow, larger than the sepals, heart-shaped, and each bears a nectary at its base. Each stamen is divided into two main portions, the filament and the anther-head ; whereas the carpels are free and each is composed of an ovary, a style and a stigma.

Half a flower should then be drawn. This shows the relative positions of the various organs of each whorl as they are set upon the receptacle. This should be followed by a true longitudinal section, which indicates only those organs through which the scalpel passes when cutting the flower longitudinally.

A floral diagram should then be drawn, and the whole study completed with a floral formula.

4. Place the anthers of a lily flower (or other large anthers) in alcohol for about a week. Cut transverse sections, mounting in glycerine, and note the structure as described in the text. Buds of marsh marigold, also hardened in alcohol, may be sectioned at the level of the anthers and a large number of anther sections thus obtained.

5. Take a just-opened flower of marsh marigold and remove the sepals and petals. Harden the group of carpels in alcohol and then cut transverse sections of these. Cut a large number of sections and pick out a few thin ones which have traversed an ovule and mount in glycerine. These will give longitudinal sections of the ovules which can be compared with the figures in the text. Transverse sections of syncarpous ovaries should also be cut, for example, those of tulip and lily.

6. Examine pollen grains from various flowers by mounting in a weak solution of methyl green in dilute acetic acid. The germination of pollen grains, for example, those of bluebell, can be observed by putting fresh pollen in a 5-10 per cent solution of cane sugar for 12-24 hours. The germination of pollen may also be observed on the styles of chickweed, taken from a flower which is just beginning to fade. The styles should be mounted in a dilute aqueous solution of methylene blue.

7. Examine stages in the development of the embryo of shepherd's purse. A flowering shoot, freshly gathered, with a few flowers at the top of the raceme and a succession of fruits below will provide the necessary material. Remove a number of ovules from a fruit which has attained a size about one third of that of a mature fruit. Place these in a drop of 1 per cent caustic potash until they become transparent ; cover and examine the general form of the campylotropous ovule. Then press gently on the coverslip. This should cause the embryos to escape from some of the ovules. Now irrigate with dilute acetic acid, a treatment which renders the embryo walls more apparent. Embryos squeezed out from the ovules of both older and younger fruits should be examined in the same way.

CHAPTER XXII

FRUITS AND SEEDS

The fruit is the complete structure formed by the ovary after fertilisation has taken place. It contains the seed or seeds, each one of which is the result of development of an ovule after fertilisation.

The fruits of flowering plants assume various forms, many of which are familiar, since they form articles of diet. The ripe fruit may open to set free the contained seeds and the latter are then dispersed by various agencies. On the other hand, the fruit may remain unopened and in such cases it is the fruit itself which is distributed. Most of the manifold modifications of fruits and seeds can in fact be related to the process of dispersing the seeds to some distance from the parent plant.

All fruits can be divided into two general groups, namely, **dry**, and **fleshy** or **succulent**.

Dry fruits can again be subdivided into two other groups, according to whether they are capable of opening by some mechanism or another, in order to allow the seeds to escape, or not. Those fruits which are capable of opening mechanically are said to **dehisce**, and are therefore described as being **dehiscent**. Those fruits which are incapable of opening, to allow the escape of seeds, are **indehiscent**.

DRY, INDEHISCENT FRUITS

The indehiscent fruits are the simplest types, and one of the simplest of all is that of the buttercup (*Ranunculus* species). Here, each carpel forms a separate fruit, since the carpels are all separate. After fertilisation, the carpellary wall undergoes no special change, so that the fruit is simply a single seed surrounded by the carpellary wall, which, after fertilisation, undergoes no change other than that of hardening, in order to protect the young seed inside. This type of fruit is called an **achene**, and it is incapable of opening of its own accord to allow the seed inside to escape (Fig. 215).

The flowers of the buttercup and the strawberry plant (*Fragaria vesca*) are somewhat similar in structure. Both contain an indefinite number of free carpels, which are very similar in shape and size. On casual observation, however, the fruits seem very different. The buttercup flower, after fertilisation, leaves a collection of fruits in the form of dry achenes, on a dry receptacle. What is usually referred to as the strawberry ' fruit ', on the other hand, is a large, red, swollen, juicy structure bearing what appear to be seeds on its surface. Now, actually, the fruit of the strawberry is very similar to that of the buttercup ; and this is so, because what is usually referred to as the ' fruit ' of the strawberry is not the fruit at all. *It is neither a fruit nor is it a berry.*

The structure of the strawberry can easily be deduced from that of the buttercup. Imagine the receptacle of the fertilised buttercup swelling to a size about a hundred times that of the normal one. Then imagine it to become red and juicy. A large, red, juicy receptacle with the achenes on its surface would result.

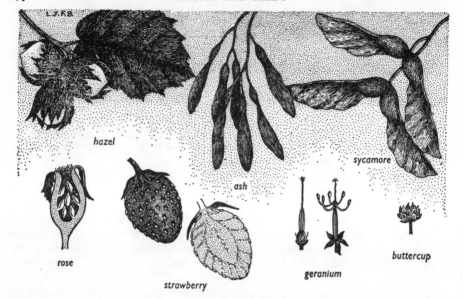

hazel

sycamore

ash

rose

strawberry

geranium

buttercup

FIG. 215. EXAMPLES OF DRY, INDEHISCENT FRUITS.

This is the case in the strawberry. The so-called 'fruit' is really a very much swollen receptacle, and the so-called 'seeds' are really fruits in the form of very small achenes (Fig. 215).

The strawberry 'fruit' is composed of about 90 per cent, by weight, of water. Its dietetic value lies chiefly in its iron content, which helps in the production of healthy blood, and its vitamin C content.

The fruit of the rose, such as dog rose (*Rosa canina*), field rose (*R. arvensis*), burnet rose (*R. spinosissima*) and cultivated roses, is also an achene. Each achene is flask-shaped and covered with tiny hook-like hairs and a collection of them is present in the large red, hollow receptacle commonly known as the **hip** (Fig. 215).

Closely related to the achene is the fruit called the **samara**. In this case, also, there is only one seed enclosed, since each fruit is formed from one carpel containing one ovule. The difference between the achene and the samara is that whereas the pericarp of the former is formed from the carpellary wall unchanged except that it becomes hardened, in the latter the carpellary wall becomes not only hardened, but also extended at the top into a long, flattened, wing-like structure (Fig. 215). A good example of a samara is seen in the fruit of the ash (*Fraxinus excelsior*), which hangs from the tree in large bunches which country-folk call 'keys'. The fruit of the maples is also a samara, but in this case often two and sometimes three samaras fuse at their swollen bases. This is well seen in the great maple or sycamore (*Acer pseudoplatanus*, Fig. 216), Norway maple (*A. platanoides*) and common maple (*A. campestre*).

Another type of fruit closely related to the achene is the **nut**. In this type, the pericarp becomes very hard and woody. The term 'nut' is often popularly applied to other types of fruit, such as the coco-nut and walnut, which are not true nuts at all, as will be seen later. A good example of a true nut is the hazel (*Corylus avellana*). The shell is the hardened pericarp, and the kernel is the seed.

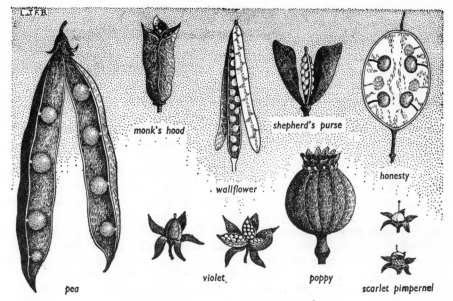

monk's hood

shepherd's purse

wallflower

honesty

violet

poppy

scarlet pimpernel

pea

FIG. 216. EXAMPLES OF DRY, DEHISCENT FRUITS.

The ovary contains two ovules, but usually only one of these develops into a seed, thus giving one kernel ; but sometimes both ovules are fertilised, and then a nut with two kernels is produced. The nut is situated in a large, green, leafy cup called a **cupule**, formed from bracts which persist after fertilisation (Fig. 215).

The hazel nut is used as a food, chiefly for the oils it contains. Many other true nuts, and other fruit erroneously classified as nuts, are used for the same purpose. Also, in many cases, the oils are extracted for commercial purposes.

The **schizocarp** is an interesting type of indehiscent fruit in that it is formed from two or more carpels, and, when fully ripe, it splits into portions, each portion containing one seed. For example, the schizocarp of the mallow (*Malva* species) and hollyhock (*Althæa rosea*) is a round, bun-shaped structure ; when ripe, it splits into its one-seeded segments in a manner similar to the way in which an ordinary round cake is cut.

The cultivated geranium and the many wild related species of *Geranium*, together with another member of the same family, the cultivated *Pelargonium*, show an even more curious type of schizocarp. Here, the ovary is composed of five, joined carpels. The style is long and tapering, and even when the fruit is ripe, the style persists as a tall spike (Fig. 215). Finally, the fruit splits at the base into its five one-seeded portions. The central portion of the persistent style remains rigid, but the outer tissues of the style also split from top to bottom into five portions, each portion being joined at the base to one of the one-seeded portions of the fruit. When splitting is complete, each ribbon-like portion of the style begins to tear away from the central rigid portion, from the bottom upwards, carrying one of the seed-containing portions with it (Fig. 215).

Although the schizocarp is a splitting fruit, it is not classified with dehiscent fruits because it does *not dehisce and expose the seeds*. Each portion formed after splitting is a seed still surrounded by a part of the pericarp wall.

DRY, DEHISCENT FRUITS

One of the simplest types of dehiscent or splitting fruits is that called the **follicle**. Examples of the follicle are seen in the marsh marigold (*Caltha palustris*), monk's hood (*Aconitum anglicum*) (Fig. 216) and larkspur (*Delphinium ajacis*). Each follicle is formed from a single, free carpel, containing several ovules. After fertilisation and the complete ripening of the seeds, the fruit or follicle splits along the inner margin, that is, the margin which bears the seeds, sometimes called the **ventral suture** as distinguished from the outer margin or **dorsal suture**. The split goes the whole length of the follicle, and thus the seeds are exposed and ready for distribution.

The **legume**, another dehiscent fruit, differs from the follicle in that it splits along both the inside margin (ventral suture) and the outside margin (dorsal suture) and the two halves of the pericarp move apart, thus completely exposing the ripe seeds. The legume, or **pod**, as it is sometimes called, is typical of peas, beans, vetches, etc.—members of the family Leguminosae (Fig. 216).

Members of the wallflower family (Cruciferae) illustrate a curious fruit structure, owing to the presence in the ovary of an exceptional tissue. The ovary is formed by the fusion, along their margins, of two carpels, as in the gooseberry ; but, as distinguished from the latter, between the two fused carpels a wall of tissue is formed, called the **false septum**. This septum persists in the ripened fruit, in which there are, therefore, two cavities, instead of one, separated from each other by the false septum. In the wallflower (*Cheiranthus cheiri*), when the fruit is ripe, the two parts of the pericarp, each corresponding to one carpellary wall, separate from the false septum, from the bottom upwards, leaving the seeds fixed to the exposed false septum which is bounded by a wiry frame formed by the persistent placentae and known as the **replum** (Fig. 216). In a closely related plant, the shepherd's purse (*Capsella bursa-pastoris*), the fruit is similar, except that whereas in the wallflower the fruit is long and flat, in the shepherd's purse it is heart-shaped; also, when the fruit is ripe, the two halves of the pericarp become completely detached, thus exposing the numerous seeds (Fig. 216).

The type of dehiscent fruit which shows the greatest number of variations is that called the **capsule**. It is impossible to consider all these variations ; but two good examples are seen in the poppy (*Papaver* species, Fig. 216) and the scarlet pimpernel (*Anagallis arvensis*, Fig. 216). In the latter, the capsule is spherical, and, when the seeds are ripe, opens by a transverse split. In the poppy, however, the capsule is of a very unique form (Fig. 216). It is formed from an indefinite number of joined carpels. The placentation is parietal, but the longitudinal parts of the ovary wall, where the carpellary margins fuse (**placentae**), project some distance into the common ovary cavity. The complete capsule is spherical or ovoid in shape, with ridges down the outside marking the lines of fusion of the carpellary margins. Stigmatic ridges also radiate at the top from the centre outwards. The top part of the ovary projects slightly beyond the edge, somewhat like the eaves of a roof. On the side of the capsule, just beneath this projection, is a series of perforations. There is one hole for each fused carpel. The reason for the perforations will be examined later (p. 283). The poppy capsule resembles a pomegranate in shape, and this is reflected in the specific name of the common red poppy (*Papaver rhoeas*) which is derived from *rhoia*, Greek for pomegranate.

The fruit of pansies and violets is a three-valved capsule (Fig. 216). This may

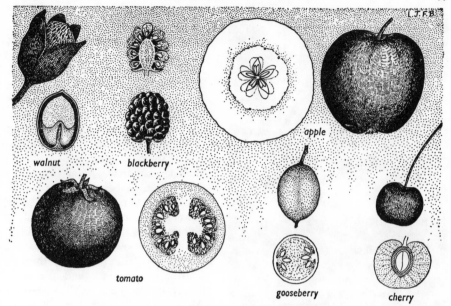

FIG. 217. EXAMPLES OF SUCCULENT FRUITS.

be seen in the dog violet (*Viola canina*), sweet violet (*V. odorata*), heartsease or wild pansy (*V. tricolor*) and others. When the capsule opens, each valve looks like a tiny boat, and the three are joined at their bows. Each valve contains a number of hard, slippery seeds. As the open valves become dry they shrink, thus pressing the seeds together until they are shot out and thus dispersed (p. 283).

SUCCULENT FRUITS

One of the simplest types of fleshy or succulent fruits is the **berry**. In this fruit, the whole pericarp becomes thick and fleshy. This fleshy pericarp is composed sometimes of more than one layer ; but all of them are formed from masses of thin-walled tissue, the cells of which usually contain plenty of food materials, especially sugar or starch.

The tomato (*Solanum lycopersicum*) is an example of a berry (Fig. 217). This fruit is formed by the simple fusion of two carpels. When ripe, the pericarp becomes very thick and juicy, and a viscous fluid is also secreted into the ovary cavities. The pericarp itself is usually divided into a thick, pale red mass of tissue, and an outer, thinner, but tougher, deep red tissue, the skin. This berry is the source of much food material and vitamins.

The fruit of the grape (*Vitis vinifera*) is also a berry. It is formed by two joined carpels, and the pericarp becomes fleshy. The so-called currants which are supplied in the dried condition, and used in the making of cakes and puddings, are not in any way to be confused with the black, red and white currants which are cultivated in gardens and fields in Great Britain (*Ribes* species). Each type belongs to a totally different family of plants. The dried currant is related to other familiar dried fruits, such as the raisin, sultana and muscatel. All these are berries, and the plants are closely related to the grape vine (*Vitis*).

FIG. 218.   BANANA IN FRUIT.
(From the collection, Royal Botanic Gardens, Kew; by permission of the late Sir Arthur Hill.)

The true currants as cultivated in Great Britain are the fruits of small deciduous shrubs. They, also, are berries. There are three types, black (*Ribes nigrum*), red (*R. rubrum*) and white, a variety of the red. The gooseberry (*Ribes uva-crispa*) fruit is also a berry (Fig. 217). It is globose or ovoid, smooth or hairy and is formed by the fusion of two carpels at their margins.

The banana (*Musa sapientum*) fruit is also a berry. The plant is indigenous to tropical regions. The flowers are borne in long, pendulous inflorescences. They are unisexual, the female flowers being borne at the base of the peduncle, that is, nearest the 'stem', and the male flowers nearer the tip of the peduncle. Although fertilisation does not occur, the female flowers develop the well-known berries, which are formed from three joined carpels. The bunches of bananas, actually the inflorescence after fertilisation, as seen in the fruiterer's shop, are usually hung upside down, for the berries in Nature grow more or less upright (Fig. 218).

Oranges (*Citrus aurantium*), grape-fruit (*C. paradisi*), lemon (*C. medica* var. *limonum*) and sweet lime (*C. medica* var. *limetta*) are sub-tropical and tropical plants, and their fruits are all berries. If any one of these fruits be cut in transverse section, they will be seen to be composed of six or more fused carpels, and, judging from the position of the seeds, the placentation of the ovary is axile. The pericarp is divided into a tough outer skin which contains many oil glands (the epicarp), an underlying zone of white tissue (the mesocarp) and a membranous endocarp, lining the cavities of the ovary. These latter are filled with large juice-containing outgrowths from the endocarp. All three of these fruits are of great importance, not only for their food value, but also for their high vitamin content (see Chap. VII).

The date (*Phoenix dactylifera*) is a one-seeded berry.

The cucumber (*Cucumis sativus*) and marrow (*Cucurbita pepo*) bear fruit very similar to the berry; but in this case the fruit is formed from an ovary of three joined carpels, *which is inferior*. It is therefore called a **pepo**.

A type of fruit known as a **drupe** is of great commercial importance. Examples of the drupe are plum, cherry (Fig. 217), peach, and three others, commercially called nuts, though botanically they are not: the coco-nut, walnut and almond.

In the drupe, the fruit is formed from either one carpel or from three joined carpels. Whereas in the berry the whole pericarp becomes swollen and fleshy, in the case of the drupe the pericarp swells and becomes divided into several distinct layers. Not all these layers are fleshy, and in some cases, for example, the coco-nut, none of them is. The layers into which the pericarp of the drupe becomes divided are usually three in number : an outer one called the **epicarp**, an inner one called the **endocarp** and a layer between these two called the **mesocarp**.

In the plum (*Prunus domestica*), blackthorn or sloe (*P. spinosa*), gean or wild cherry (*P. avium*), bird cherry (*P. padus*), cultivated cherry (*P. cerasus*), etc., the fruit is formed from one carpel (Fig. 217). The kernel inside the stone is actually the seed. It is surrounded by a thin papery testa. The epicarp is the tough skin. The mesocarp is the thick, fleshy edible portion, and the endocarp is lignified into a woody structure called the stone. Thus the cherry differs from the currant in that the seed of the former is protected by a hard layer produced by the fruit wall (endocarp), whereas the seeds of the latter are protected by a hard layer produced by themselves (testa). Apricots (*Prunus armeniaca*) and peaches (*P. persica*) are drupes which grow in warm temperate regions.

The walnut (*Juglans regia*) is not so familiar as a drupe, yet, if it is seen growing, there is no doubting its being one, for the walnut familiar as a ' nut ' is only part of the fruit, that is, the seed enclosed in the woody endocarp. The mesocarp is very fleshy and green, and is surrounded by a thin epicarp (Fig. 217). The fruit is formed from two fused carpels and contains a single seed. The lower part of the single cavity of the fruit is sub-divided by a thin partition so that as the cotyledons develop they become two-lobed and occupy the spaces on either side of the partition. Secondary partitions at right angles to the first one and a number of incomplete partitions also occur. The general result is the characteristic irregularly lobed appearance of the cotyledons which form the bulk of the kernel.

The kernel of the ripe nut contains about 18 per cent protein, 16 per cent carbohydrate and a very high percentage of fat. It therefore has a very high calorie value.

The almond (*Prunus amygdalus*) is another example of a drupe, though the portion used for food, either raw, salted, or in cake-making, is only the seed. The pericarp consists of a downy epicarp, a dry horny mesocarp and an endocarp in the usual form of a hard, woody shell surrounding the seed. The outer layers split at maturity and expose the seed surrounded by the endocarp.

There are two types of almond : the sweet, which is produced from the pink-flowered almond plant, and the bitter, which is produced from the white-flowered plant. The latter owes its bitterness to the presence of the cyanophoric glycoside, amygdalin. Almond seeds contain about 50 per cent oil, and are therefore valuable as a food supply.

One of the most important drupes, from the commercial point of view, is that of the coco-nut. These drupes are the fruits of the coco-nut palm (*Cocos nucifera*). This is a tree, attaining a height of 60 to 100 feet. The trunk is bare of leaves, but it terminates in a crown of most beautiful,

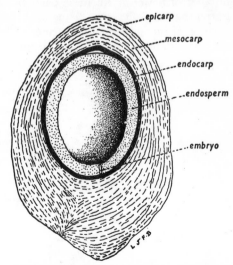

FIG. 219. LONGITUDINAL SECTION THROUGH A COCO-NUT FRUIT.

large, pinnate leaves, which are divided into many leaflets. This tree is very widely distributed in tropical countries, sometimes being so dense as to dominate the scenery for miles around. It prefers to live, however, near the sea-shore.

The drupe itself is composed of a very large pericarp, formed by the fusion of three carpels, but only one of the ovules usually develops into the seed. The pericarp is roughly triangular in cross-section, and is 15 to 20 inches in length (Fig. 219). It is divided into three layers: an outer, thin, tough layer (epicarp), a thicker, very woody endocarp, and a very thick, fibrous mesocarp between. Inside the pericarp is the seed. This is composed of a very thin testa, lined by a thick layer of white endosperm (the edible portion of the ' nut ') and, at the base, a very small embryo. The endosperm, as is well known, does not completely fill up the seed, but encloses a large cavity, in which there is a fluid called ' coco-nut milk '.

Probably the most extensively cultivated fruit in Great Britain is the apple. This is a cultivated variety of the wild or crab apple (*Malus sylvestris*). Closely allied to the apple, from the point of view of the structure of its fruit, is the pear. In both cases, the fruit is formed not only from the fertilised ovary, but also from the receptacle of the flower (Figs. 217 and 220). This very special type of fruit is called a **pome**. The flower is epigynous, and the ovary, which is formed from five fused carpels, is also fused to the surrounding receptacle. After fertilisation, this receptacle swells to form the fleshy part of the fruit, and the gynœcium becomes tough, and thus forms the core which also contains the seeds.

A special type of fruit is that found in the blackberry (*Rubus fruticosus*) and the cultivated types related to it, for example, raspberry (*R. idaeus*). In these two plants the carpels are indefinite and free. After fertilisation, each carpel forms a small drupe. The fruit is therefore said to be **aggregate**, being formed of an aggregate of drupes (Fig. 217).

The so-called ' fruit ' of the pine-apple (*Ananas sativus*) is the product of a complete inflorescence. It consists of the peduncle, bracts and the fruits of a large number of flowers. All these parts become fleshy at maturity, thus giving the barrel-shaped fruits, the diamond-shaped areas on the outside indicating the position of the individual true fruits. The peduncle continues its growth beyond the ' fruit ' and produces an apical tuft of green foliage leaves.

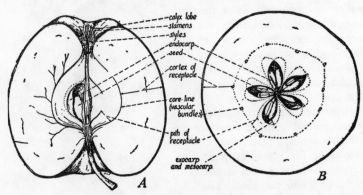

FIG. 220.   LONGITUDINAL AND TRANSVERSE SECTIONS OF THE APPLE FRUIT.
(After Robbins ; from ' Botany of Crop Plants '.)

DISPERSAL OF SEEDS AND FRUITS

The number of seeds produced by a single plant is often very large. For example, a single plant of shepherd's purse may produce 60,000 seeds ; one of tobacco, 360,000 seeds ; and some orchids may produce as many as 74,000,000 seeds. If all such seeds achieved successful germination, it is clear that the progeny of a single individual might within the course of a very few generations cover an area many times the area of the earth's surface. But most of the seeds perish, some being eaten by animals, while others will fall on ground already occupied or unfavourable to the particular species. There is therefore a severe struggle for existence.

The chances of survival in this struggle will be enhanced if the seeds are scattered some distance away from the parent plant. Even if the distance is only small, the chance of the successful germination of some of the seeds will be increased. If the distance is greater, not only will the survival of the individual be assured, but the spread of the species to new localities may be brought about.

Many seeds are so small that they are easily carried away, at any rate a short distance, from the parent plant, by even a slight breeze. Other plants, however, are more efficient in distributing their offspring. They have developed definite mechanisms for distributing their seeds.

Seeds are sometimes distributed as such, naked ; but very frequently the seeds are distributed enclosed in their fruit.

The methods whereby seeds and fruits are dispersed may be grouped under four headings : **animal**, **wind**, **mechanical**, and **water**.

DISPERSAL BY ANIMALS

The seeds and fruits of some plants are distributed by means of animals. The succulent fruit is the simplest type. The fruit of the blackberry, for example, forms a food for birds. The fleshy parts of the drupes are digested by the bird, but the seeds, protected by the hard endocarps, pass through the gut of the bird unharmed, and are then ejected with the excreta. In the case of the cherry drupe, the bird often carries it away and eats the fleshy pericarp, but the endocarp (the stone) is so hard that the bird ignores that part of the fruit and drops it. Thus the seed inside is dispersed quite unharmed.

When such seeds are ready for scattering, but not before, the fruit, etc., usually becomes very conspicuous, and thus attracts the useful animals ; for example, the cherry drupe is bright red or black, the strawberry receptacle is a bright red, etc.

In this connexion, the fruit of the mistletoe (*Viscum album*) is interesting. It is a pseudo-berry though it looks like a true berry. Actually there is a single seed, and the white sticky mass which surrounds it is not derived from the carpellary wall (as in a true berry), but is really the receptacle of the flower which, after fertilisation, has swollen and enveloped the seed. The sticky layer prevents the bird from eating the seed, so it scrapes it off its beak on to the branch of a tree which, if suitable (p. 350), acts as ground for the seed's germination.

Certain fruits, instead of becoming attractive to animals for the sake of dispersal, develop hooks so that they are enabled to cling to passing animals, and thus ' steal a ride '. These hooked fruits are sometimes called **burs**. A very familiar example is the goosegrass or cleavers (*Galium aparine*). The fruit is a more or less spherical schizocarp, and the surface of the pericarp is covered with very small

hooks (Fig. 221). By means of these hooks, the ripe fruits cling to animals such as rabbits and cows, and they easily become attached to a man's trousers if he walks through the hedgerows where the plant is growing. Thus the fruits are carried some distance before they are dropped.

In the wood avens (*Geum urbanum*) and certain cultivated species of *Geum*, the fruit is an achene, but the style does not die away after fertilisation. It remains firm, and becomes hooked at the top, thus forming an organ with which the fruit can cling to a passing animal (Fig. 221).

The inflorescence of great or common burdock (*Arctium lappa*) is a capitulum (p. 241) surrounded by a tight involucre of spiny bracts, which, after fertilisation, form brown, tough hooks. Then the entire head becomes easily detachable and forms a bur which clings to any browsing animal, thus dispersing the fruits (Fig. 221).

DISPERSAL BY WIND

As has already been seen, the samara is an achene in which the pericarp is produced into a long, thin, wing-like structure. By this wing the fruit is propelled through the air by means of wind. This is well seen in the fruit of the ash, and especially so in the double samara of the sycamore. The elm (*Ulmus* species) fruit is also winged, but the wing in this case is an extension in all directions, in one plane, instead of in one direction only. The seeds liberated from dehiscent fruits may also be winged. For example, those of the yellow rattle (*Rhinanthus major*) have the testa drawn out into a broad wing extending round most of the periphery of the flattened seed.

Plumed fruits, that is, fruits which develop hairy outgrowths, are some of the most interesting types which are distributed by wind. In the clematis, traveller's joy, or old man's beard (*Clematis vitalba*), for example, the fruits are a collection of achenes, borne on a receptacle, like those of the buttercup. But in the clematis, the hairy style is very long, and after the fruit has ripened it persists as a delicate white plume (Fig. 221).

In the flower of the dandelion the calyx is represented by a ring of white hairs called a **pappus**. The ovary is inferior and, after fertilisation, the fruit becomes an achene but the pappus persists, being pushed up on to the top of a very thin stalk, thus forming a

FIG. 221. FRUITS ADAPTED FOR WIND AND ANIMAL DISPERSAL.

Top to bottom: dandelion, clematis, avens, goosegrass, burdock.

beautiful parachute mechanism, by which the fruit travels sometimes for miles through the air (Fig. 221).

Individual seeds may also be plumed. For example, the seeds of willow, poplar and willow herb have tufts of hairs which act as parachutes.

The dehiscent capsules of many plants liberate their seeds by a pepper-pot, or **censer mechanism.** An example of this is afforded by the poppy, whose capsule has already been described (p. 276). The capsule is borne at the top of a long stalk and so is easily swayed about in the wind or by passing animals. The minute, dry seeds are shaken out, a few at a time, through the perforations in the dry peri-carp wall and may be carried still further from the parent plant by wind. During damp weather, the seeds inside the capsule are protected by small internal flaps which close the pores.

MECHANICAL DISPERSAL

In mechanical methods of seed dispersal, the fruit itself is responsible for the mechanism. The seeds are shot out, as if from a catapult. Since the mechanism is the fruit itself, it is the naked seeds that are dispersed, and not the fruit. The pods of peas, beans and vetches demonstrate this mechanical method. When the fruit is ripe and dry, the two halves of the legume or pod begin to shrivel. But the shrivelling takes place at unequal rates in different layers of the pericarp walls. The result is that the two halves first of all separate, and then each of them twists. This takes place so quickly that the action is almost explosive, thus shooting the seeds out for a considerable distance. The black or reddish brown pods of gorse (*Ulex europaeus*) and of broom (*Sarothamnus scoparius*) explode, especially on a hot day, with a very audible ' pop '. After being scattered in this way the seeds of these two plants are often carried by ants for a considerable distance. The ants are attracted by the oil-containing arils. The aril is eaten but the seed remains viable.

The capsule of pansies and violets is three-valved and distributes its seeds by mechanical means (p. 277).

The splitting of the schizocarp of the geranium and *Pelargonium*, already described, and of the meadow crane's bill (*Geranium pratense*), a closely related type, also takes place very quickly on a dry day. So sudden is this splitting that the five fruit portions, each containing a seed, are flung suddenly for some distance (Fig. 215).

DISPERSAL BY WATER

Dispersal of fruit by water is not so common, and is mainly confined to aquatic plants and to estuarine and shore plants.

The fruit of the white water-lily (*Nymphaea alba*) shows remarkable adaptation for dispersal by water. It is a berry which ripens under water. When ripe, the berry splits to release a mass of seeds which rises to the surface of the water. Each seed is covered with a spongy mass of tissue called an **aril**, and this contains many air bubbles which render the seed buoyant and capable of floating long distances in the water. Finally the air bubbles gradually disappear and then the seed sinks to the bottom where, after a period of rest, it germinates. The fruit of the yellow water lily (*Nuphar lutea*) is also a berry. Its seeds, however, are not covered with

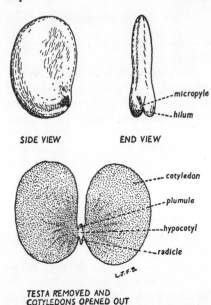

SIDE VIEW   END VIEW

- - - micropyle

- - - hilum

- - - cotyledon

- - - plumule

- - - hypocotyl

- - - radicle

L.J.F.B.

TESTA REMOVED AND
COTYLEDONS OPENED OUT

FIG. 222. SEED OF THE BROAD BEAN.

an aril of spongy tissue but the slimy testa contains air bubbles. These enable the seed to float until the testa decays.

Many of the palms growing on the shores of tropical seas and estuaries possess large fruits with fibrous, air-containing pericarps. Such fruits may be transported by ocean currents for thousands of miles and still remain capable of germination. One such palm is the coco-nut (Fig. 219). In this example, however, the seed remains viable for only a short time so that transport by ocean currents can only be effective for relatively short distances.

STRUCTURE OF SEEDS

As already described in Chapter XXI, the seed consists essentially of an embryo surrounded by a testa. The embryo usually shows a **plumule** (consisting of the stem apex and a few leaf rudiments) and a **radicle** (or first root). The stem bears either one or two seed-leaves or **cotyledons** at the base of the plumule, the number of cotyledons being characteristic of the Monocotyledons and Dicotyledons respectively. Endosperm may or may not be present in the seed, endospermic and non-endospermic types occurring in both Dicotyledons and Monocotyledons (see p. 270).

The broad bean (*Vicia faba*) will serve as an example of a non-endospermic Dicotyledon (Fig. 222). The seed is covered by a tough, light coloured testa. At one end of the seed the testa shows an elongated dark scar, the **hilum**, which indicates the point of attachment to the stalk or funicle, by means of which the seed was attached to the wall of the bean pod. A minute opening, the **micropyle**, is present at one end of the scar—this corresponds to the micropyle of the ovule from which the seed was formed. If the testa is removed, the structure revealed is the embryo. The bulk of it consists of two white, kidney-shaped structures, the cotyledons. If these are separated and opened out, the embryonic shoot and root will be seen. The root or radicle fitted into a pocket in the testa and its tip pointed to the micropyle. The plumule is sharply bent so as to occupy a protected position between the cotyledons. The cotyledons, although so highly modified, are the first leaves borne by the stem of the embryo. The short length of stem below the cotyledons is called the **hypocotyl** while the stem immediately above the cotyledons is the **epicotyl**.

The endospermic seed of the castor oil plant (*Ricinus communis*) differs from that of the broad bean in important respects (Fig. 223). The testa is hard and brittle and a warty outgrowth of the seed stalk, known as the **caruncle**, covers the hilum and micropyle. The embryo is essentially similar to that of the broad bean but the cotyledons, instead of being thick and fleshy, are thin, wafer-like structures. The embryo is embedded in a mass of endosperm, which contains food reserves in the form of proteins (in aleurone grains) and oil (castor oil).

The seeds of Monocotyledons show a similar distinction between non-endospermic and endospermic types. The water plantain (*Alisma plantago-aquatica*) seed is non-endospermic. The embryo consists of one large, food-storing cotyledon, a laterally placed plumule and a radicle. On the other hand, the small black seed of the onion (*Allium cepa*) has a similar, but less massive, embryo embedded in nutritive endosperm.

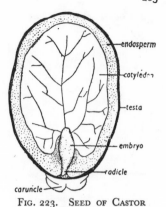

FIG. 223.   SEED OF CASTOR
OIL PLANT.
The seed has been halved and
one cotyledon removed.

The grains of wheat and other cereals are really fruits, each containing a single endospermic seed. In these plants, the ovary is formed from one carpel containing a single ovule fixed at its base. After fertilisation the testa becomes fused with the pericarp, giving a type of fruit called a **caryopsis**.

A grain of wheat is deeply furrowed down one side and the small embryo can be seen on the opposite side. A longitudinal section shows the structure indicated in Fig. 224. The embryo consists of a plumule, a radicle and one cotyledon. The plumule comprises a stem apex, surrounded by several young leaves, the outermost of which forms the plumule sheath or **coleoptile**. The radicle is surrounded by a sheath called the **coleorhiza** and in addition to this first root the rudiments of several adventitious roots (seminal roots) are already present. The single cotyledon is highly modified and forms a shield-shaped structure, the **scutellum**. The outermost layer of the scutellum, which lies in close contact with the endosperm, is glandular in nature and forms the epithelium which secretes enzymes when germination begins. The bulk of the reserve food in the endosperm is in the form of starch, but protein is stored in its outermost layer, the **aleurone layer**. This layer lies immediately inside the fused seed and fruit coats.

GERMINATION

After seeds have been liberated from the parent plant, they usually undergo a resting or dormant period before they develop further. If, at the end of this period, the seeds are provided with a supply of water and oxygen and are placed in a suitable temperature, the embryo will renew its growth and develop into a new plant. This further development of the embryo is the process of germination.

The first stage in germination is the absorption of water. The ripe seed may have as little as 10 per cent of water; but if water is available a

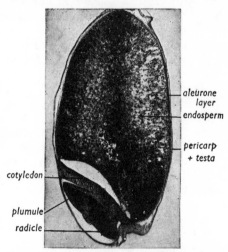

FIG. 224.   LONGITUDINAL SECTION THROUGH
A WHEAT GRAIN (× c. 8).

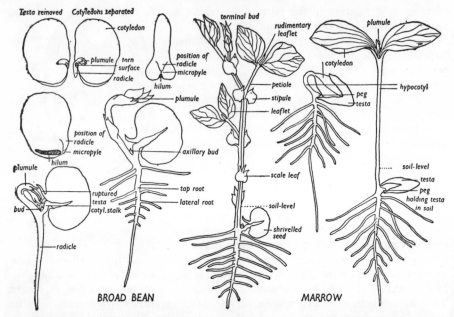

FIG. 225. GERMINATION OF BROAD BEAN (HYPOGEAL) AND OF MARROW (EPIGEAL).
(F. W. Williams)

large amount is quickly absorbed. For example, the amount of water absorbed
by germinating wheat grains is 60 per cent of their original weight, and bean seeds
absorb as much as 150 per cent. This intake of water is due in the early stages to
imbibition, first by the colloidal gels of the testa and a little later by those of the
embryo and endosperm. Intake by imbibition is followed by a further absorption
of water by the osmotic activity of the living cells of the embryo and endosperm.
The hydrated testa also acts as a general semi-permeable membrane surrounding
the whole seed. Experiment shows that water enters most rapidly at the micropyle
end of the seed but absorption does occur over the entire surface. It is for this
reason that the ground is rolled or trodden down after the seed has been sown,
since these processes ensure that the whole seed surface is brought into contact
with the moist soil.

The intake of water is followed by the renewal of all the physiological activities
of the embryonic and endosperm tissues. The enzymes in the seed are activated.
The food which had been stored in an insoluble form in cotyledons or endosperm
is changed by these enzymes into diffusible forms. Thus, starch is converted to
sugars, fats to fatty acids, and proteins to amino-acids. These food supplies now
begin to pass to the embryo, where they are used, partly as the substrate for
respiration and partly to provide new protoplasm and cell-walls in the growing
points of the embryo. Owing to the use of some of the foodstuffs for respiration,
the dry weight of a young seedling (for example, of maize) may be reduced to less
than 60 per cent of that of the original grain. The mobilisation and utilisation
of the original foodstuffs naturally also leads to complex changes in chemical
composition. A seedling of maize has about one-eighth of the amount of starch
present in the original grain, and about one-third of the fat. On the other hand,

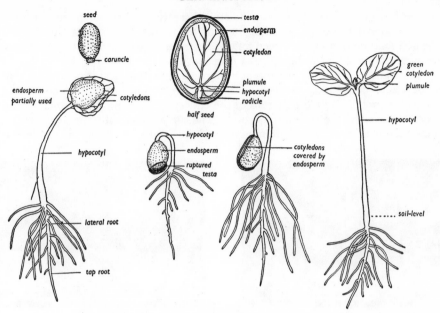

FIG. 226. SEED AND GERMINATION OF CASTOR OIL PLANT.
(F. W. Williams)

sugar, absent from the grain, is present in the seedling and the amount of cellulose in the seedling is three times that present in the grain.

With the intake of water and the mobilisation and utilisation of the foodstuffs, the embryo undergoes rapid development, accompanied by active respiration. The later stages of germination are, however, variable and will be described in relation to particular examples.

In the broad bean (Fig. 225) the radicle bursts through the testa and grows downwards into the soil. Then the stalks of the cotyledons lengthen and the plumule is thereby pushed out from between the cotyledons. The tip of the plumule is sharply reflexed so that the actual stem apex is protected from damage as the plumule now proceeds to grow up through the soil. As the plumule emerges from the soil it straightens and its upward growth is continued. The young aerial shoot bears a few scale leaves at its base and then, at higher levels, normal foliage leaves develop as photosynthetic structures. Meanwhile, the radicle has continued its downward growth and has produced lateral roots. The young plant is now self-supporting. During the whole of this development the cotyledons have remained inside the testa below ground and, by the time the young plant is established, they are shrivelled and beginning to decay. The starch and other food materials in the cotyledons were, in fact, the only source of food during the early stages prior to the development of photosynthetic leaves. This type of germination, where the cotyledons remain below soil-level, is spoken of as **hypogeal** germination. This type is characterised by the fact that the epicotyl grows in length while the hypocotyl remains short.

The single seed in the samara of the sycamore (*Acer pseudoplatanus*) is also non-endospermic. In this example, however, the hypocotyl elongates rapidly and so lifts the two green cotyledons into the air, where they carry out photosynthesis. It is only later that the epicotyl elongates and the normal foliage leaves are formed The germination of a marrow seed follows a similar course. In both these types, the germination is said to be **epigeal** (Fig. 225).

An example of epigeal germination in an endospermic seed is offered by that of the castor oil (*Ricinus communis*). When germination begins, the cotyledons absorb the mobilised food materials of the endosperm. The radicle emerges and is soon followed by the looped hypocotyl. The cotyledons themselves also increase in area and thickness. Straightening of the hypocotyl now carries the cotyledons

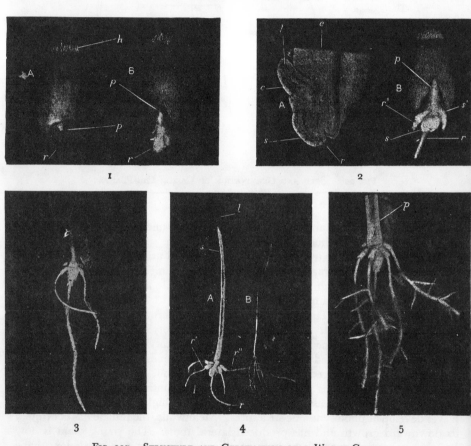

FIG. 227. STRUCTURE AND GERMINATION OF A WHEAT GRAIN.

1. The embryo with plumule (*p*) and radicle (*r*) is seen through the coat of the grain. In *B*, germination has begun.

2. *A*, longitudinal section through the lower part of the wheat grain ; *B*, the plumule and radicle have elongated and two adventitious roots have developed.

3. Germination has proceeded further than that illustrated in 2, and root hairs have formed. Another pair of roots are emerging.

4. *A* shows the first green leaf (*l*) breaking through the coleoptile (*s*) ; all roots but one have been cut off short ; *B*, a later stage.

5. Lower part of a stage later than that illustrated in 4. *A*, showing plumule (*p*) in section, and a bud enveloped by the first two green foliage leaves. Lateral roots have formed on the original adventitious roots, which have been cut off short.

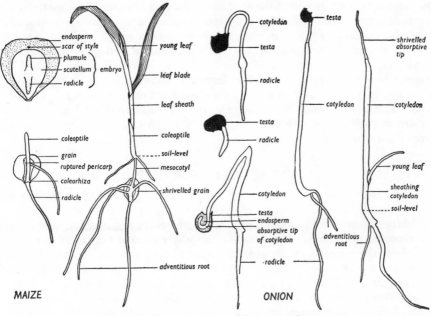

FIG. 228.   GERMINATION OF MAIZE AND ONION.
(*F. W. Williams.*)

above ground and these soon turn green and carry out photosynthesis.   By this time most of the endosperm has been absorbed, although small remnants of it may adhere to the undersides of the cotyledons.   Finally, the epicotyledonary shoot begins to elongate and produce leaves of the adult type (Fig. 226).

The germination of the seeds of Monocotyledons may also be˙ either of the hypogeal or the epigeal type.

Fig. 227, 1–5 illustrates the hypogeal germination of the wheat grain.   When germination commences, the radicle emerges first, having burst through the coleorhiza.   The radicle increases in length but is soon supplemented by the growing out of adventitious roots from the base of the plumule (the seminal roots noted in the seed); these, too, burst their way through sheaths, the remnants of which show as collars at their bases.   During these later developments of the root system the coleoptile, with young leaves inside it, elongates and eventually emerges from the soil and develops chlorophyll.   Then the first foliage leaf bursts through the tip of the coleoptile and this is followed by further leaves (Fig. 227, 4 *A* and *B*).   The stem remains very short and only elongates when, eventually, the flowering shoot is formed.   The wheat grain remains below ground and with it the scutellum or cotyledon, so that germination is hypogeal.   The sole function of the cotyledon has been to mobilise and absorb the food material from the endosperm which is reduced to a milky fluid during the process.   Further developments of the seedling result from the outgrowth of axillary shoots (the process of **tillering**) and the development of an extensive system of fibrous adventitious roots.

The germination of the onion seed is illustrated in Fig. 228.   After the emergence of the radicle, the cotyledon increases rapidly in length and becomes sharply recurved so that it first appears above ground as a green loop.   The tip of the cotyledon remains within the seed and continues to absorb food material from the endosperm. The radicle is then supplemented by adventitious roots and the cotyledon straightens itself out, often carrying the testa with it. The stem remains very short and the first true foliage leaf bursts laterally through the sheathing cotyledon, to be followed by a succession of further leaves. It is the bases of these leaves which will eventually swell to form the onion bulb. Since the cotyledon appears above ground during the germination process, the latter is epigeal in type. The epigeal condition is due here to the elongation of the basal part of the cotyledon itself and not to the elongation of a hypocotyl as in the Dicotyledon types described above.

CONDITIONS SUITABLE FOR GERMINATION

The above account of the process of germination will already have indicated the nature of the conditions under which a seed will germinate.

Water is necessary to replace that lost by the embryo and other tissues during the ripening of the seed, during which the water content falls to 10 per cent or less (from the 90 per cent or so in the general tissues of the plant). It is only when water is absorbed that the vacuoles of the embryonic and endosperm cells again expand and the cells resume their general physiological activity. Water is also needed for the activation of the enzymes and for transport to the embryo of the mobilised food reserves.

A supply of **oxygen** is necessary so that the seed may carry out aerobic respiration and thus obtain the requisite energy for the growth of the embryo. The importance of oxygen is demonstrated by the fact that seeds will not germinate in water free from air, this being true even of the seeds of aquatic plants. In Nature, seeds may fail to germinate if lying too deep in the soil since oxygen diffuses only very slowly downwards.

A **suitable temperature** is also necessary. This varies with different seeds. For example, the most suitable temperature (the **optimum** temperature) for the germination of wheat grains is $25°–30°$ C. ; they will not germinate below $0°–5°$ C. (the **minimum** temperature), nor above $31°–37°$ C. (the **maximum** temperature). Corresponding figures for maize are : optimum, $37°–44°$ C.; minimum, $5°–10°$ C.; maximum, $45°–50°$ C. For cucumber they are : optimum, $31°–37°$ C.; minimum, $15°–18°$ C.; maximum, $45°–50°$ C. Increasing the temperature from the minimum to the optimum level increases the rate of water intake, the speed of the enzyme actions and the rate of diffusion of foodstuffs. Falling off in the rate of germination at higher levels is probably due to the unfavourable effect of higher temperatures on the living cell contents. It may be noted in relation to the examples given that wheat is characteristic of temperate regions, while maize and cucumber are both plants of the tropics and sub-tropics.

The influence of **light** on germination is variable. The majority of seeds germinate equally well in light or in darkness and are said to be **light indifferent**. There are, however, some seeds which will only germinate successfully in the light. For example, the seeds of the purple loosestrife (*Lythrum salicaria*), a tall herb which grows on the banks of rivers and ponds, will not germinate unless they are exposed to light. The same is true of the seeds of the hairy willowherb (*Epilobium hirsutum*, Fig. 229), and those of the mistletoe (*Viscum album*) and the butterwort (*Pinguicula vulgaris*). Such seeds are said to be **light sensitive**. The amount of

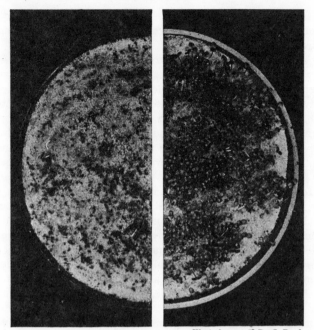

W. *Anderson and Dr.G. Bond*

Fig. 229.   Effect of Light on Germination of Hairy Willowherb
(*Epilobium hirsutum*).

Seeds were placed in moist sand in Petri dishes and the photograph taken two weeks
later.   On the left is part of the dish kept in darkness ; very few seeds have germinated.
On the right is part of a dish exposed to light, showing abundant germination (× ¾).

light needed may be very small, one hour of diffuse illumination sometimes being
sufficient, while in the purple loosestrife an exposure to light for one-tenth of a
second will stimulate germination.   The reverse reaction to light is shown by
other seeds, that is, they will not germinate except in the dark.   Such seeds are
said to be **light hard**.   Belonging to this type are the seeds of *Phacelia tanacetifolia*,
Nemophila insignis, and the familiar cultivated plant commonly called love-lies-
bleeding (*Amaranthus caudatus*).

The mode of operation of this light stimulus is not as yet fully understood.
Suggestions have been made that light influences the mobilisation of reserve
proteins or that it has an effect on the permeability of the testa to water and to
oxygen.

Before the various factors just described can bring about the germination of a
seed, the latter must itself be in a suitable condition.   Its testa must be permeable
to water and to oxygen and the embryo must be fully developed and physiologically capable of further development.   These points are dealt with in the following
paragraphs.

DORMANCY AND VIABILITY

The embryo in the developing seed ceases to grow as the seed ripens, possibly
because of the narcotising effect of the carbon dioxide retained in the seed by the

increasingly impermeable testa. All parts of the seed lose water and by the time the seed is fully ripe all the physiological activities of the embryonic and other living cells are reduced to a minimum.

In a few instances the seed is able to germinate immediately. The seeds of willows and some species of *Oxalis*, for example, are able to germinate as soon as they are liberated from the fruits if conditions are suitable. In the majority of plants, however, the ripe seed, dehydrated and physiologically inactive, will not germinate immediately even when the external conditions are entirely favourable. The seed is then said to be **dormant** and the period which has to elapse between the ripening of the seed and the time when it will germinate under suitable conditions is the **dormant period**. During this period all physiological activities appear to be almost at a standstill. It has indeed proved impossible to demonstrate that respiration is still proceeding, although on general grounds it would seem necessary for it to do so, even if extremely slowly, in order to maintain the living condition of the protoplasm. Possibly anaerobic respiration provides the small amount of energy necessary, particularly as seeds may be kept *in vacuo* for two years and then still be capable of germinating. Dormant seeds may be frozen in liquid hydrogen (–250° C.) without losing their power of germination, although it would seem that all reactions must cease in the frozen protoplasm. It is clear from these facts that dormant seeds at least approach a condition of ' suspended vitality '.

The failure of dormant seeds to germinate even under favourable conditions may be caused by a number of factors. In some examples the dormancy may be due to the fact that the embryo in the seed is not completely developed. This appears to be the case in the lesser celandine (*Ranunculus ficaria*) in which the embryo completes its development inside the fallen seed through the autumn and winter and the seed finally germinates in the spring. In the ash (*Fraxinus excelsior*) and sycamore (*Acer pseudoplatanus*) dormancy due to this cause also lasts several months. The embryos in the seeds of apple (*Malus sylvestris*) and other fruits are fully developed but have to undergo a period of **after-ripening**, involving physical and chemical changes, before germination of the seeds will take place.

Another and very frequent cause of dormancy is the impermeability of the testa or of the pericarp of indehiscent fruits. Such hard coverings prevent the entrance of the water and oxygen required for germination to take place. Even when such fruits and seeds are lying in moist soil it may take months or even years before the coverings become permeable. This is so in the case of the seeds of gorse (*Ulex europaeus*), sweet pea (*Lathyrus odoratus*), white clover (*Trifolium repens*) and many other members of the Leguminosae. Under natural conditions the testa may gradually become permeable as a result of freezing, bacterial action, or changes in the colloids of the cell-walls. Abrasion by the soil particles may also take place.

The length of the dormant period not only varies as between different species, but also as between different individual seeds of the same species. For example, if a number of clover seeds are placed under favourable conditions, some will germinate quite quickly, others more slowly, so that seedlings appear at intervals over a long period. This variation may be of biological advantage since it means that the chance of some of the seedlings developing under optimum conditions is thereby increased.

During the period of dormancy just described, the embryo remains alive and capable of eventual further growth. Such seeds which retain their ability to germinate are said to be **viable**.

The seeds of willows retain their viability for only a day or so. The seeds of the coco-nut (*Cocos nucifera*) and of Para rubber (*Hevea brasiliensis*) also retain their viability for only a relatively short time. The introduction of Para rubber to Ceylon and other countries was a difficult operation on this account. Sir J. D. Hooker had seeds collected in 1873 and again in 1876 from the forests of the Amazon. By the time these reached Kew, many of the seeds had lost their viability ; but enough seedlings were obtained from the second collection to send to Ceylon where rubber plantations were successfully established. Transport of seeds by aeroplane has now largely obviated such difficulties.

Under conditions unfavourable for germination, many seeds remain viable for periods longer than the normal period of dormancy. For example, the majority of garden and agricultural seeds will, if stored dry, remain viable for at least a year and often for longer periods. Wheat and barley grains will give full germination after storage for ten years. Many seeds of the Leguminosae, characterised by the possession of very hard seed coats, will germinate after a much longer period. **Becquerel** found that the seeds of a species of *Cassia* were capable of germination after being stored in the Paris Natural History Museum for 115 years. Another interesting example is afforded by the sacred lotus (*Nelumbo nucifera*). **Robert Brown** (1850) was able to germinate some seeds of this plant after they had remained dry in Sir Hans Sloan's collection for 120 years. Reports that wheat grains from Egyptian tombs, for example, that of Tutankamen, are still capable of germination are, however, entirely unfounded.

The above examples refer to seeds stored dry ; but seeds may remain dormant but viable for long periods when buried in moist soil, mud or peat. Here it is obviously not lack of water that prevents germination ; but it seems likely that lack of oxygen and a high concentration of carbon dioxide are the factors concerned. When pastures are newly ploughed, many weeds characteristic of arable land usually appear, charlock (*Sinapis arvensis*), a plant with bright yellow flowers, and shepherd's purse (*Capsella bursa-pastoris*) being common examples. This is due to the fact that their seeds have been buried in the deeper layers of the soil, under conditions of bad aeration, since the previous period of cultivation. Ploughing brings the seeds nearer to the surface where better conditions of aeration lead to their germination. A striking example of the retention of viability by buried seeds is afforded by those of the lotus already mentioned. In 1933 some seeds were supplied to the Royal Botanic Gardens, Kew, by a Japanese botanist who had found them in the peat of a long-dried-up Manchurian lake. They were tested and found to be still viable. In 1951, seeds having the same origin were germinated in Washington. The longevity of these seeds was then examined by **Professor R. W. Chaney** of the University of Chicago. From their investigations it seems that these still-viable *Nelumbo* seeds were at least 1040 years old.

Viability may thus be retained for very long periods ; but in the continued absence of conditions favourable for germination the seeds eventually die, possibly owing to the slow denaturing of the cell proteins. During the period of viability the seed coats appear to secrete some substance which inhibits bacterial and fungal attack. The dead seed is, however, soon subjected to the normal processes of decay.

ARTIFICIAL ACCELERATION OF GERMINATION

In examples where the seed remains in a dormant condition owing to the impermeability of a hard testa, it is possible to induce germination, that is to break the dormancy, by various methods. Filing, chipping or other mechanical abrasion of the testa of seeds such as those of the sweet pea and clover will permit the entrance of water and oxygen with resultant germination. Treatment for a limited time with sulphuric acid or with hot water may also increase the permeability of the testa.

In the case of seeds which need to undergo after-ripening, this process may be speeded up by subjecting them to a low temperature, or, in some cases, to treatment with acids. Apple seeds are often ' layered ', a process in which the seeds are embedded in layers in moist sand and kept at 5° C. Other seeds, for example those of the gooseberry, need to be subjected to − 5° C. or even lower temperatures. After-ripening does not take place in dry seeds so that seeds undergoing such treatments must be kept moist.

The germination of some garden seeds, for example, those of sweet pea, may be accelerated by a preliminary soaking in tepid water. Immersion in water should not be prolonged beyond 24 hours since anaerobic conditions soon arise and the embryo may be damaged. Runner bean seeds should not, in fact, be soaked for more than about two hours. Such soaked seeds must of course be planted in moist soil if satisfactory further development is to take place.

PRACTICAL WORK

1. By means of a hand lens, examine the collection of carpels of the buttercup, when they have ripened after fertilisation, thus forming a collection of fruit. Each fruit is an achene. Dissect one and find the single seed enclosed.

Compare and contrast a ripe strawberry with the collection of buttercup fruits.

2. Examine and draw a bunch of ash fruits. Make a detailed examination of a single fruit and cut a longitudinal section of it, thus exposing the seed.

Examine a double samara of the sycamore.

3. Draw a hazel nut, partly enclosed in its leafy cup. Examine the nut in detail and expose the seed. Compare this fruit with that of the buttercup.

Examine the schizocarp of the hollyhock, and compare this with the other types of dry, indehiscent fruits.

4. Compare and contrast the various types of dehiscent fruits.

The follicle of larkspur splits along its ventral suture. It is formed from one carpel and contains several seeds.

The legume of the pea, on the other hand, finally splits along both margins. When the pea fruit is ripe, note the manner in which both parts of the pericarp twist, thus throwing the seeds some distance from the parent plant.

Various types of siliquas and siliculas should be examined ; for example, wallflower, shepherd's purse, honesty. These fruits are formed from two fused carpels, and are distinguished by the formation of a false septum. Note how they split to expose the seeds.

The capsule shows many variations, some of which should be examined.

5. Many types of succulent fruits are of economic importance, so it is important to examine some of them.

Drawings should be made of the complete fruit, and, where possible, it should be dissected and longitudinal and transverse sections drawn.

Examples of the following should be studied : berry (gooseberry, currant, tomato, banana, grape, orange) ; pepo (cucumber) ; drupe (plum, cherry, coco-nut, walnut) ; pome (apple, pear) ; aggregate fruits (blackberry, raspberry).

6. The mechanisms adopted by seeds and fruits in order to ensure dispersal should be studied.

Apart from the fruit already examined, the following are worthy of study : burs of cleavers, burdock, and wood avens ; plumed fruit of clematis, dandelion, and thistle ; splitting schizocarp of geranium ; etc.

7. Show that seeds absorb water, both through their micropyles and through the general surface of the testa.

Take 10 dry bean seeds, weigh them, and cover their micropyles with a little rubber solution. Allow this to dry, and then place the seeds in water. Take 10 bean seeds which have not been treated thus, weigh them and place them in another vessel of water. Leave both for the same length of time (about twenty-four hours). Dry the surface of the seeds, weigh each lot separately, and calculate the percentage intake of water by the two samples.

8. Place a piece of blotting paper or cotton wool in a saucer. Then thoroughly moisten it with water. On this blotting paper, sow a hundred wheat grains. Do the same with other seeds, preferably garden seeds ; for example, cress, radish, onion, beet, carrot, etc. Then record each day the number which have germinated and calculate the percentage. More water may be added to keep the blotting paper moist. Note the diversity of time taken to germinate.

9. Examine the effect of light on seeds. Set up saucers containing seeds similar to those required for Experiment 8 ; but having duplicate sets throughout. Place one set in the light and another set in a dark room or box.

Do the same with examples of light-hard and light-sensitive seeds.

Record the results as percentages, and from them discuss the effect of light on germination.

10. Test the effect of different temperatures on seeds, by similar methods.

11. Show that oxygen is necessary for germination.

Take two wide tubes of equal length (about 1 foot). Holding them vertically, fix a wad of moist cotton wool in each, about 2 inches from the top end. Drop in 6 pea seeds or 50 wheat grains so that they lie on the wool. Insert a stopper in each. Now immerse the open lower end of one of the tubes in a beaker of water and the other in a beaker of pyrogallic acid and support in a vertical position.

Record the number of grains which have germinated each day, and, realising that pyrogallic acid absorbs the oxygen from the atmosphere, deduce the role of oxygen in germination.

12. Examine the structure of the seed of the castor-oil plant (*Ricinus communis*) and the broad bean (*Vicia faba*). In the latter case it is best to soak it for twenty-four hours before dissecting.

Dissect both seeds and draw the structures seen in each case. Note in each case the embryo, with its hypocotyl, radicle and plumule. The cotyledons of *Ricinus* are thin and tissue-like, whereas those of *Vicia* are thick and fleshy. Explain why, although in both cases there is an excellent food supply, yet *Ricinus* is albuminous whereas *Vicia* is exalbuminous.

Note the testa surrounding the seed in each case. Test the food reserves present in the seeds.

13. Plant about ten seeds of the castor-oil plant and ten seeds of the broad bean in boxes containing damp sawdust. Place in a window and allow them to germinate.

Then, after the radicles have appeared, take one of each and make a drawing of it. Do the same at intervals, thus examining the various stages of germination.

Note the hypogeal germination of the bean and the epigeal germination of the castor-oil plant.

Examine and draw the seedlings in detail.

14. Examine the seeds and seedlings of a number of Monocotyledons. The grains of wheat, barley and maize and the seeds of onion are suitable examples.

CHAPTER XXIII

SOME SIMPLE HOLOPHYTIC PLANTS : ALGAE

The simplest plants are placed in the division Thallophyta. All such plants possess a **thallus**, a plant body which has no true roots, stems or leaves. There are many different types of thallophytes, and they exhibit a great diversity of structure and mode of life. As was seen in Chapter III, they are subdivided into algae and fungi.

Most of the algae are of aquatic habit, though some are terrestrial, with a preference for moist situations.

Perhaps the most interesting feature of the algae, apart from their diversity of structure, is their diversity of mode of reproduction. There are two general methods of reproduction among such plants, namely, **asexual** and **sexual**.

ASEXUAL REPRODUCTION

This form of reproduction can take place by two methods. The first is perhaps the most simple of all methods of reproduction, that is, **fission**. This is very prevalent among unicellular plants and animals. It consists merely of the division of the unicellular organism into two daughter cells. Many unicellular algae and bacteria reproduce themselves by fission. The two daughter cells soon grow to the size of the mother cell, thus producing two new individuals.

The second method of asexual reproduction is by means of **spores**. Spores may be motile in that they possess cilia, in which case they are known as **zoospores**. These are usually naked masses of protoplasm, and they develop their cell-walls later when, after liberation, they begin to germinate. The non-motile form of spore is often just one of the plant cells. It usually has a cell-wall, which sometimes becomes thickened to withstand adverse conditions.

SEXUAL REPRODUCTION

Sexual reproduction is *fundamentally* the same in the Thallophyta as it is in the flowering plant, and in animals. It involves the production of two **gametes**, which finally fuse to produce a **zygote**. This later develops to produce a new plant. In some cases, such as the simpler plants, both gametes are alike, and are therefore known as **isogametes**. In isogamy, the fusion is usually referred to as **conjugation**, and the resulting zygote is commonly called a **zygospore**, because there is no strict differentiation of sex. As will be seen later, however, there may be a physiological difference between gametes which have exactly the same structure.

In other examples, two kinds of gametes are formed. These are distinguished by differences in both size and behaviour, though they may have the same general structure. Such gametes are described as **anisogametes** or **heterogametes** and it is usual to distinguish between smaller male gametes, which are often more active, and larger female gametes, which are often less active.

This distinction between male and female is most marked when the female gamete is non-motile while the male is small and active. In this case, the female gamete is termed an **ovum** or egg and the male gamete a **spermatozoid**. Plants whose gametes are thus differentiated are said to be oogamous and the fusion of the gametes is spoken of as fertilisation.

ALGAE

Apart from diversity of methods of reproduction, the algae exhibit great diversity of size and form, as seen in Chapter III. There are four groups of algae, namely: (1) **Green** (**Chlorophyceae**, containing chlorophyll only) ; (2) **Brown** (**Phaeophyceae**, containing chlorophyll together with the brown pigment, *fuco-xanthin*) ; (3) **Red** (**Rhodophyceae**, containing chlorophyll together with the red pigment, *phycoerythrin* and the blue pigment *phycocyanin*) ; (4) **Blue-green** (**Cyano-phyceae**, containing chlorophyll together with *phycocyanin*).

CHLAMYDOMONAS

Chlamydomonas is one of the simplest of plants. In fact, it is so simple, and so closely related to the very simple animals, that some biologists prefer to keep an open mind about it and place it in the rather heterogeneous group, *plant-animals*.

It is a motile, unicellular organism, living in stagnant water, often, especially during the summer, in such numbers as to impart a green colour to the water (Figs. 1 and 230). The cell is usually spherical or ovoid, measuring about 200µ in diameter, and is enclosed in a thin wall of hemi-cellulose which often thickens at the forward end of the cell to form a colourless ' beak '. The cell contents consist of cytoplasm, a nucleus and a single large chloroplast. The chloroplast is cup-shaped, with a thickened base in which there is embedded a protein body known as a **pyrenoid**. This latter seems to be the centre of starch deposition, for when the latter is manufactured by photosynthesis, the grains are deposited around it. The nucleus is always suspended in the cavity of the chloroplast by bridles of cytoplasm.

Towards the narrow end of the cell is a small deep-red structure known as the **eye-spot** or **stigma**. This is cup-shaped and lining it is a layer of cytoplasm sensitive to light. So the eye-spot is a photo-receptor, and under its influence the plant swims towards a source of reasonably intense light. If, on the other hand, the light is too bright, the plant swims away from it (p. 236). Nearer still to the apex are two highly refractive structures, which seem alternately to enlarge and collapse. These are therefore known as the **contractile vacuoles**, and are probably a simple means of eliminating waste products produced during the life-processes of the protoplasm.

Movement is effected by the lashing of the two fine **cilia** or **flagella** which are produced as outgrowths of the cytoplasm at the apex of the cell. These flagella pass through pores on either side of the ' beak ' and by their rapid backward strokes, they pull the plant through the water. The plant at the same time rotates on its axis.

If *Chlamydomonas* is subjected to normal conditions of temperature, nutrition, etc., it continues to grow until it attains a certain size, then it reproduces itself, asexually.

In the first place, it withdraws its cilia and comes to rest. Then the protoplast with all its contents contracts away from the wall and divides into two equal

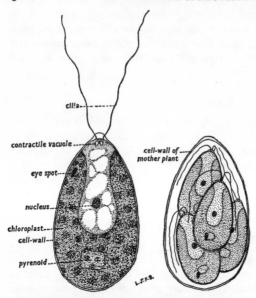

FIG. 230. ASEXUAL REPRODUCTION IN *Chlamy-domonas* (× 1500).

Left, a normal plant ; right, four young plants pro-duced by two successive divisions of the mother cell, are still enclosed in the mother cell-wall.

portions, each portion containing half of every structure present in the mother cell—nucleus, cyto-plasm, chloroplast, etc. Then each portion divides again, and perhaps even a third time. The resulting portions each develop a cell-wall and cilia, so that within the original cell-wall of the mother cell there may be two, four, or eight new, young *Chlamydomonas*-like structures called **zoospores.** After these have swum about for some time still within the mother cell-wall, the latter is dissolved or becomes ruptured, and the zoospores are liberated and quickly enlarge to form new plants. This is a very quick method of reproduction, with the result that as many as two million new *Chlamydomonas* plants can be produced from one, in a week (Fig. 230).

Under certain conditions, an individual may come to rest and its contents divide as in the formation of zoo-spores, but the daughter cells do not develop flagella and their walls become mucilaginous. Following enlargement of the daughter cells, further divisions take place so that a large number of non-motile cells remain clustered together in a large gelatinous mass. This is the **palmella** stage. Eventually, the con-stituent cells develop flagella and escape from the mucilage as normal individuals.

Yet another method of asexual reproduction is seen in a few species, for example, *C. nivalis* (the cause of red snow). Vegetative cells lose their flagella and form a thick wall around their contracted contents. Such spores, which fre-quently contain red haematochrome, can persist through an unfavourable period and then eventually give rise to one or two zoospores.

When conditions are adverse, especially when nutritive salts become depleted, *Chlamydomonas* resorts to a method of sexual reproduction. Division takes place in a manner similar to that in zoospore formation, with the exception that there are usually more divisions, thus resulting in the production of eight to sixty-four individuals, which are isogametes, each one similar to the mother cell, only smaller and naked, that is, without a cell-wall. These are soon liberated and swim about freely for a time. Then they meet in pairs and conjugate at their apices to produce the zygote (Fig. 231). This swims about for a time by means of its four cilia, then finally settles down and withdraws its cilia. Thus the zygo-spore is formed. This becomes spherical, and secretes a thick, warty membrane, while the accumulation of oil gives it a red or orange colouration. In this condi-tion, it can withstand very adverse conditions for a long time. For example, it can be safely blown about in the air or remain in dried-up mud until more water appears. Then it germinates and usually produces four zoospores, a process which is accompanied by meiosis. On liberation, these zoospores give new individuals, which by asexual reproduction rapidly produce a new colony of *Chlamydomonas* plants.

Even in isogamous species, it is only gametes from different individuals which will conjugate, so that there is a dioecious condition. Some individuals belong to what is called a plus strain and these produce plus gametes, which will only conjugate with minus gametes, produced by individuals belonging to a minus strain.

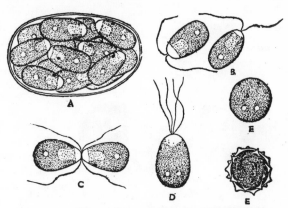

Fig. 231. Sexual Reproduction in *Chlamydomonas*.

A, gamete formation by division of one individual whose wall contains the gametes ; *B*, two liberated gametes (isogametes), having no cell-walls ; *C*, conjugation of two isogametes at their apices ; *D*, zygote ; *E*, young zygospore with the two nuclei still separate ; *F*, mature zygospore surrounded by a warty membrane.

(*Reproduced by kind permission from ' Cryptogamic Botany ' by G. M. Smith. Copyrighted, 1938, by the McGraw Hill Book Co., Inc.*)

It is impossible to distinguish between male and female gametes since they are alike in size, shape and behaviour, but the plus and minus gametes do differ in the nature of their chemical secretions. It is these latter which direct the chemotactic movements which result in the pairing of the gametes.

In other species, for example, *C. braunii* (Fig. 232), the difference between the gametes is not only physiological but there is also a difference in size between large female and small male gametes, though both are still motile. In this species the gametes have a cell-wall. When pairing occurs the contents of the male gamete pass into the female one and fuse to form the zygospore. This is the condition known as anisogamy or heterogamy and the fusion may be spoken of as fertilisation, in contradistinction to the fusion of isogametes which is called conjugation.

PLEUROCOCCUS

Pleurococcus, though in many respects simpler than *Chlamydomonas*, may be considered a true plant, whereas *Chlamydomonas* is more between the plant and animal kingdoms.

The structure of this unicellular plant may be seen in Fig. 233. One species of this plant, *Pleurococcus naegelii* (*Protococcus viridis*), is probably the commonest green alga in the world. It is found in all temperate parts of the world, growing in any damp place, such as on wooden fences, tree trunks, etc.

The cell-wall is rather thick. The protoplasm is composed of cytoplasm in which a nucleus is embedded. There is a single, lobed chloroplast, which lines the cell-wall.

There is only one method of reproduction in *Pleurococcus*,

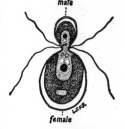

male

female

Fig. 232. Heterogamous Fertilisation in *Chlamydomonas braunii*.

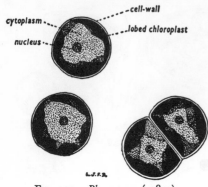

FIG. 233.   *Pleurococcus* (×840).
The specimen in the bottom right-hand is just completing division.

namely, fission. This takes place rather quickly, so that often colonies of cells may be produced, loosely joined together sometimes in three planes.

There are no sexual forms of reproduction and no motile or resting stages.

SPIROGYRA

The two types of algae so far considered are unicellular ; but there are many multicellular forms in this division of plants. The simplest multicellular form is that composed of single chains of cells, forming unbranched filaments.

One common, unbranched filamentous green alga is *Spirogyra* (Fig. 3). But the multicellular nature of this plant must be considered as very primitive, since every cell is exactly alike (with the exception that in a few species the filaments become attached to a stone, etc., by means of rhizoids borne on a colourless cell called a **holdfast**). So there is little division of labour among the constituent cells.

Spirogyra is common in stagnant pools and ponds and slow-moving rivers, and is, among other algae, responsible for the familiar green scum on their surfaces. The filaments are surrounded by a layer of mucilage which gives the plants their characteristic slimy nature.

The plant is composed of a filament of cylindrical cells arranged end on end (Fig. 234). Each cell is surrounded by a cell-wall. The cell itself is composed of cytoplasm, usually with a large central vacuole in which is suspended a nucleus. The chloroplast, which lies in the cytoplasmic lining, is very characteristic. It is ribbon-shaped with serrated margins, and lines the cell-wall in a spiral fashion. In some species there are several such chloroplasts in one cell. At intervals along the chloroplast are pyrenoids which form seats for the deposition of starch.

Growth in length of the filaments takes place by the elongation of each individual cell and by the production of new cells by mitosis. Any cell of the filament is capable of division, so that elongation takes place at any spot along the length of the filament (**intercalary growth**), not solely at the end (**apical growth**).

There are two forms of reproduction in *Spirogyra*. One is purely vegetative in that single cells or a collection of cells can become detached and grow to form a new filament. This takes place mainly during the spring and is known as fragmentation. The other is by means of conjugation. In some cases, two different filaments lie side by side and conjugate. This is known as **scalariform** or **ladder** conjugation. Sometimes, however, cells of the same filament may conjugate, in which case it is known as **lateral** or **chain** conjugation. Conjugation in *Spirogyra* can be artificially induced by giving a liberal supply of oxygen or by nitrogen deficiency. Asexual reproduction is unknown.

In scalariform conjugation, two filaments come to lie side by side as a result of slow movements and are then firmly fixed together by their mucilage. Papillae grow out from opposite cells and soon meet, their ends then becoming flattened. The papillae continue to elongate and so the two filaments are gradually pushed

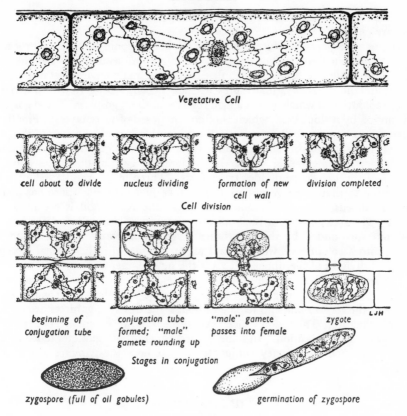

Vegetative Cell

cell about to divide   nucleus dividing   formation of new   division completed
                                            cell wall
                        Cell division

beginning of   conjugation tube   "male" gamete   zygote
conjugation tube   formed; "male"   passes into female
                    gamete rounding up
                        Stages in conjugation

zygospore (full of oil gobules)   germination of zygospore

FIG. 234. *Spirogyra.*

apart. The end wall separating the two outgrowths dissolves, thus forming a conjugation canal, in which the protoplasts of the two cells, which have extended into the outgrowths throughout their development, are now in direct contact with each other. The contents in each of the cells of one of the filaments now contract from their walls (as the result of water being removed from the central vacuole by contractile vacuoles in the cytoplasm) and form what are recognised as the male gametes. The contents of the cells of the opposite filament (which do not contract at this stage except in a few large species) constitute the female gametes. The male gametes now move through the conjugation tubes and fuse with the corresponding female gametes which now also show contraction from their walls. The fused mass in each cell is a young zygospore, the further development of which will be dealt with later. The general result of this process is that all the cells of one of the paired filaments (the male) are now empty, while those of the other one, the female, each contains a zygospore.

The process of lateral conjugation is essentially similar although only a single filament is involved. An open connexion is established between adjacent cells of the filament by means of a modified conjugation tube. The contracted contents of one cell, the male gamete, pass through the opening into the female gamete in the neighbouring cell and so form a zygospore. When this conjugation has taken place the filament shows some cells containing zygospores while cells adjacent to these are empty.

Young zygospores, produced by either method of conjugation, now proceed to ripen. Continued contraction leads to the zygospore assuming a definite ovoid form. The male and female nuclei fuse, while the male chloroplast degenerates. The abundant starch originally present soon becomes converted into a red- or orange-coloured oil which acts as a food reserve for the zygospore. The zygospore is surrounded by a thick wall, which is often composed of three layers. Finally it is released by a rupture in the mother cell-wall, and falls to the bottom of the pond, where it rests until favourable conditions arise. Then it germinates to produce a new *Spirogyra* filament (Fig. 234). Just before this germination takes place, the nucleus of the zygospore divides by meiosis. Three of the resultant four nuclei degenerate so that when germination does take place the zygospore has a single haploid nucleus. As germination commences the reserve oil is reconverted to starch. The outer layers of the wall are ruptured at one end and the contents grow out surrounded by the thin inner layer. This forms a tube which soon divides into a colourless lower cell and an upper one with the normal structure. Repeated division of the upper cell gives the new filament.

The sexual method of reproduction just described is extremely characteristic. The fact that some of the gametes are active while the others are passive justifies the use of the terms male and female although the different gametes are so much alike in structure. The fusion which takes place is, however, spoken of as conjugation and not as fertilisation. In some related genera, both gametes move into the conjugation tube where they conjugate so that here the gametes are true isogametes.

OEDOGONIUM

Another common, unbranched, filamentous alga is *Oedogonium*, several species of which are to be found in fresh, though seldom running, water, usually attached to stones, etc., or to the stems and leaves of more advanced aquatic plants. The basal cell, by means of which the plant is attached, forms a disk of finger-like processes which act as a **holdfast**.

Oedogonium may be looked upon as being more advanced than *Spirogyra*, since it demonstrates a true division of labour among its constituent cells. Whereas any cell in *Spirogyra* is capable of producing a gamete, in *Oedogonium* the gametes are produced in specialised cells known as **gametangia**. Thus, the plant is composed of two types of cells, namely, those which are able to produce gametes, commonly known as **germ cells**, and those which are purely vegetative. These latter constitute the body part of the plant or **soma**, and are therefore said to be **somatic**. This distinction between soma and germ cells is found in all the higher plants and animals.

The vegetative cell of *Oedogonium* contains a single nucleus, which is often hidden by the single large chloroplast. The latter is in the form of a cylindrical reticulum or network which is embedded in the cytoplasm lining the cellulose wall.

Growth of the filament takes place by repeated divisions in certain cells, which are usually intercalary in position and not terminal. The division of these cells is peculiar, particularly as regards wall formation. When a division takes place the lower daughter cell remains surrounded by the old wall. The wall of the upper cell is, however, largely formed by the pulling out of an invaginated ring of cellulose which had developed near the tip of the parent cell. The upper end of the wall is the bit of old wall which was above the ring and this forms a 'cap'. Division may be repeated in this upper, so-called, 'cap' cell and each time it divides a new cap is formed. (See Fig. 235 where a zoospore has been formed in such a cell.)

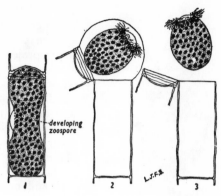

developing zoospore

FIG. 235. STAGES IN THE PRODUCTION OF A ZOOSPORE IN *Oedogonium*.

Asexual reproduction is by means of **zoospores**, or motile spores. Usually only certain cells are capable of producing zoospores. The contents of such a cell become more or less spherical, and develop a ring of cilia around the base of the colourless beak of the otherwise deep green zoospore. The zoospore thus formed is liberated by a transverse slit which appears in the cell-wall (Fig. 235). On liberation it has no cell-wall of its own. It is also sensitive to light by virtue of a small eye-spot. Then it swims about for a time (about one hour), and finally rests, withdraws its cilia, secretes a cellulose cell-wall, and by normal cell division produces a new *Oedogonium* filament.

Oedogonium also reproduces itself sexually, and since it produces large, non-motile ova and small, motile spermatozoids it is oogamous.

The spermatozoids or antherozoids are produced in cells called **antheridia**. These are disk-shaped cells which are produced in chains by the repeated division of a single cell (Fig. 236). The contents of each antheridium divide once, thus producing two sperms which are like small zoospores. These develop a ring of cilia, and are then released from the mother cell by a transverse slit. Thus they are able to swim about actively in the water. The cell which produces the egg is called the **oogonium**. This is the upper daughter cell produced by a cell division of the usual type and it thus shows a 'cap' at its upper end (Fig. 236). The lower daughter cell is termed the 'support cell'. The oogonium produces only one egg, which is formed merely by the rounding-off of the cell contents. But the egg remains passive and is not liberated from the oogonium.

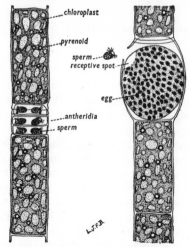

chloroplast

pyrenoid

sperm
receptive spot

egg

antheridia
sperm

The oogonium opens near one end by means of a transverse slit, and near this slit, the surface of the egg becomes devoid of chlorophyll. A spermatozoid then swims through the slit and fuses with the egg at this colourless spot, which is therefore known as the **receptive spot**. The pro-

FIG. 236. STAGES IN SEXUAL REPRODUCTION IN *Oedogonium*.

Left, production of spermatozoids; right, fertilisation about to occur.

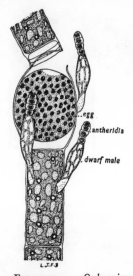

FIG. 237.   *Oedogonium*
ciliatum, SHOWING DWARF
MALES.

duct of this fertilisation process is known as an **oospore**.

The oospore develops a thick cell-wall, and the starch present becomes converted into oil and so acts as a more efficient food reserve. The presence of this oil gives a general red coloration to the oospore. Finally, the oospore is liberated from the oogonium, and after a period of rest, germinates. Its contents divide by meiosis to form four zoospores which, when they are liberated, settle down and grow into new filaments.

Some species of *Oedogonium* produce both antheridia and oogonia on the same plant (monoecious), whereas others produce male on one plant and female on another (dioecious).

In other species of this plant, there are special filaments for the production of male gametes. They are much smaller than the female filaments, and are usually made up of about four cells. Thus they are referred to as **dwarf males** and the species possessing them are said to be **nannandrous** (*nanus*, Latin, dwarf; *aner*, Greek, man). They are formed in the following manner. Certain cells of a filament divide several times transversely to give a chain of disk-shaped cells, in each of which a single zoospore is formed. This is called an **androspore,** and it is intermediate in size between an asexual zoospore and a sexual spermatozoid (antherozoid). The androspore swims about until it comes in contact with either an oogonium or a cell supporting one. Here it attaches itself and eventually germinates and developes to produce a dwarf male. Apart from the lowest cell of the dwarf male, which acts as a means of holding on to the female filament, all the cells are antheridia, two sperms being produced in each (Fig. 237).

VAUCHERIA

Vaucheria resembles *Oedogonium* in that it reproduces asexually by zoospores and sexually by oogonia and antheridia. It is, however, completely different in its vegetative structure.

There are numerous species which live in a variety of habitats. Many forms live in fresh water. Others are terrestrial, though they must live in very moist surroundings. A few species live on estuarine mud or even in sea water.

The filament of *Vaucheria* differs from that of *Oedogonium* in that, whereas the latter is actually a chain of single cells, thus having cross-walls or septa, the former has no cross-walls and is therefore said to be **non-septate**. A plant of *Vaucheria* consists of a richly branched system of such non-septate tubes, each branch continuing to grow at its tip. Most of the branches are aerial and green, but there are some narrower branches which are colourless and serve to attach the plant to the soil or other substratum ; these latter are called rhizoids (Fig. 238). An examination of one of the green tubes will show its characteristic structure. It is surrounded by a delicate wall of cellulose and pectic substances. Inside this, there

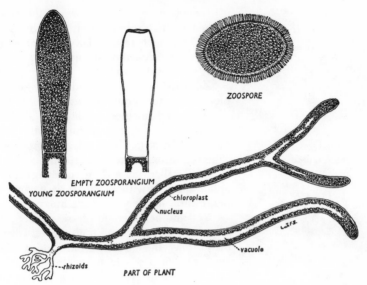

FIG. 238. VEGETATIVE STRUCTURE OF, AND ASEXUAL REPRODUCTION IN,
Vaucheria (× 120).

is a lining of cytoplasm, in the inner layers of which numerous minute nuclei are embedded. The centre of the tube is occupied by a large vacuole containing sap. The green colour of the aerial branches is due to the presence of numerous discoid chloroplasts which are embedded in the cytoplasm just inside the wall. The product of photosynthesis is not starch but oil, which appears scattered throughout the cytoplasm as minute globules.

This unusual structure is best interpreted as a collection of naked cells (each consisting of one of the nuclei and its surrounding cytoplasm), surrounded by a common wall and having a common vacuole. Such a non-septate structure is termed a **coenocyte**. It functions physiologically like a single cell in so far as water enters by the osmotic attraction of the sap in the common vacuole, the cytoplasmic lining being a semi-permeable membrane. The rigidity of the tubes depends entirely on their turgor ; but in some larger coenocytic genera related to *Vaucheria*, additional support is given by ' skeletal ' bars of wall material extending across the tube.

Vaucheria reproduces itself asexually by means of zoospores. The end of a filament becomes club-shaped and numerous nuclei and chloroplasts flow into it so that the contents become dark green. A cross-wall is then formed (Fig. 238) and the contents of the terminal portion contract to form an ovoid mass. This latter consists of a peripheral layer of cytoplasm surrounding a central vacuole. The numerous nuclei, previously more deeply seated, come to lie near to the surface and opposite to each of them a pair of flagella is protruded. Thus a compound zoospore is formed in the terminal part of the filament or **zoosporangium** (Fig. 238). The zoospore is released from the zoosporangium by the rupturing of the end of the latter. By means of its cilia, the zoospore is able to swim about for a short time ; but finally it comes to rest, withdraws its cilia and secretes a thin cell-wall.

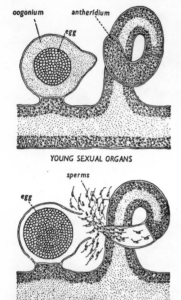

YOUNG SEXUAL ORGANS

FERTILISATION

FIG. 239. SEXUAL REPRODUCTION IN *Vaucheria* (× 300).

Germination takes the form of the production of one or two tubes which are similar to the normal filaments of the plant. Branches of these develop as colourless rhizoids fixing the young plant to the substratum.

Zoospores are only produced when there is plenty of water about, and usually only at night.

A second method of asexual reproduction sometimes occurs, but mainly in the terrestrial forms of *Vaucheria*. This is by means of what are called **aplanospores**. An aplanospore is formed in a cell called an **aplanosporangium**. Each aplanospore is without cilia ; it escapes from a perforation in the wall of the aplanosporangium.

Yet another method of asexual reproduction is resorted to when conditions are very dry. Filaments may then become partitioned into short segments, each of which becomes a thick-walled cyst capable of growing into a new plant when conditions again become more favourable.

Sexual reproduction in *Vaucheria* is effected by the production of eggs and sperms in specialised oogonia and antheridia, respectively. Both these organs are produced from side branches of the filament which develop specially for the purpose (Fig. 239). They usually arise close together at intervals along the main filament (for example, *V. sessilis*), but in other species (for example, *V. geminata*) special lateral branches bear a terminal antheridium and two or more lateral oogonia.

The antheridium is long and curved. It contains many nuclei. Then it is separated from the filament by a cross-wall. Each nucleus gradually separates and surrounds itself with some cytoplasm. Each thus produces a pear-shaped sperm. The sperm then produces two cilia, laterally, though each cilium points in opposite directions.

The young oogonium is more or less spherical, and is also multi-nucleate and packed with oil globules and chloroplasts ; but, before the cross-wall which segregates it from the filament is formed, all the nuclei except one pass back to the parent filament. A part of the oogonium wall protrudes somewhat, and this protrusion becomes perforated to allow a sperm to enter. Meanwhile, the rounded-off, uninucleate contents form a colourless **receptive spot** opposite the perforation and contracts still further to form what is called the **oosphere**, which is really the egg. Finally, the sperms are ejected from the antheridium, and though several sperms might enter the oogonium only one succeeds in penetrating the oosphere to fertilise it.

After fertilisation, the zygote forms a thick wall and rests for several months as what is called an **oospore**. Finally it germinates to produce a new *Vaucheria* filament. The single nucleus of the oospore is of course diploid as a result of fertilisation but when germination takes place, the first division of the nucleus is **meiotic** and so the new filament, like the old one, is haploid.

FUCUS

All the algae so far considered have been examples of green algae. In this book, consideration cannot be given to any form of red or blue-green algae ; but a very common and familiar brown alga is the seaweed, *Fucus*, which is therefore worthy of study.

All seaweeds are multicellular plants, and some are of extremely large dimensions. They all show a marked division of labour in their tissues. Two common types around British shores are *Fucus serratus*, the serrated wrack (Figs. 6 and 240) and *Fucus vesiculosus*, the bladder wrack (Fig. 240).

The whole **thallus** sometimes attains a length of several feet, and is composed of three outstanding parts : (1) the **holdfast**, with which the plant clings to rocks, piles of piers, etc. ; (2) the stem-like portion, known as the **stipe** ; (3) the flattened, branched portion, known as the **frond** or **lamina**. Each branch of the frond has a midrib, which is formed merely by the local thickening of the tissues. The surface of the plant is slimy to the touch owing to the presence of mucilage secreted by hairs growing in small, flask-shaped depressions (**sterile conceptacles**). The presence of these is indicated by minute, warty prominences on the surface of the frond. Branching of the frond during growth takes the simple and regular form of division of the tips into two, thus forming a pair of equal branches growing out at an angle. Later the tips of these branches split into two, and so on (**dichotomous branching**) (Fig. 241). The growth of each branch of the frond is brought about by the continued division of a large apical cell, which has the form of a truncated pyramid. This is situated at the base of a mucilage-filled, slit-like depression at the tip of the branch. It is by the division of the apical cell into two parts, each half then acting as an apical cell, that the branching just described is brought about. Segments are cut off from the sides and base of the apical cell in a very regular manner and then these, by further subdivision, give rise to the various tissues.

In *Fucus vesiculosus*, many bladders are present, which give the plant a certain buoyancy and thus prevent it from lying recumbent on the sea-floor or rocks, when submerged. Possibly they also help in respiration.

The tissues of the plant body are not highly differentiated but even so, it is possible to distinguish photosynthetic, storage and conducting regions. The anatomy of the plant is illustrated in Fig. 241.

The frond is bounded on the outside by a **limiting layer** of small, palisade-like cells, the outer walls of which are covered by a delicate, cuticle-like layer. The cells contain numerous small plastids, or chromatophores, which contain the

FIG. 240. *Fucus vesiculosus* (LEFT) AND *F. serratus* (RIGHT).

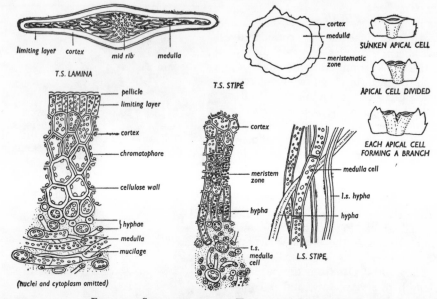

FIG. 241. STRUCTURE OF THE THALLUS OF *Fucus serratus*.
(*F. W. Williams*)

brown pigment fucoxanthin in addition to chlorophyll. Below this limiting layer there are several layers of polygonal parenchymatous cells which also contain plastids. These form the **cortex**, which may act as storage tissue for the products of photosynthesis, such as laminarin (a polysaccharide), mannitol and fats. The inner cortical cells usually show a gelatinisation of their middle lamellae and these cells merge into a central region, most strongly developed in the mid-rib, known as the **medulla**. The medullary cells are drawn out in a longitudinal direction and form filaments of cells, or **primary hyphae**. These have thin cellulose walls and are interwoven with one another. They are embedded in a mucilaginous matrix derived from the gelatinisation of the middle lamellae between the filaments. These hyphae probably serve as conducting elements and in some related seaweeds some of them have a structure very similar to that of sieve tubes.

The structure just described relates to a young part of the frond, not far below the growing point. Secondary thickening occurs in the older parts further back from the tip. The cells of the limiting layer add on to the cortex by divisions parallel to the surface and, for this reason, this layer is sometimes called the **meristoderm**. The mid-rib is strengthened by narrow, thick-walled, **secondary hyphae**, which arise from the inner cortical cells and grow downwards between the primary cells of the medulla. In the lower parts of the frond the mid-rib contains masses of these secondary hyphae, but they develop to a much more limited extent in the wings.

The structure of the stipe is not essentially different from that of the frond and it is, in fact, the mid-rib region of the oldest parts of the plant from which the wings have been worn away. In section, it is seen that the limiting layer has been lost and that the remaining tissues consist of a central medulla surrounded by cortex. The medulla contains large numbers of secondary, thick-walled hyphae

and similar hyphae may penetrate
to the outside to form an external
covering of the cortex.   The cortex
itself is similar to that of the frond ;
but, as the stipe becomes older, a
zone of its inner cells becomes meristematic and, acting like a cambium, adds to its thickness.

The holdfast also increases in
size throughout the life of the plant
and finally consists very largely of
a mass of thick-walled hyphae.

The only special form of reproduction in *Fucus* is sexual (though
vegetative reproduction takes place
by fragmentation, often by the separation of little adventitious shoots
produced as a result of injury).
Some of the branch tips are swollen
and present a dotted appearance.
Each dot indicates the position of a
flask-shaped cavity known as a **conceptacle**, which opens through the
surface of the frond by means of a
pore known as the **ostiole**.   From
the inner surface of the conceptacle numerous multicellular, unbranched hairs, known as **paraphyses**, arise, some of which grow
out through the ostiole (Fig.
242).

The sperms and eggs of *Fucus*

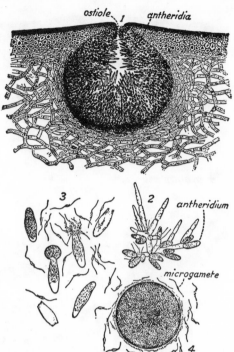

FIG. 242.   SEXUAL REPRODUCTION IN *Fucus*.

1, section through male conceptacle (× 30) ; 2,
hairs with antheridia (× 100) ; 3, sperms escaping
from antheridia (× 225) ; 4, egg surrounded by
sperm (× 225).

(*After Thuret ; from Kerner's ' Natural History of
Plants ', Blackie & Son, Ltd.*)

are produced in antheridia and oogonia respectively.   In *Fucus serratus* and *Fucus
vesiculosus*, these organs are produced in separate conceptacles on different individuals, so that there are male and female plants.   In another related species, *Fucus
spiralis*, antheridia and oogonia are produced in the same conceptacle, which is
therefore hermaphrodite.

In the male conceptacle, antheridia are produced on much-branched hairs
(Fig. 242).   Each antheridium is ovoid, and at first possesses only one nucleus.
But this nucleus soon divides to produce usually sixty-four nuclei, each
one of which surrounds itself with a small amount of cytoplasm, develops
two lateral cilia unequal in length, and thus forms a sperm.   Each sperm
also possesses a small bright orange colour-spot.   The antheridial wall is two-
layered.

The oogonia are associated with unbranched paraphyses.   They are developed
from cells on the lining wall of the conceptacle (Fig. 243).   Such a cell swells, and
divides transversely once.   The lower cell is the **stalk cell**.   The upper cell is the
oogonial cell, and its protoplasmic contents become very dense.   Its nucleus
then divides three times to produce eight nuclei.   Then each nucleus surrounds

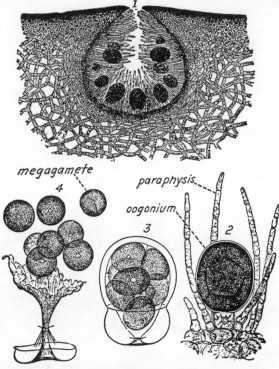

FIG. 243. SEXUAL REPRODUCTION IN *Fucus*.

1, section through female conceptacle (× 30) ; 2, an oogonium containing eggs (× 120) ; 3 and 4, oogonial wall breaking and releasing eight eggs.

*(After Thuret ; from Kerner's ' Natural History of Plants ',
Blackie & Son, Ltd.)*

itself with some of the cytoplasm, and becomes rounded off to form an egg. The oogonium wall is composed of three layers.

At low tide, the fronds of *Fucus* are subjected to a certain amount of desiccation with a consequent shrinking of tissues. This causes a pressure to be exerted on the conceptacles which contain mucilage secreted by the sterile hairs. In the case of the male conceptacles, this pressure results in the extrusion on to the surface of the frond of mucilage containing numerous packets of sperms, each of which has been liberated by the bursting of the outer wall layer of the antheridium. In the case of the female conceptacle, mucilage containing packets of eight ova (each still surrounded by the two inner layers of the oogonial wall) oozes out on to the frond surface. The mucilage extruded from the male conceptacles is coloured bright orange by the presence of millions of sperms; that extruded from the female conceptacles is olive green due to the presence of the dark green ova. When the plants are again submerged as the tide returns, the wall of each sperm packet bursts and liberates the sperms into the water. The eggs are also released, the two layers of the oogonial wall rupturing in succession, as shown in Fig. 243. Fertilisation now takes place. The sperms are attracted to the non-motile ova by a chemical secretion (see Chap. XX). A large number of sperms may swarm round an egg and each becomes attached to the latter by its forwardly pointing cilium while still lashing vigorously with its longer free cilium. The result is that the large egg is rapidly rotated. Finally, one sperm enters the egg which immediately forms a gel membrane round itself. Fertilisation having been achieved, the other sperms now fall away (Fig. 262).

The fertilised egg surrounds itself with a cell-wall and then germinates immediately to produce a new *Fucus* plant. Repeated divisions soon give a spherical body which becomes attached to a rock by means of a rhizoid. An apical cell is soon established and, by its activity, the young plant becomes first club-shaped and then, with increasing size, the terminal region becomes flattened. The holdfast increases in size, the flattened region branches repeatedly, and the main features of the adult plant are thus established.

Unlike the other algae described in this chapter, the plant of *Fucus* is diploid. Meiosis occurs in the first division in the developing antheridia and oogonia, so that both the spermatozoids and the ova are haploid. The zygote formed by fertilisation is thus diploid and it germinates to form the diploid plant.

GENERAL SURVEY

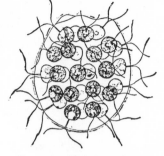

FIG. 244. *Eudorina unicocca.*
(From ' Cryptogamic Botany ', by G. M. Smith, by permission of McGraw Hill Book Co., Inc.)

The living algae are generally regarded as plants which have retained the relatively simple, but variable, structure and life-history of the earliest true plants. Even so, the elaboration of their structure and reproductive methods indicates that they must be regarded as showing a very great advance on the original living units.

It is generally accepted that the algae must represent the source from which the higher groups of plants have evolved ; but it should be realised that there is no direct evidence that this is so. The variability in structure and life-history shown by the algae does, however, give some indication of possible ways in which multicellular plants have evolved from unicellular ones. It also indicates ways in which the more uniform methods of sexual reproduction and the more standardised life-cycles of the higher groups may have been established.

One way of achieving a multicellular construction is illustrated by certain so-called ' colonial ' algae related to *Chlamydomonas*. In these, a number of cells, each similar to a *Chlamydomonas* individual, are joined together by their swollen gelatinous walls and their activities co-ordinated by means of communicating cytoplasmic strands. Such associations of chlamydomonad cells are known as colonies or **coenobia**, and these may be of varying form and complexity. For example, *Gonium* consists of sixteen chlamydomonad cells forming a flat plate ; *Eudorina* has a spherical coenobium of 16–32 cells (Fig. 244) ; while the well-known *Volvox* is a hollow sphere formed by up to 20,000 cells (Fig. 245). These colonies are motile, each cell having a pair of flagella, but in other genera the colonies are non-motile. The free-floating vegetable ' plankton ' of ponds, lakes and rivers often contains large numbers of these colonial types.

Another method of achieving a larger plant body is exemplified by the coenocyte of *Vaucheria*. This type appears to have arisen from a non-motile unicellular form by enlargement accompanied by nuclear division but without the formation of walls. *Protosiphon* illustrates an early stage in this type of development. The plant has a spherical green coenocyte, measuring only 100μ in diameter, which is fixed to the mud on which it grows by a single colourless rhizoid. Further enlargement and branching of such a plant would result in a *Vaucheria*-like structure.

Yet another way of achieving larger size and greater complexity is exemplified by the filaments of *Spirogyra* and *Oedogonium*. The origin of a filament from a non-motile unicellular form involves repeated cell divisions accompanied by wall formation, the walls always being at right angles to the long axis of the filament. Further elaboration of the plant body has taken place in other algae by the branch-

B.I.B.

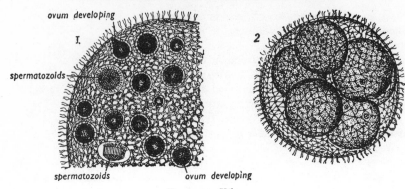

FIG. 245.   *Volvox.*
1, Part of a colony showing the formation of gametes ; 2, a colony with daughter colonies
inside it.
(*After Klein.*)

ing of the filaments, by the differentiation of basal anchoring filaments from erect photosynthetic ones, and, as in many seaweeds, by associating together a number of filaments in various ways.

Such filaments may be further elaborated by the occurrence of longitudinal as well as transverse divisions, thus giving a cell mass or parenchymatous type of plant body. This type is characteristic of many of the larger seaweeds. In some examples, the thallus is uniformly parenchymatous, while in others, for example, *Fucus* and *Laminaria*, there is considerable differentiation of tissues for special purposes.

The colonial and coenocytic types of construction are not found in any of the higher groups of plants. Very large coenobia or coenocytes would indeed be mechanically unsound and, moreover, neither type gives the possibility of any high degree of differentiation of the plant body. It seems likely that the filamentous type of construction, and particularly the parenchymatous developments of this, give an indication of the way in which the plant body of the higher groups was evolved.

The algae dealt with in this chapter also throw some light on the way in which sexual reproduction has evolved in the plant kingdom. The essential feature of such reproduction is the fusion of two nuclei which are often derived from two different individuals. Since meiosis always follows sooner or later after fusion, the progeny resulting from the fusion are different, that is they show variation, from both the parents. Such variations may be advantageous in the struggle for existence (see Chap. XXVIII). The nuclei concerned are those of gametes. Some species of *Chlamydomonas* produce isogametes, differentiated only into plus and minus types by the nature of their chemical secretions, while other species have anisogametes (or heterogametes) in the form of larger female gametes and smaller, more active male ones. In *Oedogonium*, *Vaucheria*, and *Fucus*, the distinction between female and male gametes is even more marked. The female gamete is a passive ovum, usually formed in an oogonium, while the male has retained its motility as a spermatozoid, usually produced in an antheridium. This condition of oogamy is found in all the higher groups, so that it may be inferred that it offers biological advantages over the condition where both gametes are motile.

The general life-cycle in the algae is extremely variable. In all the green algae described in this chapter, the plants are haploid and produce gametes which, on fusion, give diploid zygotes. Meiosis takes place when the zygote germinates, so that the new individuals arising from the zygospores or oospores are again haploid. The life-history of the brown seaweeds follows a number of different patterns. The *Fucus* plant is diploid and meiosis takes place in the formation of the gametes. When fertilisation takes place, the oospore is diploid and gives rise to the new diploid plant. In *Laminaria*, on the other hand, the large seaweed is diploid, but instead of producing gametes as *Fucus* does, it produces sheets of club-shaped sporangia on its surface and in each of these a large number of zoospores is produced. As meiosis takes place in the formation of the zoospores these structures are haploid. When they germinate they grow into minute filamentous plants consisting of only a few cells. Some of the filaments produce oogonia, others antheridia, and fertilisation results in oospores which grow directly into the big diploid *Laminaria* plants. There is thus a regular sequence in the life-history of a large spore-bearing plant, the diploid **sporophyte** generation, with a gamete-bearing generation, the haploid **gametophyte** generation. It is this type of **alternation of generations** which underlies the life-cycle of the higher groups of plants, as will be described in Chapter XXVI.

ECONOMIC USES OF ALGAE

The economic importance of algae is considerable, particularly that of the brown and red seaweeds. Such seaweeds have, in fact, been used since very early times as fodder for sheep and cattle, and for manuring purposes (adding humus and mineral salts to the soil). Some seaweeds, like dulse and carrageen, are occasionally used as food in Great Britain and the Republic of Ireland, and the Chinese and Japanese use a variety of seaweeds in the same way. In the past, brown seaweeds have been extensively used as a source of potash, sodium carbonate and iodine. It is, however, the organic constituents which provide the most important substances.

Agar, an important jelly-like medium used for the culture of fungi and bacteria, is derived from a number of red seaweeds. Before the Second World War, Japan provided practically the whole of the world's agar, but now the U.S.A., Australia and New Zealand all produce it. In Britain, agar is now produced from two red seaweeds, *Gigartina stellata* and *Chondrus crispus*, the pioneer work having been carried out at Millport by **Dr. Marshall** and **Dr. Orr** and at Aberystwyth by **Professor L. Newton.**

The most important seaweed product is, however, **alginic acid**. This is extracted from *Laminaria* and *Fucus* species with dilute soda solutions to give a viscous solution of sodium alginate. This is the basis of other alginates used for a wide range of purposes, such as the production of transparent, 'Cellophane'-like wrapping material, textile fibres, protective films to cover wounds and burns, and protective colloids used in the preparation of custards, ice-cream, sauces and jellies.

PRACTICAL WORK

1. Examine some living specimens of *Chlamydomonas*, and note their tactic movements. Irrigate with iodine and make fuller observations on the structure of the cell. Examine a fixed specimen and make a high-power drawing of it.

2. Examine a piece of damp wood with *Pleurococcus* growing on it. Scrape a little of the alga on to a slide, and irrigate with water and cover. Examine under the low power and look for colonies of cells.

Then study the structure of a cell under the high power.

3. Mount some filaments of *Spirogyra*, and examine under the low power. Note the chains of cells. Then study one cell under the high power and draw it, showing the detailed structure.

Examine some prepared slides of filaments of *Spirogyra*, and look for, and draw, various stages in conjugation.

4. Mount some fresh filaments of *Oedogonium*. After examining them under the low power, examine and draw a cell in detail under the high power. Look also for oogonia and antheridia.

Examine a prepared slide showing dwarf males.

5. Mount some fresh filaments of *Vaucheria*. Make a drawing to show the method of branching. Examine a portion of a filament under the high power.

Examine a prepared slide showing antheridia and oogonia.

6. Study the external features of a specimen of *Fucus*. Note the holdfast, stipe and fronds. If the specimen is *Fucus vesiculosus*, note also the bladders which give buoyancy.

Cut a transverse section across a vegetative branch, and mount and examine under the low power. Note the differentiation into limiting layer, cortex and medulla. Make a drawing under the high power.

Examine prepared slides of transverse sections cut through male and female conceptacles. Make orientation sketches of each, then enlarged drawings of antheridia, oogonia and paraphyses.

CHAPTER XXIV

IRREGULAR NUTRITION AND DISEASE: FUNGI

The majority of plants are self-supporting, or **autotrophic**, their so-called ' regular ' nutrition being based on the synthesis of organic materials from inorganic sources. There are, however, many plants which obtain all or part of their food materials in an ' irregular ' manner. These are said to be **heterotrophic**.

Many plants obtain their foodstuffs by growing on or in other living organisms, frequently making close contact with the tissues of the latter by means of suckers or haustoria. Such plants are said to be **parasites** and the living plant or animal from which they absorb foodstuffs is the **host**. Some plants, devoid of chlorophyll, are entirely dependent on the host and are **complete parasites**. Others, possessing chlorophyll, are dependent on the host for only a part of their food, usually water and mineral salts ; these are **partial parasites**. Completely parasitic plants, such as many of the bacteria and fungi, with which this chapter is largely concerned, are of great importance in agriculture, horticulture, veterinary science and medicine, since such parasitism often leads to a diseased condition of the host or even to its death. The removal of foodstuffs leads to a weakening of the host ; the secretion of toxic substances by the parasite may lead to abnormal growth and malformations of the organs of the host or to the death of the infected tissues. Death of the host may follow, particularly when the parasite is one of the less-specialised types.

Other heterotrophic plants obtain their organic foodstuffs not from a living host but from dead organic matter, such as the bodies of dead plants or animals, humus in the soil, timber, jam, cheese, leather and so on. Such plants are **saprophytes**. Some examples, devoid of chlorophyll, are **complete saprophytes**, obtaining all their food materials from the substratum upon which they are growing. Others, possessing chlorophyll, are **partial saprophytes**.

Although the bacteria and fungi offer the most numerous, and economically important, examples of parasites and saprophytes, heterotrophic methods of nutrition are also seen in some of the higher plants, particularly in the Angiosperms. Some examples of these, the mycorrhizal plants and those with root nodules, have already been described. In both of these types there appears to be a condition of symbiosis rather than of parasitism. The carnivorous plants (Chap. XVIII) are also heterotrophic in their nutrition, obtaining nitrogenous substances from the bodies of insects and so being partial saprophytes. Further examples of flowering-plant parasites and saprophytes are described at the end of this chapter.

BACTERIA

The great group of bacteria contains many parasitic types. It must *not* be imagined, however, that because many bacteria are parasitic therefore they *all* are. This is far from being the case. *Quite a large number of bacteria are free-living and harmless ; others are free-living and distinctly useful, especially from man's point of view*; for example, the nitrogen-fixing bacteria which were considered in Chapter XII. Other bacteria are used in the disposal of sewage, and also in various industries such as fermentation processes, the curing of tobacco, the retting of flax, and for numerous other industrial purposes. *Man depends on the useful bacteria almost as much as he is crippled by the harmful types.* In fact, harmful bacteria are in the minority.

All bacteria are unicellular plants. They vary greatly in shape but three principal types are recognised (Fig. 5): (1) the spherical or *coccus* form, (2) the rod-shaped form known as the *bacillus* or *bacterium*, and (3) the spiral forms, each of which might be a *vibrio, spirillum* or *spirochaete*. All of these may be further subdivided, for example, *streptococcus* or *staphylococcus*, depending upon whether the cocci occur in chains as in the former, or in packets as in the latter.

The average bacterium is about 2μ in length by 0.5μ in diameter. The largest, as a rule, belong to the spiral forms; for example, the spirochaete of relapsing fever may measure up to 40μ in length. One of the smallest is the so-called ' influenza bacillus ', which measures $0.5\mu \times 0.2\mu$.

The cell of a bacterium is a mass of cytoplasm, with particles of chromatin embedded in it; that is, there is no well-defined nucleus. Some modern workers, however, maintain that there is a definite nucleus. The cell is surrounded by a cell-wall, the outer layer of which is sometimes gelatinous and thus protects the organism from desiccation and sometimes also helps to hold the cells together in chains. The wall is seldom composed of cellulose; but is a mixture of polysaccharide with some fatty substances and proteins and sometimes a certain amount of chitin. Some bacteria possess one or many flagella which appear to be organs of locomotion.

Multiplication is by binary fission, each cell dividing into two. This may happen every hour so that in twenty-four hours about 17 million individual cells may be produced from a single one. Some bacteria produce resistant structures known as spores, which may be produced terminally (for example, tetanus bacillus) or centrally in the cell (for example, the anthrax bacillus). Such spores are able to withstand very adverse conditions, being surrounded by a very impervious wall. For example, the anthrax bacillus (*Bacillus anthracis*) can withstand boiling water for more than an hour. Normally only one spore is produced in each cell so that spore production does not lead to multiplication. Eventually the spore is released by the disintegration of the mother cell-wall.

The metabolism of the bacteria is very varied. The vast majority, being without chlorophyll, are either parasites or saprophytes. The parasitic forms causing human disease are perhaps the best known. By attacking certain tissues in the human body, and especially the blood, they are the cause of many well-known diseases, for example: erysipelas (*Streptococcus erysipelatis*), tuberculosis (*Mycobacterium tuberculosis*), lock-jaw (*Clostridium tetani*), epidemic typhus (*Rickettsia prowazeki*) scrub typhus (*Rickettsia tsutsugamushi*), bacillary dysentery (three species of *Shigella*), cholera (*Vibrio cholerae*), many forms of plague, etc.

Parasitic bacteria also cause many plant diseases leading to serious damage of a large variety of commercial crops. *Erwinia carotovora* causes the rotting of carrots, potatoes, tomatoes and many other vegetables and fruits. It secretes enzymes breaking down the middle lamellae between the cells and feeds on the carbohydrates and proteins present in the tissues. Other bacteria invade the vascular tissues of a variety of plants and cause wilting diseases. For example, *Bacterium*

FIG. 246. PEAR BLIGHT, A BACTERIAL DISEASE CAUSED BY *Bacterium amylovorum*.
1 and 3 were inoculated with the bacteria, 2 was not.
(*From Robbins and Rickett's ' Botany ', courtesy of D. Van Nostrand Co., Inc.*

solanacearum attacks tomatoes, potatoes and other members of the Solanaceae. The infection interrupts the flow of sap, so that the plant wilts, is reduced in size, and may eventually die. The fire blight of apples and pears is caused by *Bacterium amylovorum*, the bacteria penetrating mainly through the stomata. It causes much havoc among these fruit trees, especially in the United States and New Zealand (Fig. 246). Another type of disease caused by bacteria is the production of galls or other abnormal growths. The best-known example of this is the crown-gall disease of a variety of plants including chrysanthemum, apple and raspberry. This is caused by *Bacterium tumefaciens* which infects the xylem and air spaces of the host. It appears to secrete some substance which stimulates adjacent cells to rapid division so that large, warty outgrowths or galls are produced. The gall consists of abnormally large parenchymatous cells intermingled with tracheids—a type of tissue often known as tumour tissue (Fig. 247). A related type of disease due to *Corynebacterium fascians* results in the formation of masses of small leafy shoots on wallflowers, chrysanthemum, etc.

Many other bacteria are saprophytes living on dead organic matter and are important as the agents of normal decay. The bacterial breakdown of carbohydrate material in the soil is particularly important since it leads to the formation of colloidal humus, the presence of which is so important for the fertility of the soil.

A few bacteria are autotrophic. They are able to synthesise carbohydrates from carbon dioxide and water but, since they do not possess chlorophyll, the energy for the synthesis is not derived from sunlight as in photosynthesis. The requisite energy is, in fact, obtained by the oxidation of various substances. Thus, the nitrifying bacteria oxidise ammonia and nitrites (p. 140) ; the sulphur bacterium (*Beggiatoa*) oxidises sulphuretted hydrogen ; and the iron bacteria (for example, *Leptothrix*) oxidise ferrous iron. This type of carbohydrate synthesis is described as **chemosynthesis**.

Although the vast majority of bacteria are, as described, devoid of chlorophyll, it has recently been discovered that some of the purple and green sulphur bacteria are able to carry out photosynthesis. They possess **bacteriochlorophyll** (which is only slightly different from the chlorophyll of the higher plants) together with carotinoid pigments. They are thus able to use light-energy for the synthesis of

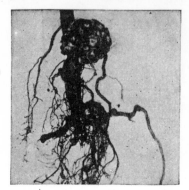 

Fig. 247. Crown Gall (due to *Bacterium tumefaciens*).
Left, on apple (*photo by Wormald*) ; right, on dahlia tubers (*photo by Dennis*).

carbohydrates from inorganic sources. The pigments absorb infra-red rays so that these sulphur bacteria, if exposed to such heat rays, are able to carry out photosynthesis even in the dark.

The respiratory processes are also varied in the bacteria. Many of them oxidise carbohydrates to obtain their energy, but, as pointed out above, other substances may be oxidised by special types of bacteria. Some bacteria, described as **aerobes**, will only grow in the presence of oxygen. Others, including many disease-causing bacteria (for example, the tetanus bacillus) are obligate **anaerobes**, that is, they will not grow in the presence of oxygen (p. 206).

Until about ninety years ago it was not really known how bacteria were formed. People believed then that bacteria were produced from putrefying plant and animal matter. In other words, they thought that living plants were generated from dead material—a belief that is now considered untenable. This sudden creation of life was called **spontaneous generation**. But the great French man of science, **Louis Pasteur**, investigated all beliefs in spontaneous generation and showed they were unproved. Pasteur did much work in fermentation, etc., and in 1864 he made his name immortal by demonstrating that bacteria are not spontaneously generated from putrefying material, but are spread about the earth either themselves or by means of spores. This immediately led to the recognition of why diseases can be infectious, thus causing a healthy person to be attacked by a disease, without necessarily coming into actual contact with a diseased person. The disease is spread by means of the bacteria floating in the air, or, even more, by the lighter bacterial reproductive spores. Recent investigations by aeroplane in the United States have revealed some interesting facts, concerning the microbiology of the upper air. For example, bacteria and moulds were found at a height of 20,600 feet, whilst yeasts and pollen grains were found 16,000 feet up.

Bacteria were seen by **Anton van Leeuwenhoek** (the Dutch microscopist who discovered Protozoa) in 1683, though he considered they were animalcules owing to their powers of locomotion. The greatest pioneer in bacteriology was the German, **Robert Koch**, who was the first to culture bacteria. His earlier work was on the bacillus responsible for anthrax, and in 1883 he was able to announce a method of inoculation against this disease. Later he did some important work on the bacteriology of tuberculosis and followed this with similar research on bubonic plague in India and cattle disease in South Africa.

The means whereby plant and animal diseases can be spread are manifold. Contact with an infected organism is the simplest, but some disease-causing organisms, especially bacteria, can be conveyed by wind, animals (especially insects), and over many miles of sea by ships. Some bacteria are actually water - borne. The aeroplane is now coming under suspicion as being a means of conveying disease into countries where it was hitherto unknown. To what extent the disease-carrying possibilities of the aeroplane may develop, unless studied and checked by science, can be gauged by progress in aviation itself.

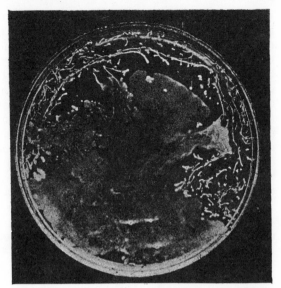

FIG. 248. COLONIES OF A BACTERIUM (*Staphylococcus aureus*) GROWING ON AN ORIGINALLY STERILE AGAR PLATE, UPON WHICH A HOUSE-FLY WAS ALLOWED TO CRAWL FOR THREE MINUTES.

(*Reproduced from Herms' ' Medical and Veterinary Entomology ', by permission of The Macmillan Company, New York.*)

The question of paramount importance is: How do the disease bacteria get into their hosts? This can take place in one of two ways. The bacteria or their spores can enter through an open wound (that is why wounds should be kept clean and well-bandaged), or they penetrate through the nose and throat. The breeding and spreading of bacteria can be curtailed by keeping homes, streets, buildings and even our own persons clean, by sneezing or coughing into a handkerchief, and by isolating disease-stricken patients. A well-known method of counteracting contagion is by putting people, disease-stricken, or suspected of a disease, into quarantine (see also Fig. 248).

Until comparatively recently, parasitic bacteria took a heavy toll of lives of people during surgical operations. Naturally a surgical operation involves a wound, and, in the early days, bacteria soon entered since they, or their spores, are constantly present in the atmosphere. The cause of death was usually a parasitic bacterium which set up blood poisoning.

Blood poisoning during an operation, so common a few decades ago, is scarcely ever known now. This is due to the great work of the British surgeon, **Joseph Lister**, who was born in Essex, in 1827. While on the staffs of the University of Glasgow and, later, King's College, London, Lister experimented with carbolic acid and found that it prevented bacteria from entering surgical wounds and setting up septic poisoning, by killing the bacteria. This work he developed to a great extent, and thus introduced what is now called **antiseptic surgery**, thereby saving millions of lives during surgical operations throughout the world. For this great work, Lister was raised to the peerage in 1897, and is therefore more familiarly known as Lord Lister.

The antiseptic surgery of Lister, introduced about ninety years ago, has since been much modified. Carbolic acid is not often used now ; but the fundamental fact is the same, that is, to prevent the parasitic activity of those dreaded disease-carrying bacteria. Antiseptic surgery means placing an antiseptic substance on the wound to prevent bacterial entry. Nowadays, surgery is not only antiseptic, but also **aseptic**. This means that all efforts are made to prevent any bacteria or their spores being anywhere near a wound, thus eliminating the necessity for killing them. This is done by keeping operating theatres scrupulously clean and sterilising (that is, so treating a thing with various solutions or heat that any living thing present on it will be killed) all instruments, dressings and even the surgeon's and nurse's hands. Rubber gloves are used for preventing any bacteria, etc., passing from the patient to the surgeon or vice versa, and masks are worn over the mouth by the nurses and surgeons, in order that they may not breathe any bacteria on to the patient or, on the other hand, inhale bacteria from the patient.

Medical practice has recently been revolutionised by the splendid researches on **antibiotics**, that is, chemical substances which prevent the growth and development of bacteria. Penicillin (p. 339) is an outstanding example.

VIRUS DISEASES

Much more recently, other infective agents of disease have been discovered. These are very difficult to investigate, and, indeed, though a great deal of research is being carried out on them at present, we know comparatively little about them. The reason why such agents of disease are so elusive of investigation is that they are so small, being much smaller than bacteria. They are called **viruses**, and are responsible for certain diseases in plants and animals, more especially the latter. Up to 1935, viruses were regarded as mysterious *living* organisms. As far back as 1922, **Twort** concluded that forms of life more primitive than bacteria or Protozoa must exist, and took the view that a form of life intermediate between simple enzymes and bacteria might account for the viruses. Then, in 1935, **Dr. W. M. Stanley** of the United States, and, in the following year, **Mr. F. C. Bawden** and others in Great Britain, showed that at any rate certain viruses were of a nucleo-protein nature. But viruses which thrive on both animals and plants vary considerably, certainly in size from about 10 $m\mu$ to 300 $m\mu$. Research on viruses is now being carried out vigorously (**Dr. K. M. Smith** being one of Britain's leading authorities on plant viruses), so that at present we can say some are very similar to living micro-organisms like bacteria, others resemble enzymes, whereas others have been obtained in crystalline form.

The first virus discovered was that responsible for the disease called **foot-and-mouth disease** of cattle. The virus which causes this disease was discovered by **Loeffler** and **Frosch** in 1898. It has been calculated that this virus is about 8 $m\mu$ in diameter ($m\mu = 0\cdot00001$ millimetre). For comparison, the diameter of the smallest bacterium is $0\cdot2\mu$ to $0\cdot5\mu$.

Much work is now being done on viruses, for they are very important, being responsible for such diseases as foot-and-mouth disease, influenza, smallpox, poliomyelitis, rabies, cowpox, dengue, smallpox, mumps, extromelia, psittacosis, etc.

It is now realised that viruses are responsible for certain plant diseases, and that viruses are probably more devastating than bacteria, in the case of plants. The first plant virus was discovered by **Iwanowski** in 1892, who showed that the **tobacco mosaic disease** (a disease which attacks the leaves of the tobacco plant

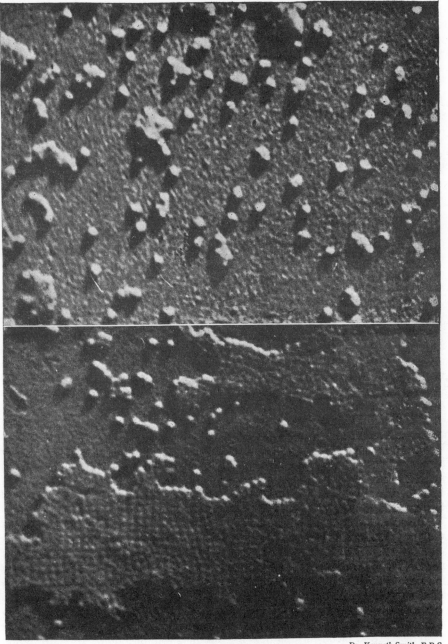

FIG. 249. A TOBACCO NECROSIS VIRUS.
Top, photographed on an electron microscope at a magnification of *c.* 20,000 and increased on the camera to *c.* 85,000, palladium shadowed. Bottom, same, showing the virus particles beginning to arrange themselves in regular order.

and causes much havoc among the crops) is caused by a virus. Another plant virus disease, **leaf roll of potato**, was recognised much earlier (from about 1770), though its cause was a mystery in spite of the fact that as far back as 1802 green aphides were suspected.

Many other plants suffer from mosaic diseases, such as the potato, sugar-cane, tomato, bean, pea, clover and so forth. These diseases are all caused by viruses, and the first symptom is sometimes the appearance of patches of various shades of green or green and yellow on the leaves. In other cases, the first symptom is the curling or crinkling of the leaves. Whereas in some ˙cases the disease scarcely

Foister

FIG. 250.   LEAF ROLL OF POTATO.

goes any further than this, and thus causes little damage, in others it causes the whole plant except the roots to be affected and sometimes destroyed. The trouble is that such mosaic diseases are usually very infectious. Research on this problem is being intensified.

Examples of plant virus diseases include leaf roll of potato (Fig. 250), spotted wilt of dahlia (Fig. 251), spotted wilt of tomato (Fig. 251), sugar beet mosaic (Fig. 251), yellow mosaic of tomato, tobacco mosaic. There are many others already known.

All virus agents will pass through a porcelain filter which is capable of retaining the smallest bacteria. Until quite recently it was customary to talk of viruses as being invisible. But to-day, by suitable staining, certain types can be seen under the microscope. With the ultra-microscope, and photo-micrography with ultraviolet light, more information concerning their size and form has been gleaned. Photographs taken on the electron microscope have proved particularly useful (Fig. 249).

Bacteria can be propagated on sterilised culture media, but in all cases viruses require the presence of *living* cells in order to multiply. It has been suggested that viruses are not necessarily invading parasites, but, by some at present unknown means, are produced by the host cells ; but this view is opposed by the regularity with which certain viruses can be induced to infect and propagate themselves in other hosts.

In animals, viruses are transmitted in various ways, for example, certain influenza-like diseases in man are carried by droplets in the breath exhaled. Bites from insects and blood-sucking ticks are also means of infection. In plants, some viruses are known to be transmitted by insects, such as aphides, others by grafting a healthy scion to an infected stock or vice versa, whereas the mode of transmission of others is unknown.

It is quite obvious that viruses are the cause of many diseases in plants and animals, so this branch of biological research is of the utmost importance.

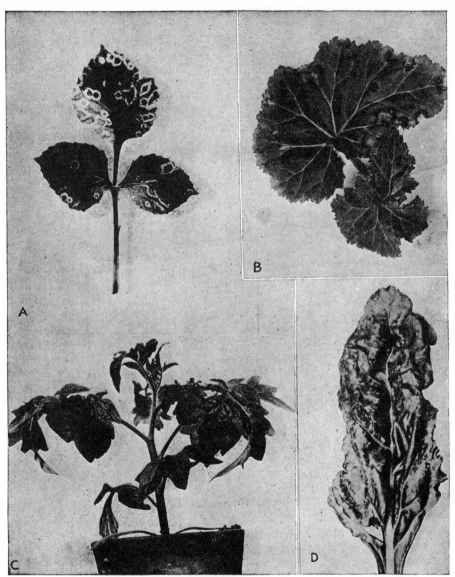

FIG. 251. PLANT VIRUS DISEASES.

A, dahlia infected with spotted wilt virus ; *B*, begonia infected with tomato spotted wilt virus ; *C*, tomato spotted wilt ; *D*, sugar beet mosaic.

(*A, C, photo by Bewley ; B, D, photo by Foister.*)

PARASITIC FUNGI

PHYTOPHTHORA

Many well-known plant diseases are caused by certain fungi. One very familiar example to all potato growers is the **potato blight**. This is caused by a parasitic fungus *Phytophthora infestans*. This plant is composed of long, colourless

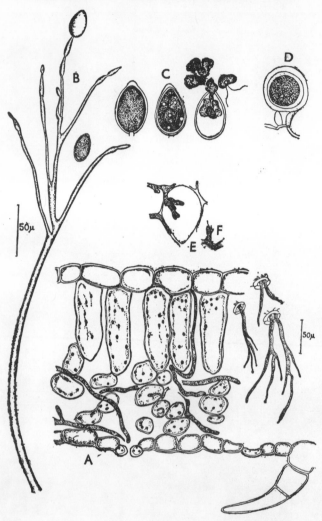

FIG. 252. POTATO BLIGHT (*Phytophthora infestans*).

A, transverse section of a leaf showing hyphae about to emerge through a stoma (to the right, young sporangiophores are shown emerging through stomata in surface view) ; B, mature sporangiophore showing development of sporangia ; C, stages in the development of zoospores in a sporangium (*after Ward*) ; D, sexual stage showing antheridium and oospore (*after Pethybridge and Murphy*) (see under *Pythium*).

threads (much finer than those of silk) known as **hyphae**. These hyphae are non-septate, but branched. Spores germinate on the leaves of the host, producing hyphae which penetrate the tissues through stomata or by boring into epidermal cells. The hyphae grow rapidly in the intercellular spaces and send fine hyphae (haustoria) into the cells from which they absorb food.

Eventually these patches turn brown, because the cells of the leaf which are attacked are killed by the activity of the parasite. The hyphae of this parasite

Foister and Noble

FIG. 253. POTATO BLIGHT (*Phytophthora infestans*).
Left, upper surface of potato leaf ; right, under surface.

grow thoughout the plant and eventually get into the tubers themselves. Here they penetrate the cells to extract foods, thus causing such cells to rot and making the tubers unfit for human consumption.

This fungus reproduces itself asexually. Certain hyphae, known as **sporangiophores**, grow through the stomata, and become branched (Fig. 252 *A* and *B*.) At the tip of each branch is a single **sporangium**. Usually a sporangium after becoming detached and blown on to a neighbouring plant, can germinate and produce new hyphae. Such fungal sporangia which germinate directly are termed **conidia**. In the presence of water, such as drops of dew or rain on the leaf, however, the contents of the sporangium break up into a number of biciliate (and therefore motile) zoospores Fig. 252, *C*. These, coming to rest, germinate to produce new hyphae. In either case, new infections are started and the disease spreads rapidly. Some of the sporangia may fall on to the soil and, in this way, tubers near the surface may also be infected. Sexual reproduction, similar to that of *Pythium* (p. 327) may occasionally take place.

Potato blight was first noticed in Europe and North America in 1840. It then became prevalent in Europe in 1845, and in 1846 it assumed widespread activity in Ireland, and caused such a failure in the potato crop that famine ensued, since that country depended so much upon potatoes for food. Wet seasons are usually favourable to the disease, probably because the sporangiophores only emerge in a very humid atmosphere and moisture is required for the further development of the sporangia.

Chemical methods for treating this disease have been devised ; but by means of up-to-date plant breeding, varieties of potato have now been produced which are either partially or almost wholly resistant.

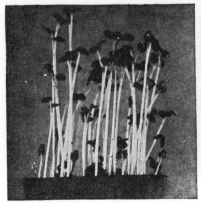

FIG. 254.   EFFECT OF DAMPING OFF BY *Pythium.*
Left, healthy cress seedlings ; right, seedlings attacked by *Pythium.*
(*From McLean and Cook, 'Theoretical Botany', Vol. I, Longmans Green & Co. Ltd.*)

PYTHIUM

Many fungal parasites live in water or very damp situations.   One is *Sapro-legnia* which attacks the gills of fish, and another, closely related to it, is *Pythium de baryanum*, which causes the common disease of seedlings, called **damping off**. This occurs especially in cress seedlings. The disease usually appears if the plants are sown too thickly and kept too moist.   It also attacks mustard and cucumber seedlings.   Other species of *Pythium* attack still further crops and other plants.

The disease usually appears on the stem just above the surface of the soil.   It attacks the cortex here to such an extent that the stem is no longer able to support the shoot (Fig. 254).

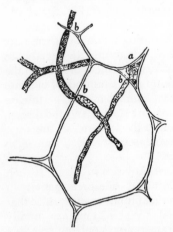

The hyphae of the fungus are non-septate, being composed of threads containing many small nuclei.   They are therefore coenocytes. The hyphae actually penetrate the cells, thus extracting nourishment from the latter (Fig. 255).   After killing the host, the parasitic *Pythium* continues to live *saprophytically* on the dead remains.

Pythium reproduces itself both sexually and asexually.   In asexual reproduction certain hyphae grow out from the tissues into the atmosphere.   The end of each hypha swells and becomes spherical. It is then cut off by a cross septum. The sporangium thus formed may, in a moist atmosphere, act as a conidium and germinate to form a hypha, capable of penetrating a new host.   In the presence of water, however, the contents of the sporangium emerge into a bladder-like sac and then become divided up into a number of biciliate (and therefore motile), uni-nucleate zoospores (Fig. 256).   The entire process

FIG. 255.   SMALL PORTION OF THE TISSUE OF A POTATO ATTACKED BY *Pythium.*

At *a*, the hyphae are in an inter-cellular space : at *b*, they have actually penetrated the cell-walls.

(*After Marshall Ward.*)

takes above a quarter of an hour. The
zoospores are capable of moving to a new
host, where they first rest, and then finally
germinate. This asexual method of reproduction is similar to the asexual method in
certain algae, for example, *Vaucheria*. On
both, the sporangia develop as terminal
structures with multinucleate contents. The
only difference is that whereas only one
multinucleate zoospore is produced in *Vaucheria*, many uninucleate zoospores are produced in *Pythium*.

Asexual reproduction in *Pythium* usually
takes place while it is living saprophytically.
Thus millions of zoospores are produced in
the soil among the dead remains of the host.
As a result, many more seedlings may be
attacked and killed.

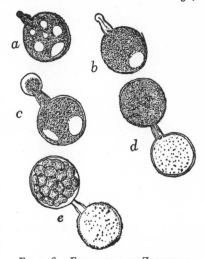

FIG. 256. FORMATION OF ZOOSPORES
BY A SPORANGIUM OF *Pythium*.
a, formation of tube ; *b*, *c*, swelling of
tube at distal end ; *d*, dilation of tube
and contents moving into it ; *e*, formation
of zoospores.
(*After Marshall Ward.*)

For sexual reproduction in *Pythium*,
oogonia and antheridia are produced, usually
within the tissues of the host. The tips of
the hyphae swell to produce spherical oogonia,
which are cut off by a cross septum (Fig. 257).
The multinucleate contents round off to form the egg, though only one nucleus
remains at its centre to be fertilised ; all the others pass to the surface of the egg.
Meanwhile another branch grows up into a smaller, more elongated swelling to
form the antheridium.   This is also cut off by a septum and its contents, at first
multinucleate, later become uninucleate.   First it comes into contact with the
oogonium, then it sends out a tube, the **fertilisation tube**, which penetrates the
latter and transmits the male gamete to the oosphere or ovum (Fig. 257). After fertilisation, the resulting zygote, called an **oospore**, rests and develops a thick wall.
Later, given suitable conditions, the zygote germinates, and the hyphae then attack
another host.   Oospores may remain viable in the soil over the winter and then
germinate in the following spring to cause a fresh outbreak of damping-off in
seedlings which may be growing in the soil.

Sexual reproduction in *Pythium* is similar again to *Vaucheria*, though there are
certain very clear distinctions.   In fact, in its methods of reproduction and in
certain aspects of its vegetative structure, *Pythium* may be said to be very similar
to certain algae.

PERONOSPORA

Closely related to the genus *Pythium* is that of *Peronospora* which contains species
living parasitically on higher plants and causing the well-known **downy mildew**
diseases.   For example, *Peronospora effusa* is responsible for the downy mildew of
spinach, whereas the disease of the same name but confined to onions, shallots,
garlic and sometimes leeks is due to *Peronospora destructor*.

In the case of spinach, the first symptom of the disease is the occurrence of

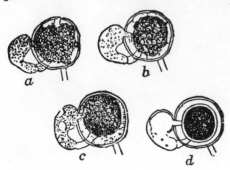

FIG. 257. SEXUAL REPRODUCTION IN *Pythium*.

a, oogonium and antheridium ; *b*, antheridium piercing oogonium by means of a fertilising tube ; *c*, transmission of contents of antheridium into oogonium ; *d*, oospore still in oogonium becoming surrounded by thick wall.

(*After Marshall Ward.*)

yellow spots on the outer leaves ; later a downy mildew appears, due to the reproductive hyphae of the fungus.

Peronospora, unlike *Pythium*, is completely parasitic. It is therefore said to be an **obligate parasite**. The hyphae of the mycelium branch between the cells of the leaves and stems ; seldom do hyphae occur in the vascular bundles, but they have been noted even in the seeds of the host plant. Unlike the hyphae of *Pythium*, therefore, those of *Peronospora* do not penetrate the cells of the host ; they are intercellular. But they do give off tiny branches called **haustoria** which pierce the cell-walls of the host and thus, enlarging and branching inside the cell, absorb food materials which are passed back into the hyphae (Fig. 258). The hyphae are non-septate, that is, coenocytic.

Peronospora can reproduce itself both asexually and sexually. The former case differs somewhat from that of *Pythium* in that the spores produced are adapted for distribution in the air. During asexual reproduction, certain hyphae in the leaf of the host grow perpendicularly through the stomata. These hyphae branch repeatedly in a dichotomous fashion (Fig. 258). Eventually the end of each branch of the exposed hyphae swells, and into the swelling a nucleus passes. Then the swelling becomes more or less oval in shape and greyish-purple in colour. The swelling forms a spore called a **conidiospore** or **conidium**, and the whole branched hypha bearing the conidiospores is called a **conidiophore**. Gradually the ripe conidiospore becomes constricted off from the end of the hyphal branch (no definite cross-wall is formed). The constriction is called the **sterigma**, and this, unlike the wall of the hypha becomes soluble in water, thus being able to assist in completing the break-away of the conidiospore, especially during damp weather. Millions of conidiospores are released ; but only a few succeed in reaching suitable ground for growth, that is, another host leaf. Once on the leaf, however, it germinates and sends out a tube which grows until it reaches a stoma through which it grows to produce a new parasitic mycelium.

Sexual reproduction in *Peronospora* is similar in essentials to that of *Pythium* (Figs. 257 and 448).

CYSTOPUS

White rust or white blister disease of many members of the flowering-plant family Cruciferae (p. 448), for example, cabbages, turnip, radish, mustard, cress and many wild members of the family, is caused by the parasitic fungus *Cystopus candidus*. It is frequently associated in its attacks with *Peronospora parasitica*, another of the downy mildews.

The diagnostic symptom of white blister disease is the appearance of white chalky pustules on most organs except the roots (Fig. 259). These pustules are followed by marked swellings of the surrounding host tissue. The results of these swellings vary according to what tissues are involved. In general, however, the result is that the infected parts turn brown and dry, and in the case of the softer parts disintegration usually occurs. The hyphae are non-septate and the mycelium grows only between cells (intercellular) ; but, like *Peronospora*, haustoria penetrate the host cells.

Asexual reproduction occurs just below the epidermal layer of the host plant. Hyphae mass together at one spot. The ends of these form club-shaped conidiophores and the mass of these causes the covering host epidermis to swell up into a blister. Conidiospores are constricted off, as in *Peronospora*. But in this case, after one conidiospore has been formed, another appears from the tip of the

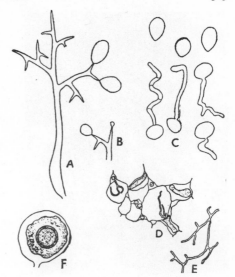

FIG. 258. REPRODUCTION IN *Peronospora effusa*.
  A, conidiophore bearing conidia ; *B*, formation of conidium ; *C*, stages in germination of a conidium ; *D*, leaf tissue showing wide hyphae and narrow haustoria ; *E*, a mycelium of hyphae ; *F*, oospore within the oogonium.
  (*After Richard, Cornell Univ. Ext. Bull., 718.*)

same conidiophore, and yet another—giving a chain of conidiospores. These eventually break away from each other (Fig. 260). Strange to say, the first conidiospore to be formed from each conidiophore is sterile. It has been suggested that its function is to aid the disintegration of the host plant's epidermis and so the final break-through of the following conidiospores. After the conidiospores have been liberated they are disseminated by wind. Then, under damp conditions, each conidiospore forms within its wall 4 to 8 zoospores, each having two cilia. After liberation, a zoospore comes to rest, surrounds itself with a cell-wall and eventually germinates on a new host. Under drier conditions, the conidia, like those of *Pythium*, are capable of growing directly into hyphae.

Sexual reproduction is similar to that of *Peronospora* and *Pythium*.

ERYSIPHE

To the genus *Erysiphe* belong certain parasitic species of fungi responsible for **powdery mildew** diseases of certain plants. They are all obligate parasites. *Erysiphe graminis* is responsible for the powdery mildew of cereals and grasses ; *Erysiphe polygoni*, the powdery mildew of clovers and peas ; *Erysiphe cichoracearum*, the powdery mildew of sunflower, etc. In fact, *Erysiphe* is a serious pest in agriculture and horticulture.

Erysiphe graminis attacks wheat, barley, oats, rye and a host of different grasses (all these—cereals and grasses—being members of the family Gramineae, p. 489). The host is seldom killed as a result of attack ; but discoloration and general weakening are the main results. The fungus develops mainly on the upper sur-

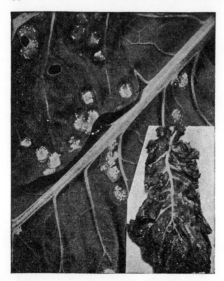

FIG. 259. WHITE RUST OR BLISTER OF CABBAGE, ETC.

Portion of a cabbage leaf showing white pustules due to sporangia (*photo by Ogilvie, by permission of the Long Ashton Research Station*); inset, same on a broccoli leaf (*photo by Foister and Noble*).

faces of the leaves and on the stems of the host. The ears are also affected, but to varying degrees (Fig. 261).

The mycelium grows superficially on the surface of the host; this eventually causes discoloration and a reduction of photosynthesis. The mycelium grips the host and absorbs food materials by means of haustoria which penetrate the epidermal cells (sometimes, but rarely, growing still deeper into the host tissues).

The hyphae are septate and composed of uninucleate cells.

Asexual reproduction is simple, taking place by the constriction of chains of uninucleate conidiospores from upright hyphae (conidiophores) (Fig. 262).

During sexual reproduction, two short hyphal branches grow erect and at the tip of each a uninucleate cell is cut off. One is large (oogonium); the other small (antheridium). The contents of the antheridium pass into the oogonium, or **ascogonium**, and fusion of the two nuclei takes place. This leads to the production of a complex fruiting body known as a **perithecium** (Fig. 262). As soon as fertilisation has occurred, the cell below the ascogonium produces numerous sterile hyphae which grow up round the sexual organs as a many-layered protective wall and some of the outer hyphae grow out over the leaf surface as anchoring filaments. Inside the protective cover, the fertilised ascogonium, after dividing into several cells, gives rise to a number of stout hyphae called **ascogenous hyphae**. The penultimate cells of these fertile hyphae develop as ovoid sacs called **asci**, inside each of which eight **ascospores** are formed (Fig. 262).

The perithecium remains intact over the winter. In the following spring, the top part of the perithecium breaks away, thus exposing the asci. These latter, which are in a very turgid condition, open at the top and the ascospores are forcibly shot out. If any alight on a suitable host, a new infection is thus started.

PUCCINIA GRAMINIS

Puccinia graminis, a member of the Basidiomycetes, causes the very serious disease of wheat and other cereals known as **black rust**. Its life-history is interesting since two different host plants are involved, part of the life-history of the fungus being spent on the wheat and part on the leaves of the barberry. Such a fungus is said to be **heteroecious**.

During the spring and summer, infected wheat leaves show orange-coloured streaks between the veins. A transverse section shows septate fungal hyphae between the mesophyll cells, food being abstracted from the latter by means of

haustoria. This mycelium
bears a large number, (a
sorus), of stalked, unicellular
spores, each with two nuclei.
These are **summer spores**
or **uredospores** (Fig. 263)
which, when liberated,
spread the disease by infecting neighbouring wheat
plants. Later in the year,
the same mycelium in the
wheat leaf produces a second
type of spore, the **winter
spore** or **teleutospore**
(Fig. 263), which is two-
celled. At first, each cell
has two nuclei but these fuse
so that at maturity the dark
coloured teleutospores have
uninucleate cells. The teleutospores are resting spores
and they lie dormant on the
stubble through the winter.
In spring, they germinate

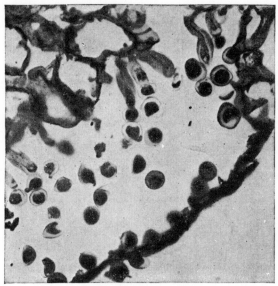

FIG. 260.   PRODUCTION OF CONIDIOSPORES BY *Cystopus*
BENEATH THE EPIDERMIS OF THE HOST PLANT.
(*From McLean and Cook, ' Theoretical Botany ', Vol. I,
Longmans Green & Co. Ltd.*)

and each cell grows out as a **basidium**, bearing four **basidiospores** (Fig. 263.)
Meiosis has taken place in the formation of the basidiospores and two of them on
each basidium belong to a plus strain and the other two to a minus strain.

The basidiospores will not germinate on a wheat plant but will do so if they
fall on the upper epidermis of a barberry leaf. Here they germinate and hyphae
enter the air spaces of the mesophyll. A plus basidiospore forms a plus mycelium
and a minus basidiospore a minus mycelium. These mycelia quickly produce
flask-shaped masses of hyphae which rupture the epidermis ; these are known as
spermagonia, some of which will be plus and others minus in type (Fig. 264).
In each spermagonium, some hyphae bud off **sporidia**, minute uninucleate
bodies which are exuded on to the leaf surface in a drop of mucilage ; other
hyphae, standing out above the leaf surface, are **receptive hyphae** (Fig. 264).
Insects, attracted by a sugary secretion from the spermagonia, may carry plus
sporidia from one spermagonium to a minus receptive hypha in another sperma-
gonium, or *vice versa*. When this happens, the nucleus of the sporidium passes
into the receptive hypha and a binucleate condition is established. Hyphae, with
two nuclei in each of their cells, now grow down through the leaf tissues to the
lower surface, where they proceed to elaborate **aecidial cups** (Fig. 265). In these,
binucleate **aecidiospores** are budded off in long chains and are eventually liberated. The aecidiospores will not germinate on the barberry, but if they are
blown on to a wheat plant they will do so. Entering the leaf through a stoma, the
germ tube establishes a mycelium in the mesophyll and uredospores are soon
produced.

Puccinia graminis, apart from the heteroecious condition, is very specialised in
other ways. It exists in a number of **physiological races** which, although alike

Foister and Noble

FIG. 261. POWDERY MILDEW OF CEREALS
(*Erysiphe graminis*).

A, on oat leaf; *B*, on ears of wheat.

in their structure, are limited to particular host plants. Thus, the race on the wheat is unable to parasitise oats, and a race growing on rye and barley will not grow on wheat.

CONTROL OF PLANT DISEASES

Fungal parasites are the cause of great financial loss since the diseases they cause, even if they are not fatal, reduce the yield of the host plants. Great efforts are therefore made to control plant diseases and, in particular, to prevent them from developing and spreading.

Fungal attack can often be prevented by spraying the surface of the host plant with substances which will prevent the germination of the fungal spores. Sulphur dusts, lime sulphur and preparations containing copper are often used for this purpose. Bordeaux mixture (for example, 5 lb. copper sulphate and 5 lb. lime in 50 gallons of water) is the most widely used and effective fungicide. High-pressure spraying and spraying from helicopters are now practised. Such sprays are used primarily to prevent infection but are also useful in reducing the severity of an attack when it has occurred.

In the case of diseases where the fungus is already present in the seeds, the latter are often treated with organic mercury compounds, cuprous oxide, or formaldehyde. In other cases, the seeds are steeped in water at a temperature high enough to kill the fungus but not to destroy the viability of the seeds.

Frequently the resting spores of the fungus persist in the soil. In this case, especially in greenhouses, the soil is sterilised by steam or by the application of formaldehyde. The incidence of disease may also be lessened by so-called field sanitation. The removal and destruction of all dead material, and particularly diseased material, will reduce the number of resting spores of disease-causing fungi. Rotation of crops is also a helpful measure, since it will result in suitable host plants not being available to the fungus for a number of years.

Efforts are made to prevent the spread of certain diseases by various regulations. Thus, notification of the occurrence of certain diseases, such as the wart disease of potatoes, is compulsory and this makes it possible for the authorities to take measures against the further spread of the disease. Measures to prevent the importation of disease-bearing plants and seeds from abroad are also taken. For example, a period of quarantine may be enforced or the entry of certain plants may be entirely prohibited.

The breeding of resistant or immune varieties is another important method of controlling disease. Thus, immune varieties of potatoes may safely be grown on land infected with wart disease (caused by the fungus *Synchitrium endobioticum*).

Varieties of wheat resistant to rust disease have also been bred by the geneticists and these are widely grown.

FUNGAL PARASITES ON ANIMAL HOSTS

The fungi considered in detail above are all parasitic on other plants. There are, however, a few fungi which are parasitic on animals, often including man, and they may be the cause of serious diseases.

Certain genera of yeasts have become parasitic. *Candida albicans*, for example, grows in the mouth, pharynx and larynx of birds and man, causing the disease known as 'thrush'. Another example is *Cryptococcus neoformans* which often causes fatal meningitis in man and also attacks horses and cattle.

Other pathogenic fungi are yeast-like in appearance while in the body of the host but, in pure culture, develop a normal mycelium and are, in fact, members of the so-called Fungi Imperfecti, whose systematic position is unknown since the zygotes, ascospores or basidiospores belonging to them have never been found. *Histoplasma capsulatum* is a fungus of this type. It causes a serious disease, histoplasmosis, of man and animals, in which the lungs, spleen, liver, etc., are attacked.

There is a group of fungi known as Dermatophytes which infect only the superficial layers of the skin, hair, nails, etc. Many of them give rise to the circular lesions known as ringworm both in man and in many animals. The fungi concerned, which may attack any part of the body surface, are members of three genera: *Microsporum* (commonly infecting puppies and kittens); *Trichophyton* (causing the common ringworm of man and other animals); and *Epidermophyton* (the cause, along with *Trichophyton*, of 'athlete's' foot').

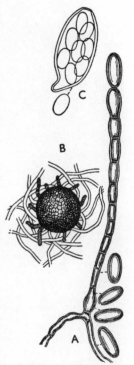

FIG. 262. REPRODUCTION IN *Erysiphe graminis*.

A, upright conidiophore producing a chain of conidiospores; B, perithecium surrounded by sterile hyphae; C, ascus containing eight ascospores.

(*A and C, after Salmon; B, after Delacroix.*)

Some common moulds, normally saprophytic, may also cause diseases in animals. For example, *Aspergillus fumigatus* infects various birds, invading the lungs and other internal cavities and causing the disease aspergillosis. Other species of the same genus, *A. flavus* and *A. niger*, infect the ears of human beings, causing the disease known as otomycosis.

Aquatic fungi may cause diseases of fish. Many sportsmen who are keen on fishing are familiar with the disease which attacks salmon, often causing so much havoc among that fish that some rivers, well-known for their salmon-fishing, are at times scarcely worth the trouble taken over the sport. This disease has been known for about a century, and is caused by a fungus called *Saprolegnia parasitica*. This fungus attacks its host, the salmon and other fish, especially goldfish, at the gills. There, the fungus grows by absorbing food from its host. Finally the growth of the parasitic *Saprolegnia* achieves such dimensions that the gills cannot carry on their work, that is, breathing. Thus the fish is killed by a form of suffocation. Excess calcium in the water favours the growth of this parasite, which is related to *Pythium* and *Phytophthora* already described.

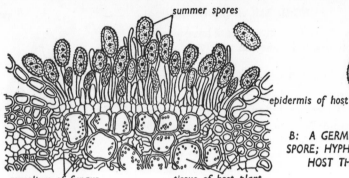

summer spores

epidermis of host

mycelium of fungus   tissue of host plant

A: FORMATION OF SUMMER SPORES

B: A GERMINATING SUMMER SPORE; HYPHA ENTERING NEW HOST THROUGH A STOMA

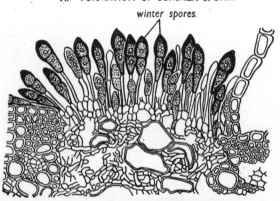

winter spores.

D: GERMINATION OF WINTER SPORE: FORMATION OF BASIDIOSPORES

C. FORMATION OF WINTER SPORES

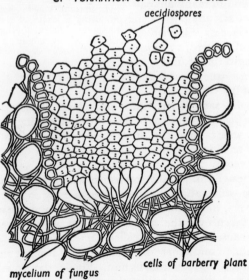

aecidiospores

E: A GERMINATING BASIDIOSPORE; HYPHA ENTERING A CELL OF A WILD BARBERRY PLANT

mycelium of fungus   cells of barberry plant

F: FORMATION OF AECIDIOSPORES

G: GERMINATION OF AN AECIDIOSPORE; HYPHA ENTERING CEREAL THROUGH A STOMA

FIG. 263. STAGES IN THE LIFE-HISTORY OF *Puccinia graminis*.
(*A, and C, after Owens ; B, E, F and G, after Holman and Robbins ; D, after Buller ; all re-drawn.*)

SAPROPHYTIC FUNGI

MUCOR

A very familiar type of saprophytic fungus is the group collectively called **moulds**. They live on different types of food-stuffs and other organic materials. A few are, however, parasitic, and live on other plants, or on animals or even on each other. One, for example, namely *Empusa muscae* is a parasite on the housefly.

One very familiar mould is *Mucor*. There are numerous species of this plant. One, *Mucor mucedo*, the **pin** or **bread mould** grows freely on damp bread, cheese, jam and horse dung.

The mycelium of *Mucor* is composed of masses of branched, non-septate hyphae (Fig. 266); though sometimes cross-walls occur in old hyphae. The mycelium ramifies on the surface of the organic substrate, and sends absorbing hyphae down into it.

FIG. 264. A THICK SECTION OF A SPERMA-GONIUM OF *Puccinia graminis* ON A BARBERRY LEAF.

(*By permission of Craigie, Dom. Canada Dept. Agric. Farm Bull. 84.*)

The plant reproduces itself by two methods, asexual and sexual. During asexual reproduction, certain hyphae grow upright. The top of an upright hypha swells to form a spherical **sporangium** (spore-container), and the upright hypha is therefore a **sporangiophore** (sporangium-bearer). The contents of the hypha become denser at the tip, and the nuclei all move towards the periphery, leaving the centre less dense and with many small vacuoles. A dome-shaped wall is now formed at the junction between these two regions, thus forming a swollen tip to the stalk of the sporangiophore, known as the **columella**. During the development of the columella, the peripheral mass of protoplasm begins to round itself off into many units, each containing several nuclei and some cytoplasm (Fig. 266). Each unit then surrounds itself with a thick wall, and forms a spore. The rest of the contents of the sporangium become mucilaginous and, when the spores are ripe, this mucilage swells. Meanwhile the sporangium wall becomes hardened and brittle. It also darkens in colour, and is covered with a layer of calcium oxalate. With the swelling of the mucilage, an internal pressure is exerted, and this pressure, together with the pressure exerted by the columella, causes the wall of the sporangium to burst, with the consequent setting free of the spores (Fig. 267). They remain for some time embedded in the mucilaginous mass, but eventually are freed by drying.

The spores of *Mucor* are small and, like those of bacteria, are capable of withstanding adverse conditions. There are millions of such spores in the atmosphere, except after a very heavy fall of rain or snow, during which they are washed down. Of course, the majority die, since they do not find a suitable place for growth. But, since they are so common in the atmosphere, it is no wonder that a suitable substrate such as exposed jam and damp bread soon becomes mouldy by the growth of this saprophyte.

FIG. 265. A COMPOUND PUSTULE OF *Puccinia graminis* ON A BARBERRY LEAF

An infection by a + sporidium has fused with an infection by a − sporidium, producing fertile aecidia.

(By permission of Craigie, Dom. Canada Dept. Agric. Farm. Bull. 84.)

Sexual reproduction of *Mucor* is of particular interest. When two hyphae from different mycelia come into contact, each begins to produce a side branch at the point of contact. The tips of these branches, which soon become flattened, are in contact with one another from the first and their further growth separates the two hyphae bearing them. Each branch becomes club-shaped and is termed a **progametangium**. Then the multinucleate tip of each is cut off by a transverse wall as a **gametangium** (gamete-container). The intervening wall between the two gametangia now breaks down and the contents fuse to form a **zygospore**. The zygospore develops a very thick, warty wall and then rests (Fig. 268). Finally it germinates to produce a short hypha, at the end of which an asexual sporangium develops. This later produces asexual spores which, on liberation, develop into new *Mucor* plants.

Sexual reproduction in *Mucor* is remarkable in that it can take place only between two *different strains* of this plant. In other words, the *Mucor* plant is unisexual; yet the two strains show no morphological difference, but there must be some physiological difference. They are referred to as the + and − strains. The only physiological difference so far recognised is that the + strain grows more vigorously. Two hyphae from a + strain cannot fuse to form zygospores, neither can two hyphae from a − strain. Sexual reproduction takes place only between a hypha of a + strain and a hypha of a − strain. Thus sexual reproduction in *Mucor* is rare, since a + strain and a − strain seldom grow near each other. This sexual difference in strains is known as **heterothallism** (Fig. 269), and is common among the fungi. Some related forms are, however, **homothallic**, that is, they possess only one sort of mycelium. In these, zygospore formation will result from the union of hyphæ belonging to a single plant.

EUROTIUM

Eurotium species are other common saprophytic fungi. At one time, the asexual stage of this genus of fungi was called *Aspergillus* (the name *Eurotium* being confined to the sexual stage), since each stage was considered to be a separate genus of fungi. Now it is known that both stages are of the same plant, so the name *Eurotium* is more favoured. This saprophyte also contaminates food, favouring those of a fatty nature. One of the most common species is *Eurotium herbariorum* which forms olive-green patches which gradually grow over the whole surface of the substrate, and also ramify beneath. The mycelium is formed of profusely branched, septate hyphae. There are two methods of reproduction, which occur at irregular phases, the asexual phase sometimes continuing for many generations without the intervention of a sexual phase.

Asexual reproduction involves the production of external spores. Certain hyphal branches grow upright and gradually swell towards their tips. Finally each branch produces asexual spores (conidia), and the branches are therefore conidiophores. After the tip of the conidiophore has swollen into a sphere, the surface develops numerous excrescences called **sterigmata** or **phialides**, and from each sterigma a chain of conidia are produced, the oldest being at the distal end (Fig. 270). Each conidium is multinucleate. The conidia rapidly develop in a humid atmosphere, and any which alight on a suitable substrate germinate and soon produce a mycelium.

The more complicated method of reproduction, that is, sexual, resembles that in *Erysiphe* (p. 329), and results finally in the production of small yellow spherical **perithecia.** These usually form when the temperature rises.

In the formation of a perithecium, a hypha coils closely to form a spiral, and another hypha de-

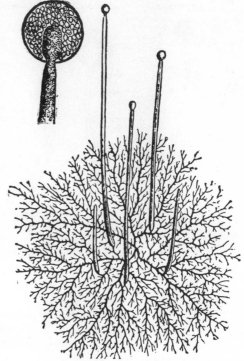

FIG. 266. *Mucor mucedo.*
A plant showing mycelium of branched hyphae and upright sporangiophores. More highly magnified is a single sporangium containing spores.
(After Brefeld.)

velops over it (Fig. 271). It seems that these hyphae are sexual. The former is female and called the **ascogonium,** and the latter is the male **antheridium.** Around these hyphae many sterile hyphae grow, and become closely interwoven into a pseudo-parenchymatous mass. Sometimes the ascogonium

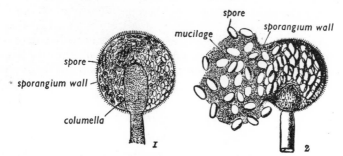

FIG. 267. SPORANGIUM OF *Mucor mucedo.*
1, a sporangium in optical longitudinal section ; 2, sporangium bursting after swelling of mucilage.
(From Strasburger.)

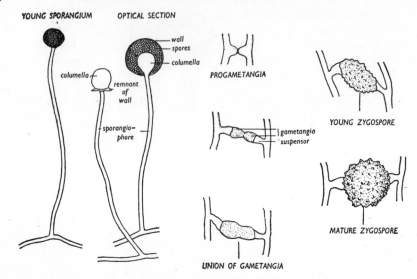

FIG. 268. ASEXUAL AND SEXUAL REPRODUCTION IN *Mucor*.
(*F. W. Williams.*)

and antheridium unite; but more frequently the antheridium withers away before any fusion can take place. Then, while the sterile hyphae are developing into a nutrifying covering, the ascogonium divides into several portions and sends out numerous branches, each of which becomes further branched. These branches are spoken of as **ascogenous hyphae**.

Near the end of each branch, which has become embedded in the pseudo-parenchymatous sphere of sterile hyphae, a sporangium is formed, each containing eight spores. Such sporangia as these are characteristic of the fungal class known as **Ascomycetes**. The sporangium is called the **ascus** and the eight spores are called **ascospores**. The ascospores derive their nutrition from the mass of sterile hyphae in which they are embedded (Figs. 272 and 273). The ascus is derived from a binucleate cell just below the tip of the ascogenous hypha (Figs. 272*A* and *B*). This penultimate cell then enlarges and the two nuclei fuse (*C*). Further enlargement takes place and the fusion nucleus divides (by meiosis) into four and then into eight. The eight ascospores now become differentiated inside the ascus (*D*).

Thus, in this mode of reproduction, sexual fusion may or may not take place; but in either case, perithecia, containing asci when ripe, are formed. This suggests that sexual fusion formerly always took place in *Eurotium*; but the process is now showing signs of degeneration.

PENICILLIUM

Penicillium is another genus of saprophytic fungal moulds closely related to *Eurotium*. Morphologically these two types are very similar, and *Penicillium* favours similar substrates for its saprophytic nutrition.

The asexual stage of *Penicillium* differs from that of *Eurotium* in that the coni-diophore does not swell at the tip and give off sterigmata all over its spherical sur-

Prof. M. Drummond

Fig. 269. Sexual Reproduction in Heterothallic *Mucor hiemalis*.
Photograph of a plate culture. + strains were grown on the left. ; – strains on the right.
Where the cultures met, sexual reproduction occurred and zygospores produced, indicated
by the dark patch down the centre.

face ; but the conidiophore branches several times dichotomously and then at the
tips of these branches conidia are constricted off in chains from modified end cells
called phialides (p. 337) (Fig. 270). The whole conidiophore of *Penicillium* is much
smaller than that of *Eurotium*. The sexual stage of *Penicillium* is closely similar to
that of *Eurotium*, but it occurs more rarely (Fig. 273).

Certain species of *Penicillium* are of considerable industrial and medical importance. For example, some are used for the ' ripening ' of certain forms of cheese
(*Penicillium camemberti* is inoculated into Camembert cheese) ; and from *Penicillium notatum* the well-known antibiotic penicillin is produced. Antibiotics are
substances, secreted by many fungi and bacteria, which are toxic to other organisms. Streptomycin (from *Streptomyces griseus*) and aureomycin (from *Streptomyces aureofaciens*) are two other well-known examples.

YEAST

Yeasts are fungi related to *Eurotium* and *Penicillium*, that is, they are Ascomycetes. They usually exist in a unicellular form, the cells being spherical or
ovoid. There are numerous different yeasts but those belonging to the genus
Saccharomyces are the best known
and the most important from the
economic point of view. The yeast
used by bakers, brewers and dis-
tillers is *Saccharomyces cerevisiae* (Fig.
274).

The single ovoid cell is surrounded by a wall of yeast-cellu-
lose. The contents consist of cytoplasm and a nucleus, the latter
lying to one side of a large vacuole,
which is sometimes regarded as
part of the nucleus. Fat globules,
glycogen in special vacuoles, and
granules of volutin are reserve food
substances which are usually found
in the cells.

Rapid vegetative reproduction
occurs by a process of **budding**.

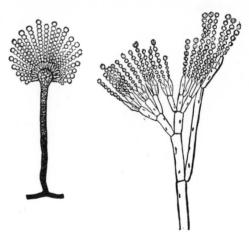

Fig. 270. Left, *Eurotium herbariorum* ; right,
Penicillium crustaceum.
Both showing upright conidiophores bearing conidia.

FIG. 271. *Eurotium*
herbariorum, SHOWING
COILED ASCOGONIUM
WITH   ANTHERIDIA
APPLIED TO IT.
(*After King and de Bary;*
from Gaumann).

One side of the cell develops a small prominence separated from the mother-cell by a very narrow collar. The nucleus, having moved to this region, becomes dumb-bell shaped and one end passes into the bud and is abstricted off as the nucleus of the new cell. The division is mitotic in character. The bud gradually enlarges but usually separates from the parent before it has attained its full size. More than one bud, however, may be formed at the same time and the daughter cells may themselves produce buds before they become detached, so that a branching group of coherent cells may be formed (Fig. 274).

When conditions become unfavourable, the yeast cells cease to form buds and enter a stage of spore production, the spores being able to withstand adverse conditions. The nucleus of a cell divides by meiosis, thus giving rise to four haploid nuclei round which the spores are organised (Fig. 274). The parent cell is an ascus and the spores are ascospores, two of which (of a plus strain) differ in sexual reaction from the other two (belonging to a minus strain). The ascospores germinate to form haploid cells, reproducing vegetatively by budding, or behaving as gametes and fusing in pairs to give diploid cells. In the yeasts of commercial value, the cells are of this diploid type. Such cells then multiply rapidly under favourable conditions by the process of budding already described. The sequence of events in the life-cycle of yeasts is, however, very varied and often complex.

The fermentation process carried on by yeast has already been described (Chap. XVII) and is, of course, the main basis of its economic importance. Yeast cells also have valuable nutritive and medicinal properties and are widely used on these accounts. *Saccharomyces ellipsoideus* is the wine yeast. It is a wild species, the spores overwintering in the soil. In the spring, they are blown on to the developing grape fruits so that the yeast is already present in the juice expressed from the ripe fruits. Numerous varieties of this yeast exist and these give the varying characteristics of the wines Burgundy, champagne, and so on.

THE MUSHROOM

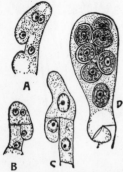

FIG. 272.   DEVELOPMENT
OF THE ASCUS OF *Pyronema*
confluens (*A*, *B*, *C*) AND
Ascodesmis (*D*). *D* SHOWS RIPE
ASCOSPORES.

Much more familiar are the saprophytic fungi which grow in meadows and on the trunks of trees. Though they never contain chlorophyll, they are not always colourless, for many are brightly coloured, often yellow or red. Familiar examples of such saprophytic fungi are the **mushrooms** and **toadstools** of meadows, etc., and **bracket** fungi which grow on trees. Many of the latter are, however, parasitic.

The familiar, umbrella-shaped portion of the mushroom (*Psalliota campestris*) is not the main part of the plant. It is the reproductive body or **fructification**. The vegetative part of the plant is composed of very fine, colourless, septate, branching hyphae which thread their way through a soil containing plenty of decaying humus, such as leaf-mould and animal manure (Fig. 275). Between two consecutive septa of the hypha are often many nuclei. In Nature, the mushroom reproduces itself by means of spores which are borne on the under surface of the well-known structure which grows above the soil. If the under surface of this fructification be examined, it will be seen that there is a large number of surfaces radiating from the centre. These are called **gills**, and on these gills the spores are produced.

The familiar fructification of the mushroom is formed entirely of inter-woven hyphae. In the stalk, or **stipe**, they are more closely packed towards the surface; whereas in the centre they are more loosely

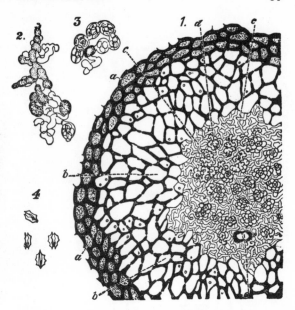

FIG. 273.   PERITHECIUM AND ASCI OF *Penicillium* (CLOSELY RELATED TO *Eurotium*).

1, section through perithecium ; *a, b,* pseudo-parenchy-matous covering ; *d,* ascogenous hyphae ; 2, 3, ascogenous hyphae with asci more highly magnified ; 4, ascospores.

(After Brefeld.)

arranged, with air-spaces between. A little more than half-way up the stipe is a ragged, membranous ring, the **annulus**, which is the remains of the membrane which formerly covered the whole lower surface of the umbrella-shaped portion or **pileus** (Fig. 276).

The gills on the under surface of the mushroom are downwardly growing extensions of the hyphae of the pileus. The structure of a gill may be seen by cutting the pileus in tangential section, when the gill will be cut in vertical section (Fig. 277). The inner part of the gill is composed of loosely packed hyphae, grow-ing downwards. This is the **trama**. Branches from these central hyphae diverge towards the two surfaces of the gill and branch repeatedly to give a more compact region known as the **sub–hymenium**. The ends of the sub-hymenial hyphae constitute the fertile surface layer known as the **hymenial layer** or **hymenium**.

The hymenial layer is composed of club-shaped hyphae growing perpendicular to the surface of the gill. Some of these are sterile and thus form **paraphyses**. Others are fertile and produce the spores. Unlike most other cases so far examined, the spores develop *externally* on the spore-bearing structure. They are therefore called **basidiospores**, and each hypha which gives rise to them is known as a **basidium**. (The mushroom therefore belongs to the fungal class known as **Basi-diomycetes**.) At the end of each club-shaped basidium, pointed outgrowths, known as **sterigmata**, are produced, four in number in the wild forms but only two in the cultivated form of the mushroom. The tip of each sterigma swells to produce the external spore or basidiospore (Fig. 277). The young basidium has two nuclei. These fuse and then meiosis occurs, giving rise to four nuclei. One

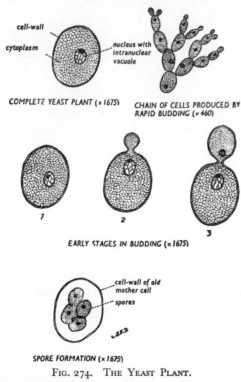

cell-wall

cytoplasm

nucleus with
intranuclear
vacuole

COMPLETE YEAST PLANT (×1675)

CHAIN OF CELLS PRODUCED BY
RAPID BUDDING (×460)

1 2 3

EARLY STAGES IN BUDDING (×1675)

cell-wall of old
mother cell

spores

SPORE FORMATION (×1675)

FIG. 274. THE YEAST PLANT.

of these passes into each of the four basidiospores of the wild form and two into each basidiospore in the case of the cultivated form. Further nuclear division takes place in the basidiospores of both types, so that they are multinucleate when ripe.

Stages in the general development of the fructification are shown in Fig. 276. Part of the vegetative mycelium develops as rope-like strands called **rhizomorphs**. It is on these that the fructifications arise. The young ovoid fructification is solid but an internal ring of hyphae in the upper part becomes modified as a hymenial primordium. This tears away from the hyphae below it so that a ring-like, or annular, cavity is formed. By this time, it is possible to distinguish the parts which are to become the stipe and the pileus respectively. Gills are formed by the young hymenium and these grow down into the enlarging cavity, as shown in the right-hand photographs in Fig. 276. Finally the floor of the cavity is ruptured, exposing the gills and leaving the annulus already described.

The number of spores that one mushroom is capable of producing is nothing short of amazing. It has been estimated by **Professor A. H. R. Buller** that one large mushroom is capable of producing 10,000,000,000 spores. It is obvious, therefore, that there must be a great mortality among these spores. The spores are shot off from the sterigmata with sufficient force to carry them away from the hymenial layer but not enough to shoot them across the interlamellar space on to the surface of the adjacent gill. The spores then fall down between the gills and when they emerge below the pileus they are readily dispersed by currents of air.

The fructification of *Psalliota campestris* is edible, and the plant is therefore cultivated for the purpose. The chief thing is to supply a soil very rich in humus, and a rather high temperature. Light is not necessary, and that is why mushrooms are frequently cultivated, not in fields, but in disused barns, cellars and tunnels. It is very seldom that the spores are planted, as seeds are, because of the difficulty of delayed germination. The old method was to collect virgin **spawn** (blocks of soil and manure containing the hyphae) and to multiply them by growing in manure and straw. This was then formed into bricks and sold as spawn. During the last few years most of the cultivated mushrooms of Great Britain have been grown by the pure culture method, where, instead of virgin spawn which has to be renewed each year, a mycelium from the inside of a young stipe is used.

In connexion with the wild varieties of saprophytic fungi closely allied to the mushroom, the so-called 'fairy rings' of our fertile meadows are of interest. These rings are usually almost perfect circles of a darker green present in the grass of certain fields, surrounded, at certain times of the year, by a ring of mushroom-like fructifications. Fairy rings are caused by several types of fungus. One is the **fairy-ring mushroom** or **champignon**. This mushroom is more delicious than the common edible mushroom, and is used extensively as an article of diet in Italy and France. It is common in Great Britain also. If one spore be imagined to develop in

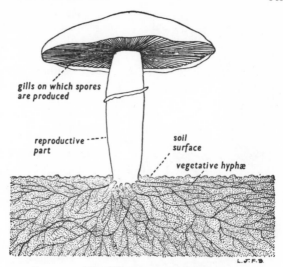

FIG. 275. THE MUSHROOM (*Psalliota campestris*).
Note the vegetative mycelium of hyphae below the soil surface, and the reproductive fructification above.

the soil, it can be seen how the hyphae will grow out in all directions from it, like the spokes of a wheel. Then new fructifications are formed, but only by the youngest parts of the hyphae ; that is, the ends of the hyphae, on the circumference of the circle. Hence the formation of the ring. The older parts of the hyphae die away, and supply nutrifying humus to the grass, and growth goes on year after year, always in an outward direction. Thus the rings are getting larger and larger. Some fairy rings are estimated to be 300 to 400 years old.

CLASSIFICATION OF FUNGI

The chief characteristics of the fungi are their lack of chlorophyll and their simple structure. There is some difference of opinion as to the plants which should be included in this group. For example, early classifications included the bacteria, as the Schizomycetes or Fission Fungi, but the bacteria are now generally recognised as a separate group not allied to the true fungi.

FIG. 276. STAGES IN THE DEVELOPMENT OF THE FRUCTIFICATION OF THE MUSHROOM.
(*From Atkinson: ' Mushrooms, Edible and Poisonous ', by permission of Henry Holt and Co., Inc.*)

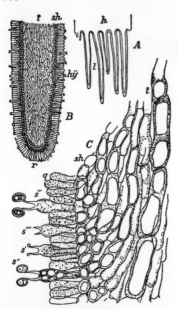

FIG. 277. STRUCTURE OF THE HYMENIUM OF THE MUSHROOM (*Psalliota campestris*).

A, vertical section through the pileus (*h*) passing longitudinally through several gills (*l*); *B*, a single gill more highly magnified, (*hy*) hymenium; *C*, portion of a gill enlarged (× 350) showing a young basidium bearing sterigmata (*s′*), an older basidium showing basidiospores developing (*s″*), a basidium bearing ripe basidiospores (*s‴*), and a basidium which has shed its basidiospores (*s⁗*).

(*After Sachs.*)

As will have been noted from the types described above, the vegetative body of the fungus usually consists of hyphae, which may be septate or non-septate. Some, however, are unicellular. Others again, the so-called slime fungi, are multinucleate masses of protoplasm without a cell wall. The following will indicate the main groups into which the fungi are divided:

Class I. PHYCOMYCETES.

Sub-class I. **Archimycetes.** Unicellular forms like *Synchitrium endobioticum* (which causes 'wart disease' of potatoes) or naked masses of protoplasm like *Plasmodiophora brassicae* (the cause of 'finger and toe' disease of cabbages and other members of the Cruciferae). Reproduction by zoospores and by motile isogametes.

Sub-class II. **Oomycetes.** Examples described here are *Saprolegnia, Pythium, Phytophthora, Cystopus,* and *Peronospora.* The hyphae are non-septate. Asexual reproduction is by zoospores or by conidia. Sexual reproduction is by oogonia and antheridia.

Sub-class III. **Zygomycetes.** The only type described here is *Mucor.* The hyphae are non-septate. Asexual reproduction is by air-borne spores. Sexual reproduction is by the union of similar gametangia to form zygospores. Some forms are heterothallic.

Class II. ASCOMYCETES.

The examples described here are *Erysiphe, Eurotium, Penicillium* and *Saccharomyces.* The last named is unicellular but the others have septate hyphae. Asexual reproduction is by means of conidia. Sexual organs, ascogonia and antheridia, may be formed but the sexual process is often modified and obscure. Fruiting bodies are formed. These may be spherical or flask-shaped perithecia, inside which the asci are borne, or disc- or cup-shaped apothecia with the asci borne on the upper surface. The fusion of two nuclei in the young ascus is regarded as a delayed fusion of the sexual nuclei.

Class III. BASIDIOMYCETES.

Examples described here are *Puccinia* and *Psalliota.* The hyphae are septate. The characteristic reproductive structure is the basidium.

Sub-class I. **Hemibasidiae.** The fungi causing the diseases known as 'smuts',

Sub-class II. **Protobasidiae.** The fungi causing 'rust' diseases, for example,

Puccinia graminis. The life-history is often complex with a number of different spore forms. In many forms, it includes a stage where there is a coming together of plus and minus nuclei leading to the production of aecidia (p. 336). These remain separate until fusion takes place in the ripening teleutospore. This can be regarded as a delayed sexual fusion.

Sub-class III. **Eubasidiae.** These are forms with large fructifications, for example, mushroom, puffballs and bracket fungi. These produce basidia of the mushroom type (p. 341). A fusion of two nuclei, which can be regarded as being sexual in nature, occurs in the developing basidia. The mushroom itself is homothallic but some forms related to it are heterothallic. In these latter, one of the two nuclei in the young basidium is of a plus type, the other is a minus one. They became associated by the fusion of a hypha of a plus strain with one of a minus strain. Fusion of the nuclei in the basidium is followed by meiosis and of the four basidiospores ultimately produced, two are plus and two are minus in type.

NUTRITITION OF FUNGI

All the fungi are devoid of chlorophyll and are therefore either saprophytes or parasites. The distinction between these two classes is, however, not always clear. Some fungi, (for example, *Mucor*) are **obligate saprophytes** which cannot attack and feed on living tissues. Others are **obligate parasites** (for example, *Puccinia*) which will not grow on dead organic material. Many fungi, however, cannot be placed in either of these groups. Some are usually saprophytic but can attack living tissues under certain conditions; these are termed **facultative parasites.** Others, which can only complete their life-cycle on a living host, are, nevertheless, capable of passing part of their existence on dead organic matter; these are **facultative saprophytes.** *Pythium* is rather unusual in that it is capable of either a completely saprophytic existence or a completely parasitic one.

The fungi require much the same sort of foodstuffs as do the higher plants. They secrete a wide range of enzymes by means of which carbohydrates, proteins and fats are converted into diffusible substances, which the fungus can absorb. Saprophytes can absorb food in solution over the entire surface of their hyphae but frequently special absorbing hyphae ramify in the substratum. The less-specialised parasites (for example, *Pythium*) also absorb food over their whole surface. Such types usually kill the tissues which they invade and often the host itself. More specialised parasites usually send fine branches into the cells from the main hyphae which remain intercellular or which ramify on the external surface of the host. These fine hyphae branch or enlarge within the cells as haustoria (such as those described in *Peronospora, Cystopus, Erysiphe,* and *Puccinia*). The haustoria secrete enzymes into the host cells to break down the foodstuffs present into soluble forms, which they then absorb and pass on to the main hyphae. Such specialised parasites usually do not kill the tissues immediately though the cells are depleted of some of their foodstuffs and the host is thus weakened.

LICHENS

In all cases of parasitic fungi considered here, the advantage seems to be with the parasite only. But there are other cases where a host has apparently taken

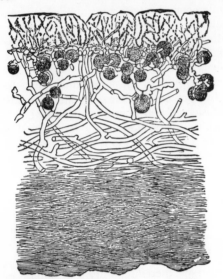

FIG. 278. *Cladonia furcata*, A LICHEN.
Vertical section of the thallus showing a cortical sheath of fungus beneath which are unicellular algae to which fungal hyphae are attached.

advantage of the presence of a parasite and uses it physiologically for its own ends. This was well seen in the case of mycorrhizal fungi (p. 135). Nitrogen-fixing bacteria are another case in point (p. 140). These are examples of symbiosis (living together for mutual benefit).

Another form of symbiosis is that strange group of plants called **lichens**. A lichen is a case of symbiosis between a green alga and a fungus. The fungal hyphae surround the alga, sometimes penetrating the algal cells with haustoria (Fig. 278). Thus the fungus obtains products of photosynthesis from the alga. On the other hand, the alga obtains water, mineral salts and perhaps some proteinaceous foods from the fungus. In most lichens, only the fungus seems able to produce spores, usually in the form of ascospores.

There are many forms of lichens, though the most familiar are those which form flattened, yellow, greenish or red halli on rocks, tree trunks or old bits of wood. They can withstand considerable desiccation, and are brittle when dry and tough and pliable when damp. Examples of common lichens are *Xanthoria parietina*, an orange lichen growing on old walls and roofs (Fig. 279), and the more shrubby type such as *Cladonia* species (Fig. 279).

Vegatative propagation often takes place by the liberation of small groups of algae surrounded by fungal hyphae.

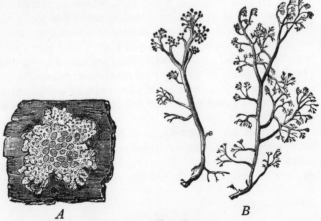

A   *B*

FIG. 279. TWO LICHENS.
A. Xanthoria parietina; B, Cladonia rangifera.
(*After Strasburger.*)

PARASITIC ANGIOSPERMS

Apart from the many parasitic fungi, there are quite a number of plant parasites among the higher groups, including the angiosperms. Complete parasitic flowering plants contain no chlorophyll, therefore they cannot manufacture their own food ; instead they depend on other living green plants for it.

A common example of a parasitic angiosperm is the **dodder**, of which there are several species, the most common being the greater dodder (*Cuscuta europaea*) and the lesser or heath dodder (*Cuscuta epithymum*). A rarer species is the flax dodder (*Cuscuta epilinum*). All are members of the convolvulus family (Convolvulaceae).

The greater dodder is parasitic on willows, stinging nettle (*Urtica dioica*) and certain vetches. It blooms during July and August in hedgerows and other places where the host plants are usually found. The sickly looking parasite is an annual which develops a yellowish, thread-like stem which twines around the stalks of the host, climbing sometimes to the very top.

At intervals, the stem of the parasite gives off suckers or haustoria which penetrate the stem of the host and by this means tap the latter of certain amounts of food supply. This being so, and since the parasite absorbs all its food supply from the host, green leaves are not necessary, and they are reduced to small, almost colourless scales. The adult parasite has no root in the soil. When its seed germinates in the soil it does give off a true root ; but this is solely for the purpose of initial anchorage. Once rooted, a prostrate shoot is formed, the tip of which grows around the stem of the host, pierces it with haustoria and then begins to nourish itself at the expense of the host (Fig. 281). Then the original root dies away. If the seed of the dodder settles away from a potential host, then the young parasite dies, for it cannot fend for itself (Fig. 280).

During July and August small rosettes of pink flowers appear on the parasite.

The hosts of the lesser or heath dodder (Fig. 281) are ling (*Calluna vulgaris*), thyme (*Thymus serpyllum*), gorse (*Ulex europaeus*) and other heath plants. (The specific name of this species of dodder, *epithymum*, is from the Greek *epi*, upon and *thymos*, thyme.) The species

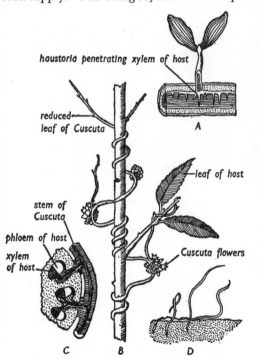

haustoria penetrating xylem of host

reduced leaf of Cuscuta

A

leaf of host

stem of Cuscuta

phloem of host

xylem of host

Cuscuta flowers

C   B   D

FIG. 280. A PARASITIC AND PARTIAL PARASITIC ANGIOSPERM.

A, young mistletoe plant showing haustoria penetrating xylem of host (*after Lowson*) ; *B*, *Cuscuta europaea* parasitic on willow ; *C*, section of the host twig showing penetrating haustoria ; *D*, seedlings of *Cuscuta* (*after Strasburger*).

FIG. 281.   LESSER DODDER (*Cuscuta epithymum*). Above, plant parasitic on gorse ; below, germinating seedling grown from starting point (right), and beginning to coil around a grass stem.

blooms later in the year—August and September. It is of smaller habit and its flowers are rose-coloured. Indeed, on the rare occasions when one meets it blooming, whole gorse bushes may be seen to be covered by its rosy, tangled masses.

The rare flax dodder is parasitic on the flax plant (*Linum usitatissimum*). Its specific name, *epilinum*, is from the Greek *epi*, upon, and *linon*, flax.

In April it is possible to find another flowering plant which is entirely parasitic —again so parasitic that it has lost its own food factories, the green leaves, and these have been reduced to insignificant scales. This plant is the **great tooth-wort** or **lungwort** and it belongs to the angiospermic family Orobanchaceae, a family which contains nothing but parasitic plants. The toothwort and the broomrapes (see below) are the only British members of this family ; all others are indigenous to North America and tropical regions.

There is only one toothwort common in Britain, that is, the great toothwort which grows in woods and copses and is parasitic on the roots of hazel and beech trees. Its botanical name is *Lathraea squamaria*. The generic name is from the Greek *lathraios*, hidden, referring to the plant's habit, because it lies hidden in the undergrowth ; the specific name is from the Latin *squama*, a scale, indicating that the creeping underground rhizome is covered with scale leaves (in four rows).

Suckers emerge from the rhizome and these penetrate the roots of the host plant and thus derive the requisite food. Periodically, erect shoots, about ten inches high, are sent up, and these bear the sickly purple flowers in one-sided racemes during April and May.

There is a much rarer toothwort— *Lathraea clandestina* (the specific name being from the Latin for hidden or secret)— which is parasitic on the roots of willows. But this plant is not indigenous to Britain.

The **broomrapes** are also members of the parasitic family Orobanchaceae. There are several species, all of the genus *Orobanche*. The clove-scented broomrape (*Orobanche caryophyllacea*) is parasitic on the large bedstraw (*Galium mollugo*) and occurs in pastures and on downs; but it is not common. The lesser broom-rape (*Orobanche minor*) grows in fields and pastures, is parasitic on clover, and blooms during June-August. The tall broomrape (*Orobanche elatior*) is parasitic on knapweeds. It flowers during June-August also. The red broomrape (*Orobanche alba*) attacks wild thyme and blooms during the same season. But the most common of the broomrapes is the greater broomrape (*Orobanche rapum-genistae*), a parasite on broom, gorse, etc., and it therefore flourishes on heaths and moorland (Fig. 282). It is a perennial, growing one to three feet high. It is a stout, fleshy plant, swollen at the base, and sending off suckers which penetrate the roots of the host. The reddish thick stem bears leaves which are reduced to drab-white scales, and at the top a long, dense spike of pink flowers.

E. G. Neal

FIG. 282. GREATER BROOMRAPE (*Orobanche rapum-genistae*).

There is an American species of *Orobanche*, namely, *O. ludoviciana*, whose stems were at one time relished as a food by the Pah Ute Indians.

One particularly interesting example of a completely parasitic flowering plant is that called *Rafflesia*. There are several species, all of which are native to Malaya. This plant is parasitic on the roots of vines of the genus *Cissus*. It is parasitic to such an extent that the whole of the plant-body, including stems, roots and leaves, are all reduced to colourless straggling threads which grow through the soil and pierce the roots of their vine host and extract food. The flower, on the other hand, is far from reduced (Fig. 283). The species known as *Rafflesia arnoldi* has the largest known flower in the world, being usually a brilliant red in colour, measuring 30 inches across and weighing 15 lb. It has a repulsive smell.

Some flowering plants attack other plant hosts but they extract only a certain portion of their food from their hosts. They cause less damage to the host than true parasites in that they do not *completely* depend upon their hosts for their manufactured food. They manufacture a great deal for themselves, since they

FIG. 283.   FLOWER OF *Rafflesia arnoldi*.
The scale represents 12 inches.
(British Museum.)

possess green leaves.   Since they extract only a certain amount of food from their hosts, they are called **partial parasites**.

A very well-known example of a partial parasite is the flowering plant **mistletoe** (*Viscum album*).   This plant belongs to the flowering plant family Loranthaceae which contains more than five hundred partial parasitic species distributed in tropical and temperate countries; yet the mistletoe is the only species indigenous to Britain.   Mistletoe is an evergreen, branches dichotomously and has opposite leaves.   It produces yellowish flowers in February and March, and eventually its well-known white fruits (p. 281) which contain a viscous, very sticky fluid. The plant lives on apple and poplar, and, more rarely, hawthorn and oak trees as hosts (Fig. 284).

The seed germinates by sending a haustorium into the bark of the host, and this finally penetrates the wood vessels (Fig. 280).   From the wood vessels of the host, it is clear that all the parasite can get is water and dissolved mineral salts.   But this is all that it requires, for, having obtained this, it can carry on food manufacture itself through its own photosynthetic leaves.   Thus the haustorium of the mistletoe in the branch of the host performs some of the physiological functions of the root of a normal plant in the soil.

The flowering-plant family Scrophulariaceae (p. 472) contains a number of partial parasites, all of which grow in fields and pastures and on heaths and moorlands with grasses as their hosts.   Casual observation would not lead one to suspect that such plants are at all parasitic, since they appear quite normal, with green leaves ; but where they grow densely, the grass itself shows relatively poor growth.   Example of these plants are : yellow rattle (*Rhinanthus major*), presenting its yellow flowers during May-July ; eyebright (*Euphrasia officinalis*), producing white or lilac flowers during May-September (Fig. 384) ; red bartsia (*Bartsia odontites*), bearing red flowers during June-September ; cow-wheat (*Melampyrum pratense*), producing yellow flowers during May-September; and lousewort (*Pedicularis sylvatica*), blooming with pink flowers during April-July.

In all these cases, there are haustoria on the roots of the parasite where these touch the grass roots. Here they penetrate the tissues of the host, and thus supplement their own food supply (Fig. 285).

SAPROPHYTIC ANGIOSPERMS

There are some, but not many, saprophytic angiosperms. These possess mycorrhiza and thus obtain their food supplies from the humus in which they grow.

One is an orchis, the bird's-nest orchis (*Neottia nidus-avis*) which grows in woods, and blooms during June and July. Being saprophytic, it needs no sunlight and therefore thrives under the densest canopy of such trees as beech (Fig. 286). The generic

H. Bastin

FIG. 284. MISTLETOE GROWING ON A YOUNG APPLE TREE.

name comes from the Greek *neottos*, nestling, for the underground stems and its branches and the roots from them form a nest-like mass among the leaf-mould substrate from which nutrients are absorbed. The Latin specific name clearly signifies the same habit (Fig. 287).

The endotrophic mycorrhiza has already been described (p. 138).

The plant is a perennial, sending up aerial shoots six to eighteen inches high. The leaves are reduced to brown scales surrounding the stem. The flowers are unattractive and crowded into pale-brown dense spikes.

Another angiospermic saprophyte is the yellow bird's nest (*Monotropa hypopithys*)—no relative of the bird's-nest orchis. This plant belongs to a small family of nothing but saprophytes, namely, Monotropaceae. It blooms during June-August (Fig. 288). Since it needs no light, it thrives among the

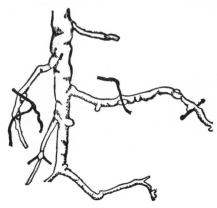

FIG. 285. ROOTS OF LOUSEWORT (*Pedicularis sylvatica*).

Fixed by suckers to grass roots (indicated in black).

After Maybrook.)

E. G. Neap

FIG. 286.   BIRD'S-NEST ORCHIS (*Neottia nidus-avis*).

undergrowth of dense beech or pine woods. (In fact, it is sometimes called fir-rape. The generic and specific name indicate this, being derived from the Greek *monotropos*, living alone, *hypo*, under, *pithys*, a pine.) The roots are embedded in the thick leaf-mould of the woods, and they are surrounded by massed ectotrophic mycorrhiza which presents the roots with a much larger absorbing surface.

The erect stem is stout and fleshy, and the leaves are reduced to tissue-like scales. The yellowish-brown flowers are tightly packed in pendulous spikes.

PRACTICAL WORK

1. Mount on a slide a little of the white scrapings from your teeth. Examine closely under the high power and look for any bacteria present. These may include cocci and bacilli.

Examine prepared slides of bacteria.

2. Examine and draw the leaf of a potato plant stricken with potato blight. Look for the lighter patches (or if the disease is advanced they will appear dark brown), where the leaf is attacked by the fungus (*Phytophthora infestans*).

Obtain a prepared microscope slide of a section of an infected leaf and look for and draw the hyphae of *Phytophthora* ramifying through the tissues of the leaf.

Prepared slides showing the method of reproduction should also be examined.

3. 'Damp off' some cress seedlings, then examine some of the *Pythium* under the low and high powers.

Prepared slides of the asexual and sexual methods of reproduction should also be studied.

4. Place some damp bread under a bell-jar, and leave it for a few days until it becomes 'mouldy'.

Then take a very small portion of the mouldy bread and examine the fungus responsible for this condition, with the high power of the microscope. Look for, draw and describe any spore-producing bodies of *Mucor* which may be discovered.

Prepared slides showing stages in sexual reproduction should also be studied.

A culture of the two strains (+ and −) should also be examined.

5. Examine a mycelium of *Eurotium* under the high power of the microscope. Note the branched, septate hyphae.

Look for condiophores producing chains of conidia.

The perithecia will be seen attached to the mycelium by means of thicker hyphae of a darker colour. By gently pressing the coverslip, the wall of sterile hyphae can be burst, and the small asci will be extruded. Examine one of these closely, and look for the eight ascospores.

H. Bastin

FIG. 287.   ROOT OF BIRD'S-NEST ORCHIS

R. Morse

FIG. 288.   YELLOW BIRD'S NEST (*Monotropa hypopithys*).

6. Examine, draw and describe the reproductive portion (fructification) of a mushroom.

Choose a ripe fructification and remove the stipe (' stem ') and place the umbrella-shaped part (pileus) on a piece of paper with the gills downwards. Leave for a few days, then gently lift the pileus, when the arrangement of the gills will be traced on the paper by thousands of small, black, reproductive spores which have been shed.

Examine a prepared slide of the longitudinal section of a gill, noting its detailed structure and especially the means of basidiospore production.

7. Examine and draw a specimen of a common lichen. Study a prepared slide showing a transverse section of the thallus revealing the presence of alga and fungus. Look for penetrating haustoria.

8. Examine plants of dodder coiling around the stems of clover, heather, gorse, nettles, etc. Make a drawing of the whole plant, noting the manner in which it coils around its host ; its reduced, scale-like leaves ; and its pink flowers. Notice the absence of chlorophyll, and explain the parasite's method of nutrition in view of this.

Make a thorough examination of the haustoria, and cut a transverse section of the stem of the host plant, in the region of a haustorium. Notice the type of host tissue which the haustorial cells penetrate.

9. Toothworts and broomrapes are not easily found in the field ; but if preserved specimens are available, these should be examined.

10. Museum specimens of mistletoe growing on a branch are sometimes available. These should be studied. So also should a prepared slide taken across a branch in transverse or longitudinal section and through the haustorium longitudinally. The members of the family Scrophulariaceae mentioned in the text should, however, be looked for during a botanical excursion at the times of flowering indicated. If slides showing method of penetrating the host are available, these should be studied.

11. Museum specimens of bird's-nest orchis might be examined, special attention being paid to the matted underground stems and the fleshy roots. Sections of the latter will show the endotrophic mycorrhiza.

CHAPTER XXV

BRYOPHYTA

In the Thallophyta which have just been described, the general life-cycle is very varied; but even in these primitive plants, fertilisation, since it involves a doubling of the chromosome number, is necessarily followed at some stage in the life-cycle by meiosis. In many Algae and Fungi the plant is haploid, producing gametes which, on fertilisation taking place, give rise to a diploid zygote. Very frequently (for example, *Chlamydomonas*, *Spirogyra*), the first division in the germinating zygote is meiotic, so that the plant or plants which develop are, like the parent, haploid. In many of the Algae and in the higher Fungi, the life-cycle is, however, more complex and varied.

Many of the Algae show a regular alternation of two phases, or two distinct types of plant, in their life-history. Some filamentous brown seaweeds (for example, *Ectocarpus*) have two types of plant regularly alternating with one another. One, a haploid plant, bears gametes and is therefore spoken of as the **gametophyte**. After fertilisation the zygote grows into a diploid plant which bears spores (the **sporophyte**), this production of spores being accompanied by meiosis. These haploid spores grow into a gametophyte plant and so the life-cycle with its alternation of generations is completed. In *Ectocarpus*, the vegetative structure of the two generations is exactly the same, but in *Laminaria* (p. 313) the big plant is the sporophyte, producing spores which grow into the minute, haploid gametophytes bearing oogonia and antheridia.

It is these Thallophyta with a regular alternation of generations which are of the greatest interest, since they give some idea of how the more standardised life-history of the higher plants may have been evolved.

In the **Bryophyta**, which include the **Hepaticae** (liverworts) and **Musci** (mosses), the plant itself is the gametophyte generation. The fertilised ovum does not grow into a new liverwort or moss plant, but forms a stalked capsule inside which spores are produced by meiosis. This capsule, which since it developed from the zygote is diploid, is the other generation of the plant, the sporophyte. The spores produced grow into the gametophyte plant. The outstanding feature of all the Bryophyta is that the sporophyte remains attached to the gametophyte throughout its development and obtains practically all its food from it. The little moss or liverwort plant bearing a capsule thus consists of both generations—the plant itself is the gametophyte, while the capsule it bears is the sporophyte.

PELLIA

Pellia is one of the commonest of British liverworts, growing in damp woods, on the sides of brooks, and wells, and sometimes even submerged in water. The plant is composed of a flattened, lobed, green thallus. Although there is little division of labour among the vegetative tissues, there is a midrib formed by the thickening of the tissues. On the under surface, long, colourless, unicellular hairs,

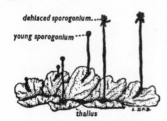

dehisced sporogonium....

young sporogonium----

thallus

FIG. 289. *Pellia epiphylla*.

known as **rhizoids**, develop and help to fix the plant to the soil and absorb water and mineral salts from it (Fig. 289).

When growing on damp soil, the thallus is broad and robust; on the other hand, when growing in water, it is long, narrow, and ribbon-like. In either case, growth is by means of a single apical cell, situated at the base of a depression at the tip of each branch. Branching is dichotomous.

The thallus is composed entirely of parenchymatous cells, all of which possess chloroplasts, though the more superficial cells of both surfaces are richer in chloroplasts than those more deeply seated. Food is stored in the form of starch. Fungal hyphae are frequently present in the rhizoids and the tissues of the mid-rib, particularly in older thalli. These may have a similar physiological significance to the mycorrhiza of higher plants (p. 135).

The thallus, as already indicated, is the gametophyte generation. It produces spermatozoids and eggs, the former in **antheridia** and the latter in organs known as **archegonia**. Both antheridia and archegonia are produced during the spring and early summer. In *Pellia epiphylla*, the common species, the plants are monœcious.

The antheridia are produced on the upper surface of the thallus, on both sides of the mid-rib. An antheridium develops from a single superficial cell. This cell enlarges and rises above the general surface of the thallus. A transverse division takes place and the lower cell (the stalk cell) undergoes a few further divisions to form the antheridial stalk. The upper cell undergoes much division, forming a mass of cells surrounded by a wall composed of a single layer of cells. The internal mass of cells forms the sperm mother cells or spermatocytes. During the formation of the antheridium, a protective wall of cells arises from the thallus around,

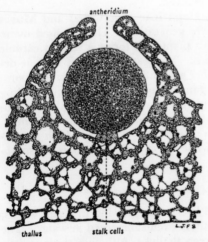

antheridium

thallus   stalk cells

FIG. 290. VERTICAL SECTION THROUGH PART OF THE THALLUS OF *Pellia*, PASSING THROUGH A RIPE ANTHERIDIUM (×90).

forming a flask-shaped antheridial cavity (Fig. 290). Finally, each sperm mother cell produces a single sperm, which is composed of a spirally coiled body with two long cilia at one end (Fig. 291).

The eggs of *Pellia* are produced in specialised organs known as archegonia. These arise just behind the growing points of branches of the thallus which have already produced series of antheridia. A whole group of archegonia is produced on the upper surface of the thallus, the latter becoming much thicker at this point. The thallus thins out again towards the apex and, at the same time, a flap of tissue (the **involucre**) grows out from the upper side of the thallus just behind the archegonia. The latter thus come to be seated in a protective pocket (Fig. 293).

A mature archegonium (Fig. 292) is a flask-shaped structure seated on a short, multicellular stalk. The basal, swollen part is the **venter**, the wall being one, or frequently two, cells thick. This contains the spherical egg and also a small cell, the **ventral canal-cell**. The upper part of the archegonium is the narrow, cylindrical **neck**, consisting of about five longitudinal rows of cells and being closed at the top by four rather large cap-cells. The central channel of the neck is occupied by a number of **neck canal-cells**. Such an archegonium arises from the segmentation of a single superficial cell. This (after cutting off a basal stalk cell) divides by three longitudinal walls in such a way that a central cell is separated from three peripheral ones. A transverse wall is now formed across these cells, thus determining the two main parts of the archegonium, the venter and the neck. The peripheral cells of the venter grow and divide to form the wall of the venter ; the central cell divides to give the large ovum and the small ventral canal-cell. In the upper region, further divisions in the peripheral cells give the neck, while the central cell divides to give the neck canal-cells.

cilia...

FIG. 291. SPERM OF *Pellia* (× 1250).

Fertilisation can only take place when the surface of the thallus is covered by a film of water, such as might be supplied by rain or dew. In the presence of such water, the wall of the antheridium is ruptured and the mucilaginous mass of sperm mother-cells is extruded on to the surface of the thallus. On reaching the surface of the film of water the sperm mother cells, each containing an active sperm, spread rapidly over the surface as a delicate film. This rapid spread of the mother-cells is due to a lowering of the surface tension at the air-water interface, by oil present in the matrix in which the cells are embedded. Many of the mother-cells are thus passively carried to the region of the archegonial group. The sperms are now liberated from the mother-cell's walls and are attracted chemotactically (possibly by soluble proteins) to the archegonia. During these events, the presence of water has also resulted in the opening of the archegonial necks. This is brought about by the swelling of mucilage derived from the breakdown of the neck-canal-cells and the ventral-canal-cell. Fertilisation can now take place. Sperms enter the open archegonial necks and wriggle down through the mucilage. Finally, in each ripe archegonium, one sperm fuses with the ovum to give a diploid zygote. Several archegonia may be fertilised, but normally only one zygote develops further.

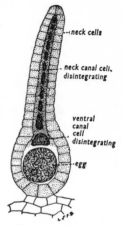

neck cells

neck canal cell, disintegrating

ventral canal cell disintegrating

egg

The zygote remains in the venter of the archegonium and ultimately gives rise, not to a new *Pellia* thallus, but to a spore-producing capsule or **sporogonium**, which is the sporophyte generation of the plant. The zygote forms a wall around itself, enlarges, and then divides transversely into three cells. The lowermost does not develop further ; the middle one gives rise by repeated divisions to a stalk or **seta** together with an organ of attachment known as the **foot** ; the top cell gives rise to the spore-containing capsule or **theca**. These developments all occur within the enlarging venter, now known as the **calyptra**.

FIG. 292. AN ALMOST RIPE ARCHEGONIUM OF *Pellia* (× 200).

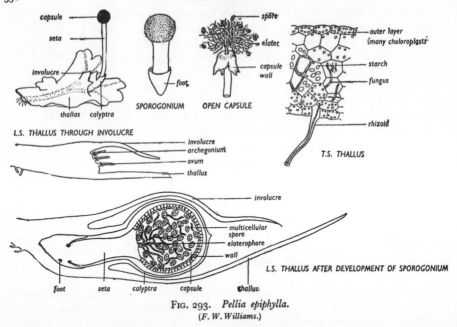

FIG. 293. *Pellia epiphylla.*
(*F. W. Williams.*)

The young capsule has a wall of several layers surrounding a mass of spore-mother-cells. Some of these latter, and particularly those in the outer and upper regions, divide (usually in July) by meiosis to give four ovoid spores, the spore-mother-cells becoming characteristically four-lobed in the earlier stages of this process. The spore-mother-cells which do not behave in this way elongate as **elaters** and develop a double spiral of thickening. Many of these are interspersed with the spores, but a mop-like tuft of elaters is fixed to a group of spirally thickened cells (known as the **elaterophore**) at the bottom of the capsule.

These developments give rise to a sporogonium (Fig. 293), now firmly fixed in the tissues of the thallus by the absorbing foot. The cells of the seta are short and packed with starch and the capsule has the structure just described. A few further changes now take place. The outer layers of the wall develop characteristic bands and rods of thickening, but four lines of thin-walled cells extend from the tip of the capsule almost down to the base, dividing the wall into quarters. These modifications are related to the opening mechanism described below. The spores themselves divide and form little oval cell groups (Fig. 294). This condition is attained by the autumn, and the sporogonium halts its development during the winter.

In the following spring, rapid developments take place. The starch in the cells of the seta is converted to sugar, water is rapidly absorbed, and the individual cells elongate to as much as forty times their original length. This results in the capsule being thrust through the calyptra and raised to a height of 2–3 in. Such sporogonia with their white, almost translucent stalks bearing aloft the black capsules, about one-eighth of an inch in diameter, give a striking appearance to patches of *Pellia* at this time of the year. As the wall of the capsule dries, it splits into four valves along the lines of dehiscence already described and the four

valves fold right back. This exposes a mass of intermingled spores and elaters supported round the basal
tuft of fixed elaters. The elaters, both fixed and free,
are now dead and empty spirally thickened cells which
are hygroscopic. They twist and turn as they dry
and thus help to scatter the dark green spores (Fig.
293).

If these haploid and already multicellular spores
alight on moist soil, they continue their development
immediately. An apical cell is formed and the young
plant is fixed to the soil by the growing-out of
rhizoids. Continued activity of the apical cell gives
rise to a young thallus or gametophyte and the
life-cycle is thus completed.

FIG. 294. SPORES (SOME
ALREADY   DIVIDED)   AND
ELATERS FROM THE CAPSULE
OF *Pellia* (× 250).

FUNARIA

Most mosses and liverworts are terrestrial plants, thus differing broadly from
the algae. The majority of mosses grow in damp situations. Nevertheless, there
are some which can withstand periods of desiccation, and grow in dry places, for
example, on rocks and the tops of walls. But, just as in the case of the liverworts,
there is one period of their life-history when they must have liquid water, namely,
for the ciliated sperm to swim to the archegonium in order to fertilise the egg.

Some common mosses which are mentioned in the following account are
Funaria hygrometrica (commonly growing on waste ground and particularly on soil
where there has been a fire), *Mnium hornum* (a woodland moss occurring in large
clumps on the floor of the wood and round the bases of the trees), *Catharinea
undulata* (another woodland moss, Fig. 295), *Polytrichum commune* (a large moss,
characteristic of moorlands), and *Sphagnum* species (the characteristic bog mosses).

In all of these, the plant is the gametophyte generation, bearing antheridia
and archegonia at certain times of the year. Unlike *Pellia*, however, the plant is
fairly highly organised and has a definite stem bear-
ing spirally arranged leaves There are no true roots,
but colourless or light brown rhizoids grow out from
the base of the stem and serve to anchor the plant and
absorb mineral solutions. Water absorption does, however, take place through the leaves as well. The anatomy of the stem is usually simple. A transverse
section shows the presence of an epidermis, a cortical
region, and, in the centre, a simple conducting strand.
This latter consists merely of elongated and thin-walled
cells, there being nothing comparable to the xylem
and phloem of the higher plants.   In the big moorland
moss *Polytrichum*, this central region is rather more
highly differentiated and shows water-conducting cells
(the **hydrome**) and food-conducting cells (the **lep-
tome**). The leaf structure of the majority of mosses
is simple and, except in the region of the mid-rib,
each is composed of a single layer of cells. The
several-layered mid-rib has a simple conducting strand.

FIG. 295.   *Catharinea undulata*
(*After Schimper.*)

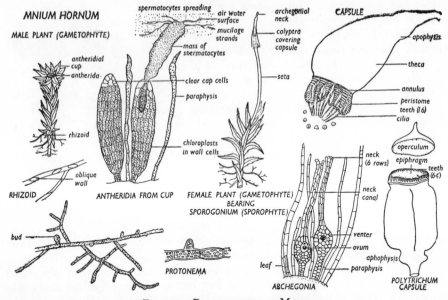

FIG. 296. REPRODUCTION IN MOSSES.

(F. W. Williams. Diagram of antheridial dehiscence modified from a figure by Prof. J. Walton)

The leaves of *Polytrichum* and *Catharinea* are more complex and have a number of longitudinally running plates of photosynthetic cells standing up from the leaf surface. These photosynthetic plates are protected in dry weather by the leaf curling into a cylinder (compare with the leaf of *Ammophila*, p. 191).

The antheridia and archegonia are borne at the tips of some of the leafy shoots. The antheridia are borne in conspicuous **antheridial cups** or ' flowers '. These consist of groups of antheridia, intermingled with sterile paraphyses, and surrounded by a rosette of large and spreading leaves, the inner ones often being reddish in colour. The **archegonial** cups are less conspicuous, since the surrounding leaves do not spread out nor are they brightly coloured (Fig. 296).

Some mosses, for example, *Funaria*, are monœcious, an antheridial cup being formed at the tip of the relatively small main shoot while the archegonial cup is formed later at the tip of a stronger lateral shoot which arises from near the base of the male shoot. Many mosses, however, are diœcious (for example, *Polytrichum*).

Moss antheridia are stalked, club-shaped structures. There is a single-layered wall. For the most part, the flat cells composing this have numerous chloroplasts, but the tip is occupied by a larger and clear-looking cap cell (in *Funaria*) or by several such cap cells (in *Mnium*). The interior is filled with extremely numerous sperm mother-cells or spermatocytes, embedded in a mucilaginous matrix containing oil. The antheridia are intermingled with paraphyses, usually of characteristic shape in the various genera, and these may serve to supply water to the developing antheridia.

The archegonia are essentially similar to those of *Pellia*, but both the stalk and the neck tend to be longer. They are also characterised by the fact that they develop by the activity of an apical cell.

Fertilisation can only take place when water is present in the antheridial and archegonial cups as a result of rain or dew. In the presence of such water, ripe antheridia discharge their spermatocytes in the following way. The walls of the cap cell become mucilaginous and first the inner and then the outer wall are ruptured, so that there is now an open pore to the exterior. Fluid collects at the base of the antheridial cavity in increasing amount, and so forces out the mass of spermatocytes through the ruptured cap cell as a worm-like mucilaginous mass. As the top of this mass reaches the surface of the water in the cup, the spermatocytes separate and spread out as a film on the surface of the water, the factors concerned being the same as in the case of *Pellia* described on p. 357. At this stage, the spermatozoids are still enclosed in the wall of the spermatocyte although they may show some movement. In the archegonial cups, any ripe archegonia will have opened their necks as the result of the mucilaginous swelling of the disintegrating neck canal-cells and ventral-canal-cell.

It is still not entirely certain how the spermatozoids reach the archegonia. Rain, dropping on the antheridial cup of a monœcious form like *Funaria*, may splash some of the film of spermatocytes down on to the archegonial cup of the same plant, which, at that time, is situated at a lower level than the antheridial cup. In the case of diœcious mosses such as *Mnium*, the male and female plants may be at some distance apart, and in this case it seems likely that insects act as the agents for carrying the spermatocytes from the male to the female plants. The legs of any insect visiting an antheridial cup will become covered with a film of spermatocytes and, if the insect then visits a female plant, some of the spermatocytes may be transferred to the water in the archegonial cup. Here, the biciliate sperms will emerge from the spermatocytes into the water and be attracted chemotactically to the open archegonial necks, from which mucilage containing cane sugar exudes. In each archegonium, one sperm will fuse with the ovum to give a diploid zygote. In the diœcious mosses, as also in *Pellia*, the transfer of the sperms to the archegonia is in two stages. The first is the passive carriage of the spermatocytes to the neighbourhood of the archegonia, and the final stage alone is accomplished by the active swimming of the liberated spermatozoids.

A number of archegonia may be fertilised but, as in *Pellia*, normally only one zygote develops further as the moss capsule or sporogonium. The zygote undergoes cell division and an actively dividing apical cell is established at each end. The lower end grows into the tissues of the stem and acts as the anchoring and absorbing foot. The upper end grows rapidly as a needle-like young sporogonium. The lower end of this forms the thin seta, while the terminal region later swells and differentiates as the spore-containing theca. During the early stages of development the whole of the sporogonium is enveloped in the enlarging archegonial venter or calyptra. When the seta elongates, however, the calyptra is ruptured round its base and carried up as a hood surrounding and protecting the developing theca (Fig. 296).

Mature capsules vary in shape in different mosses and may be spherical, ovoid, cylindrical or pear-shaped. Some are pendent from the top of the seta, while others are upright. That of *Funaria* is pear-shaped and pendent and, like those of most mosses, shows three regions. The basal part is slightly swollen and dark green in colour. This region is sterile, but actively photosynthetic, and is known as the **apophysis**. Above this is the even more swollen region in which spores are produced. The top of the capsule consists of an obliquely placed conical lid

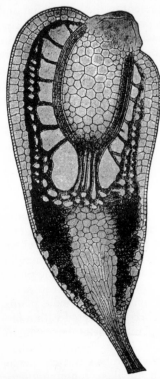

FIG. 297. LONGITUDINAL SECTION
THROUGH CAPSULE OF *Funaria hygrometrica*.

known as the **operculum**. The line of junction
between the operculum and the rest of the capsule is clearly marked, and consists of a ring of
modified cells forming the **annulus**.

The sectional view (Fig. 297) shows that
the moss capsule is much more elaborate than
that of *Pellia*. The apophysis consists of a mass
of parenchymatous tissue, the outer layers of
which are rich in chlorophyll. The epidermis
of this region has numerous stomata.

In the fertile region, the centre is occupied
by a column of parenchyma, the **columella**.
Surrounding this is the spore sac, in which, as
development has proceeded, spore mother cells
have divided by meiosis to give rise to haploid
spores. The sac itself is shaped like a barrel
with the top and bottom knocked out, and it is
bounded by an external and an internal layer
of sterile cells. No elaters are formed (this
is an important point of distinction between the
mosses and the liverworts). Outside the spore
sac, there is a wide, cylindrical air cavity which
is traversed by filaments or **trabeculae** of
chlorophyll-containing cells. These link the
spore sac with the outer, several-layered wall
of the capsule.

The distal end of the capsule is highly
modified in relation to the distribution of the
spores. A ring of epidermal cells at the junction
of the operculum with the rim of the capsule is modified as the annulus, the cells
being enlarged and filled with mucilage. A more deeply seated series of cells,
forming a cone with its base resting on the rim of the capsule and its tip pointing
upwards, becomes highly modified as the **peristome**. This lies at the junction
between the conical end of the columella and the operculum. In transverse section, the peristome appears as a ring of sixteen cells. In each of these, a strip of
thickening is laid down on both the outer and the inner tangential walls.

As the capsule ripens, a number of changes take place. The columella and
other internal tissues dry up and the spores now lie free within the dry wall of the
capsule. The peristome cells lose their living contents and the thin regions of their
walls break down, so that all that remains is a double series of teeth which are, in
fact, the patches of thickened tangential walls already mentioned. The operculum
now becomes detached by the swelling of the mucilaginous contents of the annulus
cells, and the peristome is exposed as an outer and an inner series of sixteen teeth.
The inner teeth may be split further, as in *Mnium*, into finer teeth known as **cilia**.

The spores are now dispersed. In wet weather, the peristome teeth are folded
over the mouth of the capsule so that the spores cannot escape. In dry weather,
the teeth open outwards, with jerky movements as they separate from one
another, and the spores can now leave the capsule. Many of them have become
entangled in the rough peristome teeth, and may be thrown to some little distance

by the jerky movements of the latter. Dispersal is also aided by the censer-like swinging of the whole capsule as a result of the untwisting and retwisting of the hygroscopic seta.

The peristome of the *Polytrichum* capsule (Fig. 296) consists of much smaller teeth with a membranous diaphragm stretched across the top of the teeth, so that small pores are left round the periphery of the capsule mouth, through which the spores are shaken as the capsule sways in the wind (*cf.* the capsule of poppy (p. 283)).

The small spores have a double wall and contain numerous chloroplasts. If they fall on moist soil, they germinate immediately. They do not form a new moss plant at once, but give rise to a filamentous stage known as the **protonema** (Fig. 296). In germination, the outer wall of the spore is ruptured and the contents surrounded by the inner wall grow out as a filament which soon becomes septate and branches freely. Some of the branches remain on the soil surface, giving rise to a green mat of filaments ; other branches grow into the soil as anchoring and absorbing rhizoids. Both the green filaments and the rhizoids are septate, the former by transverse and the latter by oblique walls.

Eventually, pear-shaped buds arise on the green filaments, and each of these grows, by the segmentation of an apical cell, into a moss plant of the usual form. Since the protonema may become quite extensive and bear numerous buds, a considerable tuft of leafy plants may arise from a single spore.

GENERAL

The types of Bryophyta just considered illustrate the characteristic life-cycle of this group of plants. The liverwort or moss plant is always the haploid gameto-phyte, while the diploid sporophyte which develops from the fertilised egg is always in the form of a capsule, which remains attached to the gametophyte throughout its development. It may be noted that fluid water is required by the gametophyte for the process of fertilisation, while the sporophyte liberates its air-borne spores only under dry conditions. The gametophyte is thus the generation which retains features of its aquatic ancestors, while the sporophyte is better adapted for terrestrial existence.

Although the fact has not been mentioned in relation to the individual types, most liverworts and mosses reproduce freely by vegetative means. Thus, the older parts of the thalli of *Pellia* decay and so separate the numerous branches, which grow on as separate individuals. Other liverworts have special organs of vegetative propagation known as **gemmae** (cell groups which become detached and grow into new plants). Mosses, too, show many different types of vegetative propagation. Practically any part of the plant, if damaged, may grow out as a secondary protonema, on which new plants arise. Even the seta of a capsule may do this, and, in this case, diploid moss plants are formed and, after fertilisation, these give rise to tetraploid sporogonia (that is with four instead of the usual two sets of homologous chromosomes).

The Bryophyta show great advances in complexity as compared with the Thallophyta. The gametophytes of the liverworts may be relatively simple as in *Pellia* ; but in others, known as leafy liverworts, the thallus is differentiated into stem and leaves. In yet others, as in the common *Marchantia* and *Fegatella*, the thallus, although of simple form like that of *Pellia*, is differentiated into an elaborate photosynthetic apparatus on its upper side. The moss gametophyte may likewise show considerable complexity of structure, as has been indicated in the description of *Polytrichum*. The sporogonia of both liverworts and mosses do not show any great degree of elaboration of their external form, but the internal structure may be highly differentiated. Some moss capsules, however, differ from those described by having the apophysis greatly developed as an efficient photosynthetic organ.

Another important feature of advance shown by the Bryophyta is the development of the archegonium. This gives much more adequate protection to the egg and subsequently to the zygote than does the single-celled oogonium of the Thallophytes.

In spite of the fact that the Bryophyta are, in a sense, amphibious, it is clear that they possess many features fitting them for life on land.

PRACTICAL WORK

1. Collect plants of *Pellia* in the early spring. Note the presence of dark green globular capsules just behind the growing points of some of the thallus branches. Note also small warty prominences further back from the tip and either side of the mid-rib. These are old antheridial cavities, now empty. Dissect out a sporogonium, noting the short seta. Crush the capsule into a drop of water : observe the wall with its characteristic thickenings, the spores and the elaters. Cut a transverse section of the thallus, mount in water and note the structure as described in the text.

2. Collect *Pellia* plants in the early summer. Note the presence of antheridia and cut sections through the thallus where they occur. Note also involucres just behind the tips of some of the branches and cut longitudinal sections through these in order to see the archegonia.

3. Collect various mosses, noting in each : the leafy gametophyte with rhizoids at its base and the capsule or sporogonium.

4. Collect the common woodland moss *Mnium hornum* in early summer. Note that some tufts of plants bear rosette-like antheridial cups. Dissect out the contents of one of these into water and note the structure of the antheridia and paraphyses. The liberation of the spermatocytes may be observed. The archegonial cups are less conspicuous, and a number of apices may have to be dissected before archegonia are found.

5. Examine capsules of *Funaria* or other mosses at different stages of maturity, and examine in particular the peristome and the liberation of the spores. If a cut-off capsule is placed upright in a bit of 'Plasticine', the peristome can be examined under the low power and breathing on the capsule will demonstrate the hygroscopic movements of the peristome teeth.

6. Collect protonema from hedge-rows or on the soil in flower pots. It often grows intermingled with *Vaucheria*, but is easily distinguished from the latter since it is septate. Note the green filaments with transverse septa and the brownish rhizoids with oblique septa. Observe buds on the green filaments and young plants in various stages of development.

CHAPTER XXVI

PTERIDOPHYTA

The life-history of the Pteridophyta, that is, the ferns, club-mosses and horse-tails, is essentially similar to that of the liverworts and mosses just considered, in that, in both groups, there is a regular sequence of two phases or generations. The outstanding feature of the Bryophytes is that the plant is the gametophyte generation, and the sporophyte is the sporogonium which remains dependent on the gametophyte throughout its development. In striking contrast to this is the fact that in the Pteridophytes the gametophyte is a relatively small and insignificant structure called a **prothallus**, while the sporophyte is the large fern or club-moss plant, which is completely independent of the gametophyte except in the earliest stages of its development. This alteration in the balance of the two generations as compared with the Bryophytes, is related, as will be stressed again later, to the fact that the fern is more highly adapted to terrestrial life.

THE MALE OR SHIELD FERN

The male or shield fern (*Dryopteris filix-mas*) is a very common plant in Great Britain, frequently growing in woods, hedgerows, and near streams. Though it prefers shady places, it is sometimes found in more exposed places such as hillsides. The plant presents a very robust appearance (Figs. 6 and 298).

Dryopteris has a thick and short rhizome which is covered with the remains of leaf-bases. The true roots usually arise in the regions of the leaf bases. The

H. Bastin

FIG. 298. SHIELD OR MALE FERN AMID MOSS-COVERED STONES.

365

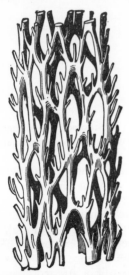

FIG. 299. DICTYOSTELE OF
Dryopteris filix-mas DISSECTED
OUT, SHOWING THE OVER-
LAPPING LEAF-GAPS.

(After Reinke.)

leaves (or fronds), which form a terminal crown, are large, often reaching a length of 2–3 ft. The main stalk or **rachis** bears two rows of leaflets or **pinnae**, and these are further subdivided into **pinnules**. Such a frond is said to be bipinnate. Other ferns have different types of frond. The hart's-tongue fern (*Phyllitis scolo-pendrium*) has an undivided lamina; the polypody (*Polypodium vulgare*) has a once pinnate leaf; while the bracken (*Pteridium aquilinum*) fronds are even more sub-divided than those of *Dryopteris*. The new leaves are developed towards the apex of the rhizome, and are covered with masses of small brownish scales known as **ramenta**. The young leaf is coiled like the spring of a watch, and as growth continues, it uncoils itself. The branches of the leaf are incurved in a similar fashion, so that the younger parts are well protected.

The rhizome has a complicated vascular system. If dissected out, this system, or stele, is seen to be in the form of a cylindrical net-work (Fig. 299). A transverse section of the stem therefore shows a ring of vascular strands, each one of which is termed a partial stele or **meristele** (Fig. 300), embedded in the food-storing parenchyma of the ground tissue. The vascular tissue supplying each large frond consists of a group of small strands, or leaf traces, which join the stele round the sides and base of a gap in the net-work, the gap itself being called a **leaf gap**. In the section, the spaces between neighbouring meristeles are the leaf gaps. The numerous adventitious roots are supplied by diarch vascular strands, which also join on to the net-work of the stele.

None of the lower plants so far considered possess vascular tissue, and even in the fern it is not so highly specialised as in the flowering plants. The sectional view of a meristele shows that it consists of a central mass of xylem sur-rounded in turn by a ring of phloem, pericycle and endodermis (Fig. 300). The xylem consists of tracheids only. These are of the scalariform type, except for the protoxylem, which consists of spiral and annular tracheids. No vessels of the angiosperm type are present. The phloem consists of sieve tubes, but these, too, are not so highly specialised as those of Angiosperms. Sieve plates occur on the longitudinal walls, as well as on the oblique end-walls between super-posed elements. There are no companion cells.

The cylindrical net-work type of vascular system is common in more advanced ferns such as *Dryopteris*; it is termed a **dictyostele** (Fig. 299). The most primi-tive ferns have a much simpler stele, a **protostele**, in which a solid central column of xylem is surrounded in turn by phloem, pericycle and endodermis. More advanced ferns have a stele in the form of a tube, which is only perforated at long intervals by leaf gaps. In these types, a transverse section of an internode shows a complete ring of xylem surrounded both internally and externally by phloem, pericycle and endodermis. This type of stele is a **solenostele**. If, as in *Dryop-teris*, the leaves are closely inserted on the stem, then the leaf gaps will overlap and the tubular stele then becomes a cylindrical net-work or dictyostele. Some

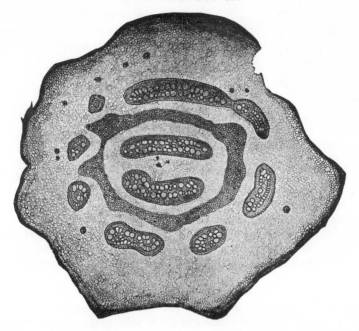

FIG. 300. TRANSVERSE SECTION OF RHIZOME OF BRACKEN (*Pteridium aquilinum*).
Showing outer and inner meristeles with a band of sclerenchyma between.

of the most advanced ferns, such as bracken (*Pteridium aquilinum*), are even more complex in the structure of their stele. The bracken rhizome has two concentric net-works, united only at the nodes, so that a transverse section shows an inner and an outer circle of meristeles (Fig. 300).

Vegetative reproduction is common among ferns. In *Dryopteris*, vegetative buds appear on the bases of older leaves which have already been shed. These buds produce new roots and leaves, and later, when the progressive rotting of the rhizome has taken place sufficiently far to leave these buds behind, they establish themselves in the soil and produce new plants. Other ferns, for example, *Asplenium bulbiferum*, produce buds or bulbils on the upper surface of their fronds. Such buds, if detached, will grow into new plants. This vegetative mode of reproduction is described as ' sporophytic budding ', and is not an essential phase in the alternation of generations. It serves merely to increase the number of sporophytes.

SPORE PRODUCTION

The sporangia of *Dryopteris* are borne on the under surface of leaves which do not differ at all from the purely vegetative ones. It is usually the leaves which are formed later in the season which thus become **sporophylls**. In other ferns, for example, the hard-fern (*Blechnum spicant*), there are two kinds of fronds, the ordinary foliage ones (known as **trophophylls**) and longer ones with narrower pinnae, the sporophylls, which bear the sporangia on their under surface.

If the under surface of a *Dryopteris* sporophyll is examined in early summer, numerous kidney-shaped structures, known as **sori**, will be seen. The charac-

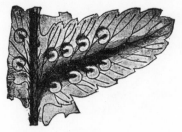

FIG. 301. PORTION OF THE UNDER-SURFACE OF THE SPOROPHYLL (LEAF) OF *Dryopteris*, SHOWING SORI.

teristic shape of each sorus is due to the presence of a membrane, the **indusium**, which covers and protects a number of sporangia (Fig. 301).

A transverse section of a pinnule passing through a sorus will give a clearer idea of the structure (Fig. 302). The indusium is connected by a stalk to a cushion-shaped structure, the **receptacle**, which is closely connected with a vein of the leaf for the provision of food, especially to the developing spores. Attached to the receptacle are numerous club-shaped organs which finally bear the spores. They are, therefore, the **sporangia**.

The head of a sporangium is shaped like a biconvex lens, the convex sides consisting of a single layer of thin-walled, flat cells. Passing around the edge of the head is a conspicuous row of cells, which stops short on one side. This is known as the **annulus**. Each cell of the annulus is shaped rather like a short Nissen hut, with its floor and end walls stoutly thickened, but with a thin curved roof and sides. In an optical section of the annulus, each cell shows golden-brown thickening on its inner and radial walls (the thickening appearing in the form of a U-shaped band), while the outer wall is thin (Figs. 302 and 303). This differential thickening of the walls is important, since it provides the mechanism for the liberation of the spores. Where the annulus stops short, the cells are thin-walled and form what is called the **stomium**. Here, dehiscence eventually takes place.

Each sporangium arises as a single cell on the placenta. Cells are cut off from the base to form the stalk, and the end cell then divides to form the head of

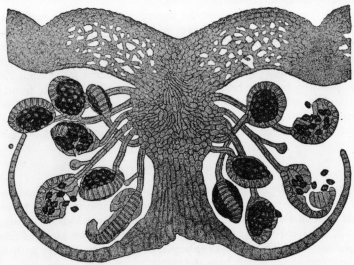

FIG. 302. VERTICAL SECTION THROUGH THE SORUS OF *Dryopteris*, SHOWING SPORANGIA IN VARIOUS STAGES OF DEVELOPMENT.

(After Kny.)

the sporangium. In this latter, a single-layered wall is soon differentiated from a central cell. Further divisions in the central cell give rise to a double layer of nutritive cells, the **tapetum,** surrounding a group of 12 spore-mother-cells. In many ferns there may be 16 or more such mother-cells. Later, as the annulus itself is reaching maturity, each spore-mother-cell divides meioti-cally to produce four spores, thus giving usually 48 spores in each sporangium. Each spore is composed of a single nucleus surrounded by some cytoplasm and a thin cell-wall.

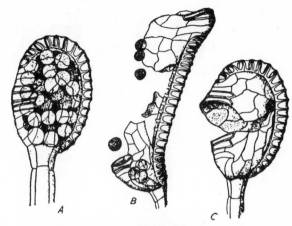

FIG. 303. FERN SPORANGIA.

A, unopened sporangium filled with spores ; *B*, the annulus has straightened and torn the sporangium into halves, freeing some of the spores ; *C*, the empty sporangium after the annulus has suddenly returned to its original position.

(*From Robbins and Rickett's ' Botany '. Courtesy of the D. Van Nostrand Co., Inc.*)

Outside this is a much thicker wall, which frequently shows a characteristic pattern of thickening.

Once the spores are ripe, they depend upon dry conditions for their dispersal. During such conditions, the indusium shrivels up, and the sporangia are exposed. Then these in turn begin to desiccate. At this stage, the annulus cells have lost their living contents, but are filled with water. As evaporation through the thin outer wall takes place, the volume of water in each cell is diminished. Owing to the cohesive power of the water and its adhesion to the walls, the volume of the cell itself is reduced by a pulling-in of the thin outer wall while the two arms of the U-shaped band of thickening are drawn together. As this happens in each cell of the annulus, the outer margin of the annulus is shortened and a considerable strain is set up, tending to straighten the annulus. This strain results in the tearing open of the sporangium at the stomium. As more water is lost by evaporation, the volume of the annulus cells is steadily reduced and the outer walls are drawn in still further, while the arms of the U-shaped band are drawn closer together. This results first in the straightening of the annulus and then in its backward curvature (Fig. 303, *B*). During this movement the annulus carries with it the top half of the sporangium still containing some of the spores. The cells of the everted annulus are still filled with water, which is now under a very considerable tension—perhaps as much as 300 atmospheres. Finally, this tension becomes so great that the cohesion of the water breaks down and the water leaves the cell-walls. With this release of tension, the elastic U-shaped band in each cell reverts suddenly to its original form This results in a rapid swinging back of the annulus to its original position, a movement which results in the jerking-out of the spores from the sling-like upper half of the sporangium still attached to the annulus (Fig. 303, *C*). It may be noted, too, that as the annulus cells suddenly regain their original volume, the drop of water which remains in each of them is

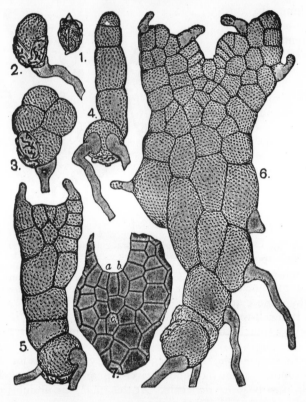

FIG. 304. STAGES IN THE GERMINATION OF A SPORE, AND DEVELOPMENT OF A PROTHALLUS
OF *Dryopteris*.
(After Kny.)

vapourised (under the resultant condition of low pressure) and appears as a black
bubble of water vapour. Although most of the spores will have been dispersed
either in the first tearing-open of the sporangium or by the subsequent catapult-
like movement, the process may be repeated. If rain falls on an opened sporangium, water is again taken in by the annulus cells and on subsequent drying
the annulus will again behave as described.

THE GAMETOPHYTE

The ripe spores of *Dryopteris* and the majority of ferns are devoid of chlorophyll and can remain viable in a dry condition for a considerable time. If,
however, they are provided with suitable conditions of warmth and moisture,
they germinate to produce the haploid gametophyte, known as the **prothallus.**

The spore takes in water, chlorophyll is developed, and the contents swell,
rupturing the outer coat. A colourless rhizoid grows out at one end, and this
serves for fixation and absorption. At the other end, the inner spore wall grows
out as a green tube. A transverse wall is formed in this latter, and the terminal
cell so formed divides repeatedly so that a filament of cells is formed. Sooner or
later (according to external conditions) intersecting oblique walls are formed in

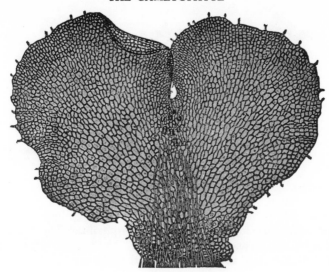

FIG. 305. PROTHALLUS OF *Dryopteris*, SEEN FROM BELOW, BEARING ANTHERIDIA AMONG ITS RHIZOIDS AND ARCHEGONIA NEAR THE APICAL INDENTATION.
(After Kny.)

the end cell, and an apical cell is thus established (Fig. 304). Eventually, the prothallus develops into a heart-shaped structure, with the apical cell, or growing point, situated at the base of the depression between the two lobes (Fig. 305). The mature prothallus is composed of a single layer of cells, except at the centre where several layers are produced to form a cushion. On the lower surface, unicellular rhizoids are produced, which carry out the physiological functions of roots.

The prothallus is the gametophyte generation of the fern plant. It therefore produces the gametes. Sperms and eggs are usually produced on the same prothallus in organs known respectively as antheridia and archegonia.

The antheridia are frequently produced on young filamentous prothalli, but on the mature prothallus they are formed on the lower surface, towards the thin marginal lobes, or among the rhizoids. They are produced from single superficial cells. Each antheridium is composed of a spherical wall of cells surrounding a collection of sperm mother-cells. The wall itself consists of two ring-shaped cells (like two quoit rings, one on top of the other) with the summit occupied by a cap-cell. The contents of each sperm mother-cell gradually becomes modified to a spirally coiled sperm which bears numerous cilia near its pointed end (Fig. 306). Some of the cytoplasm of the mother-cell usually remains as a sac attached to the lower end of the sperm.

The archegonia are also produced on the lower surface of the prothallus, but nearer the centre of the cushion just behind the apical indentation. The structure of a fern archegonium (Fig. 307) is essentially similar to that of a Bryophyte but, instead of being raised on a stalk (as in the latter), it is sunk in the tissues of the cushion. Only a short neck (consisting of four rows of cells) projects, curving sharply backwards so as to face the base of the prothallus. The canal of the neck is occupied by a neck canal-cell, which may contain several nuclei. Below this,

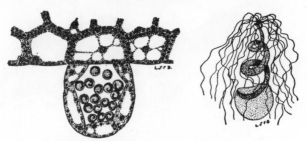

FIG. 306.   LEFT, AN ANTHERIDIUM OF *Dryopteris*, CONTAINING YOUNG SPERMS ; RIGHT, A
SPERM, BEARING NUMEROUS CILIA.

and embedded in the prothallus tissue, is a small ventral canal-cell and the large
ovum.

Fertilisation only takes place when rain or dew has resulted in the presence of
a film of water over the lower surface of the prothallus.   The presence of such
water leads to the dehiscence of the antheridia.   The cap cell is ruptured and the
ring-shaped wall cells swell to such an extent that the mass of sperm mother-cells
is extruded on to the surface of the prothallus.   On coming into contact with the
water, the wall of each sperm mother-cell dissolves and the motile sperms are set
free, usually while still near to the antheridium.   The presence of water has, at
the same time, led to the opening of the necks of any ripe archegonia.   This
opening is caused by the swelling of the mucilaginous contents of the disintegrating
canal cells.   The mucilage, some of which exudes from the open neck, contains
malic acid, a substance which acts as a chemotactic stimulus to motile fern sperms.
These latter thus swim through the film of water (with their ciliated ends fore-
most) towards the archegonial necks.   A number of sperms may enter each
archegonium.   They become somewhat straightened out and move down the
mucilage of the neck by undulatory, or corkscrew-like, movements of their bodies
rather than by the movements of their cilia.   Finally, fertilisation occurs when
one sperm enters the egg at a clear ' receptive spot ' and moves in to fuse with
the nucleus.   As in the Bryophytes, more than one archegonium may be fertilised,
but normally only one zygote develops into a new sporophyte plant.

In the early stages, the young embryo plant develops a sucker or foot which
becomes deeply embedded in the prothallus, and absorbs nourishment from it.
But growth continues rapidly, and finally a root, stem and leaf are formed.   Once
the root has become established in the soil, and the leaf unfolded, the young fern
plant is no longer dependent on the prothallus, which finally withers away (Figs.
308 and 309).

The vast majority of ferns produce only one type of spore, that is, they are
homosporous.   A few aquatic ferns, however, produce small microspores and
large megaspores ;   these are said to be **heterosporous**.   The next group to be
considered, the Lycopodiales or club-mosses, also possesses both homosporous and
heterosporous members.   Two examples will be briefly described : *Lycopodium*, a
homosporous form, and *Selaginella*, which is heterosporous.

LYCOPODIUM

Species of *Lycopodium*, or clubmosses (Fig. 6), are frequently found growing in
hilly districts, especially moorland.   Five species are indigenous to Great Britain,

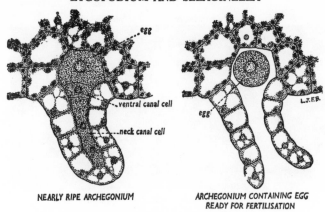

NEARLY RIPE ARCHEGONIUM

ARCHEGONIUM CONTAINING EGG
READY FOR FERTILISATION

FIG. 307. ARCHEGONIA OF *Dryopteris*.
Left, an unripe archegonium ; right, ripe archegonium from which the ventral canal and
neck canal cells have been extruded.

and about a hundred species are known altogether, spread throughout the tropics
and temperate regions. Most species of *Lycopodium* are low-growing, though
a few are epiphytic. The two commonest British species are the fir club-
moss (*Lycopodium selago*,) and the stag's-horn club-moss (*L. clavatum*), both of
which are members of the sub-alpine flora. *L. selago* is a shrubby plant with
dichotomous branching, reaching a height of six inches or more ; *L. clavatum* is
a trailing plant with monopodially branched rhizomes which reach a length of
several feet. In both, the leaves are small and simple, a feature in which they
differ markedly from the ferns (Fig. 310).

All species of *Lycopodium* are, like the fern, homosporous. Each fertile leaf
bears a single large sporangium on its upper surface near to the base. In *L. selago*,
these leaves are similar to the sterile ones, groups of fertile leaves alternating with
groups of sterile ones along the length of the stems. *L. clavatum*, however, has
definite **cones** or **strobili**. Each cone, which is borne at the tip of an upright-
growing lateral branch, consists of a central axis surrounded by closely set sporophylls. These fertile leaves are different in form from the vegetative leaves and
each bears a single large sporangium.

The sporangium when ripe splits, and the light spores are disseminated by
the wind. The prothalli which are produced from the germinating spores vary
considerably. Some are green, and others only partly green. In many species
(for example, *L. clavatum*) the prothalli develop below soil-level as colourless,
peg-top-shaped bodies about ¼ in. in diameter. The tissues contain fungal hyphae,
and the nutrition of the prothallus is closely related to the presence of this ' mycorrhiza '. The antheridia and archegonia are usually sunk deeply into the prothallial
tissue. The sperms, unlike those of the fern, are biciliate.

SELAGINELLA

Selaginella, unlike *Lycopodium* and the fern, is heterosporous. There are about
five hundred different species of this plant, most of which are indigenous to the

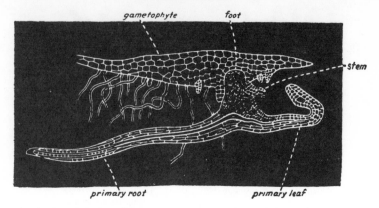

FIG. 308.　SECTION THROUGH FERN PROTHALLUS AND YOUNG SPOROPHYTE WHICH IS STILL
ATTACHED BY ITS FOOT TO THE PROTHALLUS (GAMETOPHYTE).
(From Warming's ' Systematic Botany ', published by The Macmillan Co., New York.)

damp tropical forests.　One species, namely, *S. selaginoides* (Fig. 311), is indigenous
to Great Britain.　*S. kraussiana* is commonly cultivated in greenhouses.

Selaginella is usually of a creeping habit.　Like *Lycopodium*, it branches dichotomously.　In the more advanced species, the leaves are simple and small, and are
borne on the stem in four rows, two rows of small leaves being borne on the upper
surface of the stem, and two rows of larger leaves on the lower surface.　Certain
species, however, for example, *S. selaginoides* are upright, and all their leaves are
equal in size.　The leaves of all the species possess a small membranous scale or
ligule at the base of the adaxial surface.　This ligule, which may serve to absorb
water, is also characteristic of the aquatic quill-wort (*Isoetes*), the only other
heterosporous genus of the Lycopodiales.

The rooting systems of *Selaginella* are of unusual types.　*S. selaginoides* (Fig. 311)
has a little swollen structure at the base of its stem (the ' basal knot ') from which
all the roots arise.　The dorsiventral species, such as *S. kraussiana*, produce root-
like structures, called **rhizophores**, at each forking of the stem.　These are
positively geotropic, and when their tips reach the soil they produce a number of
true roots.

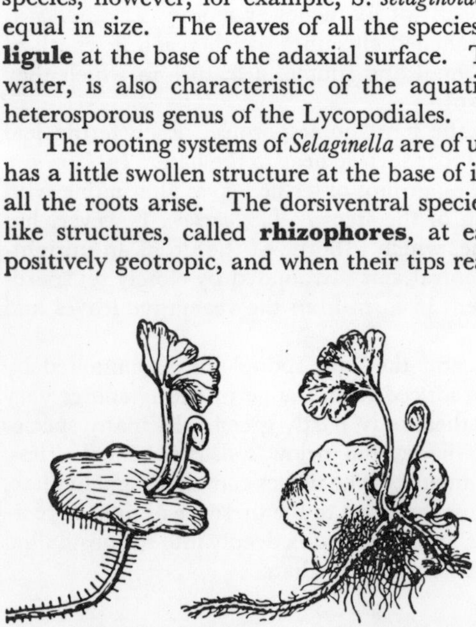

The sporophylls of *Selaginella*
are borne in distinct strobili, each
strobilus being produced at the
distal end of a vegetative branch.
In the axil of each sporophyll a
large, stalked sporangium is borne
(Fig. 312).　Since *Selaginella* is
heterosporous, there are two kinds
of sporangia, namely, microsporangia and megasporangia.　Both
types are borne in the same strobilus, though their arrangement
within the strobilus varies with the
species (Fig. 312).　The microsporangium bears numerous micro-

FIG. 309.　LEFT, VIEW OF UPPER SURFACE OF
FERN PROTHALLUS AND YOUNG SPOROPHYTE ;
RIGHT, SEEN FROM BELOW.
*(From ' Nature and Development of Plants ', by Carlton C. Curtis,
copyright by Messrs. Henry Holt & Co. Reproduced by permission.)*

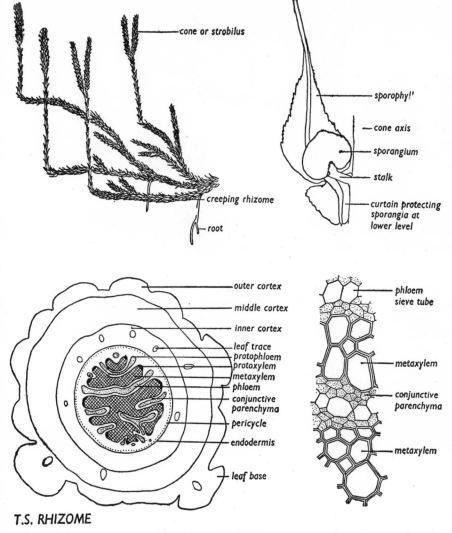

cone or strobilus

sporophyl!'

cone axis

sporangium

stalk

curtain protecting
sporangia at
lower level

creeping rhizome

root

outer cortex

middle cortex

inner cortex

leaf trace
protophloem
protoxylem
metaxylem
phloem
conjunctive
parenchyma

pericycle

endodermis

leaf base

phloem
sieve tube

metaxylem

conjunctive
parenchyma

metaxylem

T.S. RHIZOME

FIG. 310. *Lycopodium clavatum.*
(*F. W. Williams.*)

spores (Fig. 313) ; whereas the megasporangium bears only four megaspores
(Fig. 313). This condition arises from the fact that in the developing microsporangium all the spore-mother-cells divide by meiosis to give four spores ; while
in the megasporangium only one of the mother cells does so, the others serving
merely for its nutrition. Each sporophyll, like the vegetative leaves, possesses a
ligule, and this is situated just above the insertion of the sporangium.

As in *Lycopodium* and the fern, all the spores of *Selaginella* are shed from the
parent plant ; thus differing from the corresponding structures in Angiosperms,
and also, as will be seen (Chap. XXVII), in Gymnosperms. Often, however,

Fig. 311. *Selaginella selaginoides.*

the germination of the spore may begin before it is shed from the sporangium.

The microspore develops on damp soil to produce a very much reduced microprothallus. Actually only one vegetative cell and one antheridium are produced. The latter produces numerous bi-ciliate sperms. The megaspore produces, on the other hand, much more prothallial tissue. Several archegonia, too, are formed on a single megaprothallus. While the vegetative parts of the prothalli are reduced as compared with the fern prothallus (those of the microprothallus more so than those of the megaprothallus), the sexual organs, antheridia and archegonia resemble those of the fern.

The separation of male and female prothalli would seem to make fertilisation less certain than in the fern prothallus. In many cases, however, a large number of microspores and megaspores will have fallen around the plant, and if the soil is moist the bi-ciliate sperms may swim in the film of water to female prothalli in the immediate vicinity. The sperms are attracted chemotactically to the open archegonial necks, and fertilisation results in the development of a new sporophyte.

GENERAL

The ferns and club-mosses described above all show the characteristic life-cycle of the Pteridophyta, in which the sporophyte is the plant itself and the gametophyte is a relatively small prothallus. The sporophytes are much more advanced than the sporogonia of the Bryophyta, and show a differentiation of the plant body into stem, leaves and true roots. They are, in fact, well equipped for

independent existence on land. The prothallus is, however, like the liverwort or moss
plant with which it corresponds, still dependent on the presence of fluid water for the
purpose of fertilisation. To this extent, the
Pteridophyta are still amphibious.

The development of heterospory, and consequently of separate male and female gametophytes, in some of the Pteridophytes represents an important evolutionary advance which
must have been made by the ancestors of the
plants which reproduce by means of seeds, that
is, the Gymnosperms and Angiosperms.

Heterospory as seen in *Selaginella* does not
appear to be of great biological value, and,
indeed, as pointed out above, it leads to difficulties in the accomplishment of fertilisation.
Nevertheless, further modifications of this habit
have proved eminently successful in the flowering plants. One modification is the retention
of the megaspore within the megasporangium
on the parent sporophyte until after fertilisation and the development of an embryo.
Another is the use of external agencies, such
as wind and insects, to carry the male gametophytes to the neighbourhood of the female
gametophytes, while yet another important
advance is the development of a method (the
use of pollen tubes) by which the male gametes
can be carried to the female ones without the
need for fluid water. These modifications are
the characteristic features of what is known as

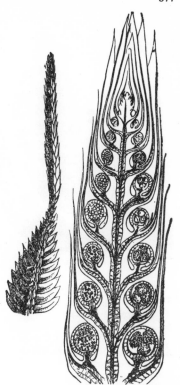

FIG. 312. *Selaginella inaequalifolia.*
Left, part of a vegetative branch
bearing a cone ; right, longitudinal
section of a cone bearing microsporangia on the left and megasporangia on the right.

the ' seed habit '. They will be examined further in relation to the life-cycle of
the Gymnosperms.

PRACTICAL WORK

1. Examine complete plants of *Dryopteris filix-mas* and *Pteridium aquilinum*. Note the
characters of the stem and fronds and examine the sori (kidney-shaped in *Dryopteris*, long
fusion sori along the margins of the pinnules in *Pteridium*). Identify other ferns by means
of a Flora.

2. Cut a transverse section of a pinnule of *Dryopteris* so as to pass through a sorus.
Note : the tissues of the leaf ; the placenta ; sporangia in various stages of development ;
the indusium.

3. Examine the dehiscence of sporangia by scraping some ripe sporangia into a drop
of glycerine on a slide. The glycerine withdraws water from the annulus cells and thus
causes the opening of the sporangia. The movements of the annulus will be slowed down
by the presence of the glycerine. Other sporangia may be scraped on to a warm slide and
the annulus movements observed under the microscope.

4. Fern prothalli may be grown in the following manner. Place a soaked flower pot
inside a larger one, packing the space between with wet sphagnum or peat. Allow a

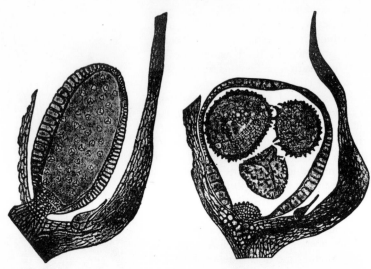

FIG. 313. *Selaginella apus.*
   Left, longitudinal section of microsporangium containing numerous microspores
(×55) ; right, longitudinal section through megasporangium showing three of the four
megaspores (×21).

(After Miss Lyon.)

mature sorus-bearing frond to dry on a piece of paper and then scatter the spores so obtained
on the inner surface of the small flower pot. Stand the pots in an inch or so of water and
cover the top of the pots with a sheet of glass. Green prothalli will soon appear, and successive stages in their development should be followed. It should be possible to see the
archegonia and also the liberation of sperms from the antheridia. If, after the prothalli
are showing archegonia, they are watered from above, young sporophytes will develop.

   5. Examine plants of *Lycopodium selago* and *L. clavatum.* Note the presence of sterile
and fertile zones in the former and of definite cones in the latter. Examine the sporangia,
both externally and by cutting sections of the cone of *L. clavatum.* Prothalli have only rarely
been found in Britain and will not be available for examination.

   6. Examine plants of *Selaginella selaginoides* and either *S. kraussiana* or *martensii.* The two
latter are commonly grown in greenhouses. Note the features described in the text and,
in particular, examine the cones by dissection and by sectioning. Note the megasporangia
each containing four megaspores and the microsporangia with numerous small microspores. Note on both types of spore the tri-radiate ridge which indicates their origin in
tetrads following meiosis. Collect the micro- and mega-spores from a ripe cone and scatter
on moist filter paper. Note the development of young sporophytes.

CHAPTER XXVII

GYMNOSPERMS

The two great characteristics of the Spermophyta are:
(1) The formation of a pollen tube.
(2) The production of seeds.

In this way the seed plants are released from their dependence on liquid water, and become truly terrestrial. By their seeds, they ensure the most efficient and sure method of reproduction.

GYMNOSPERMS

The present-day Gymnosperms are actually survivors of a much larger class of plants. They comprise four groups, namely, **Cycadales**, **Ginkgoales**, **Gnetales**, and **Coniferales**. The last-named is by far the greatest group and the most highly evolved. This group will be considered first, and then a brief review of certain other more primitive Gymnosperms will show some connecting links with the Pteridophyta.

Among the best-known members of the Coniferales are the pines (*Pinus* species), cedars (*Cedrus* and other genera), firs (*Abies* species), spruce firs (*Picea* species), Douglas firs (*Pseudotsuga* species), junipers (*Juniperus* species), cypresses (*Cupressus* and *Chamaecyparis* species), yews (*Taxus* species), redwoods (*Sequoia* species), larches (*Larix* species), and the peculiar type, the monkey puzzle (*Araucaria imbricata*).

THE SCOTS PINE

The familiar Scots pine (*Pinus sylvestris*) is the sporophyte plant, and is therefore homologous with the buttercup plant, the fern plant, and the moss sporogonium.

Young trees have an upright axis bearing whorls of branches at intervals, the distance between successive whorls representing a year's growth. The main shoot grows more strongly than the lateral branches, thus giving the familiar 'Christmas tree' effect (Fig. 314); but growth later becomes more irregular, and this symmetrical effect is lost (Fig. 315).

VEGETATIVE STRUCTURE

The Scots pine is an evergreen, native to northern Britain and the countries of northern Europe, though it is now found in much more extended areas. It is the only conifer actually indigenous to Britain. Under favourable conditions, the tree will attain a height of 100 feet, the erect trunk at this stage being bare of branches for three-quarters of its length. The bark is rough and irregularly fissured; towards the top of an old tree it becomes a glowing red. In very old trees the trunk is often gnarled and twisted.

The branches which give the main skeleton of the tree are ' **branches of unlimited growth** '. They bear only scale leaves, which are spirally arranged. In the axil of each scale leaf, however, there is a short branch of limited growth or **dwarf shoot**. Each dwarf shoot also bears scale leaves, but it also bears two foliage leaves, or pine needles, at its tip. These are bluish-green in colour, and are short compared with other members of the *Pinus* genus and have a slight twist. They remain on the tree for two or more years and then fall, along with the dwarf shoots which bore them.

Examination of a twig of unlimited growth will show that it has a terminal bud and a few lateral buds immediately below this. The terminal bud will carry on the growth of the twig and the laterals will give rise to branches, also of unlimited growth. Each bud has a number of bud scales, and in the axils of the lower scales there are minute buds which will develop as dwarf shoots next year. A few of the upper scales will have larger buds, destined to grow out as branches of unlimited growth.

In its young stages, the root system is that of a typical tap root ; but this later becomes lost in branch roots which attain equal size. The whole root system is rather shallow, and many of the finer roots show an ectotrophic mycorrhiza, such roots being short, thick, and repeatedly branched (p. 137, Fig. 86).

In marked contrast to the Angiosperms, vegetative reproduction in Gymnosperms is exceedingly rare.

The general vascular system of *Pinus* is similar to that of a woody Dicotyledon, and both stem and root undergo secondary thickening of the normal type. The

Harold Bastin

FIG. 314. SCOTS PINE GROWING UNDER FORESTRY CONDITIONS.

Lonsdale Rugg

FIG. 315.   OLD SCOTS PINE IN SOMERSET.

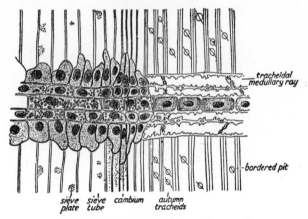

FIG. 316. PART OF A RADIAL LONGITUDINAL SECTION OF STEM OF *Pinus*, PASSING THROUGH
A MEDULLARY RAY IN THE REGION OF THE CAMBIUM (× 160).
(*After Schenck.*)

vascular tissues are, however, less specialised than those of Angiosperms. The
xylem consists almost entirely of tracheids with bordered pits on their radial
walls. The bordered pits on the spring wood are of the type described on p. 119,
each showing as two concentric circles; those of the autumn wood have a slit-
like aperture into the pit chamber, as shown in Fig. 316. No vessels are present.
The phloem consists of sieve tubes with sieve plates on their radial walls. There
are no companion cells, but some phloem parenchyma is present.

The medullary rays are numerous and differ in detailed structure from those
of the Angiosperms (Fig. 316). The central cells of a ray are usually living and
contain starch. Where the ray passes through the xylem, the marginal cells are
dead and empty and their thick walls have bordered pits; these tracheidal cells
serve for the radial transport of water. The character of the marginal cells changes
as the ray passes across the cambium into the phloem region. There, they are
living cells with abundant protein contents. These ' albuminous ' cells are drawn
out longitudinally and are closely applied to the sieve tubes. Food absorbed from
the latter may be transported radially in the stem along the central line of living
cells. The medullary rays have small radial air spaces associated with them so
that oxygen can pass radially from the cortical air spaces to living elements
embedded in the xylem mass.

The leaves are xeromorphic in their form and anatomy. The needle-shaped
leaves have a relatively small surface from which water might be lost in transpiration. There is a thick cuticle and the stomata are deeply sunken below the general
level of the epidermis (Fig. 317). The photosynthetic tissue consists of ' arm parenchyma ', that is, parenchyma with infolded walls, a modification which gives a
large surface for the display of the chloroplasts. The vein, surrounded by endodermis, has two vascular bundles embedded in a complex ground tissue, sometimes called pericycle. A band of sclerenchyma just below the bundles gives
rigidity to the leaf. A group of albuminous cells is present at the outer side of
each phloem group, while a group of tracheidal cells are similarly placed in
relation to the xylem groups. These two types of cell collectively form the trans-
fusion tissue, the tracheidal cells carrying water from the xylem to the mesophyll

and the albuminous cells carrying carbohydrates in the reverse direction to the phloem. The remaining ground-tissue consists of a mixture of tracheidal and parenchymatous cells.

A characteristic feature of *Pinus* and many other conifers is the presence of resin canals. These are long channels formed by the enlargement of intercellular spaces, and lined by an epithelium of glandular cells, which secrete resin into the cavity. Such resin canals occur in the cortex and secondary xylem of the stem, and some run radially in the larger medullary rays. They also occur in the leaf mesophyll, where they are protected by a sheath of fibres, and in the stele of the root. When one of these organs is wounded, resin is exuded at the wound surface and presumably acts as a protection against loss of water and the entrance of parasites.

THE CONES

The sporophylls of *Pinus* are borne collectively in **cones** or **strobili**. Unlike those of *Selaginella*, these are of two types, distinguished according to the type of sporophylls they bear. In one type, the carpellate or seed cone, they are all megasporophylls. In the other, the staminate or male cone, they are all microsporophylls.

FEMALE CONE

If a branch of *Pinus* is examined in June, it is possible to see three stages in the development of the carpellate or seed cones. These cones consist essentially

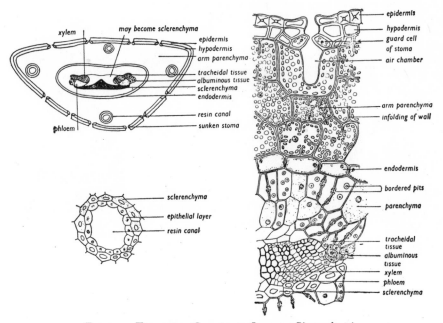

FIG. 317. TRANSVERSE SECTION OF LEAF OF *Pinus sylvestris*.
(*F. W. Williams*.)

of a central axis bearing spirally arranged megasporophylls. These latter are double structures with a small bract scale below and a larger ovuliferous scale above. Two ovules, which eventually become seeds, are present on the upper surface of the ovuliferous scale near to the axis.

The tip of the current year shoot is occupied by a terminal bud, and just below this, one or more ovoid, pinkish-green cones are present. These measure up to a quarter of an inch in length and, at this stage, they are succulent. Each replaces one of the lateral buds normally present in this position on a purely vegetative branch. These are seed cones ready for pollination.

A year's growth further back, that is, at the level of the first branches, similar but larger cones, green and still succulent, will be found. In these, fertilisation is about to take place if it has not already done so (Fig. 318).

Finally, at the level of the next lower branches, that is, another year's growth further back still, the familiar large, brown, woody cones are to be seen. These now have their scales separated, and the seeds borne on them are exposed and will soon be dispersed.

The series of cones just described gives, of course, a picture of the development of a mature seed cone. The position of the cones on the branch also gives a measurement of the time taken for this process to be completed. The young cones seen at the top of the shoot were, in fact, laid down at the end of the preceding summer in the terminal bud. They became apparent only when this bud opened in the spring to form the current year shoot. Such young cones are thus already six or eight months old when they are pollinated. A whole year elapses before fertilisation takes place, and another year before the resultant seeds are fully mature and ready to be dispersed. The process of developing ripe seeds is thus spread out over two and a half years.

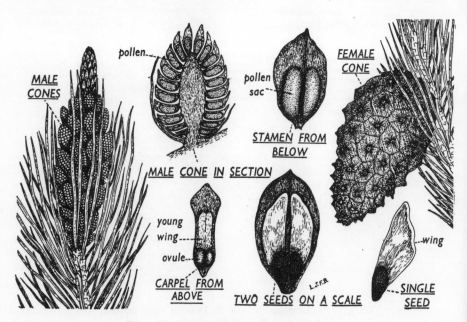

FIG. 318. REPRODUCTION IN THE SCOTS PINE.

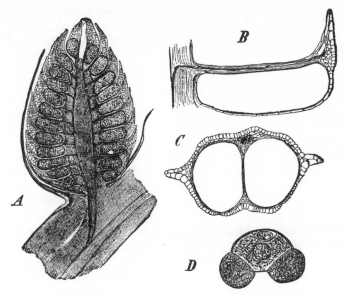

FIG. 319. MALE CONE OF SCOTS PINE.

A, Longitudinal section of the male cone (× 5) ; *B*, longitudinal section of a microsporophyll (stamen) (× 10) ; *C*, transverse section of a microsporophyll (pollen not shown)
(× 14) ; *D*, ripe pollen grain (× 200).

(After Strasburger.)

MALE CONE

Other branches of the same tree will show clusters of staminate cones (Fig.
319), yellowish, ovoid structures up to 1 cm. long. They occupy the base of the
current year shoot, and a careful examination will show that each cone is borne
in the axil of a scale leaf and thus replaces a dwarf shoot. The top of the shoot
bears normal dwarf shoots, each bearing a pair of leaves. The male cones were
laid down in the terminal bud during the preceding year, and were exposed when
the bud grew out in the spring.

The male cone consists of a central axis upon which the microsporophylls
are spirally arranged. The microsporophyll, or stamen, is composed of a stalk
which bears two pollen sacs on its ventral side (Fig. 319). After the pollen has
been shed, the withered cones fall off leaving a length of bare stem. Examination
of older parts of the branch will show similar bare patches where groups of
staminate cones were borne in previous years.

DEVELOPMENT OF MEGASPORES AND MICROSPORES

Each of the scales of the first year cone bears two ovules, the structure of
which is clearly comparable to that of an Angiosperm ovule at a corresponding
stage of development (Fig. 320). Each ovule consists of a parenchymatous
nucellus (or megasporangium) protected by a single integument. The micropyle,
which faces the axis of the cone, is rather wide, with a trumpet-shaped mouth.
Embedded deeply in the nucellus, there is a single, uninucleate embryo sac or
megaspore. This arose earlier in the development of the ovule by a single cell of

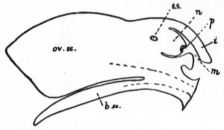

FIG. 320. LONGITUDINAL SECTION THROUGH MEGASPOROPHYLL OF SCOTS PINE AT TIME OF FERTILISATION.

b.sc., bract scale; *ov.sc.*, ovuliferous scale; *n.*, nucellus; *i.*, integument; *m.*, micropyle; *e.s.*, embryo sac; *p.*, pollen grain.

(After Coulter.)

the nucellus becoming enlarged as a megaspore mother-cell. The latter then divided by meiosis to give a chain of four megaspores, but only the innermost of these persisted as the functional embryo sac already described.

The microspores, or pollen grains, have meantime been formed in the microsporangia or pollen sacs of the staminate cone. They have developed as a result of meiotic divisions of spore mother-cells which were differentiated earlier in each sac. The microspores or young pollen grains are uninucleate.

The development of the microspores (pollen) is similar in general to the development of the spores in the fern.

DEVELOPMENT OF MALE AND FEMALE GAMETOPHYTES

The microspores or pollen grains germinate sometimes a month before they are shed. Before germination, a microspore consists of a single cell with a large nucleus. It has two balloon-like wings (Fig. 321). These are filled with air, and thus facilitate wind dispersal. The nucleus divides once, and one of the daughter nuclei forms the **first vegetative nucleus** of the microprothallus. This becomes pressed against the wall of the grain and disintegrates. The remaining nucleus divides again and produces the **second vegetative nucleus** and the **antheridial nucleus**. The two vegetative nuclei appear to have no function, and therefore may be looked upon as vestigial, vegetative, prothallial cells. Shortly before the pollen is shed, the antheridial cell divides again to form a **generative cell** and a **tube cell** (Fig. 321).

About this time, the stamen (microsporangium) dehisces by a longitudinal slit, and the pollen is shed. The large amount of pollen which is produced by a single tree is very surprising; but due allowance must be made for the wastage involved in wind pollination. Some pollen reaches the female cones of

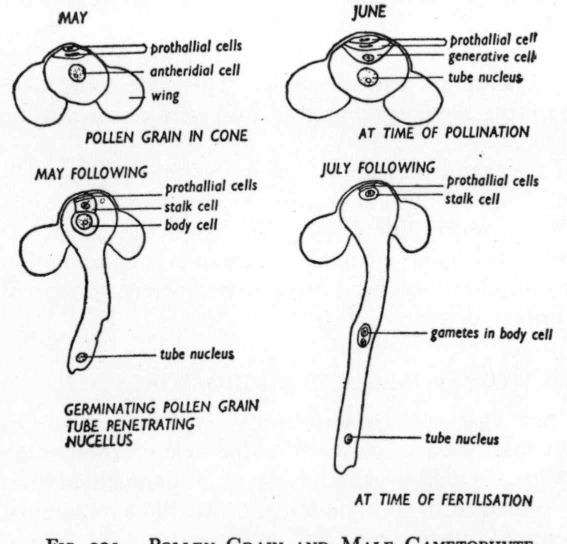

MAY
prothallial cells
antheridial cell
wing

POLLEN GRAIN IN CONE

JUNE
prothallial cell
generative cell
tube nucleus

AT TIME OF POLLINATION

MAY FOLLOWING
prothallial cells
stalk cell
body cell

tube nucleus

GERMINATING POLLEN GRAIN
TUBE PENETRATING
NUCELLUS

JULY FOLLOWING
prothallial cells
stalk cell

gametes in body cell

tube nucleus

AT TIME OF FERTILISATION

FIG. 321. POLLEN GRAIN AND MALE GAMETOPHYTE OF *Pinus*.

(F. W. Williams. After various authors.)

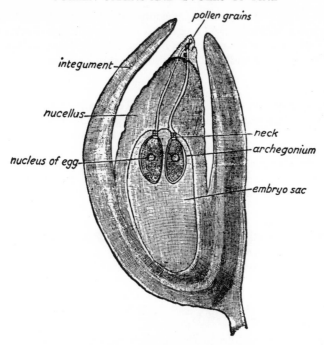

FIG. 322. LONGITUDINAL SECTION THROUGH AN OVULE OF *Picea excelsa* (SPRUCE) AT TIME OF FERTILISATION (× 9).
(*After Strasburger.*)

the first year and penetrates between the megasporophylls, which are slightly separated from one another at this stage. Some of the pollen grains reach the ovules, each of which exudes a drop of viscid fluid from its micropyle. The pollen grains are held in this fluid, which, on drying, draws the grains through the micropyle on to the surface of the nucellus (Fig. 322).

Soon after pollination occurs, the female cone closes up again, and the scales are all sealed together by the exudation of resin. Then the pollen rests inside the female cone for about twelve months. During this year, the cone enlarges and the megaspores or embryo sacs in the ovules proceed to germinate. The single nucleus of the megaspore divides repeatedly without walls being formed between the daughter nuclei (free nuclear division) and, by the end of the autumn, the enlarged embryo-sac has a number of nuclei in the peripheral layer of cytoplasm. In the following spring, development is continued. The megaspore develops a large amount of tissue which is actually the megaprothallus, and which forms endosperm (though this endosperm differs from that of Angiosperms in that it develops before and not after fertilisation). Then towards the end of the year after pollination has taken place, **archegonia** arise in the female gametophyte towards the micropylar end (Fig. 322). The number of archegonia varies. Each archegonium consists of a large ovum and small ventral canal-cell, surrounded by a layer of nutritive 'jacket cells' like a venter. At the top of the ovum there is a short neck consisting of two or three tiers of small cells ; there are no neck canal-cells.

While the development of the female prothallus is going on, the male prothallus (pollen grain) completes its development. A pollen tube is formed, and the tube nucleus passes into it. The generative nucleus divides into a **stalk cell** and a **body cell** ; then the latter passes into the tube, where it divides to produce two male gametes. The stalk cell may also pass into the tube.

FERTILISATION

Fertilisation is now ready to take place. The pollen tube grows and penetrates the canal of the archegonium and discharges its four nuclei into the cytoplasm of the egg (Fig. 323). One male gamete fuses with the egg nucleus and fertilises it. The other three cells disintegrate.

Immediately after fusion has occurred, the resulting zygote commences division. Two divisions occur quickly, and four free nuclei are produced which pass to the base of the cell, where they each divide again (Fig. 324). During this division, walls are formed. Then, by further division, four tiers, each consisting of four cells, are formed. The lowest tier but one elongates to form the suspensor, and the lowest tier of all forms the embryo. The four lowest cells forming the lowest tier may collectively form one embryo, or they may each form a single embryo. Since there are several archegonia in each ovule, all of which may be fertilised, a large number of embryos may be formed in a single seed. This is known as **polyembryony**. But usually only one embryo reaches maturity, while the rest disintegrate. When ripe, each seed bears a membranous wing which splits off from the ovuliferous scale (Fig. 318).

The number of cotyledons in the young embryo varies around fifty. The radicle is very large and robust. The seed, as already has been seen, is albuminous, but the endosperm has a different origin from that of Angiospermous seeds.

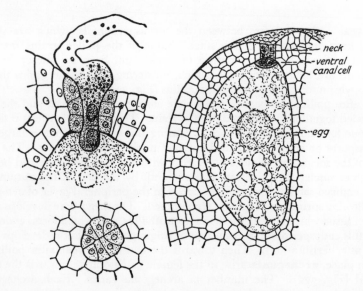

FIG. 323.   *Picea vulgaris.*

Right, longitudinal section through apex of megaprothallus, showing one archegonium (× 100) ;   bottom left, neck of archegonium seen from above (× 250) ;   top left, entry of pollen-tube into canal of archegonium (× 250).

(After Strasburger.)

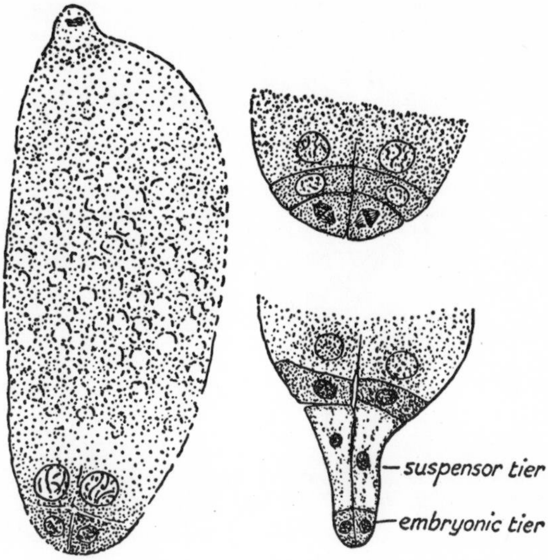

After a period of dormancy, the seed germinates, germination being of the epigeal type. The radicle emerges and then the hypocotyl elongates, thus carrying up the cotyledons (which are still embedded in the endosperm and absorbing food material from it). The cotyledons elongate and eventually become free. The plumule then elongates and produces a shoot bearing single, needle-like foliage leaves. It is not until the second year of growth that the stem produces scale leaves with dwarf shoots, bearing pairs of foliage leaves, in their axils (Fig. 325).

FIG. 324. STAGES IN DEVELOPMENT OF EMBRYO OF *Pinus laricio* (CORSICAN PINE).

Left, two tiers of four cells at base of archegonium ; top right, three tiers, showing the bottom tier undergoing further division ; bottom right, suspensor and embryo developing.

(*After Coulter and Chamberlain.*)

THE YEW

The yew (*Taxus baccata*) is one of the few members of a family which is distributed in the north temperate zone and New Caledonia. It is also a member of the Coniferales, and thus related to *Pinus* ; but it differs from *Pinus* in many respects.

The yew tree is an evergreen, and was formerly widely distributed in British forests. The general habit differs from that of *Pinus*. Some of the lower branches grow upright, thus giving several main shoots and later other branches grow erect from the base. The bases of all these shoots fuse together and thus give the short and deeply grooved ' trunk '. The leaves are borne directly on the stem and not on short shoots. They are spirally arranged but, while they are arranged radially on the erect shoots, on the horizontal branches their petioles curve in relation to a light stimulus and the leaves appear in two lateral rows (Fig. 326).

The tree is dioecious. The flowers are borne on the lower surface of the twigs in the axils of leaves of the previous year. The male flowers bear, at the base, scale leaves, and at the top, ten microsporophylls (stamens), each of which bears five to ten peltate microsporangia (pollen sacs) (Fig. 327). The pollen is distributed by wind.

The female flowers also arise as buds in the axils of leaves of the previous year. Such buds bear about ten scale leaves, and in the axil of the uppermost of these there is a very small shoot terminating in a single ovule. The ovule itself is very similar to that of *Pinus*.

Pollination (by wind) occurs in February or March. Fertilisation follows without much delay (*cf. Pinus*), the process being similar to that described for *Pinus*.

After fertilisation, the seed develops and, by the autumn, becomes surrounded

Harold Bastin

FIG. 325. SCOTS PINE.
Left, seedlings ; right, older stages.

by a fleshy red arillus which grows from the base (Fig. 326). The leaves and
the seed of the yew are poisonous, but the arillus is not. The arillus is eaten by
birds, which thus effect seed distribution.

Germination is epigeal, the seedling having two green cotyledons.

CYCADS

The group Cycadales contains only one family, namely, Cycadaceae. The
species of this family are woody plants indigenous to tropical and subtropical
regions, though some species, for example, *Cycas revoluta*, are commonly cultivated
in greenhouses in Great Britain, and are frequently used for ornamental purposes.
Cycas is a native of Asia. Some other representatives of the family are : *Macrozamia* and *Bowenia* of Australia, *Encephalartos* and *Stangeria* of Africa, and *Dioon*,
Ceratozamia and *Zamia* of America.

Cycas revoluta is a woody plant (Fig. 328). The large, pinnate foliage leaves
form a terminal crown to a secondarily thickened stem, the lower part of which
is clothed with persistent leaf-bases and scale leaves.

All the Cycadaceae are dioecious. The female plant in Fig. 328 shows a crown
of megasporophylls just below the apical bud, each sporophyll replacing a foliage
leaf. This group of megasporophylls is not a definite female cone such as occurs
in all the other living genera of the Cycadaceae. The megasporophyll (Fig. 328)
is pinnately divided and is covered with a dense growth of brown hairs. No chlorophyll is present. Towards the base, the lobes of the sporophyll are replaced by
megasporangia or ovules. These very large ovules are essentially similar to those
of *Pinus*. One characteristic feature is, however, the development of a large well-

like cavity, called a ' pollen chamber ', at the tip of the nucellus. This extends right down to the tip of the endosperm in the neighbourhood of the archegonia.

In the male plant, the microsporophylls are borne in terminal cones. The sporophylls are arranged spirally. Each consists of a hard, thick scale with an expanded head, and bears a large number of irregularly arranged microsporangia on its lower surface (Fig. 328).

The microspores germinate to form a reduced male prothallus and then, on being liberated from the microsporangia, are carried by wind to the ovules on the megasporophylls. Some of the microspores are caught in a viscid exudation from the micropyle and, on the drying of this, they are drawn into the pollen chamber. Each microspore or pollen grain then becomes anchored to the sides of the chamber by sending out a tube into the nucellar tissue. Meanwhile, in place of the two non-motile gametes seen in *Pinus*, the generative cell has divided to give two ciliated spermatozoids (Fig. 329). These are then liberated together with a drop of fluid into the bottom of the pollen chamber, where they swim to the archegonia and effect fertilisation.

This production of motile spermatozoids thus gives a link between the condition seen in Pteridophytes, such as *Selaginella*, where the male gametes are always motile, and that shown in the Angiosperms and the vast majority of the Gymnosperms, where non-motile gametes are carried to the ova by pollen-tubes.

GENERAL SURVEY

The structure and life-histories of all the main groups of plants have now been examined. The facts relating to them are of interest in a variety of ways. The

FIG. 326.   TWIGS OF YEW.
Left, female spray bearing female flowers and ripe seeds, each surrounded by an arillus ;
right, male spray bearing flowers.

structure and biology of the individual types has its own intrinsic interest, and each type will present problems to an enquiring mind. For example, the method by which the spermatozoids reach the archegonia in dioecious mosses is still only imperfectly known. So, too, is the physiology of mycorrhizal associations, such as those which occur in many tree roots. The list of unsolved problems could, indeed, be extended almost indefinitely.

FIG. 327. MALE FLOWER OF YEW.

The survey of the various groups is also interesting as giving some indication of the ways in which the complex structure of the higher plants may have evolved from simple beginnings. In the Algae, as has been pointed out, some types are unicellular, others colonial or cœnocytic, while yet others have filamentous bodies. It appears that it is the filamentous type of construction which has proved most capable of further differentiation in the course of evolution. Some of the larger Algae illustrate how such filaments may divide to give flat expanses or relatively massive parenchymatous bodies like that of *Fucus*. The Bryophyte thallus shows no fundamental advances on these latter types, nor does the sporogonium show any high specialisation of its vegetative body. It seems likely that the Bryophyte type of structure was not the one from which that of the higher plants was evolved.

It is, in fact, the sporophyte of the Pteridophyta which illustrates the possible

FIG. 328. *Cycas.*
A female plant in flower ; left, a megasporophyll of *C. revoluta* ; right, lower surface of a microsporophyll of *C. circinalis.*
(*After Richard.*)

steps in the elaboration of the highly differentiated
bodies of the Gymnosperms and Angiosperms.
The simplest Pteridophyta, the tropical *Psilotum*
and *Tmesipteris*, are rootless plants although they
have an absorbing rhizome ; they are also leafless
although their aerial stems bear leaf-like appendages. The more advanced *Lycopodium* has true
roots, and bears small leaves on its shoots. The
same is true of the ferns, though here the leaves
are large. Anatomical differentiation has gone
along with the increasing differentiation of organs
for special purposes, and the Pteridophyte type of
plant body is characterised by the presence of a
well-defined stele or vascular system with tracheidal xylem and phloem consisting of sieve tubes.

FIG. 329. MULTI-CILIATE SPERM
OF *Zamia floridana* (× 150).
(*After H. J. Webber.*)

The plant bodies of the Gymnosperms show the same differentiation into root,
stem and leaves ; but they are characterised by further anatomical elaboration,
particularly in the specialisation of their vascular elements.

In this progression of forms it will have been noted that it is the sporophyte
generation which has been elaborated in the way described, the various changes
in form and anatomy being such as to fit the plants for life on land.

This leads to a brief consideration of the types of life-cycles shown by the
various groups of the plant kingdom. That of the Algae is very variable. In
some, particularly the Green Algae, the gametophyte is the dominant plant ; in
others, it is the sporophyte. In yet others, for example, *Ectocarpus*, the two generations are alike in external form. The life-cycle of the higher groups is more
standardised. In the Bryophyta, the plant is the gametophyte and the dependent
sporogonium is the sporophyte. On the other hand, the Pteridophyte gametophyte is a small prothallus and the sporophyte is the large and independent
plant. In heterosporous Pteridophyta, reduced mega- and micro-prothalli are
formed, but the general alternation of generations is as in the homosporous forms.

THE SEED HABIT

It was apparently the Pteridophyte type of life-cycle, and particularly that
involving heterospory, which characterised the ancestors of the seed-bearing
plants, the fossil seed-ferns, and the living Gymnosperms and Angiosperms.

The further changes which took place in the establishment of the seed habit
have already been mentioned (p. 377). The number of megaspores in each
megasporangium was reduced to one, and this megaspore (or embryo sac) was
retained within the sporangium (or nucellus) until the female prothallus had
been formed, fertilisation had taken place and an embryo had been developed.
During all these stages, further protection was given by the development of an
integument growing round the sporangium. The embryo having been formed,
the whole structure, now surrounded by a testa formed from the integument, was
liberated as a seed. Correlative changes involving the microspores also took
place. They germinated to form the reduced male gametophytes before being
liberated, as structures called pollen grains, from the microsporangia. On
liberation, they were transported by wind or insects to the neighbourhood of the

megasporangia (ovules). At first, as in living Cycads and in *Ginkgo* (another primitive Gymnosperm), the male gametes eventually produced were motile spermatozoids but, as in all other Gymnosperms and Angiosperms, the gametes became non-motile and were carried passively to the female gametes.

Little is known of the actual origin of the seed habit. Gymnosperms were already numerous in Devonian times and the seed-ferns were well established in the Carboniferous floras. The seed is thus a very ancient structure, and the fossil record gives little indication of its origin. It seems likely, however, that the seed-bearing habit was first evolved by plants with a general life-history resembling that of a heterosporous Pteridophyte.

In Gymnosperms, the seeds are borne exposed on the scales of the female cone, but in Angiosperms they are always formed inside a closed fruit. The fruit, unlike the seed, appears to be a structure of relatively recent origin.

The fossil record indicates that Angiosperms first appeared in Jurassic times and that during the succeeding Cretaceous period they rapidly developed as the dominant feature of the earth's vegetation. Little is known, however, of the actual origin of the habit of bearing the ovules inside a closed cavity or ovary formed by the megasporophylls or carpels, though some of the seed-ferns (for example, *Caytonia*, from the Mesozoic rocks of Yorkshire) show some approach to it.

The development of seeds, and later of fruits, has proved to be a most advantageous innovation in the evolution of land growing plants. Those without this development, such as the Bryophytes and Pteridophytes, still depend on the presence of fluid water for the achievement of fertilisation on their gametophytes. The adoption of the seed habit, however, obviates this dependence on water, as well as giving other advantages, such as the protection and nourishment of the embryo sporophyte. Such biological advantages may well be the chief reason underlying the dominance of the Phanerogams at the present time.

SUMMARY OF LIFE-CYCLES

The main sequence of events in the life-cycles of some of the types described in earlier chapters are summarised on p. 395. It should be realised that in all of them sexual reproduction takes place so that fertilisation occurs at one stage (leading to a doubling of the chromosome number), while at some subsequent stage, the number of chromosomes is halved by the reduction division or meiosis. The part of the life-cycle between fertilisation and meiosis is the sporophyte ($2n$) generation ; that between meiosis and fertilisation is the gametophyte (n). As stressed before, the balance of these two alternating generations varies in the different groups. It should be noted in the life-cycles of the higher groups that it is the sporophyte which has become the dominant plant, well adapted for life on land. The gametophyte, on the other hand, steadily declines in size and complexity, this reduction reaching its climax in the seed-bearing plants.

PRACTICAL WORK

1. Examine, in June preferably, twigs of *Pinus* showing at least three years' growth. Note the structure of purely vegetative twigs ; the position and structure of seed cones of varying age ; the position and structure of staminate cones.

2. Dissect first-, second- and third-year seed cones and note their general structure.

PLANT LIFE-HISTORIES

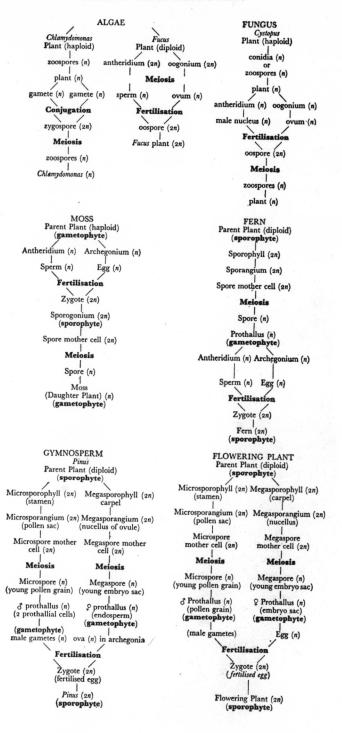

ALGAE

Chlamydomonas
Plant (haploid)
|
zoospores (n)
|
plant (n)
|
gamete (n) gamete (n)
Conjugation
|
zygospore (2n)
Meiosis
|
zoospores (n)
|
Chlamydomonas (n)

Fucus
Plant (diploid)
|
antheridium (2n) oogonium (2n)
Meiosis
|
sperm (n)   ovum (n)
Fertilisation
|
oospore (2n)
|
Fucus plant (2n)

FUNGUS

Cystopus
Plant (haploid)
|
conidia (n)
or
zoospores (n)
|
plant (n)
|
antheridium (n) oogonium (n)
|
male nucleus (n) ovum (n)
Fertilisation
|
oospore (2n)
Meiosis
|
zoospores (n)
|
plant (n)

MOSS

Parent Plant (haploid)
(gametophyte)
|
Antheridium (n) Archegonium (n)
|
Sperm (n)   Egg (n)
Fertilisation
|
Zygote (2n)
|
Sporogonium (2n)
(sporophyte)
|
Spore mother cell (2n)
Meiosis
|
Spore (n)
|
Moss
(Daughter Plant) (n)
(gametophyte)

FERN

Parent Plant (diploid)
(sporophyte)
|
Sporophyll (2n)
|
Sporangium (2n)
|
Spore mother cell (2n)
Meiosis
|
Spore (n)
|
Prothallus (n)
(gametophyte)
|
Antheridium (n) Archegonium (n)
|
Sperm (n)   Egg (n)
Fertilisation
|
Zygote (2n)
|
Fern (2n)
(sporophyte)

GYMNOSPERM

Pinus
Parent Plant (diploid)
(sporophyte)
|
Microsporophyll (2n)   Megasporophyll (2n)
(stamen)   carpel
|
Microsporangium (2n) Megasporangium (2n)
(pollen sac)   (nucellus of ovule)
|
Microspore mother   Megaspore mother
cell (2n)   cell (2n)
Meiosis   **Meiosis**
|
Microspore (n)   Megaspore (n)
(young pollen grain) (young embryo sac)
|
♂ prothallus (n)   ♀ prothallus (n)
(2 prothallial cells) (endosperm)
                      (gametophyte)
(gametophyte)
male gametes (n)   ova (n) in archegonia
Fertilisation
|
Zygote (2n)
(fertilised egg)
|
Pinus (2n)
(sporophyte)

FLOWERING PLANT

Parent Plant (diploid)
(sporophyte)
|
Microsporophyll (2n) Megasporophyll (2n)
(stamen)   (carpel)
|
Microsporangium (2n) Megasporangium (2n)
(pollen sac)   (nucellus)
|
Microspore   Megaspore
mother cell (2n)   mother cell (2n)
Meiosis   **Meiosis**
|
Microspore (n)   Megaspore (n)
(young pollen grain) (young embryo sac)
|
♂ Prothallus (n)   ♀ Prothallus (n)
(pollen grain)   (embryo sac)
(gametophyte)   **(gametophyte)**
|
(male gametes)   Egg (n)
Fertilisation
|
Zygote (2n)
(*fertilised egg*)
|
Flowering Plant (2n)
(sporophyte)

3. Remove a megasporophyll from a first-year cone and look for the two megasporangia (ovules) on the upper surface. The bract scale is on the lower surface. Examine the structure in longitudinal section under the high power of the microscope.

4. Examine a sporophyll from a second-year cone in the same way.

5. Examine a third-year cone. Remove a megasporophyll and note the seeds with their wings attached. Cut a longitudinal section through a seed and examine under the low power.

6. Dissect a staminate cone and note the form of the microsporophylls (stamens). Crush one of them into a drop of glycerine and examine the pollen grains under the high power. Examine transverse and longitudinal sections of staminate cones.

7. Examine the structure of the current-year stem and also of older stems by means of transverse and longitudinal sections. Examine in particular the tracheids, the sieve tubes, the medullary rays and the resin canals.

8. Cut a transverse section of a leaf, noting the particulars described in the text.

9. Study the external morphology of the vegetative and reproductive organs of *Taxus baccata* and species of *Cycas*, if material (of the latter) is available.

CHAPTER XXVIII

EVOLUTION AND GENETICS

The theory of evolution presents us with a picture of a long and unbroken succession of organisms, increasing in variety and complexity with the passage of vast periods of time. The plants and animals now living are the latest representatives of this succession, which began with the first substances to be endowed with ' life '.

This vast process of evolution has proceeded in the waters and on the lands of the earth's surface. The earth itself may be some 3,500 million years old. This estimate is based on the rate of the disintegration of uranium into lead and helium. The proportion of lead to uranium in a given sample of uranium ore gives the age of the ore, and thus of the rock formation in which it occurred. The oldest rocks have thus been estimated to be nearly 2,000 million years old, and more intricate calculations involving the isotopes of lead give the estimated age of the earth as stated above.

ORIGIN OF LIFE

The problem of the origin of life has interested mankind from very early times. The ancient peoples accepted ideas of the spontaneous generation of life, such as the origin of worms and frogs from mud, or the generation of putrefactive organisms in decaying organic matter. Pasteur's work in the nineteenth century disproved all such ideas, and demonstrated that living organisms can only arise from pre-existing ones.

Nevertheless, it seems that when life first appeared on the earth, perhaps about 1,000 million years ago, there must have been a synthesis of something which could be called ' living ' from non-living material. At present there is no sound evidence as to the nature of the earth's surface and atmosphere at the time when this took place, and nothing beyond the merest speculation is possible as to the origin of life. The first living matter must, however, have been able to assimilate new materials from its surroundings and to have been able to duplicate, that is, to reproduce itself. The characteristics of viruses, as already mentioned, are such as to indicate the possible nature of the bridge between non-living matter and living organisms. It must, like the viruses, have been subject to variation, for otherwise no evolutionary developments would have been possible.

These two phenomena of reproduction and variation, which must have characterised the early forms, are indeed fundamental for the whole process of evolution.

EVOLUTION

The idea of evolution was entertained by numerous early philosophers. It was, however, largely forgotten until the eighteenth century, when it was revived by several scientists, including **Jean Baptiste Lamarck** (1744–1829), whose theories will be mentioned later. It was, however, the work of **Charles Darwin** along with that of **Alfred Russel Wallace**, which first gained wide-spread acceptance of the evolution theory (p. 20). Darwin not merely stated the evidence proving that evolution has, in fact, taken place, but he also suggested the nature of the mechanism upon which the process is based. The central idea of this latter is that, as many more individuals of a species are produced than can possibly survive, there is a **struggle for existence**. Forms which vary in any way which increases their fitness for life in their environment and their capacity for successful reproduction will be ' selected ' in this struggle and will survive, supplanting those with less favourable characters. The occurrence of variations and the selection of favourable ones thus provide the mechanism by means of which the first primitive organisms gave rise to the long succession of increasingly complex organisms whose latest representatives are the plants and animals of today.

The evidence that evolution has, in fact, occurred is very varied. The main lines of evidence include the fossil record, the evidence derived from a comparison of the forms now living, and the evidence from the geographical distribution of plants and animals. This evidence may now be briefly considered.

FOSSIL PLANTS

The fossil record of plants and animals provides the only direct evidence that evolution has occurred during the passage of geological time. It displays ' a pageant of the evolutionary history of living things '.

The ancient rocks were weathered and broken up into sand or clay, which were then distributed and deposited as sediment in lakes and seas. Later, these sediments were buried by further deposits and subjected to heat and pressure, and so changed into sandstone and shale. Plant fragments were deposited along with the sand and clay and later fossilised. As geologists have been able to work out a fairly complete record of the succession of sedimentary rocks from early times, it is possible, as **Prof. J. Walton** has said, ' to draw up in chronological order a pageant of plant and animal life, which, though incomplete, is of great historical significance to the biologist.' This record is indicated in outline in the accompanying table.

The fossils themselves are in various forms. The commonest type is the ' **compression** ', formed by the plant fragment having been more or less flattened by the pressure of the surrounding sediment, the organic matter being converted into coal. Other fossils are in the form of ' **incrustations** ', giving an external mould of the plant fragment ; or of ' **casts** ', where the space occupied by the plant was subsequently filled in with mineral matter. A type of fossil called a ' **petrifaction** ' was formed when plant fragments were saturated (before compression could take place) with water containing calcium, magnesium or iron carbonates, or silica, in solution. The tissues and cells came to have a complete filling of the mineral matter as the latter separated out from the water, and the cell walls were converted to coal. Such petrifactions preserve all the details of the anatomical construction of the plant, as can be seen in thin sections.

The various types of fossils have thus provided much information about both the external form and the internal anatomy of the plants of various geological ages. Only a very brief outline of the facts so obtained can be given here.

	Geological Periods		Age in millions of years
CAINOZOIC	Quaternary	Angiosperms dominant	20
CAINOZOIC	Tertiary	Angiosperms dominant	60
MESOZOIC	Cretaceous Jurassic Rhaetic Triassic	Angiosperms appear. Gymnosperms and Pterido- phytes dominant	100 130 200
PALAEOZOIC	Permian Carboniferous Devonian	Pteridophytes dominant. Gymnosperms	280 300
PALAEOZOIC	Silurian	Earliest records of land plants	350
PALAEOZOIC	Ordovician	Algae	400
PALAEOZOIC	Cambrian	Algae	500
PALAEOZOIC	Precambrian	Algae?	900 1,800
	Formation of the Earth		3,350

TABLE SHOWING THE VEGETATION AND AGE OF THE GEOLOGICAL PERIODS.
(*Modified from Prof. J. Walton*)

The Precambrian rocks show only doubtful traces of Algae, but in Ordovician rocks, Algae, which secreted calcium carbonate over their thallus (as do some living forms which they resemble), are well preserved in limestones.

The earliest known land plants occur in the Silurian rocks of Australia. *Zosterophyllum*, for example, was a thalloid plant without differentiation into root, stem and leaves ; but it had a cuticle and vascular tissue, indicating that it lived on land. Another plant from the same rocks was more like *Lycopodium* in habit and in the way it bore its sporangia. Other primitive land plants occur in the Middle Devonian rocks of Scotland and elsewhere. For example, *Rhynia* was a rootless and leafless plant with large terminal sporangia (Fig. 330), while *Asteroxylon* had branches of its rhizome functioning as roots and its aerial shoots bore small leaves. All these early land plants were homosporous.

Great changes had taken place by later Devonian times. Ferns, club-mosses, horsetails, as well as Pteridophytes belonging to extinct groups, were abundant ;

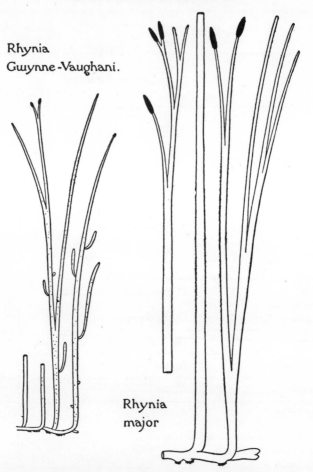

Rhynia
Gwynne-Vaughani.

Rhynia
major

FIG. 330. *Rhynia* SPECIES ; PRIMITIVE LAND PLANTS.

seed-ferns and Gymnosperms were probably also represented. In Carboniferous
times, these groups were the components of the extraordinarily rich flora whose
remains formed beds of coal (p. 203). This period was the Age of Pteridophytes,
but Algae, Fungi and Bryophytes, as well as Gymnosperms, were also present.

The Mesozoic flora was entirely different, and this period has been described
as the Age of Gymnosperms, with Conifers and Cycads spreading all over the
world. Seed-ferns and Pteridophytes were also well represented. The Cycad-
like plants were particularly abundant. They had flower-like reproductive
structures with a central receptacle bearing long-stalked ovules. At the base
of the receptacle there was a whorl of complex microsporophylls and on the
outside a number of hairy bracts. There has been much discussion as to whether
or not this group was the forerunner of the Angiosperms. This latter group
appears quite suddenly in the Cretaceous rocks, and there is, as yet, no clear
indication of its origin.

The Tertiary period had a flora much like our own, though with some forms which are now extinct. It was, in fact, the Age of Angiosperms, although all the other groups were also represented.

Even when the whole range of the fossils known to us is considered, it gives only the broadest outline of the evolutionary process. The Palæozoic period gives us a general picture of a progression from Thallophyte types on to relatively simple land plants, and so on to a flora dominated by plants with the Pteridophyte type of structure and life-history. In the later part of this period the Gymnosperm type appeared, and this became dominant in Mesozoic times. Then the Angiosperms appeared, apparently suddenly, in the Cretaceous and rapidly achieved the dominance which they have maintained to the present day.

COMPARISON OF LIVING FORMS

It was pointed out in Chapter III that the careful examination of living plants led eventually to their classification into species, genera, families and even larger groups. The fact that organisms can be classified in this way is itself evidence that evolution has taken place. The resemblance between the various species of a genus is only to be understood if we regard them as having had a common ancestor. In the same way, the resemblances between the genera composing a family indicate their origin from a common, but more remote, ancestor.

The occurrence of structures such as motile spermatozoids and the female archegonia in so many groups seems clearly to support the view that these groups must have had a common origin. The constant structure of stomata, and of tracheids and sieve tubes, throughout a series of plant groups, and a whole range of comparable facts, all support the theory of evolution. Nuclear structure and behaviour remains without basic alteration throughout the plant and animal kingdoms, clearly suggesting a common origin for them all.

A supporting argument can be based on the occurrence of modified plant organs. For example, some Leguminosae (p. 455) have pinnate leaves with all the pinnae flat and expanded, while in others some of the pinnae are modified to form tendrils. It seems reasonable to suppose that the latter were derived in the course of evolution by modification of the former. Many examples of such homologous structures have been described in earlier chapters.

GEOGRAPHICAL DISTRIBUTION

The present distribution of plants over the earth's surface also provides evidence supporting the evolutionary view. One example only can be mentioned here—the flora of the Galapagos Islands, 500 miles off the coast of South America. These islands are of volcanic origin and were formed in relatively recent geological times. Darwin found that of the 175 species of flowering plants occurring on the islands, 100 were not known elsewhere in the world. Some of these latter species, known as **endemic species**, belonged to genera which were also peculiar to these particular islands.

It seems clear that the species from the mainland which colonised the islands became so modified during the succeeding period that, by the time Darwin examined them, they had in fact evolved into new species. In some examples, the modifications had been so profound that they could not even be placed in the same genus as their relatives on the mainland.

Isolation by geographical barriers such as sea (as in the case of the islands just considered) or mountain ranges often leads to the development of new species. This result is due to the fact that favourable variations arising after isolation will have a better chance of persisting than if free inter-crossing with the parent species were possible.

GENERAL

Such is a brief and inadequate account of the evidence supporting the theory of evolution.

We have now seen that evolution has occurred and have considered the general course which evolution in plants appears to have taken. The fossil record is very incomplete and, since this is the main evidence, it will probably never be possible to construct anything like a genealogical table for the vegetable kingdom. It cannot be said that the sequence Thallophyta-Bryophyta-Pteridophyta-Gymno-sperms-Angiosperms represents a direct evolutionary line, and, indeed, we know nothing of the inter-relationships of these big groups. Nevertheless, the general trends such as those which have been described are fairly clear.

It remains to consider the mechanism by which evolution has been, and still is being, brought about. This brings us to the study of heredity and variation, for, as already mentioned, evolution depends on the occurrence of favourable variations which can be transmitted to the offspring.

GENETICS

The relatively new science of genetics is concerned with the nature and mechanism of heredity, that is, the way in which the characters of an organism are transmitted to its offspring. It is also concerned with the nature and under-lying causes of the variations, or differences in structure and behaviour, which always occur in the individuals of that progeny.

The old view was that the characters of the parents were blended in the off-spring and, at first sight, this often appears to be the case. Indeed, the first scientific experiments on hybridisation, those of **Kölreuter** in which he crossed two species of tobacco, gave offspring intermediate in most ways between the two parent species. Rigorous experiments have, however, shown that the characters of the parents are in fact transmitted as definite units, which maintain their identity from generation to generation. These units are the **genes**, the importance and nature of which has already been mentioned. The recognition of this all-important fact is the essential feature of the work of Mendel, which will now be considered.

MENDELISM

Gregor Mendel was a monk and science teacher in the Augustinian mona-stery at Brünn (now Brno, Czechoslovakia). In 1857, he began to collect and observe the numerous varieties of the garden pea, varieties which differed in height, in the characters of the seeds and in a number of other characters. He made numerous crossings between the varieties and recorded observations on the nature of the progeny. Earlier workers had made similar experiments and found great variability in the offspring. Mendel, however, unlike his predecessors, con-fined his attention to a single character at a time, for example, the height of the

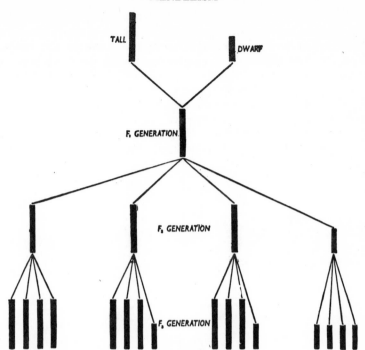

FIG. 331. DIAGRAMMATIC REPRESENTATION OF THE INHERITANCE OF TALLNESS AND DWARFNESS THROUGH FOUR GENERATIONS OF THE GARDEN PEA PLANT.

plants or the colours of the flowers. Not only so, but he counted the number of individuals in each type of offspring and kept accurate records of these in a number of successive generations. His observations were continued over a period of seven years, and he then presented his results to the Natural History Society of Brünn, together with his general conclusions which are now known as Mendel's Laws. This great contribution lay forgotten until 1900, when similar results were obtained and Mendel's paper rediscovered and proclaimed, by **de Vries**, **Correns**, and **Tschermak**.

Two of Mendel's experiments will be described. The garden pea, used throughout his experiments, is normally self-pollinated. If it is desired to cross two varieties, the immature stamens are removed from the flowers of the plant which is to be the female parent and then pollen from ripe stamens of the male parent is transferred to the stigma of the ' castrated ' flower. The artificially pollinated flowers are then covered with a small paper bag to prevent contamination by other pollen.

Mendel first crossed plants differing in a single pair of contrasting characters, such as tallness or shortness of the plants. He collected all the seed from this cross, and on sowing them in the following season obtained a generation of plants known as the **first filial** or F_1 generation. These, as indicated in Fig. 331, proved to be all tall plants. These tall plants were allowed to self-pollinate. All the resultant seeds were collected and sown to give the second filial or F_2 generation. In this generation, Mendel found tall plants and short plants in the proportion of 3 talls to 1 short (the actual number of plants in this particular experiment being 787 talls and 277 shorts). Finally, all the plants of the F_2 generation were self-pollinated, and from the resultant seeds an F_3 generation was obtained. Seeds from the short plants of the F_2 generation gave rise to nothing but short plants in the F_3 generation. But the tall plants of F_2 did not all behave in the same way. One-third of them gave rise to nothing but talls in the F_3; but the other two-thirds **segregated**, like the plants of the F_1 generation, giving rise to tall and short progeny in the proportion of 3 to 1.

Since short plants reappear in the F_2 generation, it is clear that the gametes of a short plant possess a factor or **gene**, controlling the appearance of this char-

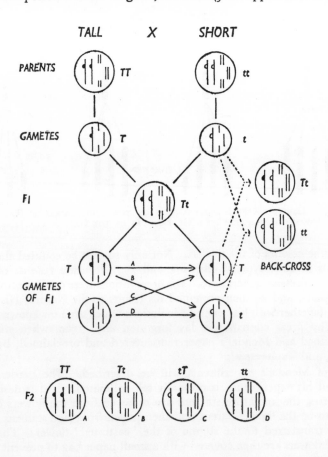

FIG. 332. RESULTS OF CROSSING A TALL WITH A SHORT VARIETY OF GARDEN PEA.
Description in text.

acter, which can be transmitted unchanged through the F_1 to the F_2 generation. Similarly the gametes of a tall plant have a gene controlling the development of tallness. Both genes are present in the F_1 plants ; but, as the experiment showed, all these plants are tall. The gene for tallness is therefore said to be **dominant** over the gene for shortness, which is said to be **recessive**.

The ratio between the different types of plant in the F_2 generation can be related to the behaviour of the chromosomes during the process of meiosis, which leads to the formation of the spores and ultimately to the gametes of the plant. This behaviour is represented diagrammatically in Fig. 332. The pure-breeding tall parent has nuclei with seven pairs of homologous chromosomes, of which only four are shown. On each of one of these pairs is the gene for tallness. Similarly, the pure-breeding short parent has the gene for shortness in the same position on the corresponding pair of homologous chromosomes. When meiosis has taken place in the formation of the pollen grains and young embryo sacs, the gametes ultimately produced will have the reduced number of chromosomes. All the gametes of the tall parent will carry the gene for tallness, all those of the short parent will carry the gene for shortness. When fertilisation takes place, the diploid progeny (F_1) will have nuclei with a pair of homologous chromosomes, one carrying the gene for tallness, the other, the gene for shortness. When these plants produce gametes, some of the male gametes will bear the tall, others the short gene, and these two types will be produced in equal numbers ; the same will apply to the female gametes. Owing to the separation of the two chromosomes of the homologous pair in meiosis, no gamete can possess both the tall and the short gene. This is, in fact, the basis for Mendel's **Principle of Segregation**. When self-fertilisation of the F_1 plants takes place, fertilisation will be brought about by the four types of union indicated by arrows in the figure. If large numbers of gametes are involved, these four types will occur with equal frequency. The result will be the plants of the F_2 generation with the nuclear constitution shown. Obviously the type A is like the original tall parent, and if selfed can only give rise to tall progeny. The type D is like the original short parent, and can only give short progeny in the F_3 generation. Types B and C, however, possess both the tall and the short gene, like the plants of the F_1 generation, and will behave exactly like the latter, giving progeny with tall and short plants in the ratio of 3 to 1.

The truth of the above interpretation may be corroborated by a further experiment. If the F_1 plants are crossed with the original short parent, we should expect to obtain equal numbers of pure shorts and tall plants bearing both the tall and short genes. This result is, in fact, obtained when such a back-cross, or test-cross, is made.

In the experiment just described, it has been seen that plants may look alike but, nevertheless, have a different genetical constitution. Two terms are used in relation to this fact. The plant as it appears is the **phenotype** ; its genetical constitution is the **genotype**. Two other terms are used for describing the genetical constitution of the plant. The true-breeding tall plants are said to be **homozygous** for the tall gene and their constitution can be written TT; the pure breeding short plants are homozygous for the short gene (tt). The F_1 generation is said to be **heterozygous** since it bears unlike genes of the pair, and its constitution can be written Tt.

Mendel worked with a number of contrasting characters which he found to

ROUND, YELLOW X WRINKLED, GREEN

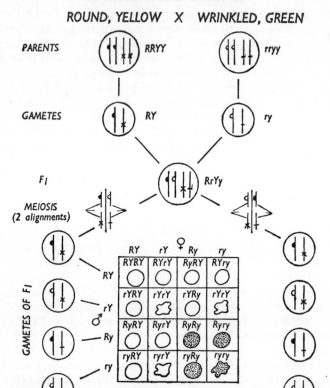

FIG. 333. RESULTS OF CROSSING A VARIETY OF GARDEN PEA WITH ROUND, YELLOW
SEEDS WITH ONE BEARING WRINKLED, GREEN SEEDS.

The resulting seeds from the four random unions between the gametes of F_1 are shown in
the checker-board diagram. Unshaded seeds, yellow ; stippled seeds, green.

behave in the same way as in the experiment just described. Thus in seeds, yellow
cotyledons were found to be dominant to green ones ; round ones to wrinkled
ones. The gene for purple flowers was found to be dominant to the one for white
flowers. In each case, there are the two contrasted genes (occupying corresponding positions on homologous chromosomes), which are described as **allelomorphs**
or **alleles**. Thus the tall and short genes are alleles; so, too, are the other pairs
of genes.

Mendel also carried out experiments in crossing plants which differed in two
contrasting pairs of characters, and found that each character is distributed
independently.

He crossed a variety of pea with round, yellow seeds (both dominant characters) with another variety having wrinkled, green seeds (recessive characters).
The F_1 generation bore only round, yellow seeds. In the F_2 generation, obtained
by self-fertilisation of the F_1 plants, plants of four different types occurred in the
proportion of 9 plants with round, yellow seeds : 3 with round, green seeds :
3 with wrinkled, yellow seeds : 1 with wrinkled, green seeds. It will be seen that
for each character considered separately, the proportions are the same as in the

first experiment. There are 12 plants with round seeds to 4 plants with wrinkled ones and 12 plants with yellow seeds to 4 plants with green seeds ; in each case the proportion is 3 to 1. Each gene has behaved independently, and the experiment illustrates Mendel's **Principle of Independent Assortment**. The interpretation of this experiment is given in diagrammatical form in Fig. 333.

The different pairs of characters only behave in this way if the genes controlling them are situated on different chromosomes. Thus, if the genes controlling the seed characters described above had been situated on one pair of homologous chromosomes instead of on a different pair, then the ratio 9 : 3 : 3 : 1 would not have been obtained, for the round and yellow genes would always have been inherited together, as would the wrinkled and green ones. We should, therefore, not have obtained round and green or wrinkled and yellow seeds, unless the phenomenon of crossing-over described below had occurred.

Any genes which are grouped on one chromosome are, since they tend to be inherited together, said to form **linkage groups**. All the characters of an organism are, in fact, arranged in such groups, the number of groups being equal to the haploid number of chromosomes. There are seven such groups in the pea.

Linkage has, however, been frequently found to be incomplete. This is due to the interchange of parts of chromatids at the chiasmata formed during meiosis (p. 265). The effects of crossing-over are illustrated diagrammatically in Fig. 334, which shows a pair of homologous chromosomes (bearing the genes A, B and C and their alleles a, b and c) and the behaviour of the chromatids in meiosis. It will be seen, as on the left, that if crossing-over occurs between A and B, some gametes having the constitution Abc and aBC will be formed, while if crossing-over is between B and C, gametes with the constitution ABc and abC will be formed. If the linkage had been complete, that is, if crossing-over had not taken place, only gametes with the constitution ABC or abc would have been formed.

An elementary account of Mendelism gives the impression that each of the thousands of characters of an organism is controlled by a single gene in a definite position on one of the chromosomes. This, however, is far from the true position, for the matter is infinitely more complicated than is indicated by such a view. The fact is that each gene may have effects on a number of characters, though one of these effects is usually the most obvious, and is the one by means of which

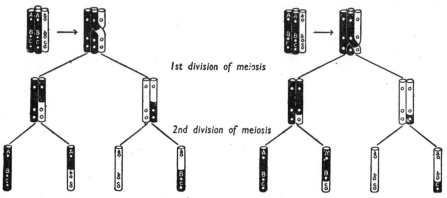

FIG. 334. DIAGRAMS SHOWING CROSSING-OVER AT DIFFERENT POINTS ON THE CHROMOSOME.
(From Murray : ' Biology.')

the gene is known. It is also the case that, in some examples, a number of genes acting together are responsible for a single character, such a series of genes being what is termed a **multiple factor**. It is the operation of such multiple factors which often give rise to a great range of variation in the offspring, which may give the appearance of gradual and continuous variation.

It must be realised, too, that dominance is not always so complete as in the examples described. For example, red-flowered antirrhinums crossed with white-flowered ones give an F_1 generation with pink flowers, and the F_2 generation consists of plants bearing red, pink and white flowers in the proportion of $1 : 2 : 1$. Dominance here is incomplete and gives an apparent blending in the heterozygous condition.

The characters controlled by the genes may vary in their nature under the influence of external conditions, and it must be realised that the characters of the organism as we see it (that is, the phenotype) are determined not merely by the hereditary factors (that is, the genotype), but by the interaction of the latter with the environment.

VARIATION

The progress of evolution depends upon the origin of variations, some of which, being favourable to the species, may be selected so that ultimately a new species may arise.

Variations are of two main types, those which are heritable and those which are not. Organisms possessing the same genetical constitution may vary in many characters, such as size and vigour, variations which are due to the action of the environment. Such variations, or **modifications**, are not heritable. Nor are so-called **acquired characters**, that is, characters acquired during the lifetime of the individual, often as a result of the use or disuse of organs, or as a result of growth under abnormal conditions. For example, a bean plant grown under shade conditions will show a straggly habit and yellowish leaves ; but the seed of such a plant will give normal progeny. The etiolated condition is not passed on to the next generation. Nor, so far as numerous experiments go to show, are any other acquired characters or modifications. Lamarck based his theory of evolution on the idea that the effects of use and disuse are, in fact, inherited ; but no experimental evidence has ever been obtained to support this view. Charles Darwin also believed in the inheritance of acquired characters, and this belief is indeed the major difference between Darwin's views of evolution and the modern views.

It seems probable that the only variations which can be passed on to the offspring are those which are related to the actual mechanism of heredity.

In Mendel's experiments with peas bearing different types of seeds, it was seen that in the F_2 generation there appeared seed types differing from those of the parents (round, green seeds, for example). This example indicates the result of associating genes already in existence in various **recombinations**. This phenomenon of recombination is, in fact, one of the commonest sources of variation. It always occurs when sexual reproduction is followed by meiosis. This latter process, too, is always associated with crossing-over and, as we have seen, this leads to new combinations of characters.

Changes in the chromosomes may occur. Parts of a chromosome may become detached and join on to another chromosome, or join up again in an inverted position, or (as in the evening primrose) join with other detached segments to form a ring. All such changes result in the origin of variations, since the position of genes relative to their neighbours has an influence on the functioning of the gene.

The number of chromosomes in the nucleus may be increased so that three, four or more sets of homologous chromosomes may be present instead of the normal two. Such nuclei are said to be **triploid** or **tetraploid**, or more generally they are described as being **polyploid**, a term covering all the conditions where more than two sets of chromosomes are present. This increase in chromosome numbers is brought about in a variety of ways, which involve abnormalities in mitosis and meiosis. The phenomenon of polyploidy is of very widespread occurrence and of great importance in the evolution of new species. For example, many genera of flowering plants contain a series of species characterised by varying degrees of polyploidy. In *Chrysanthemum*, the basic number is 18 chromosomes in the vegetative cells ; other species have 36, 54, 72 and 90 chromosomes. Polyploidy is important, too, from a practical point of view, since plants with this character are often very vigorous, and may be more resistant to frost and the attacks of parasitic fungi.

The term **mutation** is applied to any heritable variation which arises from a change in the genotype. The examples so far discussed have involved changes in the chromosomes themselves, either in their structure or in their number. Mutations, usually relatively inconspicuous in character, also occur as the result of changes in a single gene. They may be induced by irradiation with X-rays, and they occur naturally but apparently only at long intervals. They must, nevertheless, have been very important factors in the evolutionary process, and they have certainly been so in the development of cultivated plants. In the case of the sweet pea, for example, the original wild species was introduced in 1699. Since then, a series of gene mutations have occurred at intervals, the Spencer type of flower, for example, having arisen in 1900.

The mutations arising in the various ways mentioned, together with the recombinations of characters mentioned earlier, give the variations required before evolution can proceed by the selection of those increasing the fitness of the organism. Much is known about the way in which such variations spread in the natural population and of how new species arise. A major problem to which no adequate solution has been found is presented by the manifold examples of **adaptation** in living organisms. For example, the origin of the complex modifications of the leaves of carnivorous plants, or of all the modifications leading up to the adoption of the seed habit, are processes about which we can only conjecture.

PRACTICAL APPLICATIONS OF GENETICS

The science of genetics has important practical applications. The underlying principles of genetics were, in fact, utilised by mankind long before the origin of the science as such. Many examples of this are provided by the history of the origin of our domestic plants and their subsequent improvement.

The main work of plant breeders is to obtain and select genotypic variants which will give greater yield or show other features improving the economic value of the crop. The important factor may be sugar content, as in the sugar beet ;

or protein and oil content, as in the maize ; or greater grain size and greater number of grains per ear, as in wheat. Plant breeders have, in fact obtained improved varieties of such crops by the selection of favourable mutants and by hybridisation.

Hybridisation is an extremely useful method of obtaining improved varieties. For example, if inbred strains of certain crops are crossed, then the resultant seed produces plants of greatly increased vigour and productivity, that is, plants showing what is termed ' **hybrid vigour** '. By growing hybrid maize, the annual crop in the United States has been increased by 700 million bushels. Another example is the development in Sweden of wheat-rye, or *Triticale*, by the crossing of these two plants followed by a doubling of the chromosome number. This plant may become a useful cereal. New varieties able to grow under a new set of environmental conditions have also been produced by hybridisation and selection, for instance, outdoor varieties of tomatoes.

Another important aspect of the geneticist's work is the development of varieties of various plants able to resist the attack of prevalent diseases. Thus, potato varieties immune to wart disease are available ; so are rust-resistant varieties of antirrhinum. New varieties of wheat resistant to stem rust have been bred in the United States, thus giving an increase of some 40 million bushels annually.

Finally, brief mention may be made of the fact that the science of genetics also provides a tool for certain types of physiological investigation. For example, our knowledge of the pigments concerned in flower coloration is in part due to the investigation of the genes controlling their development. The elucidation of some biochemical problems in the nutrition of fungi such as the Ascomycete, *Neurospora*, has been aided by studying so-called ' biochemical mutants ', obtained by X-ray irradiation of cultures of this fungus.

CHAPTER XXIX

THE PLANT AND ITS SURROUNDINGS

To know the relations which exist between a plant and its surroundings (or **environment**), both below the soil and above it, is a great help in appreciating why the plant grows best in a certain locality and on a certain soil. Such knowledge is an absolute necessity in the cultivation of plants ; whether extensively as in agriculture, or intensively as in horticulture.

SOIL

In the normal land-growing (**terrestrial**) plant—and this includes nearly all the important plants from man's point of view—the soil forms the environment of the roots.

In many localities, the soil has been formed from the rocks on which it is immediately situated. In its original state, the land was composed of nothing but bare rock. Of course, some such rocks are seen exposed even to-day. By various means, which will be examined shortly, most rocks have crumbled or have been **eroded**, thus producing the soil particles. Naturally, various soils are produced according to the various rocks from which they are formed.

FIG. 335. A SOIL PROFILE.
Showing soil, sub-soil and rocks.

(From ' The Geology of South London ', by permission of the Controller of H.M. Stationery Office.)

411

As one would expect, there is no hard and fast line between the top soil and the rocks beneath from which it has been formed. However, between the real top soil and the rocks there is usually a more or less definite layer composed of stones intermediate in size between the large boulders of the rock and the very much smaller soil particles. This layer is referred to as the **subsoil**. The three layers, (a) rock, (b) subsoil, (c) soil, can be easily distinguished where it is possible to examine the soil in profile, such as on the edge of a quarry or the top of a cliff (Fig. 335).

The soil and the subsoil layers vary considerably in different localities, the thickness of soil being dependent upon the hardness of the rocks and the amount of erosion that has taken place. Usually the thicker the layer of soil, the more fertile it is from the point of view of plant production, wild or cultivated. The soil is generally rich in humus and in the mineral elements essential for the nutrition of the plant ; its texture, too, is more favourable for the growth and functioning of the roots. In these characters it differs from the infertile subsoil.

Erosion of rocks, resulting in the formation of soils, is dependent upon several factors. The main factors are those concerned with the weather ; and their effect is often referred to as **weathering**. Rain, sun and frost all play their part. For example, the rain gets into the crannies of the rocks ; then frost may occur. This causes the water to freeze, which results in an increase in volume. This has the effect of pushing the pieces of rock apart.

Chemical weathering may also take place as a result of the solution, oxidation, or hydrolysis of the rock components. Biological weathering also occurs ; for example, growing plants and animals help in the making of soil. Charles Darwin showed that earthworms are very effective in bringing up soil particles from the lower layers of the surface, thus turning the soil just as it is turned in digging and ploughing operations.

The formation of all soils depends ultimately on the processes of weathering just described. Many soils have been formed, as described, directly from the underlying rocks. In many cases, however, the soil is not related to the underlying rock and has, in fact, been transported from another locality. An example of this is the boulder clay, formed by the action of glaciers on rocks, and carried as glacial drift for long distances. In regions with such boulder clay, the soil which has been formed on top of this material has no relation to the underlying rocks. Alluvial deposits offer another example of such transported soils.

Weathering processes thus give the mineral skeleton of the soil. Further changes, however, take place. The most important of these is the addition of humus. This is derived from fallen leaves on the surface of the soil and from dead roots in the soil. These are acted upon by a very varied microfauna and microflora, the latter consisting of algae, fungi and, in particular, bacteria. There may be as many as 1,000 millions of these latter in a gramme of soil. They are concerned with the process of nitrification and with the fixation of atmospheric nitrogen. They are also the main agents in converting organic debris into humus. This is a black, colloidal substance derived largely from the cellulose and lignins of the plant remains, and it is of the greatest importance in the formation of soil suitable for plant life. It gives a colloidal-gel coating to the soil particles which holds water, adsorbs calcium amd other bases, and aggregates the soil particles into larger particles called **soil crumbs**.

Under natural conditions, the soil thus formed from mineral particles and humus becomes stratified by climatic agencies, mainly by the downward percolation of water, the water carrying fine particles and dissolved salts down to lower levels. This stratification gives what is called a **soil profile**, consisting of a number of characteristic layers overlying the parent material. Soil types may be classified on the basis of their profiles, which can be investigated by digging a trench down to the parent material. In Britain, there are two main types. One is the **brown earth**, formed in regions with moderate rainfall and temperature, such as in most of our deciduous woodland areas. In this type, the washing out, or leaching, from the upper layers is not excessive. The surface layer is rich in bases and in humus and the soil is very fertile. The other type is the **podsol**, characteristic of the cooler and wetter regions in the north and west of Britain. Under these conditions, the humus-forming bacteria are absent and the plant-remains collect as acid peat on the surface of the soil. Acid water, percolating downwards through the soil, leaches the upper layers, which thus become deficient in basic salts and are often grey in colour as a result of the removal of the sesqui-oxides of iron. Such soils support coniferous forests, moors or heaths.

It should be noted, however, that cultivation destroys such stratification, and the soils of fields and gardens are more or less uniform down to the subsoil.

COMPOSITION OF SOIL

The average garden soil which may be considered a good one should contain the following ingredients, the numbers representing percentage by volume:

Rock particles - - - -	40
Water - - - - -	25
Air - - - - - -	25
Humus - - - - -	10

The rock particles constituting the actual soil vary in size according to the rocks from which they have been formed. The three most common types of soil formations are those in which sand, clay or calcareous material predominates, namely, sandstones, clays and limestones. The average soil, of course, is composed of a mixture of these, the percentage of each varying with different localities. Then, in many cases, the soil is further characterised by the presence of other substances. For example, the very white, limy (or calcareous) soils of Kent and Sussex are due to the presence of chalk. The red soils of Devonshire and certain parts of Somerset owe their colour to the large amount of iron salts in the soil, and the fact that they are formed from red sandstone. Soils very rich in humus are usually dark in colour. Soils containing a great deal of lime are alkaline in reaction, whereas those which contain excess water and humus, such as bogs and marshes, are very acid in reaction. All this is important from the point of view of the plants which will colonise these soils, for certain plants must have very acid soils, whereas others must have alkaline soils, etc.

The largest particles in the soil are the **stones** and **gravels**. With too many such large particles, the soil is very poor, since it cannot retain water. **Sand** forms very coarse soil particles. **Clay**, on the other hand, is composed of very fine soil particles. Intermediate between sands and clays are those particles called **silt**.

A typical soil contains a certain percentage of all these various soil particles, the percentages varying with different localities. A soil which contains chiefly sand and silt, with about 6 per cent of clay, is called a **sandy soil**. If it contains more than 25 per cent of clay it is said to be **clayey**. Between the two, we have the more common type of soil, with plenty of sand and clay. Such a soil is called a **loam**. **Sir John Russell** gives the following as being a typical loam : clay, 6–15 per cent ; silt, 40–60 per cent ; and sand, 20–50 per cent. If an exceptional amount of calcium be present in the soil in the form of calcium carbonate, such as limestone or chalk, the soil is called a **marl**. A marl is usually a rich soil, since the calcium it contains serves a double purpose. First, calcium is one of the raw elements required in the plant's nutrition. Secondly, calcium carbonate is an alkali, and thus prevents the soil from becoming acid or sour.

It is comparatively easy to tell whether a soil is acid or alkaline. In the case of marls, the reaction will clearly be alkaline. The presence of the limestone can easily be detected by adding a few drops of concentrated acid, such as hydrochloric, to some of the soil. The reaction with the excess amount of calcareous alkali can easily be seen by the effervescence which takes place. A very acid soil, which clearly contains no limestone, can often be detected by collecting some soil water and testing with litmus. A more precise method of measuring the reaction of the soil is to determine the hydrogen ion concentration in a suspension of the soil in distilled water. This concentration is expressed as a number indicating the logarithm of the reciprocal of the hydrogen ion concentration, this number being termed the pH value. A pH value of 7 indicates neutrality ; figures below this indicate the degree of acidity, while figures above it indicate alkalinity. Measurements are carried out by means of dyes or indicators which give a variation of colour related to the pH. The B.D.H. Soil Indicator, for example, gives a red colour for a pH of 4, orange 5, yellow 6, green 7 and blue 8. The pH may also be measured by electrometrical methods.

Clay is colloidal in nature, and therefore has a very high water-retaining capacity. Drainage from clayey soils, therefore, is very difficult. Naturally, holding the water so tenaciously as a clayey soil does, it is very badly aerated and is heavy. Also, since it is saturated with water, it tends to remain cold. Therefore, a very clayey soil is heavy, badly aerated and water-logged. Crops grown on such soils are usually poor and slow to ripen. In agriculture and horticulture such soils are very often improved by adding larger particles such as sand or coal ash.

Sandy soils are the reverse. They allow an easy percolation of water, and therefore remain very dry, sometimes too dry. They are also very well aerated, and therefore extremely light. Crops grown on very sandy soils sometimes suffer from the want of water, etc., so a sandy soil is really not much better than a clayey soil, although it is the reverse in composition. Such soils are usually improved by adding heavy farmyard manure, for with such easy water drainage through a sandy soil the leaching of mineral salts goes on at an excessive rate.

It is quite clear that, on the average, loams are the best soils for crops, etc. A rough mechanical way of showing the constituents of a loam is to stir a little of such soil in a beaker or glass of water. The heaviest sand particles will soon settle to the bottom of the glass. Closely following the sand will be the silt which will settle above it. The clay particles, on the other hand, being very small and colloidal, will remain suspended in the water, whereas some of the humus will float. The alkalinity or acidity can then be examined by testing the reaction of the water.

The colour of the soil is an important physical property, for a dark soil absorbs more heat from the sun's rays than a light soil does. Thus, the temperature of a dark soil is higher than that of a light soil under similar conditions. The temperature of the soil is also affected in two other important ways : (a) degree of slope towards or away from the sun ; (b) amount of evaporation of water from the soil. Evaporation from a surface has a cooling effect on the surface, therefore the more the evaporation takes place, the cooler does the soil become. Naturally, a very loose, sandy soil has a much higher degree of evaporation than a compact clayey soil.

The rate of passage of water through the soil is also of great importance. As was seen in Chapter XII, the soil water is usually present on the surface of the soil particles. Passage through the air channels of the soil is due to **capillarity**. It can easily be shown that the smaller the bore in a glass tube, the greater is its capillary attraction. Place several tubes, with very small but varying bores, standing in water. By capillarity, the water will pass up the tubes, but the height to which it rises varies inversely as the diameter of the bores of the tubes. Therefore, the smaller the diameter, the greater the capillarity. Thus, in soils, passage of water will be greatly affected by the soil particles. The smaller the particle (for example, clay) the smaller the air spaces, and therefore, the greater the capillarity. The result is that water moves upwards by capillary attraction much more quickly in a clayey soil than in a sandy one.

The physical properties (such as size of particles), chemical properties (such as water-content, mineral-content), and biological properties (such as the soil bacteria) of soils are all of the utmost importance to plants.

PLANT ENVIRONMENT

In the case of wild plants, it is largely a matter of chance whether the seeds or spores reach a suitable habitat, that is, a situation favouring their germination and healthy development. If they do reach a favourable habitat, they will establish themselves ; otherwise they will perish or develop feebly. With cultivated plants it is different, for cultivation means that the plant is assured of a suitable habitat ; then, once it is developed, the conditions are kept suitable, artificially by man so far as he is able. For example, soils are manured, they are ploughed, hoed and harrowed, and water is supplied when there is insufficient rain. It is not so easy to supply suitable temperatures and light intensities outside, but this can be done in the greenhouse ; and even soil temperatures are controlled in some cases by heating the soil by electricity or steam pipes. Ligh intensity, too, is not easily controlled, though in the case of some valuable plants this is done by a system of artificial electric lighting to raise the intensity, or the use of green glass to lower it. In controlling conditions, man has been able to make plants grow in situations which otherwise would be unsuitable for them.

In Nature, there is a vast range of habitats, varying in the nature of the soil and in climatic conditions. Each habitat tends to have its own distinctive vegetation, or plant community, consisting of species which, in the course of evolution, have become adapted to the particular habitat conditions. The study of this relationship between plants and their environment is called **ecology**.

MESOPHYTES

Many habitats provide the conditions usually regarded as those favourable for plant growth, for example, plentiful water, favourable soil factors, and the absence of conditions which might lead to excessive transpiration. The plants growing in such habitats have what we regard as ' normal ' structure, and they are termed **mesophytes**.

On the other hand, many habitats present conditions deviating in some way from the favourable ones described above. The plants of such habitats tend to show special adaptations, as in the types which will now be described.

EPIPHYTES

Some plants, instead of growing on the soil, that is, instead of leading a **terrestrial** mode of life, grow attached to other plants, most frequently growing on the trunks of trees, or in the axils of their branches.

All such plants are termed **epiphytes**. In Great Britain and other temperate countries, epiphytes are represented only by certain algae, such as *Pleurococcus* on the bark of trees, many lichens and some mosses, though certain mesophytes, such as dandelions, sometimes grow epiphytically in the forks of trees. It must be remembered that epiphytes are not parasites, like the mistletoe, for they only use the actual tree, etc., as a support, and not as a source of nutrition.

In tropical countries, especially in the luxuriant vegetation of the jungle, many ferns and flowering plants are epiphytic. Certain orchids are. They cling to the supporting branch of the tree by means of short **clinging roots** which twine round it. They absorb nutrition from the humus (chiefly dead leaves) which collects around them, by means of **absorbing roots**. Many of these plants have special methods of obtaining water. The commonest one is by means of long **aerial roots** (see Chap. V). In the Tropical House at the Royal Botanic Gardens at Kew, many of these long aerial roots can be seen suspended from the epiphytic plants growing on the tropical trees there. Other epiphytes have some of their leaves modified as water-collecting pitchers, into which their adventitious roots grow. Many epiphytes have water-absorbing hairs and scales, particularly on their leaves.

XEROPHYTES

Xerophytes are plants growing in habitats where water is scarce or difficult to absorb. The majority of desert plants, such as cacti, are xerophytic. Pine trees also are subjected to physiologically xerophytic conditions, since the water supply is often frozen and then unavailable. The various methods of adaptation, both with regard to the storage of water and the prevention of excess transpiration, have already been examined.

HYDROPHYTES

Special peculiarities in structure are found in plants which grow in water (called **hydrophytes**). The majority of them, such as the bladderwort (p. 212) and spiky water milfoil (*Myriophyllum spicatum*), are able to absorb water all over their surfaces. The requisite gases, oxygen and carbon dioxide, are absorbed from the surrounding water in solution, just as in the case of seaweeds, etc., which are submerged in the sea.

Many hydrophytes have their leaves floating on the surface of the water, such as in the case of the yellow water-lily (*Nuphar lutea*), white water-lily (*Nymphaea alba*) (Fig. 336), some species of pond-weed (*Potamogeton* species), frogbit

(*Hydrocharis morsus-ranae*), etc. The adaptation of the leaves to this habit, especially with regard to the position of the stomata, has already been considered.
Floating leaves carry out gaseous interchanges through their stomata as in land
plants and not in solution. All their stomata are on the upper surface of the
leaf. But such leaves run the risk of being flooded with water, especially when the
level of the lake or river rises. Certain adaptations prevent this. For example,
the floating leaves of *Potamogeton* are borne on very long stems which thus allow
for changes in water-level. The petioles of the floating leaves of certain tropical
hydrophytes are shaped like a corkscrew, and therefore, in order to remain floating, stretch or contract according to the rise or fall in water-level.

A very exceptional type is seen in the tremendous leaves of the tropical hydro-
phyte, *Victoria regia*, the giant water-lily of the Amazon. This plant is often
cultivated in botanic gardens. There is a special *Victoria regia* house at the Royal
Botanic Gardens, Kew, where the plant can be seen. It can only be seen at
certain times of the year, since it is an annual. In this case, the floating leaves
are so large as to be able to support a baby. Flooding over the top of the leaf is
prevented by a vertically growing margin, which is two to three inches high,
thus making the leaf look like a floating tray.

Certain hydrophytes grow rooted in shallow water, in which case some leaves
grow below the surface of the water, whereas others grow upright above it. The
latter are normal in structure. The former, on the other hand, are modified in
relation to their submerged position. For example, stomata are absent. Also,
they are subjected to currents of water which might easily tear them. How com-

T. Edmondson

FIG. 336. WHITE WATER-LILY (*Nymphaea alba*).

pletely submerged leaves adapt themselves to this contingency was seen in the case of the bladderwort in Chapter XVIII. In this case, the leaves become finely divided. In the case of plants with submerged leaves and also leaves above water, the submerged ones *only* become divided. Thus, in such plants, there are two types of foliage leaves in one and the same plant. Common examples of this in Great Britain are the arrowhead (*Sagittaria sagittifolia*) (Fig. 337), with upright aerial leaves bearing an arrow-shaped lamina and submerged leaves, long and ribbon-shaped ; and the water crowfoot (*Ranunculus aquatilis*) (Fig. 338), with lobed floating leaves and finely divided submerged leaves. One species of pond-weed, the various-leaved pond-weed (*Potamogeton gramineus*), has oblong float-ing leaves and narrow, linear submerged leaves.

In relation to the difficulty that hydrophytes experience in obtaining air for respiratory purposes, the tissues of the stems, roots and leaves often have very large intercellular spaces, usually in the cortex of the stem and roots. These large intercellular spaces act as air chambers, and this aerated tissue is usually called **aerenchyma**. Such tissue is common in the petioles and roots of the water-lily, and in the stems of *Potamogeton*, and marestail (*Hippuris vulgaris*) (Fig. 339).

Plants growing in swamps may be considered as semi-hydrophytes, since their roots are subjected to water-logged conditions. Many such plants produce respiratory roots, which grow up into the air (Chap. V) in order to allow gaseous exchange.

HALOPHYTES

Quite a number of flowering plants are capable of living near the coast where the soil is periodically swamped with sea-water, and several flowering plants can

Ernest G. Neal

FIG. 337. ARROWHEAD (*Sagittaria sagittifolia*).
Frogbit and duckweed are also present.

grow in shallow sea-water. Such
plants thus adapted to salt water
are called **halophytes**. Between
high-tide mark and the sand dunes
further inland may be found the
sea rocket (*Cakile maritima*), salt-
wort (*Salsola kali*), etc. ; all are
halophytes with succulent leaves,
and in some cases succulent stems.
On tracts of tidal mud, where the
soil is constantly saturated with
sea-water, may be found the glasswort, which has very minute leaves,
but thick, fleshy, jointed stems (Fig.
340). The glasswort (*Salicornia
stricta*) is often found growing
practically alone in salt marshes
just out of reach of the neap tides.
The succulence of halophytes is not
fully understood. Their root-hairs
have an extremely high osmotic
pressure, so that the plants can
absorb water even from a soil
saturated with sea-water. The significance of the storage of water in
the succulent tissues of the aerial
parts is, however, not understood.
The branched stems of the glasswort can absorb dew and rain-
water readily. The eel-grass or

FIG. 338. WATER CROWFOOT (*Ranunculus aquatilis*).
Showing floating leaves, submerged leaves,
and flowers.

grass wrack (*Zostera marina* and *Zostera nana*) is a land plant which has become
totally adapted to marine aquatic conditions. The entire plant, including flowers, is submerged. It spreads by means of a creeping stem, and bears ribbon-
shaped leaves sheathed at their bases. *Zostera marina* suddenly mysteriously disappeared from the coasts of Britain ; but there are localised signs of its return.

Of great value to man are certain halophytic grasses. These are capable of
growing in salt marshes, and they grow so prolifically that finally the marshes
become dried up, and their levels raised, owing to the great activity of the plants
in forming humus from their own dead bodies. In Europe, one halophytic grass,
Spartina maritima, occurs in the salt flats of southern England, and in the Mediterranean coasts. In the salt marshes around Southampton is another species,
Spartina alterniflora, where it was introduced by ships from America. There it is
colonising the marshes so much that certain parts around Southampton, which
were once marshes, are now dry land.

The most efficient halophyte in colonising salt marshes is the rice grass,
Spartina townsendii, a hybrid of *maritima* and *alterniflora*. This halophyte originally
grew around Southampton, but now it has spread along the mud flats of the
southern coast, and is choking up certain parts of Poole harbour, in Dorset. Rice
grass is being used for the reclamation of new land from the sea. This is the

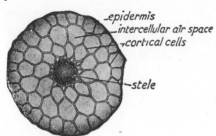

epidermis
intercellular air space
cortical cells
stele

FIG. 339. PHOTOMICROGRAPH OF TRANS-
VERSE SECTION OF MARESTAIL
(T. D. T. Hall.)

case in Holland, where, since 1924, when the plant was definitely introduced for the purpose, it has grown so prolifically that acres of land, previously salt marshes or actually under the sea, have been reclaimed and now used for agricultural purposes.

COMPETITION AND COLONISATION

The study of the social life of plants can be made a very extensive one, taking in the examination of the distribution and inter-relationships of plants throughout the world. This type of study, dealing mainly with plant distribution, is described as **plant geography.** Ecology, however, is concerned with the relation of plants, either as individuals or as communities, to their habitat. Communities of plants behave in many ways like individuals, and thus have an origin and a developmental history.

One very important thing to realise is that the seeds of plants in their wild or native state are distributed largely by chance. On any one piece of ground, thousands of plant seeds must fall in a year, yet it is clear that only a few actually take to the soil and grow. If it is a barren, infertile soil, with a low mineral content, bad exposure and unsuitable climate, few plants will succeed. The rest of the seeds perish.

One excellent method of studying colonisation by plants is to watch the gradual invasion of a newly exposed piece of soil. Of course, in Great Britain, this is not very easy; yet one can gather much information concerning plant colonisation by studying the newly made banks of roads and railway cuttings. The study may extend over several seasons, and the type of plant, the time that it arrives, the percentage of different plants, etc., duly noted.

In such cases, plants which develop quickly from spores, and not seeds, are the first arrivals. Spores are produced by many of the simpler types of plants, such as algae,

Ernest G. Neal

FIG. 340. GLASSWORT (*Salicornia herbacea*).

fungi, mosses, liverworts and ferns. These usually arrive first, since spores are very small and are easily carried in the air. Next come the ephemeral and annual flowering plants common to the neighbourhood. A **community** of plants thus established is still not very closely packed and it is still open to new-comers. It is therefore called an **open** community.

Hardy perennials such as certain grasses, thistles, plantains, dandelions, etc., then become established, and competition is set up between the plants, because, by now, the community is becoming overcrowded. Then the weaker plants are choked out and die, and the hardier ones become more firmly established. Finally, only those plants which are suited to that soil and climate remain, and then, when the ground is fully occupied, the community is said to be **closed**. This does not necessarily complete the development of the vegetation, for further changes may occur. Shrubs may invade the community of herbaceous perennials and, even-tually, trees may become established. Thus, the habitat shows a **succession** of plant communities. If no other factors intervene, the succession will end with the development of the highest type of plant community which the climate will permit. In this country, woodland is what is termed the ' climatic climax ', and much of our vegetation would revert to woodland were it not for the action of man and his domestic animals.

At all stages in the development of a plant community there is competition between the component plants for light and air. But there is also competition among their roots for a place in the soil. It is amazing sometimes how far the roots of trees will penetrate the soil ; some roots go down more feet than the topmost shoots are high. The root competition is chiefly due to the search for an adequate water supply. It has been shown experimentally that apple trees planted 40 feet apart yielded 43 bushels more fruit to the acre than similar trees planted 30 feet apart. This is interesting from the point of view of plant accommodation in both air and soil.

The largest and most comprehensive type of plant community is the plant **formation**. Such formations are, in fact, the main world types of vegetation. Some are determined by climate, as for example, tropical rain-forests, temperate grasslands (like prairies and steppes), and deciduous forests (such as oak and beech woods). Others are determined by soil factors, for example, reed swamps and salt marshes. Although the general type of vegetation in each formation is the same all over the world, the species composing the formation vary in different regions.

So far as British vegetation is concerned, sown crops and plantations occupy a great part of the country. The remaining vegetation is largely semi-natural, that is, it consists of native plants, but has been more or less interfered with by man's activity. It is practically only the vegetation of the high mountains and some parts of the sea-coast which is truly natural.

The actual units of the semi-natural and natural vegetation are called **associa-tions**. Each is composed of a group of characteristic species, one or several of which have a great effect on the community as a whole and are therefore called the **dominants**. For example, the oak is the dominant plant in an oak-wood ; the heather may be dominant on a moor. Smaller component units of an associa-tion are termed **societies**. For example, in an oak wood, there may be a society of primroses, or one of bluebells.

PLANT COMMUNITIES

The examination of a plant community makes interesting study. There are several methods of study. One is to examine a rather large area. Then obtain a blank map and shade in with various colours the different types of vegetation, such as woodland, scrub, meadow, hedgerows, moor, marsh, water, cultivated land, etc. Then this should be correlated with the physiography, the geological formations, the weather and types of animals, etc., all of which have definite effects on the community.

A more intensive study is that referred to as the **quadrat chart** method. This can only be applied to a much smaller area. In this case a rectangular area is marked out on the land by tapes. This is then divided into smaller squares by tapes or string, then the distribution of the individual plants inside the quadrat is represented on squared paper.

A second method applicable especially to a hedgerow, edge of a wood, river bank, etc., is that called the **transect method**. This involves drawing a section through the habitat to be examined, then representing the various types of plants, along the section.

It must be remembered that no plant community is absolutely stable. There are certain factors which decide what type of plant shall constitute it in the first place, such as nature of the soil, climate, etc. ; but then there are other factors such as disease epidemics, useful and harmful animals, which come along afterwards and modify it. All the factors which influence the nature and development of the community are known collectively as the **habitat factors**.

Such factors are so numerous that it would be difficult to name them all. Also, whereas one certain factor may be very potent in one plant formation, it may be entirely absent in another. However, certain factors are rather general. **Climatic** factors, for example, are very important. These comprise chiefly conditions of rainfall, temperature, light, and prevailing winds. In Great Britain, rainfall is heaviest in the west, and there wet moorlands and marshes are found ; in the east of England, where the rainfall is much less, dry heaths are common. In the north, especially Scotland, it is much colder, and this greatly affects vegetation.

Physiographic factors depend upon the shape of the country, mountainous or flat, etc. The slope of the land, too, helps to determine the degree of warmth and light and amount of exposure to the prevailing winds that the vegetation will receive.

The chemical and physical nature of the soil is a very important factor. Such factors associated with the soil are called **edaphic** factors.

Finally, a very potent factor in nearly all cases is the **biotic** factor. This involves the influence of other forms of life, plant and animal. For example, a beech tree casts a very dense shade. The result is that in beech woods very little ground vegetation may be found (Fig. 341). Therefore the beech trees are a biotic factor. Many animals form a biotic factor. Some give manure and therefore form a useful biotic factor ; others, such as herbivorous animals, for example, rabbits, slugs and snails, eat the plants of the community and are thus a harmful one. Animals such as those used in the dispersal of seeds and fruits, and insects used in pollination, clearly are important factors in plant distribution. Man himself is a biotic factor even where wild plants are concerned. This is well seen in the forests where lumbering goes on.

An intensive study will involve listing all the plants in the community, with notes on the frequency with which each occurs. Since communities show seasonal variations, several studies should be made at different times of the year. First of all one should begin with a note of the type of country. Dates when examined should be stated, and a detailed study of the factors, especially edaphic and climatic, should be made. At all times the ratio and distribution of plants within the association should be thoroughly investigated.

For example, one, two or perhaps even three plants form the most important members of a plant association. These, as mentioned earlier, are called the **dominant** plants. The dominants very often help to answer the question of why are the others such as they are, for dominants are often biotic factors to the other members of the association. This may be seen in a beech-wood association. Here, few herbs are found growing on the soil. Very often, closely associated with the dominant plants are others which are almost as common. They are then said to be **subdominant**. Less common again, but evenly distributed throughout the association, are other plants. These are said to be **distributed**. Much less common again are those plants said to be **occasional**, and the sparsest of all are classified as **rare**.

WOODLANDS

Very familiar plant associations in Great Britain are the **woods**. There are several types of wood associations in this country. In the highest altitudes are

T. Edmondson

FIG. 341. BEECHES IN SUSSEX.
Note the sparsity of undergrowth.

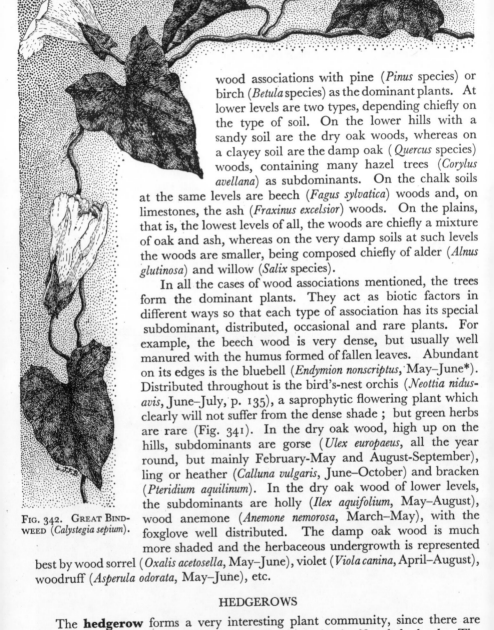

wood associations with pine (*Pinus* species) or birch (*Betula* species) as the dominant plants. At lower levels are two types, depending chiefly on the type of soil. On the lower hills with a sandy soil are the dry oak woods, whereas on a clayey soil are the damp oak (*Quercus* species) woods, containing many hazel trees (*Corylus avellana*) as subdominants. On the chalk soils at the same levels are beech (*Fagus sylvatica*) woods and, on limestones, the ash (*Fraxinus excelsior*) woods. On the plains, that is, the lowest levels of all, the woods are chiefly a mixture of oak and ash, whereas on the very damp soils at such levels the woods are smaller, being composed chiefly of alder (*Alnus glutinosa*) and willow (*Salix* species).

In all the cases of wood associations mentioned, the trees form the dominant plants. They act as biotic factors in different ways so that each type of association has its special subdominant, distributed, occasional and rare plants. For example, the beech wood is very dense, but usually well manured with the humus formed of fallen leaves. Abundant on its edges is the bluebell (*Endymion nonscriptus*, May–June*). Distributed throughout is the bird's-nest orchis (*Neottia nidus-avis*, June–July, p. 135), a saprophytic flowering plant which clearly will not suffer from the dense shade ; but green herbs are rare (Fig. 341). In the dry oak wood, high up on the hills, subdominants are gorse (*Ulex europaeus*, all the year round, but mainly February-May and August-September), ling or heather (*Calluna vulgaris*, June–October) and bracken (*Pteridium aquilinum*). In the dry oak wood of lower levels, the subdominants are holly (*Ilex aquifolium*, May–August), wood anemone (*Anemone nemorosa*, March–May), with the foxglove well distributed. The damp oak wood is much more shaded and the herbaceous undergrowth is represented best by wood sorrel (*Oxalis acetosella*, May–June), violet (*Viola canina*, April–August), woodruff (*Asperula odorata*, May–June), etc.

FIG. 342. GREAT BIND-WEED (*Calystegia sepium*).

HEDGEROWS

The **hedgerow** forms a very interesting plant community, since there are several levels to examine, comprising chiefly the hedge itself and the bank. The

* Throughout this chapter, times of flowering are indicated in brackets.

hedge itself varies in different parts, being composed of some, or all, of the following: hawthorn, hazel, maple, bramble, blackthorn, willow, holly, etc. The bank of the hedge shows more variations. At the top are plants which have to contend with a certain amount of shade. These may be climbers which can scramble to the light, such as the bramble (*Rubus fruticosus*, July–October), clematis (*Clematis vitalba*, July–September), great bindweed (*Calystegia sepium*, June–August) (Fig. 342), honeysuckle (*Lonicera periclymenum*, June–September), black bryony (*Tamus communis*, May–June), white bryony (*Bryonia dioica*, May–September) and various vetches. They have different means of climbing, which are worthy of study. Other plants which can endure the shade of the top of the bank are those with very long, erect stems, such as stinging nettle (*Urtica dioica*, June–September), white deadnettle (*Lamium album*, May–September) but frequently much earlier and much later), certain grasses, hedge mustard (*Sisymbrium officinale*, June–July), garlic mustard (*Alliaria petiolata*, May–June,) greater stitchwort (*Stellaria holostea*, June–August), etc. Finally, there may be plants which prefer the shade, such as certain ferns, broad-leaved garlic or ramsons (*Allium ursinum*, May–July), wild arum (*Arum maculatum*, August–October), enchanter's nightshade (*Circaea lutetiana*, June–August), foxglove (*Digitalis purpurea*, June–September), primrose (*Primula vulgaris*, February–May), ground-ivy (*Glechoma hederacea*, March–June), germander speedwell (*Veronica chamaedrys*, May–June), violets, and a host of others.

On the slope of the bank the light is abundant. Here, therefore, we find prostrate plants such as creeping buttercup (*Ranunculus repens*, May–September), wild strawberry (*Fragaria vesca*, May–July), etc., and rosette plants such as daisy (*Bellis perennis*, March onwards), dandelion (*Taraxaum officinale*, March–September, sometimes earlier, sometimes later), great plantain (*Plantago major*, June–August), etc.

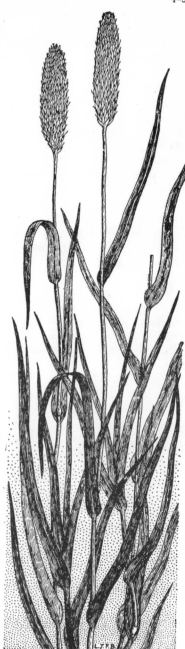

FIG. 343. MEADOW FOXTAIL
(*Alopecurus pratensis*)

MEADOWS

In the majority of **meadow** associations the dominants are usually grasses, such as perennial rye grass (*Lolium perenne*, June–July), meadow foxtail (*Alopecurus pratensis*, April–June, Fig. 343) and meadow fescue (*Festuca pratensis*, June–August), together with white or Dutch clover (*Trifolium repens*, May–September) and red clover (*Trifolium pratense*, May–September). In various meadows, certain sub-dominants may be found according to the soil structure and composition. These may be daisy, buttercup, cowslip (*Primula veris*, May–June), common sorrel (*Rumex acetosa*, May–July), ox-eye or moon daisy (*Chrysanthemum leucanthemum*, June–August), etc.

Lawns and **arable land** are associations closely related to meadows, with the exception that in these the dominant plant or plants are cultivated. Usually, all other plants, subdominant, distributed, occasional or rare, in such associations, are **weeds**. A weed is any kind of plant which grows where it is not wanted or is a nuisance. Certain meadows have their weeds, since they are required for the production of hay. Any plant growing in such a meadow which will spoil the hay is therefore a weed. Such weeds are chiefly the ox-eye daisy, common daisy, common sorrel, common buttercup (*Ranunculus acris*, June and August), spear plume thistle (*Cirsinm vulgare*, July—August), etc.

On cultivated lawns the dominants are various grasses. The most troublesome weeds of such associations are the rosette plants such as dandelion, daisy and plantain. The yarrow (*Achillea millefolium*, July–August) and creeping buttercup (*Ranunculus repens*, May–June) also form troublesome occasionals. In gardens,

Ernest G. Neal

FIG. 344.   A TYPICAL MARSH.
Plants present include osier willow, reed mace, rushes, willowherb, bulrush, water plantain, iris and gypsywort.

the chief weeds are groundsel (*Senecio vulgaris*, February–December), small or lesser bindweed (*Convolvulus arvensis*, June–August), small chickweed (*Stellaria media*, February–October), various thistles, and dandelion. In cornfields they are the same at the early stages of cultivation. Later, along come the charlock or wild mustard (*Sinapis arvensis*, June–August), various buttercups, and, later still, common red poppy (*Papaver rhoeas*, June–July) and corncockle (*Agrostemma githago*, July–September).

AQUATIC AND SEMI-AQUATIC ASSOCIATIONS

Marshes, ponds and **streams** have very characteristic plant associations. The diagnostic features of some of these plants have already been examined with relation to their hydrophytic habit.

In a marsh (Fig. 344) the soil only is water-logged, therefore only the roots and rhizomes of the plants are affected by the excess water. In some marshes the dominant is the osier willow (*Salix viminalis*, April–June), whereas in others it is the common rush (*Juncus conglomeratus*, July–August) ; more rarely it is the great reed mace (*Typha latifolia*, July–August) or common reed (*Phragmites communis*, July–August). Subdominant and distributed include branched bur-reed (*Sparganium ramosum*, June–July), unbranched bur-reed (*Sparganium simplex*, June–July), meadowsweet (*Filipendula ulmaria*, June–September), ragged robin (*Lychnis flos-cuculi*, April–July), wild iris or flag (*Iris pseudacorus*, June–August), marsh marigold (*Caltha palustris*, April–June), marsh mallow (*Althaea officinalis*, August–September), water mint (*Mentha aquatica*, August–October), water forget-me-not (*Myosotis palustris*, June–August), etc.

Between the marsh and the pond or stream proper comes an intermediate formation, commonly called the **marginal** association. This formation can often be divided into three zones or associations, which merge into each other, provided the slope of the land towards the open water is a gentle one. The first zone is that nearest dry land. Here, only the roots are subjected to water-logged conditions. The members of this zone are similar to those of the marsh, together with others that are dominant or subdominant, such as reeds (*Phragmites* species, July–August), bulrushes (*Schoenoplectus lacustris*, July–August), common rushes, sedges (*Carex acutiformis*, May–July), horsetails (*Equisetum* species), watercress (*Lepidium sativum*), etc. This shore association is often called a **reed-swamp** association.

The next association is that intermediate between this reed-swamp association and the complete water association. In it, plants which are rooted in the soil, with parts of their stems submerged but leaves and flowers held above the water or floating on it, may be found. Such plants are the arrowhead (July–August), water crowfoot (May–August), water-lilies (July–August), *Potamogeton* species (p. 418), etc.

In the third association, the plants are entirely submerged. The most common types found here are Canadian pondweed (*Elodea canadensis*), and tape grass (*Vallisneria spiralis*). *Elodea* is of particular interest owing to its prolific growth. It can reproduce itself by a very simple means of vegetative reproduction. Any small branch, if broken off, will reproduce a new plant. So prolific is this, that once *Elodea* establishes itself in a suitable pond, it soon chokes up the pond, unless it is periodically cleared out. That is why this plant is such a nuisance in waterworks, etc. The great rate at which it can vegetatively reproduce itself can be realised when one learns that before 1841 it was quite unknown in England. Now it is one of the most common and prolific water plants throughout the country, and also Scotland and Wales. Actually it is a native of North America and Canada. Then it was mysteriously introduced into Co. Down in Ireland in 1836 and in England in 1841. To-day it is common in nearly every lake, pond, stream and ditch of the islands. Vegetative reproduction is the only means of multiplication of *Elodea* known in Britain. The flowers are unisexual, and in this county the male plant is very rare ; so fertilisation is unlikely.

HEATHS AND MOORLANDS

The **heath** association may be found on dry, sandy soils with a thin covering of acid peat. The dominant is ling or heather (*Calluna vulgaris*, July–October, Fig. 345). In certain areas other dominants are cross-leaved heath (*Erica tetralix*, July–October) and bilberry or whortleberry (*Vaccinium myrtillus*, May–June). Most heath plants are much exposed and therefore have certain xerophytic adaptations. The heather, for example, has rolled leaves. A local subdominant is the gorse which, as we have already seen, is xerophytic. Other local dominants are the mat grass (*Nardus stricta*, June–August) and bracken. Occasionals are represented by the Scots pine.

Real **moorland** associations differ from the heath in that they have a great deal of peat in the soil, which is usually very acid. Peat is formed chiefly by the accumulation through hundreds of years of *Sphagnum*, the peat moss, common in

Harold Bastin

FIG. 345. COMMON LING OR HEATHER (*Calluna vulgaris*).

the north and west of Britain and also in Ireland. The dried peat forms a kind of fuel. The formation of this deep, acid peat is due to the prevailing low temperatures, lack of calcium, and lack of soil aeration (due to high rainfall). Under these conditions the aerobic, humus-forming bacteria are largely absent. Higher up, wet moors and associations often have cotton grass (*Eriophorum angustifolium*, June–August) as the dominant. This cotton grass association can be found in the higher moors of Yorkshire and on Exmoor (Somerset and Devon) and Dartmoor (Devon) (Fig. 346). Cotton grass is not a member of the grass family (Gramineae), but of the sedge family (Cyperaceae).

Harold Bastin

FIG. 346. COTTON GRASS (*Eriophorum angustifolium*).

The very low-lying moors contain not only peat, but also much water and are very acid. Here may be found certain carnivorous plants such as the sundew (p. 212, July–August), butterwort (p. 212, May–July), and, in the free water, bladderwort (p. 213, June–July). These make up for the lack of available nitrogen in such soils by developing mechanisms for catching small animals which have a high nitrogen content. Large rushes also abound on such soils.

The wet peat soils of the east of England, especially in Cambridgeshire and Norfolk, vary from this in that they are very rich in mineral salts, especially of calcium, and are therefore alkaline. They are commonly called **fens**, and the plant associations of the Fen District vary accordingly. The formation of peat is here due entirely to the lack of soil aeration. The dominant tree is the alder. The Fen District of East Anglia, right up to and including the Wash, has been the subject of intensive drainage schemes ever since the time of the Roman invasion. Consequently, soil conditions, especially from the point of view of water content, are constantly undergoing changing conditions. Thus has man become a very potent biotic factor in this area. Its detailed ecological study has therefore offered great opportunities for botanists and zoologists.

With changes in conditions, great changes in the flora are taking place. New plants come, old-established ones disappear, dominants become rare, and rare become dominant. Thus the ecological aspect of the country is constantly changing. The study of this is known as dynamic ecology, as opposed to static ecology. So important is such an area to the plant and animal ecologists that one part of it, Wicken Fen in Cambridgeshire, has been purchased for the nation by the National Trust. Thus has it been secured, for all time, for further studies in ecology, etc.

SAND-DUNES

An important plant formation is that of the **sand-dune**, so common around our seashores. Here, conditions are very exceptional. The almost pure sandy soil, consisting of sand blown up from the shore, is very loose and consequently the surface layers are abnormally dry, although water (formed by condensation) occurs at a lower level. The surface is exposed to bright light, heat and strong wind. These are factors tending to increase the rate of transpiration.

Sand-dunes usually occur as a series of ridges running parallel to the shore line. Moving inland, one comes first to low hummocks of sand colonised by the sea couch grass (*Agropyron junceiforme*) or by the sea lyme grass (*Elymus arenarius*).

Behind these are the mobile or shifting dunes on which the dominant plant is the marram grass (*Ammophila arenaria*, Fig. 309). The sea holly (*Eryngium maritimum*) is also characteristic.

Further inland still are the fixed dunes. These are covered with a more or less complete carpet of grasses, and there may be patches of shrubs like the sea buckthorn (*Hippophaë rhamnoides*) and the elder (*Sambucus nigra*). Characteristic leguminous plants such as rest harrow (*Ononis reclinata*) and bird's-foot trefoil (*Lotus corniculatus*) are usually present.

Between the lines of mobile and fixed dunes there are dune valleys where the vegetation is often dominated by the creeping willow (*Salix repens*). The flora is frequently very rich, including the wintergreen (*Pyrola rotundifolia*), grass of Parnassus (*Parnassia palustris*), etc.

The series of dunes is interesting, since it represents a definite succession of vegetation. The mature, fixed dunes have, in fact, all passed through the stages of the low couch grass hummock and the mobile *Ammophila* dune. The building up of the dunes is brought about by the sand-binding capacity of the tangled rhizomes and roots of the dominant plants.

COASTAL AREAS

Coastal areas offer splendid opportunities for intensive ecological studies, such as that of the sandy seashore, or of cliffs and rocks, salt marshes, etc. Space will not allow full descriptions here. Many plants to be seen in such areas are not confined solely to the coast; but there are some plants which almost are, though even some of these do not show obvious halophytic adaptation. A brief account of those to be looked for, given in order of flowering throughout the year, follows.

Harold Bastin

FIG. 347. SAMPHIRE (*Crithmum maritimum*).

On muddy seashores, the thrift or sea pink (*Armeria maritima*) is one of the first plants to bloom, presenting its rose-coloured flowers in April, and so on until September. This symbolic plant is sometimes also known as sea gilliflower ; but it is not confined to the seashore, being also an inhabitant of mountainous areas, especially in Scotland. It is compact and has a strong rootstock and grass-like leaves. It is therefore prized by gardeners as a rock and border plant.

A very small member of the cabbage family (Cruciferae, p. 448), namely, scurvy grass (*Cochlearia officinalis*) is displaying small white flowers during May, and continues to do so until August. It grows four to ten inches high, and betrays its halophytic habit by the fleshiness of the vegetative organs.

The sea pearlwort (*Sagina maritima*) is a tiny member of the pink family (Caryophyllaceae, p. 442). It blooms during May to September. The flowers are white and the leaves fleshy.

Another member of the pink family, the sea sandwort (*Honkenya peploides*) grows in tangled masses on the seashore, a pro-

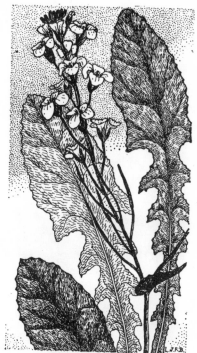

FIG. 348. SEA CABBAGE (*Brassica oleracea*).

strate plant showing its white flowers during May to September. Its leaves also are fleshy.

Of special interest is the samphire (*Crithmum maritimum*), a member of the family Umbelliferae (p. 464). It flourishes on coastal rocks and cliffs. This plant is much prized by local inhabitants, who like to gather it just before it blooms, that is, late May and early June, sprinkle the fleshy leaves with salt and pickle them. They make good eating—a fact known by seashoremen for centuries. This tufted plant has a strong rootstock, which grows well into the crevices of the rocks. The umbels of yellowish-green flowers are displayed mainly during June to August. The stems and leaves are thick and very fleshy, and the latter are glaucous, that is, covered with a ' bloom ' like a plum or black grape (Fig. 347).

The progenitor of all cultivated members of the. cabbage tribe is the sea or wild cabbage (*Brassica oleracea*), which grows on cliffs and sometimes still further inland, though it is found mainly in the south and west of England. Its large, lemon-coloured flowers appear during June to August. Though the roots are somewhat fleshy, the leaves are not ; but they are glaucous. The lower leaves are lobed and have wavy margins ; the upper leaves are sessile and not lobed (Fig. 348).

Another rarity, and again a progenitor of a garden plant, is the wild asparagus (*Asparagus officinalis*), which grows on sea cliffs and sometimes inland on waste areas. It belongs to the lily family (Liliaceae, p. 480). It attains a height of eight to eighteen inches, differing from the cultivated form only in size. Wild asparagus

FIG. 349.   TREE-MALLOW (*Lavatera arborea*).

grows abundantly in some parts of Europe ; in fact, on the steppes of the U.S.S.R. it is so prolific that cattle eat the young shoots as fodder. The young thick and fleshy shoots bear green or purplish scale-leaves. During June and July taller and more graceful shoots emerge from the axils of the lower scale-leaves. These bear needle-like phylloclades (p. 51) and yellowish-green pendulous flowers.

On the dunes, the marram grass blooms during June to August (p. 191). So also does the common hound's tongue (*Cynoglossum officinale*).

In salt marshes, the parsley dropwort (*Oenanthe lachenalii*), of the family Umbelliferae (p. 464), blooms during June to August. Its roots are fleshy, the leaves finely divided and the small white flowers borne in loose umbels.

Among the sea shingle it should not be difficult to discover the sea catchfly or sea campion (*Silene maritima*) blooming with fairly large white flowers during June to August. It is closely related to the bladder campion of the pink family (Caryophyllaceae, p. 442).

Seaside bindweed or seaside convolvulus (*Calystegia soldanella*) favours hard sand. It is a near relative of the great bindweed (p. 425). Its leaves are small and heart-shaped, borne on a creeping stem ; they are somewhat fleshy. The flowers are pink and are typical of the family, appearing during June to August.

A still further progenitor of a cultivated plant is the sea or wild beet (*Beta maritima*). It grows on the seashore, one to three feet high. The leaves are large and glossy, the ground-level (radical) ones being broad and spear-shaped, those growing up on the stems (cauline) being smaller and lance-shaped. The insignificant green flowers appear on long spikes during June to October.

During June to August, the submerged grass wracks or eel-grasses (p. 419) are in bloom.

The month of July sees the beginning of flowering of several outstanding seashore plants.

On maritime cliffs and rocks the tall, handsome tree-mallow grows, though it is not very common. It is a member of the mallow family (Malvaceae, p. 460) ; but it is more nearly related to the ornamental garden plant *Lavatera* than to the mallows themselves (*Malva* species). It is *Lavatera arborea*. It grows six to eight feet high. The pale rose-purple flowers appear during July to October (Fig. 349).

On dunes, especially in the south-west, the bushy tamarisk (*Tamarix gallica*) blooms during July to September. This plant is native to Europe and south-east Asia where the plant assumes the habit of a tree ; but in Britain it is never anything but of bushy habit. There are several different species of *Tamarix*, and several different varieties of *Tamarix gallica*, including *Tamarix gallica* var. *mannifera* of Egypt to Afghanistan, which exudes a white, honey-like secretion—the manna of the Bedouins, but not of the Israelites of the Bible. Tamarisk bears tufted branches covered with blue-green leaves, which are so reduced in size that the evaporating surfaces are cut down to a minimum—a valuable asset to a plant exposed to high winds. The plant becomes very conspicuous during July to September, when the small flowers appear in long, rose-pink spikes so closely arranged as to give a plumed effect.

The rice or cord grass, *Spartina townsendii* (p. 490), begins flowering during July.

Yet another progenitor of a cultivated plant is to be found, though not easily, in salt marshes. This is the wild celery (*Apium graveolens*), and it blooms during July and August. The greenish-white flowers are borne in umbels. The stalks of the leaves are not fleshy as those of the cultivated forms are.

On the seashore some particularly interesting plants begin blooming during July. One is the yellow horned poppy (*Glaucium flavum*) of the poppy family (Papaveraceae, p. 447). (There is also a much rarer scarlet horned poppy,

Ernest G. Neal

FIG. 350. YELLOW HORNED POPPY (*Glaucium flavum*).
Showing flowers and fruits.

Glacium corniculatum.) The leaves of the plant are glaucous, due to a covering of wax which reduces evaporation. The large yellow flowers appear during July to September ; but it is the fruit which is most interesting in that, unlike that of the common red poppy (p. 276), it takes the form of a horn-like pod six to twelve inches long. Inside are two longitudinal chambers containing seeds (Fig. 350).

The strangest of all umbelliferous plants (Umbelliferae, p. 464) is the sea holly (*Eryngium maritimum*), which grows on the seashore and blooms during July and August. The plant attains a height of one to two feet and bears curious leaves. The radical leaves are round, but tough and spiny ; the cauline leaves are lobed, and also have tough, spiny margins. These leaves, too, have a waxy bloom. The pale blue flowers appear in heads which look more like those of teasel than of an umbellifer, yet they really are umbels in which the separate flower-stalks have been suppressed. The roots are long and fleshy, and at one time they were candied as a sweetmeat—the 'kissing comfits' of Elizabethan days.

Sea lavender (*Limonium vulgare*) grows on the seashore. It is no relative of the garden lavender, but is closely related to thrift (p. 431) and more closely to garden *Statice*. All leaves are radical, being narrow at the base then broadening towards the distal end and suddenly ending in a pointed apex. The inflorescences of purple flowers appear during July and August. The flower stalks are repeatedly branched, and these bear panicles of small flowers, all of them arranged along the upper surfaces only of the branches.

In August, glasswort (*Salicornia herbacea*) begins blooming, and the insignificant green flowers appear from then until September.

The only indigenous member of the genus *Aster* (Family Compositae, p. 476) —a genus to which Michaelmas daisies but not cultivated asters belong, is the sea aster or starwort (*Aster tripolium*). This is an inhabitant of salt marshes and marine cliffs. Its leaves are fleshy and the flower-heads (capitula, p. 241) appear during August and September. The tubular disk flowers are yellow ; the fairly widely spread ligulate ray flowers are bluish-purple.

CHAPTER XXX

CLASSIFICATION OF ANGIOSPERMS

Throughout the ages, different bases have been chosen for plant classification ; but to-day we use the most natural of them all, that is, according to evolutionary sequence so far as that can be recognised. During evolution, living organisms have undergone progressive changes, mainly involving an increase in efficiency and complexity, though at times one comes across examples of degeneration and degradation. So it does not always follow that the more simple a plant organism is, therefore the lower it is in the scale of evolution ; but it is usually so.

In a classification which is based on evolutionary sequence and change the more variable characteristics are the most valuable as clues. The root, for example, is practically useless, for it is subject to the least changeable of all environments and has, therefore, suffered little change throughout the ages. Foliage leaves, too, are not sufficiently diagnostic as a basis for plant classification, though sometimes they are useful for making specific distinctions. In fact, the most changeable organ of the plant is the flower, so it is on this structure mainly, though not entirely, that flowering-plant classification is based.

The classification of plants (and animals) to-day follows fairly closely that made by the Swedish biologist Carl von Linné (usually known as Linnaeus). His whole scheme for the classification of plants was set out in his *Species Plantarum* published during 1753, though he had actually announced his system in *Systema Naturae* in 1735. But, of course, many thousands of different plants have been discovered since the days of Linnaeus, and botanical research has also revealed much more about them, so that though present-day classification does follow the technique devised by Linnaeus, it has become much modified.

Angiosperms may first of all be divided into two great groups—the Monocotyledons and the Dicotyledons.

The Dicotyledons form the larger group. They are characterised by the presence of two cotyledons or seed-leaves in their embryos. Their foliage leaves are usually narrow at the base, and may be stalked or sessile. Most of them are net-veined. The stems and roots of many become secondarily thickened. The flowers are sometimes composed of whorls with an indefinite number of segments, though most dicotyledonous flowers have either five, or a multiple of five, or four, or a multiple of four, members to each whorl. Dicotyledons are either annual, biennial, ephemeral or perennial, the last-named often taking the form of shrubs or trees. In some Dicotyledons the petals are free, whereas in others they are joined. Those having free petals are looked upon as being lower in the scale of evolution.

The Monocotyledons comprise those families of Angiosperms characterised by the presence of only one cotyledon in the embryo. Their leaves are usually (but not always) parallel-veined with almost entire (smooth) margins, whereas the leaves of Dicotyledons are net-veined and frequently serrated, lobed or compound. In Monocotyledons the number of parts of the flower is usually three or multiples of three. The internal anatomy, especially of stems and roots, also differs from that of Dicotyledons. For example, in very few Monocotyledons is there any mechanism for secondary thickening, so there are few monocotyledonous trees or shrubs.

PLANT NOMENCLATURE

Every type of plant is a species. For example, among the buttercups or crow-foots, there are several types, such as creeping buttercup, bulbous buttercup, water crowfoot, and so forth. Each is a separate **species**. Creeping buttercups belong to one species, but bulbous buttercups belong to a different species from that of the creeping. So there are different species of buttercups, but they are all closely related to each other so they are grouped together in what is called a **genus**. Therefore different buttercups belong to the same genus but different species.

In order to make the necessary distinction in nomenclature, Linnaeus suggested each plant having two names, one to designate the genus, the other the species. This is the **binomial system** adopted to-day for all plants. For example, all buttercups belong to the genus *Ranunculus*. This is the generic name. Then each species is assigned a second or specific name. Thus the botanical name for the water crowfoot is *Ranunculus aquatilis* ; that for the bulbous buttercup is *Ranunculus bulbosus* ; creeping buttercup, *Ranunculus repens* ; the lesser celandine, *Ranunculus ficaria*, and so on for the rest of the genus. The generic name is usually chosen to indicate some character of the whole genus, and the specific name to indicate an outstanding character of the species. Sometimes the names are chosen to celebrate the name of some well-known botanist, and sometimes the names are chosen for other reasons.

Just as species differ from each other, though some resemble each other sufficiently to be grouped under the same genus, so do genera themselves differ from or resemble each other. For example, closely related to the buttercup is the paeony. Thus, though the paeony is sufficiently different from the members of the genus *Ranunculus* to warrant another genus (*Paeonia*), the two genera resemble each other closely and are kept together in classification. Other genera also resembling these two are the following (in brackets after each genus is the common name of one plant belonging to it) : *Caltha* (marsh marigold), *Nigella* (love-in-a-mist), *Aquilegia* (columbine), *Delphinium* (larkspur), *Aconitum* (monk's hood), *Clematis* (clematis or traveller's joy), *Thalictrum* (meadow rue), *Anemone* (wood anemone), *Adonis* (pheasant's eye), *Myosurus* (mouse-tail), *Trollius* (globe flower), *Helleborus* (hellebore), *Eranthis* (water aconite), *Actaea* (bane-berry). All these genera are so closely related to each other that they are placed in a still bigger group called a **family**. In this cases the family is called RANUNCULACEAE. Examples of this family are given in the table on p. 437.

The classification is carried still further in that many families are closely related to each other. Those so related are grouped into what is called a **cohort** or **order**. For example, other families closely related to RANUNCULACEÆ are NYMPHÆACEÆ, to which the white water-lily (*Nymphæa alba*) and the yellow water-lily (*Nuphar luteum*) belong ; CERATOPHYLLACEÆ, to which the hornwort (*Ceratophyllum submersum*) belongs ; and BERBERIDACEÆ, to which the barberry (*Berberis vulgaris*) and other genera belong. Thus all these families, together with Ranunculaceæ, are grouped in the order **Ranales**.

THE BUTTERCUP FAMILY

RANUNCULACEAE

Genus	Species	Common Name
Caltha	*palustris*	marsh marigold
Helleborus	*viridis*	green hellebore
	foetidus	stinking hellebore
	niger	Christmas rose
Trollius	*europaeus*	globe flower
Aquilegia	*vulgaris*	columbine
Delphinium	*ajacis*	larkspur
Aconitum	*anglicum*	monk's hood
Paeonia	*sinensis*	Chinese paeony
Ranunculus	*aquatilis*	water crowfoot
	ficaria	lesser celandine
	acris	common buttercup
	bulbosus	bulbous buttercup
	repens	creeping buttercup
	sceleratus	celery-leaved buttercup
Anemone	*nemorosa*	wood anemone
	pulsatilla	pasque flower
Thalictrum	*minus*	lesser meadow rue
	flavum	common meadow rue
Clematis	*vitalba*	traveller's joy or old man's beard
Adonis	*annua*	pheasant's eye
Myosurus	*minimus*	mouse-tail
Eranthis	*hyemalis*	winter aconite
Actaea	*spicata*	bane-berry

N.B.—There are 48 genera and about 1,300 species belonging to this family altogether.

Then all the orders may be grouped finally into the Monocotyledons and Dicotyledons.

Thus Angiosperms may be divided into two groups, Monocotyledons and Dicotyledons ; then these groups into orders or cohorts ; the orders into families ; families into genera ; and genera into species.

By natural variation, plant breeding and hybridisation, species have been even still further subdivided into what are called **subspecies** and **varieties.** For example, the potato is a single species called *Solanum tuberosum*. This species is then subdivided into many varieties, some of which are familiar to the gardener.

HOW TO STUDY PLANT CLASSIFICATION

A study of plant classification cannot be made with either profit or interest from a book alone. The only interesting way to learn the various families, etc., of flowering or, indeed, non-flowering plants, is to go out into the country and examine the plants, first growing in their habitats, then more closely at home **or** in the laboratory. This, of course, to the novice, and even the trained naturalist, is impossible without some form of guide. The present book could not hope to function satisfactorily in this respect, unless it were expanded to many more pages. The best guide, on the other hand, is a good Flora. This is one which has all the most familiar, and many of the unfamiliar, plants classified in their orders, families, genera, species, and variations. Many elementary Floras ignore orders and begin with families. This does not matter very much to a beginner. A Flora helps in the description of a plant, and not only classifies it but usually gives adequate reasons for so doing. It is also a means of identifying unfamiliar plants by means of a key. The most up-to-date Flora, and therefore the one using the generally accepted nomenclature, is *Flora of the British Isles* by **A. R. Clapham**, **T. G. Tutin** and **E. F. Warburg** (Cambridge University Press).

Merely to identify a plant and see how it is classified is not enough to those who are really interested. A written description is highly to be desired, and drawings, scientifically clear, though perhaps not necessarily of very great artistic merit, should *always* be made. Special attention should be paid to the flower, since it is chiefly upon this, especially its reproductive organs (stamens and carpels), that the plant is classified. A member of each floral whorl should be drawn separately, and a longitudinal section of the flower and a floral diagram constructed. The whole should be finished off with a floral formula.

A few specimens of each family (each specimen being, if possible, a member of a different genus) will soon help the student in an appreciation of the study of plant classification. Since cultivated plants often undergo considerable ' artificial ' modifications under cultivation, they chould be avoided in this connexion so far as possible.

As a general guide, the following résumé of the classification of Angiosperms, with the outstanding features of certain families (many are omitted), should prove useful ; but it cannot be too much emphasised that practical work with the help of a Flora is essential. (The usual months of flowering are inserted after each plant mentioned ; but it must be borne in mind that such dates vary within days or even weeks according to latitude and prevailing climatic conditions.)

DICOTYLEDONS

The Dicotyledons are characterised by the presence of two cotyledons in their embryos. The leaves are usually narrow at the base, and may be stalked or sessile. They are net-veined. The stem and root contain cambium, and therefore are often secondarily thickened. The flowers are sometimes composed of whorls with an indefinite number of segments, though most dicotyledonous flowers are either **pentamerous** (five, or a multiple of five, members to each whorl) or **tetramerous** (four, or a multiple of four, members to each whorl). Dicotyledons are either annual, biennial, ephemeral, or perennial, the last-named often taking the form of shrubs or trees.

The Dicotyledons are subdivided into groups according to whether their petals are separate, that is, **polypetalous (Polypetalae)**; joined, or **sympetalous (Sympetalae)**; or without petal (**Apetalae**). Polypetalae and Apetalae are now usually grouped together under **Archichlamydeae,** while the term **Metachlamydeae** is used instead of Sympetalae.

ARCHICHLAMYDEAE

The Archichlamydeae are more primitive than the Sympetalae, especially in view of the fact that the latter are more specialised in their pollination mechanisms. The Archichlamydeae contain many familiar families.

SALICACEAE

This a cosmopolitan family which contains only two genera, namely, *Salix* (the willows) and *Populus* (the poplars). All are trees or shrubs which often reproduce themselves profusely by means of suckers. The flowers are borne in inflorescences known as catkins (Fig. 180). The plants are dioecious (p. 239). Each flower is borne in the axil of a hairy bract. The margin of the bract is entire in *Salix*, but toothed in *Populus*.

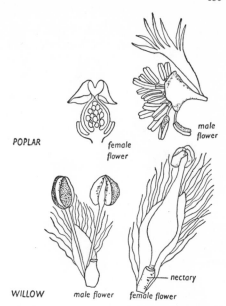

Salix species flower very early (March and April). Self-pollination is naturally impossible, since these plants are dioecious. Cross-pollination is affected by insects which visit flowers for the nectar or pollen. The male flower consists of two to five stamens, the number varying with the species. The female flower is composed of two syncarpous carpels, with a bilobed stigma, and an ovary enclosing one chamber. The ovules are borne parietally (Fig. 351). Both male and female flowers have a small, green, club-shaped nectary at the base. The fruit is a hard capsule, and each seed bears a tuft of hairs which proves useful for wind dispersal.

FIG. 351. FLOWERS OF POPLAR AND WILLOW.
(*F. W. Williams*.)

The genus *Salix* is not difficult to recognise, though the separation into species is not so easy, and there are many wild hybrids. Examples of *Salix* are : *S. caprea*, goat willow sometimes also called pussy willow, sallow, or ' palm ' (April–May) ; *S. fragilis*, crack willow or withy (April–June) ; *S. alba*, white or Huntingdon willow (May) ; *S. alba* var. *coerulea*, cricket bat willow (May–June) ; *S. viminalis*, common osier or withy (April–June) ; *S. repens*, creeping willow (April–May) ; *S. babylonica*, weeping willow (May–June).

The flowers of the genus *Populus* are also borne in catkins (Fig. 180). They are wind-pollinated, so an abundance of pollen is necessary in order to ensure pollination—there is a great wastage. The male flowers bear four to forty stamens in a cup-shaped structure representing the perianth. The female flowers have a small, cup-shaped perianth and a large stigma (Fig. 351). The fruits and seeds are similar to those of *Salix*.

Examples of *Populus* are *P. alba*, white poplar (March–April) ; *P. nigra*, black poplar (March–April) ; *P. tremula*, aspen (March–April) ; *P. italica*, Lombardy poplar (March–April) ; *P. canescens*, grey poplar (March–April).

JUGLANDACEAE

A family of half a dozen genera of north temperate and tropical Asiatic trees.

The commonest genus is *Juglans*, examples of which are : *J. regia*, English or European walnut (April–May) ; *J. nigra*, black walnut (April–May) ; *J. cinerea*, North American butternut. A genus common in America is *Carya*, embracing *C. alba* and *C. tomentosa*, the hickory nuts, and *C. olivaeformis*, the pecan nut.

BETULACEAE

A north-temperate family of trees and shrubs, though a few genera are indigenous to South America. The most common genera are *Betulus* (birch) and *Alnus* (alder). All

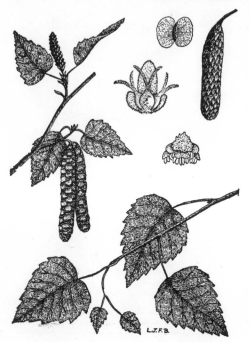

leaves are simple and flowers unisexual, borne in male and female catkins on the same plant (Fig. 352).

Examples of *Betula* are : *B. verrucosa,* silver birch (March–April) ; *B. pubescens,* the silver birch of more northerly habitat and more pendulous habit with downy twigs. *B. papyrifera* is the paper or canoe birch of North America.

The only common species of *Alnus* is *A. glutinosa,* alder or aller (March–April).

CORYLACEAE

The north-temperate family CORY-LACEAE contains the genera *Carpinus* (hornbeam) and *Corylus* (hazel). (Some Floras include these genera in the family BETULACEAE.)

The genus *Carpinus* contains a large number of eastern Asiatic species. The only common European species is *C. betulus,* hornbeam or yoke-elm (April–May).

The genus *Corylus* is native to north temperate regions in both Old and New Worlds. The only common example is *C. avellana,* hazel (January–February), bearing its male flowers in the familiar catkins and the female flowers in erect structures superficially resembling winter

FIG. 252. BIRCH (*Betula verrucosa*).
Top left, male catkins ; bottom right, spray of foliage ; top right, female catkin, winged fruit, scale bearing female flowers, scale bearing male flowers.

buds with several pairs of red styles protruding. Three bracteoles form the capsule around the nut.

FAGACEAE

The FAGACEAE is a family of noble trees such as *Quercus* (oaks), *Fagus* (beeches) and *Castanea* (sweet chestnut). The family is distributed throughout tropical and temperate regions, though it is absent in the wild state in tropical and South Africa.

Examples of *Quercus* are : *Q. robur* and *Q. petraea,* common oaks (April–May), the latter being sometimes known as durmast oaks, having sessile female flowers, whereas those of the former are stalked ; *Q. cerris,* Turkey oak (April–May) ; *Q. ilex,* evergreen, holm or holly oak (April–May) ; and a number of cultivated species.

Castanea is represented in Britain only by *C. sativa,* sweet, Italian, French, Spanish or European chestnut (July–August). The fact that this tree flowers so late in the year explains why in these regions it seldom bears *mature* fruit.

Fagus is represented by *F. sylvatica,* beech (April–May). The copper beech is not a separate species, but is believed to have been derived from the purple beech (*P. sylvatica* var. *purpurea*). There are several varieties of cultivated beech.

ULMACEAE

This is a family of trees and shrubs distributed throughout the north temperate zone. The only common genus in Britain is *Ulmus* (elms). (In some Floras, *Ulmus* is placed in the related family URTICACEAE, p. 441.)

Examples of *Ulmus* are : *U. procera*, common or English elm of hedgerows (February–March) ; *U. glabra*, wych elm of more open spaces (February–March). Less common are *U. diversifolia*, East Anglian elm ; *U. hollandica*, Dutch elm ; *U. stricta*, Cornish elm ; *U. plotii*, Plot elm. The fruits are winged, with a more or less central seed.

MORACEAE

No member of the family MORACEAE is indigenous to Britain ; but several are well known and some are cultivated. The genus *Morus* contains *M. nigra*, black mulberry, commonly cultivated ; *M. alba*, white mulberry, rare in Britain, but cultivated elsewhere to feed the silkworm. The perianth becomes succulent and edible as the fruit ripens. The genus *Ficus* contains *F. indica* and *F. benghalensis*, the banyans ; *F. elastica*, india-rubber tree ; *F. sycomorus*, the sycomore fig of Biblical lands ; *F. carica*, common fig. The genus *Artocarpus* contains several well-known tropical species, including *A. incisa*, bread-fruit.

URTICACEAE

A tropical and temperate family of herbs and small shrubs containing some species which yield tough fibres of economic importance. The flowers are usually unisexual with a sepaloid perianth and the plants either monoecious or dioecious. Three genera are common to Britain, namely, *Urtica*, *Parietaria* and *Humulus*. (In some Floras, the last named is placed in the closely related family MORACEAE, and in others it is included in the small family CANNAEINACEAE.)

The genus *Urtica* contains the three stinging nettles : *U. dioica*, common stinging nettle (June–September) ; *U. urens*, small stinging nettle (June–September) ; *U. pilulifera*, Roman stinging nettle (June–September), a rarer species with a more painful sting.

Parietaria is represented in Britain by *P. diffusa*, pellitory-of-the-wall (June–September).

Humulus contains two species only, namely, *H. lupulus*, hop (July–September), and *H. americanus*, American hop. Both grow wild (the former in Europe, the latter in North America) and both are cultivated either for ornamental purposes or for their fruiting catkins used for flavouring beer.

POLYGONACEAE

POLYGONACEAE is mainly a temperate family of plants bearing very inconspicuous, hermaphrodite flowers. The leaves have stipules which unite to form a ring around the stem (an ochrea). It contains the well-known genera *Rumex* (docks and sorrels), *Polygonum* (bistorts and persicarias), *Rheum* (rhubarb) and *Fagopyrum* (buckwheat).

Common examples of the genus *Rumex* are : *R. acetosa*, common sorrel (May–July) ; *R. acetosella*, sheep's sorrel (May–July) ; *R. crispus*, curled dock (June–August) ; *R. obtusifolius*), broad-leaved dock (July–September) ; *R. conglomeratus*, sharp dock (July–September) ; *R. sanguineus*, red-veined or blood dock (July–September).

Polygonum is also a common genus, being represented by *P. bistorta*, bistort (June–September) ; *P. convolvulus*, climbing persicaria (July–September) ; *P. lapathifolium*, pink persicaria (July–September) ; *P. persicaria*, spotted persicaria (July–October) ; *P. aviculare*, knotgrass or knotweed (July–September) ; *P. nodosum*, large persicaria (July–October) ; *P. amphibium*, amphibious persicaria (July–October) ; *P. hydropiper*, biting persicaria (July–October).

The genus *Rheum* is represented by *R. rhaponticum*, cultivated rhubarb ; and *Fagopyrum* by *F. esculentum*, buckwheat (July–September).

CHENOPODIACEAE

This is a small but economically important family of plants, since it contains the sugar beet, spinach and mangel-wurzel. In many ways the family resembles POLYGONACEAE,

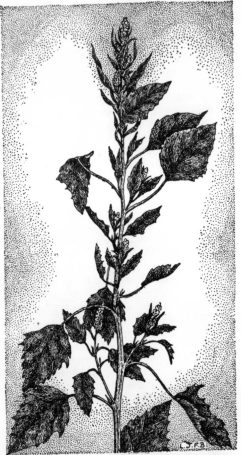

FIG. 353.   WHITE GOOSEFOOT (*Chenopodium album*).

having similar leaves, but no stipules, and inflorescences, with flowers bearing no petals. The main genera are *Atriplex, Salicornia, Chenopodium, Beta* and *Spinacia*.

Atriplex is represented by *A. patula*, orache (July–September), and others.

Salicornia is represented by *S. stricta*, glasswort or marsh samphire (August September), the salt-marsh plant (p. 434).

The genus *Chenopodium* has several representatives, some of which are common weeds in gardens and on arable land, for example, *C. bonus-henricus*, Good King Henry (June–October); *C. rubrum*, red goosefoot (July–September); *C. album*, white goosefoot or fat hen, a common weed (August–September, Fig. 353); *C. urbicum*, upright goosefoot (August–September); *C. murale*, nettle-leaved goosefoot (August–September).

The genus *Beta* is represented in the wild state by *B. vulgaris*, the sea beet (June–October) of coastal areas (p. 432). The garden beetroot, sugar beet and mangel-wurzel (mangold) are all varieties of this.

Spinacia has no wild representative in Britain, but *S. oleracea*, spinach, is widely cultivated.

AMARANTACEAE

A family closely related to CHENO-PODIACEAE, but not indigenously represented in Britain, is AMARANTACEAE. In tropical and other areas certain members of this family are cultivated for food; in Britain and elsewhere some are popular ornamental favourites, for example, *Amaranthus caudatus*, love-lies-bleeding; *A. hypochondriacus*, prince's feathers; and *Celosia cristata*, cock's-comb.

PORTULACACEAE

This is a family which is mainly native to America, where many of its members are of fleshy habit and therefore thrive in desert areas. In Britain, some of these plants have been imported and are now cultivated, the most common being *Portulaca* species, *Calandrinia umbellata*, *Claytonia* species and *Lewisia* species, the last named having handsome rosettes of fleshy leaves. *Portulaca oleracea* and *Claytonia perfoliata* and *C. alsinoides* are locally abundant as garden escapes. Species of blinks (*Montia*) are the only native members of the family.

CARYOPHYLLACEAE

This is a cosmopolitan family confined to north-temperate and cold regions. It comprises about sixty genera, many of which are indigenous to Britain. Most of them are herbs, though there are a few shrubs among them. None is of economic value, apart from the many cultivated for ornamental purposes. There are diagnostic features of this family. The leaves are simple and opposite. The stems are usually swollen at the nodes. The inflorescence is also characteristic in that it is a dichasial cyme (Fig. 245).

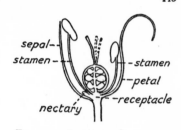

The flower is actinomorphic and hypogynous. It has five inferior sepals, five free petals, usually ten hypogynous stamens in two whorls of five each, and a superior gynoecium with two to five free styles and a one-chambered ovary with incomplete partitions. The flower is protandrous and the outer stamens ripen first. The fruit is usually a capsule, opening by ten teeth.

FIG. 354. LONGITUDINAL SECTION THROUGH FLOWER OF GREATER STITCHWORT (*Stellaria holostea*).

A typical longitudinal section is shown in Fig. 354.

The main genera are : *Stellaria, Cerastium, Sagina, Arenaria, Spergularia, Silene, Lychnis, Melandrium, Agrostemma, Dianthus* and *Gypsophila*.

The following are examples :

Stellaria : *S. media*, small chickweed, a common garden weed (February–October) ; *S. holostea*, greater stitchwort, adder's meat or satin flower (April–August) ; *S. nemorum*, wood stitchwort (May–July) ; *S. graminea*, lesser or heath stitchwort (May–August) ; *S. alsine*, bog stitchwort (May–July).

Myosoton : *M. aquaticum*, water chickweed (July–August).

Cerastium : *C. vulgatum*, common mouse-ear chickweed (April–September) ; *C. arvense*, field mouse-ear chickweed (April–September).

Sagina : *S. apetala*, annual pearlwort, a wall plant having no petals (May–September) ; *S. procumbens*, procumbent pearlwort (May–September) ; *S. maritima*, sea pearlwort (May–September) ; *S. subulata*, heath pearlwort (June–August) ; *S. nodosa*, knotted pearlwort or knotted spurrey (July–September).

Arenaria : *A. serpyllifolia*, thyme-leaved sandwort (May–September).

Spergularia : *S. rubra*, seaside sandwort spurrey (July–September).

Silene : *S. cucubalus*, bladder campion or white bottle (June–August) ; *S. acaulis*, moss campion, confined to high mountainous areas (May–August) ; *S. maritima*, sea campion (July–August).

Lychnis : *L. flos-cuculi*, ragged robin (April–July).

Melandrium : *M. rubrum*, red campion, rose campion, cock robin or red robin, having unisexual flowers (May–August) ; *M. album*, white or evening campion (June–September, Fig. 355), also having unisexual flowers.

Agrostemma : *A. githago*, corn cockle (July–September).

Dianthus : *D. gratianopolitanus*, Cheddar pink, a plant of calcareous soils (June–July) ; *D. deltoides*, maiden pink (June–July) ; *D. armeria*, Deptford pink (June–July) ; *D. caryophyllus*, clove pink or wild carnation (June–July)—all very rare plants in the wild state.

The family CARYOPHYLLACEAE is very popular among gardeners, the most commonly cultivated plants being : *Dianthus caryophyllus*, varieties of pinks and carnations, together with those species of *Dianthus* mentioned above : *D. barbatus*, sweet William ; and *Lychnis, Silene, Arenaria* and *Gypsophila* species and varieties.

NYMPHAEACEAE

The NYMPHAEACEAE is a well-known tropical and temperate family of marsh and water plants, represented indigenously in Britain by the genera *Nymphaea* and *Nuphar*.

The following are worthy of note :

Nymphaea alba, white water lily (July–August, p. 416) ; *Nuphar lutea*, yellow water lily or brandy bottle (June–August, p. 416) ; *Victoria regia*, Victoria water lily of the Amazon

FIG. 355. WHITE OR EVENING CAMPION
(*Melandrium album*).

(p. 417) ; *Nelumbo nucifeca*, sacred lotus of the East (p. 293). In ornamental lakes and ponds, many differently coloured varieties of *Nymphaea alba* may be seen.

CERATOPHYLLACEAE

This is a small family of water plants comprising one genus only, namely, *Ceratophyllum*, the hornworts of which two species are indigenous to Britain—*C. demersum*, hornwort, a submerged freshwater plant having unisexual, very insignificant flowers (June–July) ; and *C. submersum*, which differs from the former in that its fruits are smooth instead of spiny.

RANUNCULACEAE

This large, mainly north-temperate family is well known in Britain, both wild and cultivated. Many of its members are herbaceous perennials. The general and floristic characters of the family have been described elsewhere in this book (pp. 242, 436). The main genera are : *Paeonia, Caltha, Trollius, Helleborus, Nigella, Eranthis, Actaea, Aquilegia, Delphinium, Aconitum, Anemone, Clematis, Ranunculus, Adonis, Thalictrum* and *Myosurus*.

Wild members to be looked for include the following :

Caltha : *C. palustris*, marsh marigold, kingcup, golden cup, soldier's button, brave celandine, may-blob, mare-blob, horse-blob, Mary-bud, publicans and sinners, luckan gowan (in Scotland) and cowslip (in parts of the United States) (March–June).

Trollius : *T. europaeus*, globe flower of rocky mountain pastures (witch's gowan in Scotland) (May–July).

Helleborus : *H. viridis*, green hellebore of limestone areas in the south and east (January–June, Fig. 356) ; *H. foetidus*, stinking hellebore or setterwort.

Eranthis : *E. hyemalis*, winter aconite (January–April).

Actaea : *A. spicata*, baneberry in copses of northern England.

Aquilegia : *A. vulgaris*, wild columbine or granny's bonnet, very rare in woods and heaths (May–June).

Aconitum : *A. anglicum*, monk's hood or wolf's bane, rare habitué of river banks, having zygomorphic flowers (p. 247). The entire plant is poisonous (June–August).

Ernest G. Neal

FIG. 356. GREEN HELLEBORE (*Helleborus viridis*).

Anemone : *A. nemorosa*, wood anemone or wind flower, of woodlands (March–May) ; *A. pulsatilla*, pasque flower, a rarity of calcareous soils (April–May). The foliar appendages on the flower stalk are bracts.

Clematis : *C. vitalba*, wild clematis, old man's beard, traveller's joy, virgin's bower, a common climber of woodlands and hedgerows in the calcareous regions of the south, seldom seen north of the Midlands (July–September).

Ranunculus : *R. ficaria*, lesser celandine (April–May) ; *R. auricomus*, rare goldilocks or wood crowfoot (April–July) ; *R. repens*, creeping buttercup (May–August) ; *R. bulbosus*, bulbous buttercup (May–August) ; *R. parviflorus*, small-flowered buttercup (May–July) ; *R. arvensis*, corn buttercup (May–July) ; *R. sceleratus*, celery-leaved buttercup (May–September) ; *R. aquatilis*, water crowfoot—divided into several sub-species in some Floras (May–August, p. 418) ; *R. acris*, common buttercup (June–August) ; *R. lingua*, greater spearwort of marshes and bogs (June–September).

Adonis : *A. annua*, pheasant's eye, having red flowers and confined to cornfields (June–September).

Thalictrum : *T. flavum*, common meadow rue (July–August) ; *T. minus*, small meadow rue (July–August) ; *T. alpinum*, alpine meadow rue (July–August).

Myosurus : *M. minimus*, mousetail, a small weed of cornfields (April–June).

Among the many cultivated members of the RANUNCULACEAE are : *Paeonia sinensis* and other species and varieties of paeony ; *Trollius europaeus*, globe flower ; *Helleborus niger*, black hellebore or Christmas rose ; *Nigella* species, love-in-a-mist ; *Aquilegia* species, columbine ; *Delphinium* species, delphiniums and larkspurs ; *Aconitum anglicum*, monk's

FIG. 357. *Magnolia grandiflora.*
Note simple structure of flower.

hood ; *Anemone* species and varieties ; *Clematis* species and varieties ; *Eranthis hyemalis,* winter aconite.

Members of the RANUNCULACEAE show interesting modifications of the petals for the production of nectaries. In the globe flower (*Trollius*) the petals are represented by narrow honey leaves. The Christmas rose (*Helleborus*) possesses small trumpet-shaped petal nectaries, the sepals being petaloid and attractive. The petals of the buttercup (*Ranunculus*) have small basal nectaries. All these flowers are visited by short-tongued insects. Many flowers require long-tongued insects to effect pollination. The petals of columbine (*Aquilegia*) are prolonged into spurs containing nectar. The monk's hood (*Aconitum*) shows two posterior petals elongated into stalked glandular spurs covered by the hooded sepal. Larkspur (*Delphinium*) has a similar elongation of the petals fitting into a spurred calyx.

Those members of the family which are devoid of petals are usually wind-pollinated, for example, wood anemone (*Anemone*) and meadow rue (*Thalictrum*), or the nectar is secreted elsewhere, for example, carpels of marsh marigold (*Caltha*).

BERBERIDACEAE

A family of shrubs indigenous mainly to South America, but represented in Britain by *Berberis vulgaris*, common barberry and other species and varieties in parks and gardens. The leaves may be simple or compound, and are usually prickly. In *Berberis*, the leaves on the main shoots are reduced to spines, and the normal foliage leaves are borne in short branches in the axils of the spines. The flowers usually hang in pendulous inflorescences, white, yellow or orange. The berries are bright yellow, orange or red. Barberry acts as one of the hosts to the fungal parasite *Puccinia graminis* (p. 330), which causes black rust disease of wheat, the other host.

MAGNOLIACEAE

A family of trees and shrubs (a few of which are climbers) all indigenous to tropical and sub-tropical regions, but well known in Britain through introduced exotics. Most bear large flowers of very simple structure. The best-known genera are *Magnolia* (magnolias) and *Liriodendron* (tulip tree).

Species of *Magnolia* cultivated are many and varied (Fig. 357).

Liriodendron contains one species only, namely, *L. tulipifera*, the tulip tree, whose yellowish-green flowers superficially resemble tulips (hence the common name), though basically they are quite different (the tulip is a Monocotyledon). The tulip tree is cultivated in Britain, though it flowers (June–July) usually only in the Midlands and South. The curious leaf has a re-entrant notch at its distal end. In the United States, the tulip tree is very common, and there it is sometimes also known as yellow poplar or whitewood. The timber is commercially known as white poplar, yellow poplar, tulip wood and canary white wood.

LAURACEAE

Another tropical and subtropical family of trees and shrubs, but these bear tough, evergreen foliage. The most important genera are : *Cinnamonum, Persea* and *Laurus*. These and other genera contain trees yielding medicinal and aromatic bark.

Best-known examples are :

Cinnamonum : *C. zeylanicum*, cinnamon ; *C. cassia*, yielding cassia bark ; *C. camphora*, camphor.

Persea : *P. gratissima*, avocado pear.

Laurus : *L. nobilis*, laurel or sweet bay. This is the true laurel cultivated in Britain since the sixteenth century—the laurel of victors and conquerors. It must not be confused with the Japanese laurel, which usually bears green leaves splashed with yellow (*Aucuba japonica*, CORNACEAE) or the cherry laurel (*Prunus laurocerasus*, ROSACEAE).

PAPAVERACEAE

A fairly small family, mainly of herbs, almost entirely confined to north temperate regions. Chief genera are : *Papaver, Eschscholtzia, Chelidonium* and *Glaucium*.

Among those indigenous to Britain are :

Papaver : *P. rhoeas*, common red poppy (June–July) ; *P. argemone*, rough-head poppy, rarer and of paler red (June–July) ; *P. somniferum*, opium poppy, with white petals having purple patches at their bases—more at home in Mediterranean regions (July–August).

Glaucium : *G. flavum*, yellow horned poppy (July–September, p. 433 and Fig. 350) ; *G. corniculatum*, scarlet horned poppy (July–September).

Meconopsis : *M. cambrica*, Welsh yellow poppy, a rarity of rocky areas of Wales, West of England and Westmorland (June–August).

Chelidonium : *C. majus*, greater celandine (May–August).

PAPAVERACEAE is another family popular among gardeners. *P. somniferum* is cultivated in various parts of the world for the opium it yields. Among ornamental plants are : *Papaver rhoeas* varieties ; *P. orientalis* varieties of oriental poppies ; *P. nudicale*, yellow Iceland poppy ; *Eschscholtzia* species and varieties of Californian poppy ; *Meconopsis cambrica*, Welsh yellow poppy ; *M. baileyii*, Tibetan blue poppy ; *Romneya* species, perennial Californian tree poppy.

Members of the PAPAVERACEAE are noted for their latex canals. *Papaver* species usually contain a white latex which stains the hands a dark colour ; *Meconopsis*, yellow latex ; *Glaucium*, yellow ; *Chelidonium*, yellow.

FUMARIACEAE

In some Floras, the members of this family are placed with the PAPAVERACEAE. There are two wild genera native to Britain, namely, *Fumaria* and *Corydalis* ; others are cultivated. Examples of indigenous plants are :

Fumaria : *F. capreolata*, rampant fumitory, a climber of hedgerows (June–September) ; *F. officinalis*, common fumitory, a weed of gardens and arable land (June–September).

Corydalis : *C. claviculata*, white climbing corydalis, rather rare (June–September).

Among cultivated ornamentals are : *Dicentra* (or *Dielytra*) *spectabilis*, bleeding heart or lyre flower ; *Corydalis cheilanthifolia*, in rock gardens and dwarf borders.

CRUCIFERAE

This is a cosmopolitan, though mainly north-temperate, family of great economic importance, since it contains mainly edible plants such as cabbages, radish, turnip, watercress, etc. The family is so named from the cross-shaped or cruciform arrangement of the petals. There are well over two hundred genera.

The hypogynous flower has a calyx of four sepals and a corolla of four petals. The androecium is composed of two short and four long stamens (Fig. 247), the superior gynoecium comprises two syncarpous carpels. The ovary is at first one-chambered ; but this eventually becomes divided by the growth of a false septum (p. 276). Placentation is parietal. The petals are long and narrow, though the distal ends often are expanded. The sepals hold them up, pressed against the stamens. The two lateral sepals are often pouched to hold nectar from the nectary round the bases of the lateral stamens. Pollination is clearly only possible by long-tongued insects. A few species are self-pollinated.

The flowers of the CRUCIFERAE vary only slightly, and that chiefly in size and colour. The fruit is a siliqua or silicula (a short and broad form of siliqua).

The floral formula for most species is :

$$\oplus K2 + 2\ C4\ A2 + 4\ G(\underline{2}).$$

A representative floral diagram is given in Fig. 358.

Only a few of the many examples indigenous to Britain can be mentioned :

Cheiranthus : *C. cheiri*, wallflower, usually yellow in the wild state, and possibly really a garden escape (April–May).

Capsella : *C. bursa-pastoris*, shepherd's purse, a very common weed in almost every part of the world (January–December, p. 276, Fig. 359).

FIG. 358. FLORAL DIAGRAM OF CRUCIFERAE.

Erophila : *E. verna*, whitlow grass, a tiny plant of banks and walls (March–May).

Draba *D. aizoides*, very rare yellow whitlow grass (March–May) ; *D. incana*, hoary whitlow grass (June–July) ; *D. rupestris*, rock whitlow grass (June–July).

Cardamine : *C. pratensis*, cuckoo flower, lady's smock, meadow bitter cress or milk maids (April–June) ; *C. hirsuta*, hairy bitter cress (April–August) ; *C. amara*, large-flowered bitter cress (April–August).

Arabis : *A. hirsuta*, hairy rock cress (June–August).

Nasturtium : *N. officinale*, water-cress (June–July). In some Floras this plant has been renamed *Radicula nasturtium* ; it must not be confused with the garden flowering nasturtium (*Tropaeolum* species, of the South American family TROPAEOLACEAE), August.

Rorippa : *R. sylvestris*, creeping yellow cress (June–July) ; *R. islandica*, marsh yellow cress (April–September).

Alliaria : *A. petiolata*, garlic mustard, Jack-by-the-hedge or sauce alone (May–June).

Erysimum : *E. cheiranthoides*, treacle mustard (June–August).

Sinapis : *S. arvensis*, charlock (June–August).

Isatis : *I. tinctoria*, woad, a rarity of river banks and chalk pits (July–September).

Among the many members of the CRUCIFERAE cultivated in flower gardens are : *Cheiranthis cheiri*, wallflower ; *C. allionii*, Siberian wallflower ; *Erysimum* species, alpine wallflowers ; *Lunaria annua*, honesty, bearing large, flat, oval fruits, the central partition of which persists as a white tissue-like plate against which the seeds are pressed (Fig. 216) ; *Matthiola incana* and varieties of stock ; *Iberis amara*, candytuft ; *Arabis*, *Aubretia* and *Alyssum* species and varieties.

Those members cultivated for food include : varieties derived from *Brassica oleracea*, the wild sea cabbage (p. 431 and Fig. 348), such as cabbages, brussels sprout, cauliflower, kale or borecole, and kohl-rabi ; *B. rapa*, turnip ; *B. napus*, swede and rape ; *B. nigra*, black mustard ; *Sinapis alba*, white mustard, sometimes found growing wild ; *Nasturtium officinale*, water-cress ; *Lepidium sativum*, garden cress ; *Raphanus sativus*, radish, derived from the rather rare wild radish, *R. rhaphanistrum* ; *Armoracia rusticana*, horse radish.

The flowers of all the cultivated food crucifers are typical of the family, and may be white, yellow, mauve or purple.

RESEDACEAE

RESEDACEAE is a small family, members of which grow in Europe, Asia, Africa and the southern parts of the United States. In Britain it is represented by the genus *Reseda* only.

Examples are : *R. lutea*, yellow wild mignonette of chalky and sandy soils (June–August) ; *R. alba*, white wild mignonette, a rarer species (July–August) ; *R. luteola*, dyer's rocket or weld, from which at one time a green dye was extracted (July–August). The scented garden species, *R. odorata*, is of Egyptian origin.

DROSERACEAE

A family of carnivorous herbs, represented in Britain by the genus *Drosera* (p. 211), the species being : *D. rotundifolia*, round-leaved sundew (July–August) ; *D. anglica*, long-leaved sundew (July–August) ; *D. intermedia*, lesser long-leaved sundew or narrow-leaved sundew (July–August). In all cases the flowers seldom open, but are pollinated while still in the bud. *Dionaea muscipula*, Venus' fly-trap of Carolina (p. 214), is also a member of this curious, carnivorous family.

CRASSULACEAE

This is the stonecrop family, a very large one, widely represented in South Africa. Genera represented in Britain are : *Sedum*, *Sempervivum* and *Umbilicus*.

Examples are :

Sedum : *S. acre*, biting stonecrop, wall pepper or poor man's pepper of walls and stony places (June–August) ; *S. telephium*, orpine or live-long (July–August) ; *S. rosea*, rose-root (May–August). There are other rarer stonecrops.

Umbilicus : *U. rupestris*, wall pennywort or navelwort (June–August).

Sempervivum : *S. tectorum*, houseleek (July–August, p. 50 and Fig. 32).

Many of the above wild species are cultivated in gardens, mainly rock-gardens and on old walls and cottage roofs.

SAXIFRAGACEAE

An important family, mainly temperate and mostly herbs. A large number are alpine or arctic and of xerophytic character. The main genera are : *Saxifraga*, *Chrysosplenium*, *Parnassia*. In this family may also be included certain genera that are placed in other families by some schemes of classification, namely, *Hydrangea*, *Deutzia* and *Philadelphus* (all HYDRANGEACEAE) and *Ribes* (GROSSULARIACEAE).

Wild examples of SAXIFRAGACEAE include the following :

Saxifraga : *S. granulata*, meadow saxifrage, of meadows and banks, which multiplies vegetatively by means of basal axillary bulbils (May–June, Fig. 360) ; *S. tridactylites*, rue-leaved or three-fingered saxifrage (April–July) ; and other rarer species.

Chrysosplenium : *C. oppositifolium*, golden saxifrage (April–May) ; *C. alternifolium*, alternate-leaved saxifrage (April–May).

Parnassia : *P. palustris*, grass of Parnassus, of moorland marshes (July–September). (Some authorities place this genus in a separate family—PARNASSIACEAE.)

Among members of this family cultivated for their ornamental beauty are : *Saxifraga* species, including *S. umbrosa*, London pride ; *Astilbe* species, a popular pot

FIG. 359.   SHEPHERD'S PURSE (*Capsella bursa-pastoris*).

plant (often erroneously called *Spiraea*);
Escallonia species; *Hydrangea* species;
Philadelphus species (sometimes called
orange blossom and often erroneously
named *Syringa*, which is the generic name
for lilac of the family OLEACEAE).

Those cultivated for their fruit (sometimes placed in a separate family, GROSSU-
LARIACEAE) are : *Ribes* : *R. sylvestre*, red
currant and its white variety ; *R. nigrum*,
black currant ; *R. uva-crispa*, gooseberry.
All three of these species grow wild in
the North. Some *Ribes* shrubs are cultivated as ornamental plants, for example,
R. sanguineum, flowering currant ; *R. speci-
osum*, a wall climber ; *R. aureum*, Buffalo
currant.

PLATANACEAE

A very small, mono-generic family of
trees. The only genus is *Platanus*, which
contains the true plane trees, a well-
known character of which is that their
bark periodically peels off in patches,
which prevents the lenticels becoming
choked with soot, etc. Moreover, the
winter buds are fully protected by the
base of the leaf stalk, which covers it
like a helmet. These characters render
the trees suitable for growing in towns.
These trees bear unisexual flowers (June–
July) ; but they are monoecious. Of

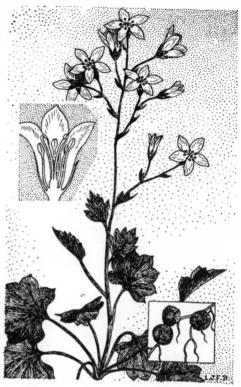

FIG. 360. MEADOW SAXIFRAGE (*Saxifraga granulata*).
Left, section of flower ; bottom right, bulbils.

the several species, three are to be found in Britain, namely, *Platanus orientalis*, oriental
plane ; *P. occidentalis*, western plane ; *P. acerifolia*, London plane, the last named being
characteristic of streets and parks in London and other towns.

ROSACEAE

ROSACEAE contains many familiar plants, both wild and cultivated. It is a cosmopolitan
family containing about a hundred genera. Vegetative reproduction is common in this
family, such as by runners in the strawberry, suckers in the raspberry, etc. Artificial
budding of roses and grafting of apples and other trees are common horticultural practices.
All forms of plants are represented—trees, shrubs and herbs. The leaves are usually
alternately arranged, simple or compound, and in many cases they bear conspicuous
stipules.

The flowers are borne in varying forms of inflorescences, and the flowers themselves
show well the different stages from perigny to epigyny (p. 250 and Fig. 195). They are
regular, with four or five sepals and the corresponding number of petals. The stamens are
indefinite in number. The carpels vary from one in the cherry to indefinite in the strawberry. They are generally apocarpous ; rarely syncarpous.

The stages from perigyny to epigyny are interesting. In the perigynous *Potentilla* the
receptacle is practically a flattened disk ; in the perigynous peach (*Prunus persica*) the
receptacle becomes incurved ; in the cherry (*P. cerasus*, etc.) perigyny is even more strongly

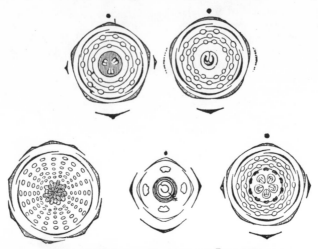

FIG. 361. FLORAL DIAGRAMS OF ROSACEAE.

Top left to bottom right: True Service (*Sorbus domestica*) ; bird cherry (*Prunus padus*) ;
rose (*Rosa canina*) ; great burnet (*Sanguisorba officinalis*) ; meadowsweet (*Filipendula ulmaria*).
(*After Eichler.*)

marked with the sepals, petals and stamens borne on the rim of a cup-shaped receptacle ;
whereas in the apple (*Malus sylvestris*) the flower is entirely epigynous.

The genera of ROSACEAE can be classified into several groups.

Group 1. This contains the genus *Filipendula* and several other less well-known genera.
Wild examples are : *F. ulmaria*, meadowsweet or queen of the meadow, a common plant
in damp meadows and along river and lake banks (June–September) ; *F. vulgaris*, dropwort of chalky pastures and downs (June–August).

Group 2. This contains the genera *Pyrus, Malus, Mespilus, Sorbus, Crataegus*, etc. The
flower is epigynous. The receptacle grows around the carpels and unites with them. The
walls of the carpels in *Pyrus* and *Malus* are horny, hence the characteristic toughness of
the 'core' of the apple and pear fruit ; while in *Crataegus* the wall of the carpel is stony.

Characteristic floral diagrams of Groups 2 and 3 are given in Fig. 361, and characteristic floral formulae of Group 2 are :

$$Malus\ sylvestris\ \text{(apple)} : \ \oplus K_5 C_5 A\infty\ G(\overline{5}),$$
$$Crataegus\ monogyna\ \text{(hawthorn)} : \ \oplus K_5 C_5 A\infty\ \overline{G\mathrm{I}}.$$

Wild examples are :

Malus : *M. sylvestris*, wild crab apple (April–August).

Pyrus : *P. communis*, wild pear, a rarity of southern counties (April–May).

Crataegus : *C. monogyna*, may or hawthorn, usually forming hedge-bushes with one
(sometimes two) carpels (May–June) ; *C. oxyacanthoides*, may or hawthorn, usually growing as small, isolated trees, having two or three carpels (May–June).

Mespilus : *M. germanica*, wild medlar tree of southern counties, probably a garden
escape (May–June).

Sorbus : *S. aria*, white beam, a bush or tree of limy soils (May–June) ; *S. latifolia*, rare
Cornish white beam (May–June) ; *S. intermedia*, rare Scottish white beam (May–June) ;
S. torminalis, wild service, rather rare tree of the south, having lobed leaves (May–June) ;
S. aucuparia, rowan or mountain ash, with many other more localised names (June–Sep-
tember) ; *S. domestica*, true service or chequer tree, a rarity (June–September).

Cultivated examples include plants derived from all the above wild plants, either for
their fruit or for ornamental purposes.

Group 3. This contains the genera *Rubus, Fragaria, Potentilla, Geum, Agrimonia, Alchemilla, Poterium, Rosa* and others.

The receptacle varies from a flat disk to a slightly cup-shaped structure. In *Fragaria* (wild and cultivated strawberry) the receptacle becomes succulent after fertilisation (p. 273 and Fig. 215). In *Rubus* (blackberry, dewberry, etc.), the receptacle remains dry after fertilisation, but the carpels are fleshy and form a collection of drupes (Fig. 362). Very often in certain genera the calyx has stipular structures which form what is called the epicalyx. This is well seen in *Fragaria*.

A typical floral formula of this group is :

$$\oplus K_5 C_5 A\infty \underline{G\infty}.$$

Agrimonia, however, has only two carpels, and certain other genera have no petals.

Wild examples are :

Rubus : *R. fruticosus*, bramble or blackberry (July–September) (actually this is a general specific name for a very large number which are now treated as separate species) ; *R. caesius*, dewberry (July–September) ; *R. idaeus*, raspberry, a heath plant ; *R. chamaemorus*, cloudberry of high moors (June–September) ; *R. saxatilis*, stone bramble of mountainous districts (June–September).

Fragaria : *F. vesca*, wild or wood strawberry (May–July) ; *F. moschata*, wild ' hautboy ' strawberry, probably a garden escape (May–July).

Potentilla : *P. sterilis*, barren strawberry, having prostrate runners, a dry receptacle, and dry, non-splitting achenes (January–May) ; *P. anserina*, silverweed (June–August) ; *P. reptans*, creeping cinquefoil (June–September) ; *P. erecta*, tormentil (June–September).

Geum : *G. urbanum*, wood avens, common avens or herb Bennet (June–August) ; *G.*

Fig. 362. Ripe and Unripe Fruits of Blackberry (*Rubus fruticosus*).
Inset, a collection of fruits in section.

FIG. 363. BURNET OR SCOTCH ROSE (*Rosa spinosissima*).
Showing flowers and fruits.

rivale, water avens (June–August) ; *G. intermedium*, a rare hybrid between the two former species.

Agrimonia : *A. eupatoria*, agrimony (June–July).

Alchemilla : *A. vulgaris*, lady's mantle (June–September).

Poterium : *P. sanguisorba*, lesser or salad burnet (June–August).

Sanguisorba : *S. officinalis*, great or common burnet (June–August).

Rosa : *R. canina*, wild or dog rose (June–July) ; *R. arvensis*, trailing or field rose (June–July) ; *R. rubiginosa*, sweet briar or eglantine (June–July) ; *R. spinosissima*, burnet or Scotch rose (May–June, Fig. 363).

Among cultivated examples of this group are : **ornamental**—many species and varieties of rose (*Rosa*) ; *Geum* species ; *Potentilla* species ; *Poterium obtusum* : **food, etc.**—*Rubus, fruticosus, R. caesius, R. idaeus*, and loganberry (a species of *Rubus* of doubtful origin) ; *Fragaria vesca*, etc.

Group 4. This contains the genus *Prunus* and a little-known genus *Nuttallia*.

The genus *Prunus* contains several well-known cultivated and wild plants, including plum, peach, cherry, apricot, etc. The receptacle is cup-shaped, and the flower perigynous. There is one carpel, and the pericarp of the fruit becomes succulent. The fruit is therefore a drupe.

The floral formula is :

$$\oplus K_5 C_5 A\infty\ G_1.$$

Wild examples include :

Prunus : *P. spinosa*, sloe or blackthorn (March–May) ; *P. domestica*, wild plum, rather rare (April–May) ; *P. avium*, gean (April–May) ; *P. padus*, bird cherry (April–May) ; *P. cerasus*, wild or dwarf cherry (April–May).

Cultivated plants include : *P. domestica*, cultivated varieties of plum ; *P. domestica*, sub-species *insititia*, bullace or damson ; *P. amygdalus*, sweet and bitter almonds ; *P. serrulata*, Japanese cherry and a host of ' flowering ' cherries ; *P. armeniaca*, apricot ; *P. persica*, peach and its variety nectarine ; *P. laurocerasus*, cherry laurel, and many other ' flowering ' *Prunus*.

LEGUMINOSAE

Leguminosae is one of the largest of the flowering-plant families, comprising about six hundred genera. The family is very cosmopolitan. In nearly all cases, the fruit is characteristic, being derived from a single carpel. Most genera have highly zygomorphic, butter-fly-like flowers. These are placed in the sub-family PAPILIONACAE, and this is largely temperate in distribution ; in fact, all members of the family LEGUMINOSAE indigenous to Britain belong to this sub-family. Certain exotic genera bear flowers which are less irregular, and these are placed in the sub-family CAESALPINIOIDEAE— largely tropical. A few exotics bear quite regular flowers, and these are placed in the sub-family MIMOSO-IDEAE, also tropical.

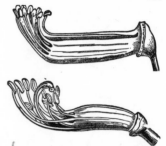

FIG. 364. ABOVE, DIADELPHOUS STAMENS OF GARDEN PEA (*Pisum sativum*) ; BELOW, MONADELPHOUS STAMENS OF BROOM (*Sarothamnus scoparius*).

(*After Oliver.*)

The family LEGUMINOSAE is of considerable economic importance in supplying food and floral beauty. Not only does it cover most parts of the world, but also it favours many habits and consequently contains plants of diverse habit such as trees, shrubs and herbs, climbers, hydrophytes, xerophytes, mesophytes, etc. Above all, most of them have root nodules (p. 138).

All familiar British genera belong to the PAPILIONACAE, for example, garden pea (*Pisum sativum*). The flower is highly zygomorphic, based on a practically flattened receptacle and therefore perigynous (Fig. 250). The calyx is either two-lipped formed by two united sepals, one with two small teeth, the other with three small teeth, for example, gorse (*Ulex europaeus*), or with five teeth produced by five united sepals, for example, garden pea (*Pisum sativum*). The corolla is polypetalous, and consists of five petals—a large standard at the back, two side wings and two smaller petals forming the keel. Bulges on the wings fit into the hollows in the keel. There are ten stamens ; in some, all ten are united for about two-thirds of their length (**monadelphous**), for example, broom (*Sarothamnus scoparius*), whereas in others, nine stamens are so united but the tenth (upper) is free (**diadelphous**), for example, pea, broad bean (*Vicia faba*) and bird's-foot trefoil (*Lotus corniculatus*) (Fig. 364). The gynoecium is formed from one carpel and is superior. Most diadelphous species are herbs ; most monadelphous species are trees or shrubs ; moreover, in the latter, the calyx is usually two-lipped. Most diadelphous species secrete nectar on the inner sides of the bases of the filaments of the stamens. The majority of monadelphous species do not secrete nectar, but in *Laburnum*, tissue rich in nectar is to be found at the base of the standard.

Representative floral formulae are :

$$\text{Pisum sativum (garden pea)} : \quad \uparrow K(5)C5A(9) + 1G\underline{1},$$

$$\text{Ulex europaeus (gorse)} : \quad \uparrow K(2)C5A(10)G\underline{1}.$$

Examples of floral diagrams are given in Fig. 365.

In LEGUMINOSAE there are several interesting examples of floral mechanisms which ensure cross-pollination by insects. The stamens (and therefore the pollen) and the nectar (if present) are protected by the keel ; therefore only strong insects can get at them. Bees are the usual agents of pollination. The bee alights on the wings of the flower and thus, by virtue of its own weight, it presses the wings apart and the keel downwards, thus forcing the stamens and the stigma to expose themselves and rub against the ventral

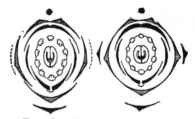

FIG. 365. FLORAL DIAGRAMS OF LEGUMINOSAE.

Left, broad bean (*Vicia faba*) ; right, laburnum (*Laburnum anagyroides*).

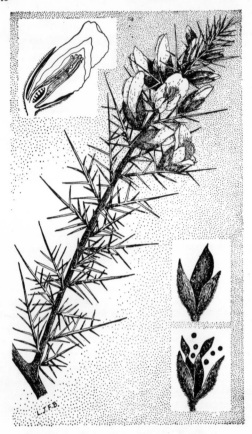

FIG. 366. GORSE (*Ulex europaeus*).
Top left, section of flower ; bottom right, pod
opening to release seeds.

part of the insect's body. Thus the pollen is transmitted to another flower, where it is rubbed on to its stigma. This simple mechanism is seen in clover and laburnum. In the pea, the stamens are not forced out. They shed their pollen on the inside of the keel ; then as the stigma is forced out, the pollen is swept on to the insect's body by hairs present on the style. In the broom, the stamens and style and stigma are held in the keel under tension, and the depression of the keel results in explosive liberation of the style and pollen. The stigma hits the back of the bee, receiving pollen from a previously visited flower, and the pollen of the present flower descends on to the back of the insect. These flowers can only be exploded once, as the stamens do not return to their position in the keel.

There are several other more complicated methods of cross-pollination in this family.

The methods of fruit dispersal of the more common species are discussed in Chapter XXII. The various types of leaf modifications are described in Chapter V, and the exceptional leaf movements in *Mimosa* and *Amicia* on p. 235. The presence of root nodules, their structure and significance are considered in Chapter XII.

Wild examples of the family, indigenous to Britain, include the following (all members of the sub-family PAPILIONACAE) :

Sarothamnus : *S. scoparius*, broom, a shrub of mountains and moorland (May–June).

Ulex : *U. europaeus*, gorse, furze or whin, a shrub of heaths, hills, etc. (all the year round, but mainly February–March and August–September). The line in Goldsmith's *The Deserted Village*, ' With blossom'd furze unprofitably gay,' is probably explained by the fact that for months gorse (furze) flowers are blooming when no bees are available and they therefore die unpollinated (Fig. 366).

Genista : *G. anglica*, needle or petty whin, a shrub of moorlands and heaths (May–July) ; *G. tinctoria*, dyer's green weed, a shrub of fields and meadows (July–September).

Astragalus : *A. glycyphyllos*, milk vetch, cream-coloured (June–September) ; *A. danicus*, purple milk vetch (June–August).

Anthyllis : *A. vulneraria*, kidney vetch, a herb of dry pastures (June–August).

Ononis : *O. repens*, common rest harrow, a dwarf shrub of moorlands and heaths, also characteristic of sand dunes (June–September) ; *O. spinosa*, spiny rest harrow of similar habitats (June–September).

Medicago : *M. lupulina*, black medick or nonsuch, a common herb of fields and waste places, with indehiscent fruits (May–September).

Melilotus : *M. altissima,* tall melilot, a herb of waysides and waste places (June–August) ; *M. alba,* a rare white species (June–August) ; *M. indica,* rarer small yellow species (June–August).

Trifolium : *T. dubium,* yellow-flowering trefoil, the true shamrock, a small herb (June–August) ; *T. pratense,* red clover (May–September) ; *T. repens,* white or Dutch clover (May–September) ; *T. campestre,* hop trefoil (June–September) ; *T. micranthum,* least yellow trefoil or slender clover (June–July) ; *T. striatum,* knotted clover or trefoil of the sandy hinterland of coastal regions (June–July) ; *T. fragiferum,* strawberry clover or trefoil (July–September) ; *T. arvense,* hare's-foot trefoil (July–September).

Lotus : *L. corniculatus,* bird's-foot trefoil or Tom Thumb, fingers and thumbs, shoes and stockings, of fields (July–September) ; *L. tenuis,* narrow-leaved bird's-foot trefoil (July–September) ; *L. uliginosus,* march bird's-foot trefoil.

Vicia : *V. angustifolia,* narrow-leaved vetch (April–July) ; *V. sepium,* hedge or bush vetch (May–August, Fig. 367) ; *V. sativa,* common vetch (May–July) ; *V. cracca,* tufted vetch (June–August) ; *V. hirsuta,* hairy tare (June–August).

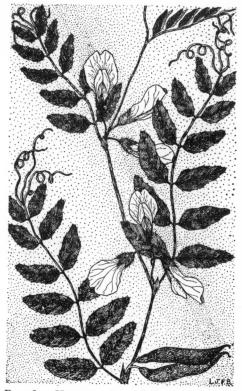

FIG. 367.   HEDGE OR BUSH VETCH (*Vicia sepium*).

Lathyrus : *L. montanus,* tuberous bitter vetch (May–August) ; *L. nissólia,* grass vetch (May–June) ; *L. pratensis,* meadow vetchling (June–September).

Among cultivated leguminous plants, the sub-family PAPILIONACAE is best represented ; many are familiar as food or ornamental plants.   They include :

Food, etc. *Pisum sativum,* varieties of garden pea ; *Phaseolus multiflorus,* varieties of runner and haricot bean : *P. vulgaris,* varieties of French, kidney or dwarf beans ; *P. mungo,* pulse or gram of India ; *Trifolium* species, clover for fodder ; *Medicago lupulina,* medick, and *M. sativa,* lucerne or alfalfa, for fodder ; *Onobrychis viciifolia,* sainfoin for fodder ; *Glycine max* and *G. hispida,* soya beans, mainly in the United States and Japan ; *Lens esculenta,* lentils of Africa and Mediterranean regions ; *Arachis hypogaea,* groundnut, peanut or monkeynut of Africa, America and Asia.

Ornamental. *Lathyrus odoratus,* varieties of sweet pea ; *L. latifolius,* everlasting pea ; *Lupinus,* species and varieties of lupin ; *Genista* species ; *Sarothamnus* species and varieties of broom ; *Laburnum anagyroides,* varieties of laburnum ; *L. alpinum,* alpine laburnum ; *Wistaria chinensis,* *W. floribunda* and *W. venusta,* species of wistaria ; *Robinia pseudoacacia,* common false acacia or locust, a street tree with spiny stipules (p. 53) ; *Sophora japonica,* Japanese pagoda tree.

Among cultivated CAESALPINIOIDEAE are *Gleditschia triacanthus,* honey locust ; *Cercis siliquastrum,* Judas tree ; *Caesalpinia pulcherrima,* peacock flower or Barbados pride : and among cultivated MIMOSOIDEAE are *Mimosa,* the sensitive plant, sometimes grown in greenhouses ; *Acacia dealbata,* the so-called ornamental ' mimosa ' or silver wattle of Australia.

GERANIACEAE

GERANIACEAE is a cosmopolitan family, though most genera are indigenous to temperate regions. The fruit is characteristic, taking the form of a splitting schizocarp (p. 275). The flowers are very conspicuous. The most important genera are : *Geranium*, *Erodium* and *Pelargonium*.

Among indigenous examples are :

Geranium (ten stamens) : *G. pratense*, meadow crane's bill (June–July) ; *G. molle*, dove's-foot crane's bill (April–September) ; *G. robertianum*, herb Robert, which also has a host of local names (May–August) ; *G. sanguineum*, bloody crane's bill (July–August).

Erodium (five fertile and five sterile stamens) : *E. cicutarium*, stork's bill (May–August) ; *E. moschatum*, rare musky stork's bill (May–August).

Among cultivated members of this family are several crane's bills (*Geranium*) and some species of *Erodium* ; but the most popular are the bedding, greenhouse and pot plants frequently called geraniums but actually species and varieties of *Pelargonium*.

OXALIDACEAE

A mainly tropical and sub-tropical family, but represented indigenously in Britain by the beautiful yet common wood sorrel, a member of the best-known genus, *Oxalis*. The common wild example is *Oxalis acetosella*, wood sorrel, alleluia or green sauce. At one time, this plant was looked upon in England especially as the shamrock ; but now we know that the true shamrock is *Trifolium dubium* (p. 457), though wood sorrel has trifoliate leaves. Its white flowers appear during May–June. In the West of England a rare yellow species, *O. corniculata*, grows, blooming during June–September.

Several species and varieties of *Oxalis*, in colours yellow, pink and white, are cultivated for their ornamental beauty.

LINACEAE

A cosmopolitan family, represented in the wild state in Britain by the two genera *Linum* and *Radiola*.

Wild examples are :

L. usitatissimum, common flax, not really indigenous but an escape from cultivation (July–September) ; *L. bienne*, narrow-leaved flax, of the south-eastern counties (July–September) ; *L. anglicum*, perennial flax (June–July) ; *L. catharticum*, purging flax (June–August).

Radiola : *R. linoides*, allseed or thyme-leaved flaxseed of heaths, one of Britain's tiniest flowering plants (July–August).

The most important of the cultivated examples is *Linum usitatissimum*, the flax of commerce, supplying fibres for textiles, linseed oil from the seed, and oil-cake from the compressed seeds. Several species of *Linum* are cultivated for their beautiful flowers—bright blue, pale blue, deep blue, bright yellow, pale yellow, red and white. The red species has short- and long-styled flowers.

POLYGALACEAE

This is a small family, but it is worthy of mention since it is represented in Britain by the interesting *Polygala vulgaris*, the milkwort (June–September). This graceful little perennial grows on cliff-tops, chalk downs, railway sidings, and is very characteristic of sub-alpine grasslands. It shows an unusual range of colours for a wild flower, because not only are there varying shades of white, deep blue, purple, pink and mauve, but also two differently coloured flowers may be seen in the same raceme. The colour is due to the calyx, which is irregular. There are three insignificant green sepals and two large petaloid sepals which are coloured. The petals themselves are much reduced.

In the same chalky habitats, a rarer milkwort (*P. calcarea*) grows, but it blooms earlier (May–June). There are other rare species.

EUPHORBIACEAE

A strange family, better known in tropical and subtropical areas, but represented by a couple of common herbaceous genera in Britain. In tropical regions, members are of varying habit, some being of outstanding xerophytic character, the trees and shrubs especially resembling cacti. Some are of considerable economic importance. The two British genera are *Mercurialis* and *Euphorbia*.

Wild examples are :

Mercurialis : *M. perennis*, dog's mercury, a common hedgerow and wood plant, being unisexual and dioecious (March–April) ; *M. annua*, annual mercury, a less-common weed (July–November).

Euphorbia : Unlike *Mercurialis*, the members of this genus contain a milky juice (latex). *E. amygdaloides*, wood spurge (March–May) ; *E. exigua*, dwarf spurge, a weed (July–November) ; *E. helioscopia*, sun spurge, a weed (July–November) ; *E. peplus*, petty spurge (July–November). Seventeen different species of spurge are indigenous to Britain.

Among members of the EUPHORBIACEAE of economic importance are : *Hevea brasiliensis*, Para rubber ; *Manihot glaziovii*, Ceara rubber ; *M. utilissima*, bitter cassava ; *M. aipi*, sweet cassava ; *Ricinus communis*, castor oil.

BUXACEAE

A small temperate and tropical family of evergreen shrubs, represented in Britain only by *Buxus sempervirens*, the box tree, native to beech woods and the chalk and limestone areas of the south (February–May). Other species and varieties are cultivated.

AQUIFOLIACEAE

Another small temperate and tropical family of trees and shrubs, mainly evergreen, represented in Britain only by *Ilex aquifolium*, holly (May–August). The flowers are usually unisexual, but sometimes they are hermaphrodite. *I. paraguensis* is the yerba maté or Paraguay tea of South America.

CELASTRACEAE

A somewhat larger family of tropical and temperate trees, shrubs and woody climbers ; but again represented in Britain by only one species—*Euonymus europaeus*, spindle tree, pegwood or skewerwood (May–June). During autumn, its leaves present brilliant tints, and the pink fruits are divided into four lobes which open out to reveal seeds surrounded by orange-coloured arils. This and other species and varieties are cultivated.

ACERACEAE

A family of temperate and tropical trees and shrubs (mainly of mountainous regions) represented in Britain by one genus only, namely, *Acer*, the maples and sycamores. The common species are *A. pseudoplatanus*, sycamore or great maple (the plane in Scotland) (May–June) ; *A. campestre*, common or field maple (May–June, Fig. 368) ; *A. platanoides*, Norway maple (May–June). Other species of maple are cultivated for ornamental purposes, and *A. saccharum* is the sugar maple of the United States and Canada.

HIPPOCASTANACEAE

A family of mainly tropical trees, shrubs and woody climbers, represented in Britain by *Aesculus hippocastanum*, the well-known horse-chestnut (May–June). The pink horse-

FIG. 368. FIELD MAPLE (*Acer campestre*).
Left, twig bearing leaves and flowers; top right, twig bearing fruit; bottom right,
male flower and hermaphrodite flower.

chestnut (*A. carnea*) arose as a hybrid between *A. hippocastanum* and an American species *A. pavia*. The species of the genus *Aesculus* are cultivated for ornamental purposes in many parts of the world.

RHAMNACEAE

Still another family of trees, shrubs and climbers, cosmopolitan in distribution. To it belong the two British buckthorns—the purging buckthorn (*Rhamnus cathartica*) and the alder buckthorn (*Frangula alnus*). Both are small trees which thrive on limy soils, bearing black fruit in the form of drupes. The fruit of the purging buckthorn was at one time used in the preparation of a purgative. The purgative cascara sagrada is prepared from the bark of the American buckthorn (*Rhamnus purshiana*).

TILIACEAE

TILIACEAE is a fairly large family mainly of trees and shrubs, distributed chiefly in Brazil and south-east Asia; but it is represented in Britain by the well-known limes or lindens (*Tilia* species). The most familiar is the common lime (*T. vulgaris*), which is now considered to be a hybrid of the other two rarer species—*T. platyphylla*, large-leaved or red-twigged lime, and *T. cordata*, small-leaved lime. A curious feature of limes is the manner in which the flowers are borne (July). They grow in clustered cymes, each flower growinig at the end of a short stalk and then all the stalks joining at the end of a longer stalk which, for half its length, is fused with a curious, oblong, light-green bract.

MALVACEAE

A large tropical and temperate family of herbs, trees and shrubs, represented in Britain by the three genera *Malva*, *Lavatera* and *Althaea*.

Wild examples include :

Malva : *M. sylvestris*, common mallow (June–September) ; *M. neglecta*, dwarf mallow (June–September) ; *M. moschata*, musk mallow (July–September).

Lavatera : *L. arborea*, tree mallow (July–October, Fig. 349).

Althaea : *A. officinalis*, marsh mallow (August–September) ; *A. hirsuta*, hispid or hairy mallow (August–September).

Among examples cultivated outdoors or in greenhouses for their floral beauty are : *Malva moschata* ; *Althaea rosea*, hollyhock ; *Lavatera* species ; *Hibiscus* species ; *Abutilon* species

Gossypium species are the cotton plants.

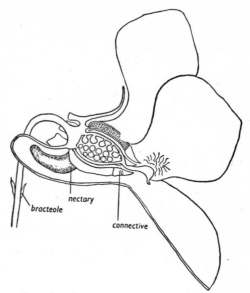

FIG. 369. LONGITUDINAL SECTION THROUGH FLOWER OF VIOLET.
(*F. W. Williams.*)

CISTACEAE

This small, mainly temperate family is represented indigenously in Britain by the rock roses (*Helianthemum*) only. The common rock rose (*H. chamaecistus*) is easily found displaying its yellow flowers on chalk downs during July–September. There are other species, but they are rarer. Rock roses are cultivated favourites in tints of white, pale yellow, orange, pink, amber and crimson.

VIOLACEAE

A well-known cosmopolitan family of herbs, with a few trees and shrubs, yet the only indigenous genus is *Viola*.

Of outstanding interest is the highly zygomorphic flower (Fig. 369). It is composed of five, backwardly projecting sepals, five petals (two at the top, one on each side and one at the bottom), the last-named having the characteristic nectar-containing spur, five stamens and three united carpels. The floral diagram is depicted in Fig. 370.

The stamens are almost sessile and have the connective, the tissue between the anther lobes, prolonged into a thin orange-coloured triangular flap (Fig. 369). These connectives form a pollen-holding chamber surrounding the style. Any movement of the style by the visiting insect while sucking nectar from the spur disturbs the flaps and results in a shower of pollen descending on the insect's head. The two anterior stamens have long, green projections which fit into the petal spur. The nectar-secreting tissue is at the tip of these outgrowths. There are hairs on the lateral petals guarding the entrance to the spur and honey-guides on the petals pointing towards it. The hairs are probably to keep out small insects. The stigma is placed in the centre of the opening. Any visiting insect when it alights is bound to touch the stigmatic surface first, and leave on it some previously gathered pollen, so effecting cross-pollination. As the bee leaves the flower the stigma is pushed up and self-pollination is avoided. In some species, a flap of tissue helps to protect the stigmatic surface. The shape and position of the stigma and stigmatic surface vary slightly in the different species.

FIG. 370. FLORAL DIAGRAM OF *Viola*.
(*After Strasburger.*)

Cleistogamic flowers occur in some species (p. 254) ; the fruit is described on p. 276.

Wild examples of *Viola* include : *V. odorata*, sweet violet (February–April) ; *V. canina*, dog violet (April–August) ; *V. hirta*, hairy violet (April–June), and several other species of violet ; *V. tricolor*, heartsease or wild pansy (May–September), the plant that yielded its juice to the jealous Oberon (*Midsummer-Night's Dream*) ; *V. lutea*, rarer yellow heartsease (May–September) ; *V. nana*, dwarf heartsease (May–September).

Species of *Viola* are very popular among gardeners, for example, varieties of sweet violet (*V. odorata*) in shades of white, lilac, blue and deep violet and single and double ; *V. tricolor* varieties in the form of pansies and violas, etc.

LYTHRACEAE

A very cosmopolitan family represented in the wild state in Britain by the genera *Lythrum* and *Peplis*. The flowers of *Lythrum* are heterostyled (p. 256).

Wild examples include :

Lythrum : *L. salicaria*, purple loosestrife of river banks (July–September) ; *L. hyssopifolia*, hyssop-leaved loosestrife—very rare (July–September).

Peplis : *P. portula*, water purslane of the margins of lakes and of river banks (July–August).

Species of *Lythrum* are used in water gardening, so also are species of the exotic genus *Cuphea*.

ONAGRACEAE

ONAGRACEAE contains the familiar willowherbs. It is a large temperate and tropical family comprising mainly herbs. The two most common genera found growing wild in Britain are *Epilobium* and *Chamaenerion* (willowherbs) and *Circaea* (enchanter's nightshade) ; but there are others.

Examples of British genera and species include the following :

Epilobium : *E. hirsutum*, great hairy willowherb, or codlins and cream of river banks (July–August) ; *E. parviflorum*, small-flowered willowherb (July–August) ; *E. adnatum*, square-stemmed willowherb (July–August) ; *E. montanum*, broad-leaved willowherb ; *E. palustre*, marsh willowherb ; and others.

Chamaenerion : *C. angustifolium*, rose-bay willowherb or fireweed, the familiar willowherb of bombed sites and other burnt-out areas (July–August, Fig. 13).

Circaea : *C. lutetiana*, enchanter's nightshade (June-August) and other species and hybrids.

Among ornamental cultivated members of this family are *Oenothera biennis*, evening primrose, often found in the wild state—possibly an escape from cultivation ; other species of *Oenothera* ; *Fuchsia* species and varieties ; *Godetia* species and varieties ; *Clarkia* species and varieties. *Fuchsia magellanica* was introduced as a hedge plant in south-west England, the Isle of Man and north-west Ireland where, especially in Ireland, it has spread as an escape.

HALORAGACEAE (HALORAGIDACEAE)

A small cosmopolitan family of herbs, many of which are marsh or aquatic plants, represented in Britain by the water milfoils—popular aquarium plants, namely, *Myriophyllum verticillatum*, whorled water milfoil (June-August) ; *M. spicatum*, spiked water milfoil (June-August) ; *M. alterniflorum*, alternate-flowered water milfoil (June-August).

HIPPURIDACEAE

Another small family of aquatic herbs represented indigenously in Britain by *Hippuris vulgaris*, mare's tail (June-July).

CALLITRICHACEAE

Another family of aquatics containing the indigenous, unisexual water starworts, the commonest of which is *Callitriche verna* (April–September).

LORANTHACEAE

A tropical and temperate family of semi-parasitic shrubs, represented in Britain by the unisexual, dioecious *Viscum album*, mistletoe (March, p. 350).

HYPERICACEAE

This is the family of the St. John's worts, most members of which are characterised by large yellow flowers. The family is mainly temperate in distribution, and comprises herbs, shrubs and trees. Some inhabit tropical mountainous regions.

Of the eight genera, the only one indigenous to Britain is *Hypericum*, the most common species of which is *H. perforatum*, the common or perforated St. John's wort. The leaves of this species contain small oil glands which appear like small, paler green pin-pricks. The flower has five small sepals and five oval and pointed, bright yellow petals. The stamens are numerous, longer than the petals, and are joined to form three bundles—a curious character. Like many other introductions, this plant is proving to be a bad weed in Australia. It blooms during June–September.

Among other British members of the genus *Hypericum* are :

H. dubium, imperforate St. John's wort (June–August) ; *H. pulchrum*, small or slender

FIG. 371. POLYMORPHISM IN LEAVES OF THE IVY (*Hedera helix*).

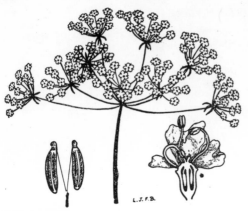

FIG. 372. A TYPICAL UMBELLIFER.
Showing inflorescence, fruit (left) and flower
(right) with stamen pollinating stigma.

St. John's wort (June–August) ; *H. humifusum*, trailing St. John's wort (June–September) ; *H. androsaemum*, tutsan (June–August) ; *H. hirsutum*, hairy St. John's wort (July–August) ; etc.

Several species of *Hypericum* are cultivated for their ornamental beauty. Most are introduced species. The best known is *H. calycinum*, the rose of Sharon or Aaron's beard, having flowers three to four inches in diameter. *H. polyphyllum* is a charming trailing species seen in some rock gardens. *H. moserianum* and *H. prolificum* are taller shrubs.

ARALIACEAE

A mainly tropical family of trees and shrubs, represented in Britain solely by the late-flowering, evergreen ivy, *Hedera helix* (October–November). This plant shows exceptional variation in leaf form (polymorphism, Fig. 371).

UMBELLIFERAE

A large family of annual and perennial herbs (a few shrubs and trees), containing many familiar British plants. The main characteristic is the grouping of the insignificant flowers into characteristic umbels (p. 241). The flower is usually cross-pollinated by flies ; but if this fails, self-pollination is effected by the stamens turning inwards towards the short styles which turn outwards (Fig. 372). The fruit is characteristically a splitting schizocarp (Fig. 372), and in many species contains oil canals.

A typical floral formula is :

$$\oplus K_5 \, C_5 \, A_5 \, G(\overline{2}),$$

and a floral formula is shown in Fig. 373.

Of the many members indigenous to Britain may be mentioned :

Sanicula europaea, wood sanicle (June–July).

Eryngium : *E. maritimum*, sea holly (July–August, p. 434).

Chaerophyllum : *C. temulum*, rough chervil (June–July).

Anthriscus : *A. sylvestris*, cow parsley, wild beaked parsley or keck (April–June) ; *A. cerefolium*, chervil (May–July) ; *A. neglecta*, bur chervil or common beaked parsley (May–July).

Scandix : *S. pecten-veneris*, shepherd's needle (June–September).

Myrrhis : *M. odorata*, sweet Cicely (May–June).

Smyrnium : *S. olusatrum*, alexanders (May–June).

Conium : *C. maculatum*, hemlock (June–July).

Bupleurum : *B. rotundifolium*, hare's ear or thorow-wax (June–July).

Apium : *A. graveolens*, wild celery or smallage (July–August).

Cicuta : *C. virosa*, cowbane or water hemlock (June–July, Fig. 374).

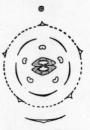

FIG. 373. FLORAL DIAGRAM OF UMBELLIFERAE.

Pimpinella : *P. major*, large burnet saxifrage (July–September) ; *P. saxifraga*, small burnet saxifrage (July–September).

Ernest G. Neal

Fig. 374. Cowbane or Water Hemlock (*Cicuta virosa*).

Crithmum : *C. maritimum*, samphire (June–August, p. 431).

Oenanthe : *O. fistulosa*, water dropwort (July–September) ; *O. aquatica*, fine-leaved water dropwort (July–September) ; *O. lachenallii*, parsley water dropwort (June–August) ; *O. crocata*, hemlock water dropwort (July–August).

Aethusa : *A. cynapium*, fool's parsley (July–August).

Angelica : *A. sylvestris*, wild angelica (July–August).

Pastinaca : *P. sativa*, wild parsnip (July–August).

Heracleum : *H. sphondylium*, cow parsnip or hogweed (July–August).

Daucus : *D. carota*, wild carrot (June–August).

Few members of the UMBELLIFERAE are cultivated in ornamental gardens, though the

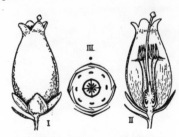

FIG. 375. CROSS-LEAVED HEATH
(*Erica tetralix*).

I, flower ; II, longitudinal section
of flower ; III, floral diagram.

gigantic *Heracleum giganteum* may often be seen growing on the edges of ornamental lakes and ponds, attaining a height of nine or more feet and having enormous leaves and umbels as big as umbrellas.

Among those of economic value are : *Daucus carota*, carrot ; *Pastinaca sativa*, parsnip ; *Apium graveolens*, celery ; *Petroselinum crispum*, parsley ; *P. carvi*, caraway ; *Pimpinella anisum*, aniseed ; *Coriandrum sativum*, coriander ; *Foeniculum vulgare*, fennel ; etc.

CUCURBITACEAE

This large family favours tropical regions, and comprises a large number of tendril-climbing herbs. It is represented in the indigenous British flora by one species only, namely, *Bryonia dioica*, white bryony (May–September). But the family is also familiar through its members of economic importance. These include *Cucurbita pepo*, pumpkin and its varieties the vegetable marrow and squash ; *C. maxima*, giant pumpkin of North America and its varieties the gourds ; *Cucumis sativus*, cucumber ; *C. anguria*, the true gherkin of the West Indies ; *C. melo*, various melons ; *Citrullus vulgaris*, water melon ; *Lagenaria vulgaris*, the flask-shaped calabash ; *Ecballium elaterium*, squirting cucumber of Mediterranean regions ; *Luffa cylindrica*, of which the woody vascular network of the fruit supplies the loofah of the bathroom.

CORNACEAE

A temperate and tropical family of mainly trees or shrubs. The only important British member is *Cornus sanguinea*, dogwood or cornel (June–July). Several varieties of this, and other species, are cultivated in parks and gardens. *Aucuba japonica*, Japanese laurel (p. 447), is also a member of this family.

METACHLAMYDEAE

ERICACEAE

A fairly large cosmopolitan family of plants, many of xerophytic habit and inhabitants of dry moorlands, for example, heather (Fig. 428) and cross-leaved heath (Fig. 375).

Native examples are :

Erica : *E. tetralix*, cross-leaved heath (July–October) ; *E. cinerea*, fine-leaved heath or bell heather (July–October), and other more localised species.

Calluna : *C. vulgaris*, common ling or heather (July–October).

Pyrola : *P. minor*, common or lesser wintergreen (June–July) ; *P. rotundifolia*, larger or round-leaved wintergreen (July–September). (In some Floras, this genus is placed in a separate family—PYROLACEAE.)

Vaccinium : *V. vitis-idaea*, cowberry or red whortleberry (June–August) ; *V. myrtillus*, bilberry, blaeberry or whortleberry (May–June).

Oxycoccus : *O. palustris*, crowberry (May–June).

Among cultivated members of the ERICACEAE are : *Rhododendron* species, *Azalea* species, *Arbutus unedo* (strawberry tree), and many species of *Erica* and *Calluna*.

PRIMULACEAE

This family contains several well-known wild and cultivated species. There are twenty-eight genera in all. Most are herbaceous perennials with rhizomes or tubers.

The flower is actinomorphic with a calyx of five gamosepalous inferior sepals, a corolla of five gamopetalous, hypogynous petals, an androecium of five epipetalous stamens fixed opposite the petals, and a single-chambered, superior ovary with free-central placentation (p. 250, and Figs. 193 and 194. The fruit is a capsule. This opens when ripe, but in the scarlet pimpernel the seeds are exposed by a transverse split (p. 276 and Fig. 216).

Certain genera of the PRIMULACEAE show heterostyly, that is, different lengths of style as compared with the insertion of the stamens (p. 248 and Fig. 202). This is a mechanism for ensuring insect-pollination.

A typical floral formula is :

FIG. 376. FLORAL DIAGRAM OF PRIMU-LACEAE.

$$\oplus K(5) \overset{\frown}{C(5)} A5 \, G(\underline{5}).$$

A typical floral diagram is shown in Fig. 376.

Certain wild examples of this family are very common ; others are not so. Examples are :

Primula : *P. vulgaris*, primrose (March–May) ; *P. veris*, cowslip (May–June) ; *P. elatior*, oxlip (April–May).

Hottonia : *H. palustris*, water violet (May–July).

Cyclamen : *C. neapolitanum*, cyclamen or sowbread, a rarity of woods and meadows (August–September).

Lysimachia : *L. nemorum*, yellow pimpernel (May–August) ; *L. nummularia*, creeping Jenny or moneywort (June–July) ; *L. vulgaris*, yellow loosestrife (July–August).

Anagallis : *A. arvensis*, scarlet pimpernel or poor man's weather glass (June–November) ; *A. foemina*, an exceedingly rare blue form (June–November) ; *A. tenella*, bog pimpernel (July–August).

Among the cultivated favourites of this family may be mentioned such primulas as *Primula sinensis*, *P. japonica*, *P. obconica*, *P. stellata*, *P. malacoides*, *P. auricula* and *P. elatior*. Then there are the many forms of *Cyclamen*, and cultivated varieties of other wild genera, especially *Lysimachia nummularia*.

PLUMBAGINACEAE

This family is mainly of ecological interest, since it contains a high percentage of plants of salt marshes and coastal areas, represented in Britain by the genera *Armeria* and *Limonium*.

Examples indigenous to Britain are :

Limonium : *L. vulgare*, sea lavender (July–August, p. 434), and several other rarer species.

Armeria : *A. maritima*, thrift, sea pink or sea gilliflower (April–September, p. 431).

Among cultivated plants are species of *Limonium* and *Armeria*, together with *Statice*, *Plumbago* and *Ceratostigma*.

OLEACEAE

A tropical and temperate family of trees and shrubs, represented in Britain indigenously by the ash and under cultivation by the lilac, jasmine, etc.

Most familiar examples are :

Fraxinus : *F. excelsior*, ash (March–April).

Cultivated members include *Syringa vulgaris*, lilac ; *Ligustrum vulgare*, privet ; *Jasminum* species, jasmine ; *Forsythia* species ; etc.

GENTIANACEAE

This is a large and very versatile family both in habit and habitat, widely represented in the wild state in Britain and popular among flower gardeners.

Indigenous examples include the following :

Gentiana : *G. pneumonanthe*, marsh gentian (August–September) ; *G. verna*, spring gentian (April–June) ; etc.

Gentianella : *G. campestris*, field gentian (August–September) ; *G. amarella*, autumnal gentian or felwort (August–September) ; etc.

Centaurium : *C. minus*, common centaury (June–September) ; etc.

Blackstonia : *B. perfoliata*, yellow-wort or yellow centaury (June–September).

Menyanthes : *M. trifoliata*, buckbean or bogbean (May–September).

Certain members of this family, especially the gentians, are cultivated.

APOCYNACEAE

This family is quite large, but it is in the main tropical, though represented in Britain by the familiar periwinkles, in fact these are the sole representatives, being members of the genus *Vinca*—*V. minor*, lesser periwinkle (March–May, Fig. 377) ; *V. major*, greater periwinkle (March–June).

The greenhouse favourite, the oleander, *Nerium oleander*, is also a member of this family.

CONVOLVULACEAE

The CONVOLVULACEAE, a family of temperate and tropical plants, contains many woody climbers and a large number of herbs and shrubs, but rarely trees. It is well represented in the native flora of Britain. In most cases, the flowers are unmistakable, having five petals joined to form a five-angled cornet-shaped tube, which in many cases closes up at night.

Examples from the British native flora include :

Convolvulus : *C. arvensis*, small bindweed, cornbine (June–August), a pernicious weed.

Calystegia : *C. sepium*, great bindweed or bellbine (June–August, p. 425) ; *C. soldanella*, seaside bindweed (June–August, p. 432)

Cuscuta (parasites) : *C. europaea*, greater dodder (July–August, p. 347) ; *C. epithymum*, common or lesser dodder (August–September, p. 347) ; *C. epilinum*, flax dodder (July–August, p. 348).

Only plants of ornamental beauty belonging to this family are cultivated in Britain, for example, *Ipomaea purpurea* and other species of morning glory (climbers) ; *Convolvulus althaeoides* (a creeper) ; *Nolana grandiflora* (a creeper) ; but among plants of economic value cultivated elsewhere are : *Ipomaea batatas*, the sweet potato of tropical and subtropical areas ; and *I. purga*, the Mexican jalap, which yields a strong purgative.

FIG. 377.   LESSER PERIWINKLE (*Vinca minor*).

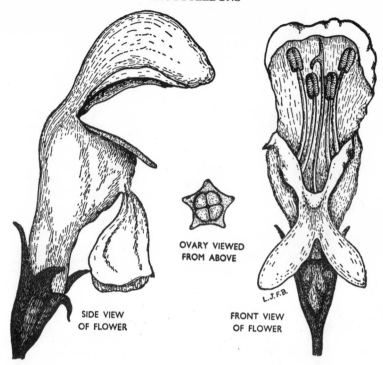

OVARY VIEWED
FROM ABOVE

SIDE VIEW
OF FLOWER

FRONT VIEW
OF FLOWER

FIG. 378. FLOWER OF WHITE DEADNETTLE (*Lamium album*).

BORAGINACEAE

This is another family well-known in the British indigenous flora, though there are many tropical and temperate members unknown in Britain. A curious character of many members of this family is the frequent tendency to change the colour of the flowers during the life of the individual plant. For example, the forget-me-not often changes from pink in its early stages to blue in the later. Moreover, the inflorescences are frequently scorpioidal cymes, and look like the coiled tail of a scorpion.

Indigenous British examples include :

Myosotis : *M. arvensis*, common forget-me-not or field scorpion grass (June–July) ; *M. caespitosa*, water forget-me-not (June–July) ; *M. hispida*, early forget-me-not (April–July) ; *M. palustris*, another water forget-me-not (June–August) ; *M. discolor*, yellow-and-blue or partly-coloured forget-me-not (April–July).

Cynoglossum : *C. officinale*, hound's tongue (June–July) ; *C. germanicum*, green-leaved hound's tongue (June–July).

Symphytum : *S. officinale*, comfrey (May–July) ; *S. tuberosum*, tuberous comfrey (May–July).

Borago : *B. officinalis*, borage (June–July).

Lycopsis : *L. arvensis*, small bugloss (June–August).

Pulmonaria : *P. officinalis*, lungwort (June–July).

Lithospermum : *L. arvense*, corn gromwell (May–July) ; *L. officinale*, common gromwell or grey millet (July–August).

Echium : *E. vulgare*, viper's bugloss (June–July).

Certain members of the BORAGINACEAE are popular among flower gardeners, especially

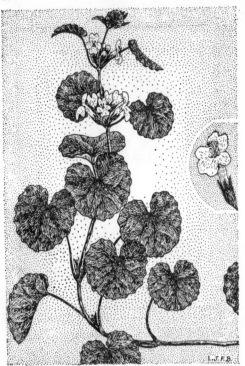

FIG. 379.   GROUND IVY (*Glechoma hederacea*).

those of the genera *Mysotis* (forget-me-not), *Heliotropium* (cherry pie and heliotrope), *Anchusa*, *Echium* and *Cynoglossum*.

LABIATAE

The Mediterranean regions form the centre for this important family, which is composed of about two hundred genera. Certain genera are, however, limited in their distribution in such regions as Australia, Tasmania, Malaya, etc. Many species have an aromatic smell, and in certain cases are therefore of economic importance.

The majority of genera are herbs with stems square in cross-section, and simple, decussate leaves. The flowers are gamopetalous and highly zygomorphic. The calyx is gamosepalous, the corolla united and two-lipped, the stamens epipetalous and four (rarely two) in number, and the gynoecium composed of two syncarpous carpels. The stigma is usually forked and the superior ovary is composed of four lobes, each loculus being divided by a false septum. In each lobe there is one ovule.

In this family there are several beautiful mechanisms for ensuring cross-pollination by insects. That of the sage has already been examined (pp. 257 and 258 and Fig. 204). In the white deadnettle (*Lamium album*), pollination is by bees. The nectar is at the base of the long corolla tube (Fig. 378). The bee alights on the lower lip of the corolla which acts as a kind of landing stage. The bee's back thus rubs the stigmas and stamens; but since the stigma protrudes further than the stamens, it touches the insect's back first and catches any pollen already present from another flower. Thus, although the stamens and stigma ripen at the same time, and although self-pollination is possible, cross-pollination is made more likely. Other species show peculiar adaptations for cross-pollination by insects.

A typical floral formula for the family is :

$$\text{\emph{Lamium album} (white deadnettle) :}\quad \uparrow K(5)\ \widehat{C(5)}A4\ G(\underline{2}).$$

That of the rarer types of flower is :

$$\text{\emph{Salvia officinalis} (sage) :}\quad \uparrow K(5)\ \widehat{C(5)}A2\ G(\underline{2}).$$

Twenty-five genera are represented in the native flora of Britain, the most important of which are :

Lamium : *L. album*, white deadnettle (mainly May–September, but often throughout the year) ; *L. purpureum*, red deadnettle (April–October) ; *L. amplexicaule*, henbit deadnettle (May–September).

Galeobdolon : *G. luteum*, yellow archangel (May–June).

Mentha : *M. pulegium*, pennyroyal (August–October) ; *M. arvensis*, corn or field mint (August–October) ; *M. aquatica*, water mint (August–October) ; *M. rotundifolia*, round-leaved mint (August–October) ; *M. longifolia*, horse mint (August–October) ; *M. piperita*, peppermint (August–September) ; *M. spicata*, spearmint (August–September).

Lycopus : *L. europaeus*, gipsywort (June–September).

Origanum : *O. vulgare*, marjoram (July–September).

Thymus : *T. serpyllum*, wild thyme (June–August).

Acinos : *A. arvensis*, basil thyme (June–September).

Clinopodium : *C. vulgare*, wild basil (July–September).

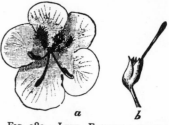

FIG. 380. LEFT, FLOWER ; RIGHT, CALYX AND STYLE OF MULLEIN (*Verbascum*).

Salvia : *S. verbenaca*, wild sage or clary (May–September) ; *S. pratensis*, meadow clary, very rare (May–September).

Prunella : *P. vulgaris*, self-heal (July–September).

Stachys : *S. arvensis*, corn or field woundwort (April–November) ; *S. officinalis*, betony (June–September) ; *S. sylvatica*, hedge woundwort (July–August) ; *S. palustris*, marsh woundwort (July–August).

Ballota : *B. nigra*, black horehound (June–October).

Galeopsis : *G. tetrahit*, common hempnettle (July–August) ; *G. speciosa*, large-flowered hempnettle (July–August) ; *G. ladanum*, red hempnettle (July–October) ; *G. dubia*, downy hempnettle (July–October).

Nepeta : *N. cataria*, catmint (July–September).

Glechoma : *G. hederacea*, ground ivy (April–June, Fig. 379).

Marrubium : *M. vulgare*, white horehound (July–October).

Scutellaria : *S. galericulata*, greater skullcap (July–September) ; *S. minor*, lesser skullcap (July–October).

Teucrium : *T. scorodonia*, wood sage (June–October).

Ajuga : *A. reptans*, bugle (blue) (May–July) ; *A. chamaepitys*, ground pine or yellow bugle, rather rare (May–September).

Among cultivated members of this family are : *Lavandula* species, lavender ; *Thymus serpyllum*, thyme ; *Salvia* species ; *Nepeta* species ; and *Coleus* grown for its gaily variegated

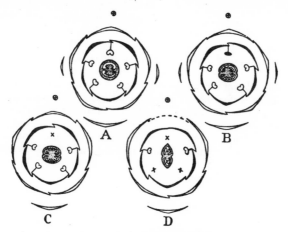

FIG. 381. FLORAL DIAGRAMS OF SCROPHULARIACEAE.

A, mullein (*Verbascum*) ; *B*, figwort (*Scrophularia*) ; *C*, Foxglove (*Digitalis*), snapdragon (*Antirrhinum*), and toadflax (*Linaria*) ; and *D*, speedwell (*Veronica*).

2 H

FIG. 382.   YELLOW TOADFLAX (*Linaria vulgaris*).

Bottom left, section of flower ; bottom right, fruit.

leaves—all grown for ornamental purposes : and *Rosmarinus officinalis*, rosemary ; *Mentha piperita*, peppermint ; *M. pulegium*, pennyroyal ; *M. viridis*, mint ; *Origanum majorana*, sweet marjoram ; *O. onites*, pot marjoram ; *Salvia officinalis*, sage ; *Thymus vulgaris*, thyme ; *Satureia hortensis* and *S. montana*, summer and winter savories—all grown for their perfume or as pot-herbs and savouries.

SCROPHULARIACEAE

This is a rather large cosmopolitan family of about two hundred genera. Most are herbs or small shrubs, though a few, such as *Paulownia imperialis* (having large, blue flowers), a native of Japan, attain the habit of trees.

Many members of this family demonstrate interesting features in their vegetative organs. Certain genera are climbers. Others are xerophytic, and others, such as *Bartsia*, are semi-parasitic on the roots of grasses, etc. (p. 350).

The flowers are gamopetalous, hypogynous, and in most cases zygomorphic. The majority are two-lipped. The stamens are usually four in number, although in others there are two or five. They are epipetalous. The ovary is two-chambered and superior. The fruit is a capsule.

Many species are beautifully adapted for cross-pollination by insects. The two-lipped nature of the corolla makes this possible, and especially in view of the fact that the anthers and stigmas can move in succession, in order to come into contact with the visiting insect.

In the snapdragon and toadflax, the two lips of the corolla are in close contact with each other. They can be opened only by very strong insects such as bees. In the snapdragon the nectar is stored in a pouch, whereas in the toadflax it is present in a long spur (Fig. 382). In the figwort the stigma ripens before the stamens, and the former occupies a position near the lower lip of the corolla before the latter do, and thus comes into contact with the under surface of the insect's body. Thus is self-pollination prevented. In the mulleins, the flower is only very slightly zygomorphic, this being due chiefly to a difference in size of petals (Fig. 380). In the species of *Veronica* there are several deviations from the general characteristics of the family. For example, there are only four petals (one large, two medium and one small), four sepals, and two stamens. Great diversity of vegetative structure is shown in this family. Most British species are herbs, but other species are shrubs with thick, xerophytic leaves.

A typical floral formula is :

$$\uparrow K(5) \; \overset{\frown}{C(5)} A4 \; G\underline{(2)},$$

but there are many deviations from this, for example :

Verbascum sp. (mullein) $\uparrow K(5) \; \overset{\frown}{C(5)} A5 \; G\underline{(2)}$

Veronica sp. (speedwell) : $\uparrow K(4) \; \overgroup{C(4)} A_2 \; G(\underline{2})$

Deviations from the typical scrophulariaceous flower may be seen in Fig. 381, showing especially a gradual reduction in number of stamens.

Of the many indigenous members of the SCROPHULARIACEAE, the following are worth examining :

Verbascum : *V. thapsus*, great mullein, Aaron's rod, high taper, Adam's flannel, Adam's blanket (June–August), and others.

Antirrhinum : *A. orontium*, lesser snapdragon, weasel's snout or calf's snout (July–October) ; *A. majus*, snapdragon (July–September).

Linaria : *L. vulgaris*, yellow toadflax (July–October, Fig. 382).

Cymbalaria : *C. muralis*, ivy-leaved toadflax (May–September).

Kickxia : *K. spuria*, fluellen or round-leaved toadflax (July–October) ; *K. elatine*, sharp-leaved toadflax (July–October).

Scrophularia : *S. nodosa*, knotted figwort (July–September) ; *S. aquatica*, water figwort or water betony (July–September).

Mimulus : *M. guttatus*, monkey-flower (July–August) ; *M. moschatus*, musk, a garden escape (July–August).

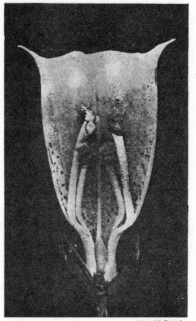

Harold Bastin

FIG. 383. SECTION OF FOXGLOVE FLOWER.

Digitalis : *D. purpurea*, foxglove (June–September, Fig. 383).

Veronica : *V. agrestis*, field speedwell (April–September) ; *V. arvensis*, wall speedwell (April–July) ; *V. chamaedrys*, germander speedwell or bird's eye (May–June) ; *V. officinalis*, common speedwell (May–July) ; *V. scutellata*, marsh speedwell (July–August) ; *V. anagallis*, water speedwell (July–September) ; *V. beccabunga*, brooklime (May–September).

Pedicularis (semi-parasitic) : *P. palustris*, marsh lousewort or marsh red-rattle (May–September) ; *P. sylvatica*, field or heath lousewort (April–July).

Melampyrum (semi-parasitic) : *M. pratense*, common cow-wheat (May–August).

Euphrasia (semi-parasitic) : *E. officinalis*, eyebright (May–September, Fig. 384). There are many sub-species of this.

Odontites (semi-parasitic) : *O. verna*, red bartsia (June–September).

The SCROPHULARIACEAE has yielded many of its members for cultivation as ornamental plants ; but none is of value as food, though some, such as the foxglove, produce valuable drugs. Among the most popular cultivated members are : *Antirrhinum* species, snapdragon or chatterbox ; *Digitalis purpurea* varieties, many colours of foxglove ; *Verbascum* species and varieties, the mulleins ; *Pentstemon* varieties derived mainly from the American *P. hartwegii* and *P. cobaea* ; *Mimulus guttatus*, monkey flower ; *M. moschatus*, musk plant ; *Calceolaria*, sometimes called the slipperworts ; *Linaria* species ; *Veronica* species (herbs and shrubs) ; *Torenia*, a greenhouse annual ; *Rehmannia* species ; *Paulownia imperialis*, a tree.

SOLANACEAE

This tropical and temperate family is centred chiefly in Central and South America, from which the most important members cultivated in this country originated, for example, potato and tomato ; but there are some familiar members of the family native to Britain, all of which are either mildly or strongly poisonous.

Examples of wild members are :

FIG. 384. EYEBRIGHT (*Euphrasia officinalis*).

Atropa : *A. belladonna*, deadly nightshade or dwale (June–August).

Hyoscyamus : *H. niger*, henbane (June–September).

Solanum : *S. dulcamara*, woody nightshade or bittersweet (June–August) ; *S. nigrum*, black nightshade (July–November).

Datura : *D. stramonium*, thorn-apple, a rarity originally introduced (June–September).

PLANTAGINACEAE

This small cosmopolitan family is familiar to British botanists mainly through the very common plantains (*Plantago*). No member is of economic or ornamental value.

The commonest British examples are :

Plantago : *P. major*, great plantain (June–August) ; *P. media*, hoary plantain (June–October) ; *P. lanceolata*, ribwort plantain (June–August) ; *P. maritima*, sea plantain (June–September ; *P. coronopus*, buck's-horn plantain (June–July).

Littorella : *L. uniflora*, shoreweed (June–August)

RUBIACEAE

The RUBIACEAE is one of the largest of flowering plant families, comprising about 450 genera, and well represented in the tropics (certain are of economic importance), though some very common plants in the British native flora are also members. All the British members have very characteristic leaves. They are borne in circular whorls of what appear to be four or more. Actually only two have axillary buds. These are the true leaves (opposite each other) ; the others, though morphologically similar, are really stipules of these two.

Among those members indigenous to the British flora are :

Galium : *G. aparine*, goosegrass (June–September, p. 281 and Fig. 385) ; *G. verum*, lady's bedstraw (June–September) ; *G. mollugo*, hedge or large bedstraw (July–September) ; *G. palustre*, marsh bedstraw (July–August) ; *G. uliginosum*, fern bedstraw (July–August).

Sherardia : *S. arvensis*, field madder (April–October).

Asperula : *A. odorata*, sweet woodruff (May–June).

CAPRIFOLIACEAE

A family mainly of shrubs and trees, distributed chiefly in temperate and mountainous tropical regions.

The best-known examples native to Britain include :

Sambucus : *S. nigra*, elder (June–July) ; *S. ebulus*, dwarf elder or danewort (June–August).

Viburnum : *V. opulus*, guelder rose (June–July) ; *V. lantana*, wayfaring tree (May–June).

Lonicera : *L. periclymenum* (June–September).

The only British herbaceous member of this family, according to most Floras, is *Adoxa moschatellina*, moschatel or townhall clock (April–May) ; but recently this plant has been transferred to a new, monogeneric family, ADOXACEAE.

Among members of this family which are cultivated for their ornamental effect are : *Viburnum opulus*
var. *roseum*, snowball tree, and *Symphoricarpos rivularis*,
snowberry.

VALERIANACEAE

A small, well-distributed family of herbs represented indigenously in Britain by the following genera :

Valeriana : *V. officinalis*, valerian (June–September) ;
V. dioica, marsh valerian (May–September).

Valerianella : *V. locusta*, lamb's lettuce or corn salad
(April–June).

Kentranthus : *K. ruber*, red-spurred valerian (June–
September) ; but probably a garden escape.

Some species of *Valerian* are cultivated for their
floral beauty ; *V. officinalis* yields the medicine valerian. From the thick underground stems of the Hima-
layan member of this family, *Nardostachys jatamansi*,
spikenard is extracted.

DIPSACACEAE

A small temperate and tropical family, familiar
to all British field botanists. It is different from any
family so far considered in that the small flowers are
crowded together in a single head, which, at a casual
glance, looks like one large complicated flower.

Among British indigenous examples are some conspicuous types, such as :

Scabiosa : *S. columbaria*, small scabious (July–
August).

Succisa : *S. pratensis*, devil's-bit scabious (July–
October).

Knautia : *K. arvensis*, field scabious (July–August).

Dipsacus : *D. sylvestris*, wild teasel (August–September) ; *D. fullonum*, fuller's teasel (August–September) ; *D. pilosus*, small teasel (August–September).

Dipsacus fullonum, the fuller's teasel, is still cultivated in a few localities and the dead heads used for
' teasing ' or combing wool and for raising the nap
on cloth. The genus *Scabiosa* contains some very
attractive species and varieties for the flower garden.

CAMPANULACEAE

A large family of temperate and sub-tropical
regions, containing some especially beautiful flowering herbs.

Native British examples include the following :

Campanula : *C. rotundifolia*, harebell or bluebell of
Scotland (June–September, Fig. 386) ; *C. latifolia*,
large campanula, large bellflower, giant bluebell or
throatwort (July–September) ; *C. trachelium*, nettle-leaved bellflower or bats-in-the-belfry
(July–September) ; and others.

Wahlenbergia : *W. hederacea*, ivy-leaved bellflower (July–August).

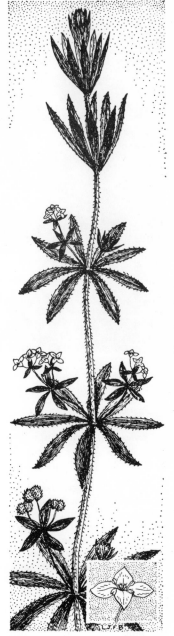

FIG. 385. GOOSEGRASS (*Galium aparine*).

FIG. 386.   HAREBELL   (*Campanula rotundifolia*).

Phyteuma : *P. tenerum*, round-headed rampion (July–August) ; *P. spicatum*, spiked rampion (July–August).

Jasione : *J. montana*, sheep's-bit scabious (June–September).

COMPOSITAE

This is the largest family of the flowering plants.   It comprises about nine hundred genera and thirteen thousand species, which is about ten per cent of the total of the flowering plants.   The family is very cosmopolitan, and some species are found in almost every conceivable habitat.   The majority are herbs, though certain tropical species of *Senecio* are trees, while some species of the same genus are xerophytic, having fleshy stems and leaves.   Many contain latex, others oil canals.

All British species are herbaceous.   The inflorescence is very characteristic, being a capitulum (Fig. 186).   The flowers are small and are usually referred to as florets.   The calyx is sometimes absent, but in most cases it is very much modified into an epigynous collection of hairs known as a pappus used in the dispersal of the fruit (pp. 282 and 283 and Fig. 221).   The corolla is composed of five, or more rarely, three, petals which are united either in the form of a tube (tubular) or a strap (ligulate).   There are five epipetalous stamens, which are very characteristic in that they are joined to each other by their anthers to form a tube.   The gynoecium is formed of two syncarpous carpels with one chamber containing a single ovule.   The fruit, in most cases, is an achene.

The Compositae are subdivided into two sub-families.   The **Liguliflorae** possess capitula which are composed solely of ligulate flowers.   The dandelion is an example.   A diagram-

matic representation of a longitudinal section through
such a capitulum is shown in Fig. 387. The petals are
joined to form a strap-shaped corolla with five teeth.
The lower part of the corolla forms a tube to which the
stamens are attached. All the florets are hermaphrodite.

The **Tubuliflorae** possess either all tubular florets,
for example, the thistles, or some ligulate and some
tubular florets, for example, the daisy. In the genus
Centaurea(knapweeds and cornflowers), all the florets
are tubular, but the inner florets are small and hermaphrodite, whereas the outer ones are neuter that is,
they possess neither stamens nor carpels and serve
merely to attract insects (Fig. 388). In many Tubuliflorae only the disk florets are tubular. The ray florets
are ligulate with three teeth, and these are often highly
coloured and large. In most cases, for example the daisy,
the disk florets are hermaphrodite, whereas the ray florets
are female since they possess carpels, but no stamens.

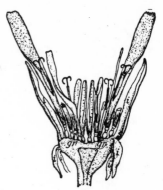

FIG. 387. LONGITUDINAL SECTION THROUGH CAPITULUM OF DANDELION *Taraxacum officinale*).
(*After Thompson.*)

There are many different methods of cross-pollination in the Compositae. The florets
possess an abundance of nectar which is sought by numerous types of insects. The hermaphrodite florets are protandrous. All stages of development may sometimes be seen in the
same capitulum. Some species produce their seeds parthenogentically.

The following are typical floral formulae for the Compositae :

Taraxacum officinale (dandelion) : $\uparrow K$ pappus $\widehat{C(5)}\ A(5)\ G(\overline{2})$.

Bellis perennis (daisy) : Ray floret : $\uparrow Ko\ C(3)\ Ao\ G(2)$.

Disk floret : $\uparrow Ko\ \widehat{C(5)}\ A(5)\ G(2)$.

A typical floral diagram is shown in Fig. 389.

So very well represented is the COMPOSITAE in the indigenous British flora that space
will not allow a comprehensive list. The following are not necessarily the commonest,
but those having very typical or perhaps some peculiar characters.

Tubuliflorae.

Bellis : *B. perennis*, daisy (all the year round, but mainly March–October).

Chrysanthemum : *C. leucanthemum*, moon daisy, ox-eye daisy, or marguerite (June–
August) ; *C. parthenium*, feverfew (July–September) ; *C. segetum*,
corn marigold (August–September).

Tussilago : *T. farfara*, colt's foot (March–May).

Senecio : *S. jacobaea*, ragwort (June–September, Fig. 390) ;
S. vulgaris, groundsel (February–December) ; *S. aquaticus*,
marsh ragwort.

Bidens : *B. cernuus*, nodding bur-marigold (July–October) ;
B. tripartitus, trifid bur-marigold (July–October).

Petasites : *P. hybridus*, common butterbur (January–May) ;
P. fragrans, winter heliotrope (January–May) ; *P. albus*, white
butterbur (January–May).

Solidago : *S. virgaurea*, golden rod (July–September).

Aster *A. tripolium*, sea aster or starwort (August–September).

Erigeron : *E. acris*, blue fleabane (July–August) ; *E.
canadensis*, Canadian fleabane (now naturalised in Britain)
(July–August).

Eupatorium : *E. cannabinum*, hemp agrimony (July–October).

FIG. 388. LONGITUDINAL
SECTION THROUGH CAPITULUM OF LESSER KNAPWEED
(*Centaurea nigra*), SHOWING A
NEUTER RAY FLORET AND
A HERMAPHRODITE DISK
FLORET.

(*After Oliver.*)

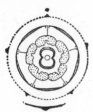

FIG. 389. FLORAL
DIAGRAM OF COM-
POSITAE (*Carduus*).

Anthemis : *A. cotula*, stinking mayweed (June–September) ; *A. arvensis*, corn chamomile (June–August) ; *A. tinctoria*, yellow chamomile (July–August) ; *A. nobilis*, common chamomile (July–October).

Achillea : *A. millefolium*, yarrow or milfoil (July–August) ; *A. ptarmica*, sneezewort (July–August).

Matricaria : *M. maritima*, scentless mayweed (June–October, Fig. 391) ; *M. chamomilla*, wild chamomile (June–August).

Tanacetum : *T. vulgare*, tansy (August–September).

Artemisia : *A. vulgaris*, mugwort (July–September) ; *A. absinthium*, wormwood (August–September).

Carlina : *C. vulgaris*, carline thistle (June–October).

Arctium : *A. lappa*, great burdock (July–August, Fig. 221) ; *A. minus*, small burdock (July–August).

Carduus : *C. crispus*, welted thistle (June–August).

Cirsium : *C. vulgare*, common spear thistle (July–August).

Centaurea : *C. scabiosa*, giant knapweed (June–September) ; *C. nigra*, black knapweed or hard-head (June–September, Fig. 388) ; *C. cyanus*, cornflower or corn bluebottle (June–August) ; *C. solstitialis*, yellow star thistle or St. Barnaby's thistle.

Liguliflorae.

Cichorium : *C. intybus*, chicory (July–October).

Lapsana : *L. communis*, nipplewort (July–September).

Hypochaeris : *H. radicata*, long-rooted cat's ear (June–September, Fig. 392).

Leontodon : *L. autumnalis*, autumnal hawkbit (July–October) ; *L. hispidus*, rough hawkbit (June–September).

Harold Bastin

FIG. 390. RAGWORT (*Senecio jacobaea*) invading Meadowland.

Tragopogon : *T. pratensis*, goat's beard (June–July) ; *T. porrifolius*, salsify (May–July).

Sonchus : *S. oleraceus*, milk or sow thistle (August–September).

Hieracium : There are ten to twenty thousand species or subspecies of this genus, about 260 of which are British. Examples are : *H. pilosella*, mouse-ear hawkweed (July–August) ; *H. murorum*, wall hawkweed (July–August) ; *H. umbellatum*, umbellate hawkweed (July–August).

Crepis : *C. biennis*, rough hawk's beard (June–July) ; *C. capillaris*, smooth hawk's beard (July–September).

Taraxacum : *T. officinale*, common dandelion (throughout the year, but mainly during March–September).

As with the native plants, those members of the COMPOSITAE cultivated for various reasons are too numerous to list in detail. but the following are worthy of note :

Ornamental (most of which belong to the TUBULIFLORAE).—*Ageratum* ; *Bellis perennis*, single and double varieties of daisy ; *Aster*, perennial asters and Michaelmas daisies ; *Callistephus hortensis*, varieties of single and double summer asters; *Solidago virgaurea*, varieties of golden rod ; *Ammobium*, *Helichrysum*, *Rhodanthe* and *Xeranthemum*, 'everlasting' flowers ; *Helianthus annuus*, varieties, etc., of sunflower ; *Dahlia variabilis*, varieties and hybrids of single and double dahlias ; *Cosmos*; *Coreopsis* ; *Zinnia* ; *Calendula*, English marigold ; *Tagetes*, African or French marigold ; *Gaillardia*, *Pyrethrum*, *Chrysanthemum leucanthemum*, varieties of marguerite ; *C. sinense* and *C. indicum*, single and double varieties of chrysanthemum ; *Senecio* species and varieties ; *S. cineraria*, varieties of cineraria ; *Dimorphotheca*, star of the veldt ; *Echinops*, blue globe thistle ; *Centaurea*, cornflowers.

Economic.—*Lactuca sativa*, lettuce ; *Cichorium intybus*, chickory ; *C. endivia*, endive ; *Helianthus tuberosus*, Jerusalem artichoke ; *Cynara scolymus*, true artichoke of Mediterranean regions ; *H. annuus*, sunflower, for the oil in its seeds ; *Pyrethrum*, for insecticides.

Harold Bastin

FIG. 391. SCENTLESS MAYWEED (*Matricaria maritima*).

MONOCOTYLEDONS

The Monocotyledons comprise those families of Angiosperms characterised by the presence of only one cotyledon in the embryo. Their leaves are usually parallel-veined with almost always entire margins, whereas the leaves of Dicotyledons are net-veined and frequently serrated, lobed, or compound (composed of leaflets). In Monocotyledons, the number of parts of the flower (sepals, petals, stamens and carpels) is usually three or a multiple of three (**trimerous**). In the stem of Monocotyledons, the vascular bundles are numerous and irregularly arranged and, with few exceptions p. 155, there is no cambium or secondary thickening. In Dicotyledons, the vascular bundles are usually arranged in

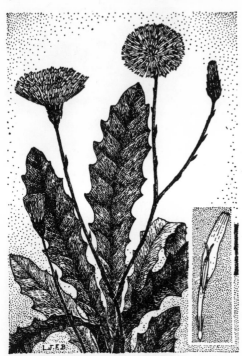

FIG. 392. LONG-ROOTED CAT'S EAR
(*Hypochaeris radicata*).
Inset, single flower.

a single ring, and cambium is present with a resulting secondary thickening in most cases.

ALISMATACEAE

A cosmopolitan family of water and marsh herbs, many possessing perennial rhizomes. The best-known examples native to Britain are :
Alisma : *A. plantago-aquatica*, water plantain (June–August).
Sagittaria : *S. sagittifolia*, arrow-head (July–August, Fig. 337 and p. 418).

BUTOMACEAE

A small tropical and temperate family, also of water and marsh herbs, some containing latex. Represented in Britain by only *Butomus umbellatus*, flowering rush (June–July).

HYDROCHARITACEAE

A still further example of a family of water-loving plants—both fresh water and sea water, some having floating leaves, others submerged ribbon-shaped leaves.

Among British examples are the following :

Hydrocharis : *H. morsus-ranae*, frog-bit (July–August, Fig. 337.)
Stratiotes : *S. aloides*, water soldier (July–August).
Elodea : *E. canadensis*, Canadian pondweed, water weed or water thyme, a unisexual plant, usually bearing only female flowers in Britain. Reproduces vegetatively (p. 427).

Some of the members of the three foregoing families are frequently grown in ornamental ponds and lakes, especially *Alisma, Sagittaria, Butomus, Hydrocharis* and *Stratiotes*.

LILIACEAE

This is one of the largest families of flowering plants, being composed of about two hundred genera. It is cosmopolitan in distribution. A few show exceptional secondary thickening, for example, *Yucca* and *Dracaena* (pp. 155–6).

The majority are herbs which perennate by means of bulbs or rhizomes. In most cases the inflorescence is a raceme ; but in the tulip (*Tulipa*) it is solitary, and in the onion (*Allium cepa*) and garlic (*A. ursinum*) it is a cymose umbel.

The flower is actinomorphic, hermaphrodite, hypogynous, and the parts alternate in whorls of three. The sepals and petals are similar and form a perianth of two whorls of three, and are sometimes joined at the base, for example, hyacinth, or are free, for example, tulip. The stamens form two whorls of three each, and are hypogynous as in the tulip (Fig. 393) or epiphyllous as in the bluebell (Fig. 192). The gynoecium is composed of three syncarpous carpels, and the ovary is superior and three-chambered. The fruit is a capsule in the lilies, tulips, bluebell, etc., opening by three valves at the tip ; or a berry, as in the lily-of-the-valley, butcher's broom, Solomon's seal. Pollination takes place chiefly by means of insects.

Typical floral formulae are :

$Tulipa$ sp. (tulip) : $\oplus P3 + 3\ A3 + 3\ G(\underline{3})$.

$Endymion\ nonscriptus$ (hyacinth) : $\oplus \overset{\frown}{P(3+3)}A\ 3 + 3\ G(\underline{3})$.

A floral diagram is given in Fig. 394.

Examples of a number of genera are to be found growing wild in Britain ; but they are not all indigenous—some have either been introduced into the wild flora by man or were originally garden escapes. Some native genera are confined to very small areas.

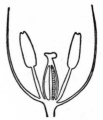

FIG. 393. LONGI-TUDINAL SECTION THROUGH FLOWER OF TULIP (*Tulipa*).

The following are the most important, some having certain peculiarities discussed elsewhere in this book.

Endymion : *E. nonscriptus*, bluebell or wild hyacinth (May–June).

Convallaria : *C. majalis*, lily-of-the-valley (May).

Narthecium : *N. ossifragum*, bog asphodel (July–August).

Polygonatum : *P. multiflorum*, Solomon's seal (June–July) and other species.

Asparagus : *A. officinalis*, asparagus (June–July). Very rare.

Ruscus : *R. aculeatus*, butcher's broom (February–April).

Fritillaria : *F. meleagris*, fritillary (May–June, Fig. 395).

Gagea : *G. lutea*, yellow star of Bethlehem (March–April).

Ornithogalum : *O. umbellatum*, white star of Bethlehem (March–April) ; *O. pyrenaicum*, spiked star of Bethlehem or Bath asparagus (June–July) ; *O. nutans*, drooping star of Bethlehem (April–May).

Allium : *A. ursinum*, ramsons or broad-leaved garlic (May–July) ; *A. vineale*, crow garlic (June–July).

Colchicum : *C. autumnale*, autumn crocus or meadow saffron (August–October)—not to be confused with the genus *Crocus* of the IRIDACEAE (p. 483).

Although the two genera *Lilium* and *Tulipa* are very popular among gardeners, neither is indigenous to Britain, though some species occur locally in the wild state, having thus originated as garden escapes. They are : *Lilium martagon*, the martagon lily or Turk's head, *L. pyrenaicum* and *Tulipa sylvestris*, the wild tulip.

The family LILIACEAE is a great favourite among flower gardeners, having given us such popular flowers as : *Endymion hispanicus*, garden bluebell ; *Hyacinthus orientalis*, varieties of hyacinth ; *Muscari racemosum*, grape hyacinth (also found wild in certain parts of the east of England) ; *Scilla* species ; *Fritillaria imperialis*, the crown imperial and other species of fritillary ; *Kniphophia*, species of red-hot poker ; *Gloriosa*, greenhouse climbers ; *Erythronium dens-canis*, dog-toothed violets ; *Tulipa*, tulips ; *Lilium auratum*, golden-rayed lily ; *L. candidum*, madonna lily ; *L. regale*, regal lily ; *L. giganteum* ; *L. longiflorum*, white trumpet lily ; *L. bulbiferum*, orange lily ; *L. tigrinum*, tiger lily ; *L. martagon* and *L. chalcedonicum*, Turk's cap or martagon, and a host of other species of *Lilium* ; *Funkia* ; *Aspidistra* ; *Asparagus officinalis*, for its tall, graceful late-summer feathery shoots bearing phylloclades ; *A. plumosus*, asparagus ' fern ' ; *Convallaria majalis*, lily-of-the-valley.

The thick shoots of *Asparagus officinalis*, which are formed earlier in the year, are much relished as a vegetable.

Apart from this, the only genus of economic value is *Allium*, which includes *A. cepa*, onion ; *A. ascalonicum*, shallot ; *A. schoenoprasum*, chives ; *A. porrum*, leek ; *A. sativum*, edible garlic.

TRILLIACEAE

This small family is represented indigenously in Britain by a very curious, though fairly common woodland plant, *Paris quadrifolia*, herb Paris or true love knot. Aerial stems shoot up from

FIG. 394. FLORAL DIAGRAM OF POLY-PHYLLOUS LILIACEAE.

Fig. 395.   Fritillary (*Fritillaria meleagris*).

a thick underground rhizome, and near the top of each stem is a whorl of four or more large, broad leaves, having net venation (unusual in Monocotyledons). At the top of the stem is a flower, also atypical in that it has four long, green sepals, usually four lance-shaped, greenish-yellow petals, six to ten stamens, and an ovary with three to five chambers with a corresponding number of styles. The fruit is a black, poisonous berry.

JUNCACEAE

A small family of herbs favouring damp, cold areas, containing certain rushes. British examples include the following :

Juncus : *J. conglomeratus*, common rush (July–August) ; *J. effusus*, soft rush (July–August) ; and about two dozen other species.

Luzula : *L. campestris*, field wood-rush or sweep's brush (April–June) ; *L. sylvatica*, greater woodrush (May–July) ; *L. pilosa*, broad-leaved hairy woodrush (May–July).

AMARYLLIDACEAE

This family contains about sixty genera, most of which are tropical or subtropical, though certain species are very familiar in Great Britain. The family closely resembles Liliaceae in its diagnostic features, except that it has an inferior ovary. In certain genera the perianth tube is prolonged beyond the region where its six segments diverge, to form a tube—the **corona** (Fig. 396). Most members are herbs having pronounced bulbs.

Among wild examples in Britain are the following :

Narcissus : *N. pseudonarcissus*, daffodil (March–April, Fig. 396) and other more localised species and hybrids.

Galanthus : *G. nivalis*, snowdrop (February–April).

Both these genera supply some of our most beautiful cultivated spring flowers, such as *Galanthus nivalis*, snowdrop ; *Narcissus pseudonarcissus*, daffodil, and a host of hybrids and varieties ; *N. majalis*, poet's narcissus, pheasant's eye or primrose pearl ; *N. jonquilla*, varieties of jonquil, etc.

IRIDACEAE

The chief centres of this family are South Africa and tropical America. The yellow flag (*Iris*) and cultivated *Iris*, *Crocus* and *Gladiolus* are common in Great Britain.

Most species are herbs which perennate by means of either corms (*Crocus*) or rhizomes (*Iris*). The flowers are distinguished by having only three stamens. These are epiphyllous.

The leaves are alternate and folded in half longitudinally. The flowers are hermaphrodite and either actinomorphic (*Crocus* and *Iris*) or zygomorphic (*Gladiolus*). The perianth is petaloid and made up of two whorls of three each. The stamens are opposite the outer perianth whorls, and thus are actually the outer whorl of the two whorls of stamens which are usually found in Monocotyledons. The gynoecium is formed from three syncarpous carpels. The ovary is inferior and three-chambered, with axile placentation. The fruit is a capsule, which opens by three valves.

The flowers are cross-pollinated by insects. In the *Iris* the three lobes of the style are large and petaloid. They arch over the stamens and then turn upwards (Fig. 397). Just on the outside of the bend each style lobe bears a stigmatic surface (Fig. 398). The stamens dehisce outwards. Thus the bee, in seeking the nectar, first brushes its head against the stigmatic surface and there deposits any pollen it may be carrying from another flower, then later it brushes its head against the dehisced anthers and thus collects more pollen.

A typical floral formula for the family is :

$$\oplus P\overbrace{(3+3)}A3 + 0\ G(\overline{3}).$$

A floral diagram is shown in Fig. 397.

Among British wild representatives are some interesting but rare examples. The following are worthy of mention :

Iris : *I. pseudacorus*, yellow iris or flag (June–August), common in certain localities ; *I. foetidissima*, gladdon, gladwin, stinking iris or roast beef plant (May–July), rather rare ; and others which are probably garden escapes.

Crocus : *C. nudiflorus*, naked flower crocus (September, Fig. 399), very rare ; *C. purpureus*, purple crocus (March–April), very rare ; *C. biflorus*, purple and mauve crocus (March–April), very rare ; *C. flavus*, golden crocus (March–April), very rare. All these species were probably originally garden escapes.

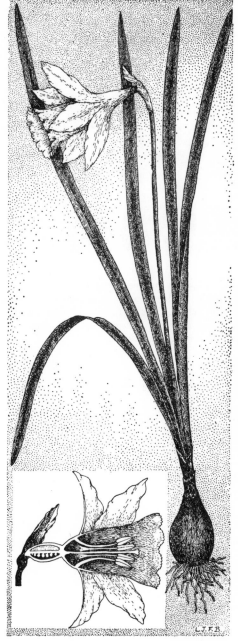

Fig. 396. Daffodil (*Narcissus pseudonarcissus*). Inset, section of flower.

Gladiolus : *G. illyricus*, wild gladiolus (June–July), indigenous but confined to Hampshire and the Isle of Wight. Other genera may be found growing wild, but they were either originally introduced or garden escapes.

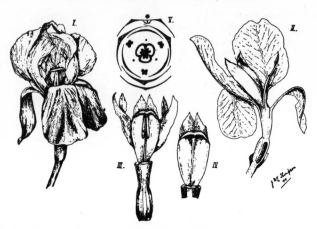

FIG. 397. FLOWER OF *Iris pseudacorus*.
I, complete flower ; II, longitudinal section ; III, perianth removed ;
IV, single lobe of petalloid style ; V, floral diagram.
(*After Thompson.*)

IRIDACEAE supplies some very well-known garden flowers, for example : *Iris germanica*, large iris and its many varieties such as bearded irises ; *I. kaempferi*, clematis-flowered iris ; *I. reticulata*, *I. bucharica* and *I. orchioides*, bulbous or Spanish irises ; *Crocus purpureus*, *C. flavus*, *C. versicolor*, spring crocuses ; *Gladiolus*, varieties of gladiolus ; *Crocosmia*, montbretia; *Sisyrinchium*, satin flower ; *Ixia* ; *Tigridia pavonia*, tiger flower ; *Freesia*.

DIOSCOREACEAE

Climbing plants are rare among the Monocotyledons, but this family is an exception. It contains a large number of dioecious climbers, many of which have thick tuberous rhizomes or stem-tubers ; but it is confined mainly to tropical and warm temperate regions, and is represented in Britain by only one species, namely, *Tamus communis*, the very common black bryony, a hedge climber (May–June).

Among the tropical and warm temperate members of this family are the several species of the genus *Dioscorea*, which yield the edible yams, which in some cases are stem-tubers, in others swollen rhizomes.

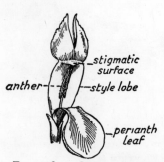

FIG. 398. OUTER PERIANTH LEAF, STAMEN AND STYLE LOBE OF *Iris pseudacorus*.
(*After Lord Avebury.*)

ORCHIDACEAE

This cosmopolitan family is very abundant in the tropics, but rare in arctic regions. It is composed of about 450 genera. The genera agree in certain diagnostic features, but vary considerably in details owing to their diversity of habit—terrestrial, epiphytic (pp. 43, 416) and saprophytic (pp. 351–353).

The flower of even a relatively simple type like *Orchis maculata* (Fig. 401) shows very specialised adaptations to insect pollination. It is highly zygomorphic. The perianth is composed of two whorls of three polyphyllous, superior segments. The three outer ones are small and about equal in size. The three inner ones are very unequal (Fig. 400). The front segment forms a large platform (**labellum**), and is extended downwards into a spur. The single stamen fuses with the short style to form what is called the **column**, and the other two segments of the inner perianth whorl, together with the anterior outer segment, form a hood over this column.

There is only one stamen in most; two in some. Each of the two anther lobes is covered by a thin membrane and when this latter splits, a club-shaped **pollinium** is revealed in each lobe. Each pollinium has an expanded head made up of little packets (**massulae**) of pollen grains; a stalk or caudicle; and a small, sticky, basal disk. The disks of the two pollinia are covered by a **rostellum** cup. This latter stands immediately above the entrance to the spur.

The inferior ovary consists of three united carpels. Very numerous minute ovules are inserted on the parietal placentae. Two stigmatic surfaces are placed left and right of the entrance to the spur; the third stigma may be represented by the rostellum. During its development, the ovary twists through 180° so that the labellum, which appears to be anterior in the adult flower, is really posterior in position (see floral diagram).

After alighting on the platform the insect, in trying to force its proboscis down the spur, pushes the rostellum aside. On being moved in this fashion the rostellum exposes two adhesive disks of the pollinia which stick to the insect's head. On withdrawing, the insect thus pulls the coherent contents of the anther lobes from their sheaths and the two pollinia stand erect on the insect's head. After a few seconds, they gradually curve forwards and soon stand at right-angles to their stalks, at the same time diverging from one another. In this position they are carried to the

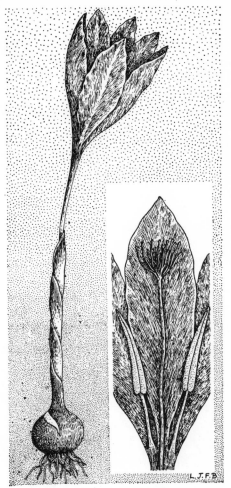

Fig. 399. NAKED-FLOWER CROCUS (*Crocus nudiflorus*).

Inset, part of flower in section showing two stamens and long style with much-branched stigma.

next flower and, as the insect moves to the entrance of the spur, the pollinia strike the stigmatic surfaces. Some of the pollen packets adhere to the sticky stigmas and cross-pollination takes place.

The ORCHIDACEAE is well represented in the native flora of Britain, but only by terrestrial forms—some very common, others very rare. Examples are as follows:

Orchis: *O. mascula*, early purple orchis (May–June), common; *O. maculata*, spotted orchis (June–September, Fig. 401), common.

Anacamptis: *A. pyramidalis*, pyramidal orchis (July–August).

Ophrys: *O. apifera*, bee orchis (June–July), fairly common on chalk and limestone; *O. sphegodes*, spider orchis (April–May), rare.

Platanthera: *P. chlorantha*, greater butterfly orchis (June–August), common locally; *P. bifolia*, lesser butterfly orchis (June–August), less common.

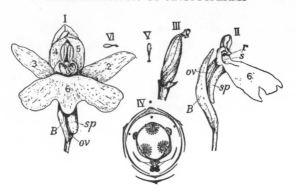

FIG. 400. FLOWER OF SPOTTED ORCHIS (*Orchis maculata*).
I, Front view ; II, lateral view (perianth partly removed) ;
III, young fruit ; IV, floral diagram ; V and VI, pollinia erect and
curved. *B*, bract ; *sp*, spur ; *ov*, ovary ; *s*, stigma ; *r*, rostellum.
(*After Thompson.*)

Neottia : *N. nidus-avis*, bird's-nest orchis (June–July, p. 350).

Listera : *L. ovata*, twayblade (May–June), common ; *L. cordata*, lesser or heart-leaved twayblade (June–August), locally common.

Spiranthes : *S. spiralis*, lady's tresses (July–August), rather rare.

Cephalanthera : *C. damasonium*, white or large-flowered helleborine (May–June), fairly common ; *C. longifolia*, long- or narrow-leaved helleborine (May–June), locally common.

Epipactis : *E. palustris*, marsh helleborine (July–August), locally common ; *E. helleborine*, broad-leaved helleborine (July–August), rather rare.

The number of orchids cultivated both outdoors and in greenhouses defy description in a short space. Some of the most popular are : *Cypripedium calceolus*, lady's slipper orchis (once a native of Britain, but now probably extinct in the wild state) ; and such exotic genera as *Cymbidium*, *Dendrobium*, *Cattleya*, *Epidendron*, *Laelia*, *Odontoglossum*, etc. Many of these are epiphytes. Among outdoor cultivated species, one of the most popular is *Orchis ericetorum*.

ARACEAE

This is a large but very curious family of temperate and tropical plants, though more than ninety per cent of the genera are tropical. In most cases the flowers are small and are crowded on an axis called a **spadix**, which is

Harold Bastin

FIG. 401. SPOTTED ORCHIS (*Orchis maculata*).

enclosed in a foliose **spathe**. They may be hermaphrodite or unisexual.

Large though the family is, there is only one genus indigenous to Britain, namely, *Arum*, and one that was originally introduced, namely, *Acorus*.

The best-known British representative is very common. It is *Arum maculatum*, the wild arum, whose curious form has inspired many local names such as cuckoopint, lords and ladies, wake robin, cows and calves, Jack-in-the-pulpit, parson-in-the-pulpit, and so forth. Its pale green spathes stand out against the darker green of hedgerows, bushes and woods during April–May.

There is a persistent root-stock.

The large handsome leaves are arrow-shaped, dark green with dark-red blotches (Fig. 402).

The floral shoot is extraordinary. It bears unisexual flowers, both sexes being on the same shoot. It takes the form of a succulent axis known as a spadix. At the base of the latter is a mass of female flowers, each of one carpel only and little else. Above this zone comes a ring of strong wavy hairs. Then above this is a zone of male flowers, each being singly composed of two to four stamens. Then above this is a second zone of wavy hairs, which morphologically are sterile male flowers. These point downwards. Terminating the spadix is a long, club-shaped fleshy part, which varies in colour from biscuit through pink to crimson or deep purple.

Inserted on the floral stalk just below the base of the spadix is a large pale green bract called the spathe. This is about twice as long as the spadix. It enfolds the spadix entirely during the early stages, but later opens out from the second zone of hairs and upwards. Below this zone, the spathe is completely wrapped round the flower-bearing part of the spadix, thus forming an urn-shaped surround for the floral parts. The second zone of hairs, therefore, are inserted at the level of the neck of the urn.

This entire floral arrangement forms a very effective mechanism for ensuring insect pollination. The foetid smell of the plant attracts flies, especially midges, which crawl past the downward-curling hairs and eventually reach the ripe carpels at the bottom where nectar is to be found. The insects are probably already covered with pollen from another

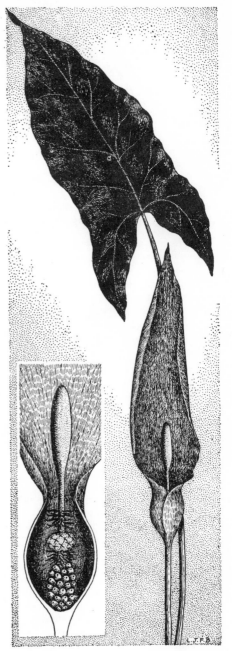

FIG. 402. WILD ARUM (*Arum maculatum*). Inset, spadix and inflorescence exposed.

FIG. 403.   WHEAT (*Triticum*).

1, ear, and 2, whole plant of bearded wheat ; 3, ear of beardless wheat ; *a*, axis of ear
b, spikelet ; *c*, fruit ; *d*, flower.

(*After Baillon.*)

cuckoopint, and thus the carpels are pollinated.   Now the flies are kept prisoner by the
rampart of downward-pointing hairs above the zone of female flowers until all the flowers
have been fertilised.   Then these hairs wither and the flies are free to crawl as far as the
male flowers, where they collect the ripe pollen.   After this, the upper zone of hairs withers
away and the flies can then escape.

The fertilised carpels ripen into large berries, first green then bright red, and while
this is happening the rest of the floral shoot, including both spathe and spadix-tip, withers
away, leaving the cluster of berries standing out prominently in their vivid redness against
the green of the hedgerow.   The berries are poisonous, as also is the rest of the plant, thus
protecting it from browsing animals.

The other genus of the ARACEAE found in the wild state in Britain, namely, *Acorus*, is
represented by *A. calamus*, sweet flag (June–July).

Some tropical genera of this extraordinary family produce organs of astounding dimensions.   For example, *Amorphophallus titanum*, of tropical Asia, bears leaves ten feet high and
a spadix three feet long which is dirty-red in colour and has an offensive smell.

The only cultivated members of this family are : *Zantedeschia aethiopica*, arum lilies, and
Calla palustris, water arum.

TYPHACEAE

A very small monogeneric family of shallow-water herbs, represented in Britain by *Typha latifolia*, great reed-mace (July–August), and *T. angustifolia*, lesser reed-mace (July–August).

CYPERACEAE

A very large, widely distributed family of perennial herbs, represented in Britain by fifteen genera. The most important or interesting are :

Carex, the sedges, of which there are more than seventy recognised native species, including *C. acutiformis*, lesser pond sedge (May–July) ; *C. riparia*, great pond sedge (May–June) ; *C. sylvatica*, wood sedge (May–June) ; *C. flacca*, carnation grass or glaucous sedge (May–June).

Eriophorum : *E. angustifolium*, common cotton grass (June–August, p. 429, and Fig. 346).

Scirpus : *S. sylvaticus*, wood clubrush (July–August) ; *S. maritimus*, sea clubrush (July–August).

Schoenoplectus : *S. lacustris*, bulrush (July–August).

FIG. 404. MEADOW FES-CUE (*Festuca pratensis*).

Below, spikelet with two open flowers ; above, flower, from which the outer pale has been removed.

(After Schenck.)

GRAMINEAE

This is one of the largest families of flowering plants, being composed of 600 genera and about nine thousand species, growing in nearly all regions of the globe. In temperate regions, species of this family form the most important features of the vegetation.

A few are annual, such as oats, wheat, and meadow grass, but most are perennials, and perennate by means of rhizomes (couch grass) or runners (common bent). In most cases the stem has hollow internodes (in maize they are solid). The leaves are alternate, two-ranked, and have split sheaths and ligules. The flowers are hermaphrodite, and are borne in spikelets which are enclosed in bracts called **glumes**. There is no perianth. The stamens are usually three in number and are hypogynous. Their filaments are usually very long and slender (Figs. 403 and 404). The gynoecium is formed from one carpel. The two stigmas are feathery. The ovary is superior, one-chambered, and contains one ovule. The fruit is a caryopsis (p. 285 and Fig. 224).

The appearance of the inflorescence of a grass varies considerably according to whether the stalks of the spikelets are long and spreading outwards from the stem as in oats (Fig. 405), short as in the meadow fescue, or absent as in wheat (Fig. 403) and perennial rye grass (Fig. 407). Each spikelet is subtended by two bracts called glumes (Figs. 403 and 404). Each flower in the spikelet is protected by two bracts called **pales** (Fig. 403). In some cases the outer pale bears a long bristle called an **awn**. This is very evident in oat (Fig. 405), barley, and the ' bearded ' wheat (Fig. 403).

In most cases the flowers are wind-pollinated, though in the majority of cereals self-pollination usually takes place. Stamens and stigmas are exposed when two scales of *lodicules* (seen in front of the ovary in Fig. 103) swell and force the pales apart.

FIG. 405. INFLORESCENCE OF OAT (*Avena sativa*).

(After Figuier.)

A typical floral formula is :

$$\uparrow \text{Po } A3 + 0 \, \underline{G1}.$$

A typical floral diagram is shown in Fig. 406.

The GRAMINEAE is a family well represented in Britain, the following being the most common or of peculiar interest (Fig. 407) :

FIG. 406. FLORAL DIAGRAM OF GRA-MINEAE.

Poa : *P. pratensis*, common meadow grass and several sub-species ; *P. nemoralis*, wood meadow grass ; *P. annua*, annual meadow grass (June–July).

Anthoxanthum : *A. odoratum*, sweet vernal grass (June–July).

Hordeum : *H. secalinum*, meadow barley (June–July) ; *H. murinum*, wall barley (June–July).

Lolium : *L. perenne*, perennial rye grass (June–July).

Agrostis : *A. stolonifera*, white bent grass or fiorin (June–August).

Festuca : *F. pratensis*, meadow fescue (June–August) ; *F. arundinacea*, tall fescue (June–August).

Phleum : *P. pratense*, timothy grass or cat's tail grass (June–July).

Dactylis : *D. glomerata*, cock's-foot grass (June–September).

Briza : *B. media*, quaking grass, or doddering dillies (June–July).

Alopecurus : *A. pratensis*, meadow foxtail (April–June).

Glyceria : *G. maxima*, reed grass (of marshes) (May–August).

Agropyron : *A. repens*, couch grass (June–August).

Avena : *A. fatua*, wild oat (June–July).

Nardus : *N. stricta*, mat grass (June–August).

Phalaris : *P. canariensis*, canary grass (June–July).

Ammophila : *A. arenaria*, marram grass (June–August, p. 430).

Molinia : *M. caerulea*, purple moor grass (July–August).

Phragmites : *P. communis*, common reed (July–August).

There are, of course, a large number of other genera and a number of additional species in some of the genera already mentioned.

Apart from lawns, the GRAMINEAE supplies few ornamental plants. Those worthy of mention are : *Briza maxima*, a large species of quaking grass ; *Stipa pennata*, feather grass ; *Phalaris arundinacea*, variety of ribbon grass having green leaves striped with white ; *Gynerium argenteum*, pampas grass.

But economically, the GRAMINEAE is of the utmost importance, being unrivalled by any other family. It includes :

Cereals.

Triticum vulgare and varieties, wheat ; *T. durum*, hard wheat of Mediterranean areas from which macaroni is made ; *Hordeum vulgare* var. *distichum*, cultivated barley ; *Avena sativa*, cultivated oat ; *Secale cereale*, rye ; *Zea mais*, maize or Indian corn ; *Oryza sativa*, rice ; *Panicum* species, millet ; *Sorghum vulgare*, Mediterranean millet or guinea corn.

Phalaris canariensis yields the grain on which many cage birds are fed.

Sugar.

Saccharum officinarum, sugar cane, of Cuba, India, Java, West Indies, etc. (Sugar is also obtained from the sugar maple and sugar beet.

Building and Construction.

Bambusa species, bamboos ; *Spartina townsendii*, rice grass used in reclaiming land from the sea (p. 433).

Paper and Cordage.

Lygeum spartum of Mediterranean regions ; *Stipa tenacissima* of North Africa, the esparto grasses used for making cordage and paper.

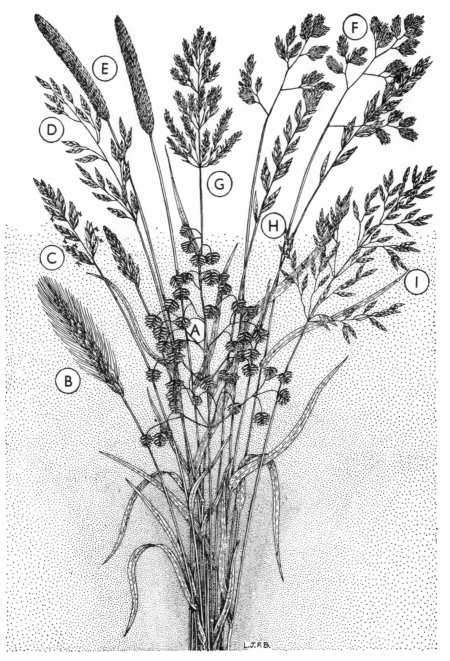

FIG. 407. SOME GRASSES.

A, quaking grass ; *B*, meadow barley ; *C*, sweet vernal grass ; *D*, meadow fescue ;
E, timothy grass ; *F*, cock's-foot ; *G*, white bent ; *H*, perennial rye grass ; *I*, smooth
meadow grass.

INDEX

Numbers printed in thick type indicate pages on which illustrations appear.
Examples of plants mentioned in Chapter XXX are not indexed.